WASTES – SOLUTIONS, TREATMENTS AND OPPORTUNITIES II

T0174100

SELECTED PAPERS FROM THE 4TH EDITION OF THE INTERNATIONAL CONFERENCE WASTES: SOLUTIONS, TREATMENTS AND OPPORTUNITIES, PORTO, PORTUGAL, 25–26 SEPTEMBER 2017

WASTES – Solutions, Treatments and Opportunities II

Editors

Cândida Vilarinho & Fernando Castro
University of Minho, Guimarães, Portugal

Maria de Lurdes Lopes
Faculty of Engineering of the University of Porto, Porto, Portugal

CRC Press
Taylor & Francis Group
Boca Raton London New York

CRC Press is an imprint of the
Taylor & Francis Group, an **informa** business

A BALKEMA BOOK

Published by:
CRC Press/Balkema
P.O. Box 447, 2300 AK Leiden, The Netherlands
e-mail: Pub.NL@taylorandfrancis.com
www.crcpress.com – www.taylorandfrancis.com

First issued in paperback 2020

© 2018 by Taylor & Francis Group, LLC
CRC Press/Balkema is an imprint of the Taylor & Francis Group, an informa business

No claim to original U.S. Government works

Typeset by V Publishing Solutions Pvt Ltd., Chennai, India

ISBN 13: 978-0-367-73586-9 (pbk)
ISBN 13: 978-1-138-19669-8 (hbk)

This book contains information obtained from authentic and highly regarded sources. Reasonable efforts have been made to publish reliable data and information, but the author and publisher cannot assume responsibility for the validity of all materials or the consequences of their use. The authors and publishers have attempted to trace the copyright holders of all material reproduced in this publication and apologize to copyright holders if permission to publish in this form has not been obtained. If any copyright material has not been acknowledged please write and let us know so we may rectify in any future reprint.

Except as permitted under U.S. Copyright Law, no part of this book may be reprinted, reproduced, transmitted, or utilized in any form by any electronic, mechanical, or other means, now known or hereafter invented, including photocopying, microfilming, and recording, or in any information storage or retrieval system, without written permission from the publishers.

For permission to photocopy or use material electronically from this work, please access www.copyright.com (http://www.copyright.com/) or contact the Copyright Clearance Center, Inc. (CCC), 222 Rosewood Drive, Danvers, MA 01923, 978-750-8400. CCC is a not-for-profit organization that provides licenses and registration for a variety of users. For organizations that have been granted a photocopy license by the CCC, a separate system of payment has been arranged.

Trademark Notice: Product or corporate names may be trademarks or registered trademarks, and are used only for identification and explanation without intent to infringe.

Visit the Taylor & Francis Web site at
http://www.taylorandfrancis.com

and the CRC Press Web site at
http://www.crcpress.com

WASTES – Solutions, Treatments and Opportunities II – Vilarinho, Castro & Lopes (Eds)
© 2018 Taylor & Francis Group, London, ISBN 978-1-138-19669-8

Table of contents

Preface   ix

Energetic recovery of municipal solid waste scenario in São Paulo State, Brazil   1
F.C. Dalmo, N.M. Simão, S. Nebra & P.H.M. Sant'Ana

Recycled synthetic waste fibres for the reinforcement of concrete   9
K. Bendjillali, M. Chemrouk & B. Boulekbache

Leaching characteristics of co-bearing glasses obtained from spent Li-ion batteries   17
B.J. Forero, J.V. Díaz-Salaverría, P. Delvasto & C.X. Gouveia

Valorisation of different wastes: A sustainable approach in the design of new products   23
A. Teixeira, D. Monteiro, R. Ribeiro, V. Canavarro, B. Rangel & J.L. Alves

Belgian and Portuguese apple tree bark and core: Comparison of antioxidant content   31
M.M. Moreira, M.F. Barroso, S. Morais, C. Delerue-Matos, A. Boeykens & H. Withouck

Process development for a combined treatment of EAFD and jarosite   37
S. Wegscheider, S. Steinlechner, G. Hanke & J. Antrekowitsch

Mechanical damage of a nonwoven geotextile induced by recycled aggregates   45
J.R. Carneiro, M.L. Lopes & A. da Silva

Smart biofilms produced from fish filleting wastes   51
R.S. Brito, C.A. Araújo, L.F.H. Lourenço, E.J.G. Pino-Hernádez,
G.S. Souza Matos & M.R.S. Peixoto Joele

PAC and tannin as coagulants in swine slaughterhouse wastewater treatment   57
M.C. Bongiovani, T. Werberich, R.M. Schneider & A.G. do Amaral

Waste to energy as a complementary energy source in Abuja, Nigeria   63
O.M. Aderoju & G.A. Dias

Study on properties related to energy recovery from waste streams in Finland   69
E. Sermyagina, M. Nikku, E. Vakkilainen & T. Hyppänen

Chars from co-gasification of rice wastes as Cr(III) removal agents   75
D. Dias, W. Ribeiro, N. Lapa, M. Bernardo, I. Matos, I. Fonseca & F. Pinto

Anaerobic digestion sludge composting—assessment of the star-up process   81
M.E. Silva, S. Araújo, I. Brás, G. Lobo, A. Cordeiro, M. Faria,
A.C. Cunha-Queda & O.C. Nunes

Diagnosis and assessment of the management of sanitary landfill leachates in Portugal   87
A. Fernandes, L. Ciríaco, M.J. Pacheco, A. Lopes & A. Albuquerque

Truck tire pyrolysis optimization using the self-produced carbon black as a catalyst   93
N. Akkouche, M. Balistrou, M. Hachemi, N. Himrane, K. Loubar & M. Tazerout

Waste cooking oils: Low-cost substrate for co-production of lipase and microbial lipids   99
S.M. Miranda, A.S. Pereira, I. Belo & M. Lopes

Proposal for MSW management facilities location in a state of Brazil 105
D.A. Colvero, A.P.D. Gomes, L.A.C. Tarelho, M.A.A. Matos & K.A. Santos

Production of tannin-based adsorbents and their use for arsenic uptake from water 113
H.A.M. Bacelo, C.M.S. Botelho & S.C.R. Santos

Damage induced by recycled C&D wastes on the short-term tensile
behaviour of a geogrid 119
P.M. Pereira, C.S. Vieira & M.L. Lopes

Lean-green synergy awareness: A Portuguese survey 125
M.F. Abreu, A.C. Alves & F. Moreira

Efficiency of regeneration by solvent extraction for different types of waste oil 133
C.T. Pinheiro, M.J. Quina & L.M. Gando-Ferreira

Diethylketone and Cd pilot-scale biosorption by a biofilm supported on vermiculite 139
F. Costa & T. Tavares

Removal of wastewater treatment sludge by incineration 147
S. Dursun, Z.C. Ayturan & G. Dinc

Double benefit biodiesel produced from waste frying oils and animal fats 153
M. Catarino, A.P. Soares Dias & M. Ramos

Alkali-activated cement using slags and fly ash 161
S. Rios, A. Viana da Fonseca, C. Pinheiro, S. Nunes & N. Cristelo

Waste of biodiesel production: Conversion of glycerol into biofuel additives 167
S. Carlota, J.E. Castanheiro & A.P. Pinto

Geotechnical characterization of recycled C&D wastes for use as trenches backfilling 175
C.S. Vieira, M.L. Lopes & N. Cristelo

Recycling of MSWI fly ash in clay bricks—effect of washing and
electrodialytic treatment 183
W. Chen, E. Klupsch, G.M. Kirkelund, P.E. Jensen, L.M. Ottosen & C. Dias-Ferreira

Processing of metallurgical wastes with obtaining iron oxides nanopowders 191
I.Yu. Motovilov, V.A. Luganov, T.A. Chepushtanova, G.D. Guseynova & Sh.S. Itkulova

A step forward on cleaner production: Remanufacturing and interchangeability 197
F. Moreira

Review of potential ways for resource recovery from human urine 203
J. Santos, E. Cifrian, T. Llano, C. Rico, A. Andrés & C. Alegría

Portugal lacks refuse derived fuel production from municipal solid waste 209
P.C. Berardi, M.F. Almeida, J.M. Dias & M.L. Lopes

Removal of Cr(III) from aqueous solutions by modified lignocellulosic waste 215
A.L. Arim, D.F.M. Cecílio, M.J. Quina & L.M. Gando-Ferreira

Valorization of residues from fig processing industry by anaerobic digestion 221
*D.P. Rodrigues, C.I. Alves, R.C. Martins, M.J. Quina, A. Klepacz-Smolka,
M.N. Coelho Pinheiro & L.M. Castro*

Hazards identification in waste collection systems: A case study 227
B. Rani-Borges & J.M.P. Vieira

Outlining strategies to improve eco-efficiency and efficiency performance 235
E.J. Lourenço, A.J. Baptista, J.P. Pereira & C. Dias-Ferreira

Enzymatic esterification of pre-treated and untreated acid oil soapstock
J. Borges, C. Alvim-Ferraz, M.F. Almeida, J.M. Dias & S. Budžaki
241

Formulation of waste mixtures towards effective composting: A case study
M.J. Fernandes, F.C. Pires & J.M. Dias
245

Selective extraction of lithium from spent lithium-ion batteries
N. Vieceli, F. Margarido, M.F.C. Pereira, F. Durão, C. Guimarães & C.A. Nogueira
251

The main environmental impacts of a university restaurant and the search for solutions
M.C. Rizk, D.B. Nascimento, B.A. Perão & F.P. Camacho
259

Treatment of food waste from a university restaurant added to sugarcane bagasse
M.C. Rizk, I.P. Bonalumi, T.S. Almeida & F.P. Camacho
265

Biosolids production and COD removal in activated sludge and moving
bed biofilm reactors
R.A. Dias, R.C. Martins, L.M. Castro & R.M. Quinta-Ferreira
271

Anaerobic digestion impact on the adaptation to climate change in São
Tomé and Príncipe
J.F. Pesqueira, M.F. Almeida, J.M. Dias, D. Carneiro, A. Justo & M.J. Martins
277

Garden waste quantification using home composting on a model garden
T. Machado, B. Chaves, L. Campos & D. Bessa
283

Acid esterification vs glycerolysis of acid oil soapstock for FFA reduction
E. Costa, M. Cruz, C. Alvim-Ferraz, M.F. Almeida & J.M. Dias
287

Sweet potato bioethanol purification using glycerol
J.O.V. Silva, M.F. Almeida, J.M. Dias & M.C. Alvim-Ferraz
293

Methodology for the assessment of non-hazardous waste treatment areas
V. Amant, A. Denot & L. Eisenlohr
299

MAESTRI efficiency framework as a support tool for industrial symbiosis
implementation
*A.J. Baptista, E.J. Lourenço, P. Peças, E.J. Silva, M.A. Estrela, M. Holgado,
M. Benedetti & S. Evans*
305

Activated carbons from Angolan wood wastes for the adsorption of
MCPA pesticide
E.F. Tchikuala, P.A.M. Mourão & J.M.V. Nabais
311

Separate collection of packaging waste: Characterization and impacts
V. Oliveira, J.M. Vaz, V. Sousa & C. Dias-Ferreira
317

Improvement of a clayey soil with alkaline activation of wastes
M. Corrêa-Silva, T. Miranda, N. Araújo, J. Coelho, N. Cristelo & A. Topa Gomes
323

Gasification of RDF from MSW—an overview
F.V. Barbosa, J.C.F. Teixeira, M.C.L.G. Vilarinho & J.M.M.G. Araújo
331

Extraction of copper from dumps and tails of leaching by hydrochloric acid
*K.K. Mamyrbayeva, V.A. Luganov, Y.S. Merkibayev,
Zh. Yesken & S.D. Orazymbetova*
339

Construction wastes application for environmental protection
A.S. Sakharova, L.B. Svatovskaya, M.M. Baidarashvili & A.V. Petriaev
345

Design of a laboratory scale circulating fluidized bed gasifier for residual biomass
D.A. Tibocha, D.C. Guio-Pérez & S.L. Rincón
351

Efficient activated carbons from chars of the co-pyrolysis of rice wastes 359
D. Dias, M. Miguel, N. Lapa, M. Bernardo, I. Matos, I. Fonseca & F. Pinto

Recovery of the polymer content of electrical cables for thermal and acoustic insulation 365
J. Bessa, C. Mota, F. Cunha & R. Fangueiro

Recovery of wood dust in composite materials 371
J. Bessa, C. Mota, F. Cunha & R. Fangueiro

Modified biological sorbents from waste for the removal of metal ions from the water system 377
L. Rozumová, P. Kůs & I. Šafařík

Phytoremediation of soils contaminated with lead by *Arundo donax* L. 383
S. Sidella, S.L. Cosentino, A.L. Fernando, J. Costa & B. Barbosa

Employment of industrial wastes as agents for inclusion modification in molten steels 389
F.A. Castro, J. Santos, P. Lacerda, R. Pacheco, T. Teixeira, A. Silva, E. Soares,
J. Machado & M. Abreu

Olive pomace phenolics extraction: Conventional *vs* emergent methodologies 395
M.A. Nunes, R.C. Alves, A.S.G. Costa, M.B.P.P. Oliveira & H. Puga

Incorporation of metallurgical wastes as inorganic fillers in resins 403
A. Oliveira, C.I. Martins & F. Castro

Sustainability and circular economy through PBL: Engineering students' perceptions 409
A.C. Alves, F. Moreira, C.P. Leão & M.A. Carvalho

Suitability of agroindustrial residues for cellulose-based materials production 417
D.J.C. Araújo, M.C.L.G. Vilarinho & A.V. Machado

Pyrolysis of lipid wastes under different atmospheres: Vacuum, nitrogen and methane 425
L. Durão, M. Gonçalves, A. Oliveira, C. Nobre, B. Mendes, T. Kolaitis & T. Tsoutsos

Effect of temperature in RDF pyrolysis 431
A. Ribeiro, J. Carvalho & C. Vilarinho

Potential of exhausted olive pomace for gasification 437
C. Castro, A. Mota, A. Ribeiro, M. Soares, J. Araujo, J. Carvalho & C. Vilarinho

Analysis of foundry sand for incorporation in asphalt mixtures 443
L.P. Nascimento, J.R.M. Oliveira & C. Vilarinho

Author index 451

WASTES – *Solutions, Treatments and Opportunities II – Vilarinho, Castro & Lopes (Eds)*
© *2018 Taylor & Francis Group, London, ISBN 978-1-138-19669-8*

Preface

Wastes: Solutions, Treatments and Opportunities II contains selected papers presented at the 4th edition of the International Conference Wastes: Solutions, Treatments and Opportunities, that took place on 25–26 September 2017, at the Faculty of Engineering of the University of Porto, Portugal.

The Wastes conference, which takes place biennially, is a prime forum for academics and industry representatives from the waste management and recycling sectors around the world to share their experience and knowledge with all in attendance.

The papers included in this book focus on a wide range of topics, including: Wastes as construction materials, wastes as fuels, waste treatment technologies, MSW management, recycling of wastes and materials recovery, wastes from new materials (nanomaterials, electronics, composites, etc.), environmental, economic and social aspects in waste management and circular economy.

All the articles were individually reviewed by members of the Scientific Committee of the Conference, who also contributed to the editing of the 69 articles presented. The editors wish to thank all reviewers, namely:

Ana Cavaleiro
Ana Luísa Fernando
Anabela Leitão
André Mota
André Ribeiro
António Brito
António Roca
António Roque
Benilde Mendes
Carlos Bernardo
Carlos Nogueira
Castorina Vieira
Cristina Cunha Queda
Fernanda Margarido
Gerasimus Lyberatos
Isabel Ferreira
Javier Viguri
Joana Carvalho

Joana Dias
João Labrincha
Joel Oliveira
Jorge Araújo
José Carlos Teixeira
Madalena Alves
Margarida Gonçalves
Margarida Quina
Maria Alcina Pereira
Mário Costa
Miguel Brito
Nídia Caetano
Nuno Lapa
Paulo Brito
Regina Monteiro
Rosa Quinta-Ferreira
Susete Martins-Dias
Victor Ferreira

With this book we expect to contribute to the dissemination of state of the art knowledge, as well as present the results of real application studies in the waste management and treatment field towards a circular economy perspective.

The Editors,

Cândida Vilarinho
Fernando Castro
Maria de Lurdes Lopes

Energetic recovery of municipal solid waste scenario in São Paulo State, Brazil

F.C. Dalmo, N.M. Simão, S. Nebra & P.H.M. Sant'Ana
Universidade Federal do ABC, Santo André, Brazil

ABSTRACT: This work aims to present a scenario of the energetic recovery potential of the Municipal Solid Waste (MSW) of the São Paulo State, Brazil. The landfills that receive Urban Solid Residues (USR) from two or more counties are considered individually, calculating the potential of energy generation according to the methodology proposed by the International Panel on Climate Change—IPCC. The other landfills throughout the state, which receive MSW from their own counties, were considered together. It is also presented the energy recovery plants in operation, as well as those that have the concession term issued by the National Agency of Electric Energy. It was calculated that, by this mean, the energy potential of the State is 464 MW, with 341.7 MW referring to landfills that receive MSW from two or more counties and 122.3 MW referring to the others. It is suggested for future studies that the cost analysis carried out.

1 INTRODUCTION

The generation of municipal solid waste (MSW) in the world was estimated in 1.3 billion tons per year with an annual increase projection, reaching 2.2 billion in 2025 (Hoornweg & Bhada, 2012). The Organization for Economic Co-operation and Development (OECD) counties in which 34 member counties participate make up nearly half of the current global generation of waste, or 44% of the total, corresponding to 572 million tons per year.

Latin America and the Caribbean, which is Brazil's region, account for 12% of the world's MSW generation, which corresponds to 160 million tons/year. In Brazil, the daily generation targeted was 13,616 tons/day with a projection of 15,886 tons/day for 2025 with a population of 144,507,175 and 206,850,000 inhabitants respectively (Hoornweg & Bhada, 2012).

In Brazil, landfill disposal represents 52.4% of the local total. Besides that, 13.1% in controlled dumps and 12.3% in dumps; 3.9% are sent to sorting and composting units and on 18.3% does not exist information (SNSA, 2016).

In summary, the problem in Brazil is still discussed with regard to the disposal of MSW, and discussions on energy recovery are, for this reason, incipient, even considering that the National Solid Waste Policy (NSWP) through Law no. 12,305/2010 provides the residues energy recovery as one of its objectives (BRAZIL, 2010).

Considering the scenario of the MSW disposal in Brazil, the best way of dealing with the Brazilian reality is therefore the use of the biogas generated spontaneously in the landfills, even considering the fugitive emissions of methane by the landfill surface, which compromises the efficiency of the biogas capture system (Silva & Freitas & Candiani, 2013).

In order to carry out the proposed study of evaluating the energy potential of MSW, the State of São Paulo was chosen. This state has an area of 248,222 km² (IBGE, 2016), and although it is not the most extensive in Brazil—which has a territorial extension of 8,515,767 km² (IBGE, 2016) -, it is the most populous and therefore the largest RSU generator. It is also the state with the highest participation in the national economy, representing 1/3 of the total Brazilian Gross Domestic Product (GDP) between 2010 and 2014 (SEADE, 2016).

In this way, the main objective of this work is to present the energy recovery potential of MSW scenario in the State of São Paulo. The amount of MSW from landfills receiving residues from two or more counties in the State of São Paulo was calculated, and the biogas utilization potential of each one of them was calculated according to the IPCC methodology (IPCC, 1996). The MSW of the other landfills were added to carry out this calculation. It will also be pointed out the locations where there are power generation plants in operation, beyond those that already have the application granted at the National Electric Energy Agency (ANEEL).

2 WASTE DISPOSAL SCENERIO IN THE STATE OF SÃO PAULO

From the 645 counties belonging to the State of São Paulo, 36 have landfills that receive their own waste and from other counties, which in total, attend 277 counties. The others dispose in their own headquarters or in other states and are presented as "others" in the Table 1. Table 1 shows the counties that have landfills, the number of counties depositing, and the total sum of tons of USR received per year.

From the landfills presented in Table 1, there are thermal plants (UTE) installed in four of them: Caieiras (UTE Termoverde Caieiras), São Paulo (UTE São João Biogás), Santana de Parnaíba (UTE Tecipar) and Guatapará (UTE Guatapará). The fifth plant is located in the county of São Paulo, at the Bandeirantes landfill, which no longer receives MSW. Table 2 shows the operation start dates of these plants, as well as the installed capacity; it reports also two other plants already authorized by ANEEL. As showed in Table 2, the energy generation of the Barueri plant is the only one that will be implanted through the incineration tecnology of MSW.

Table 1. Landfills, number of counties disposing on them and tons of USR received per anuum.

Landfill	Number of depository counties	Total MSW (ton/year)	Landfill	Number of depository counties	Total MSW (ton/year)
Botucatu	3	49,965	Oscar Bressane	2	1,106
Cachoeira Paulista	13	120,158	Pariquera-Açu	2	6,150
Caieiras*	17	2,993,073	Paulinia***	32	1,091,898
Catanduva	16	93,283	Piacatú	2	2,431
Cesario Lange	5	41,581	Piratininga****	13	108,087
Coroados	2	2,409	Quatá ****	13	87,501
Guará	6	42,581	Rio das Pedras***	5	63,302
Guatapará	19	521,326	Sales Oliveira	3	5,676
Indaiatuba	3	93,805	Santa Fé do Sul	2	9,435
Iperó	12	327,383	Santa Salete	2	548
Itapevi	8	287,708	Santana de Parnaíba	5	387,458
Jambeiro**	11	258,387	Santos	7	532,780
Jardinópolis	5	44,289	São Carlos	2	79,172
Julio Mesquita	2	6,796	São Paulo**	11	2,080,577
Leme	2	29,171	São Pedro	4	16,991
Mauá	7	740,297	Taquaral	2	1,997
Meridiano	14	57,915	Tremembé	9	159,907
Mesópolis	2	1,263	Others	368	3,773,312
Onda verde	15	192,866			
			Total*****	641	14,312,581

*The São Paulo County also deposits in Caieiras.
** The counties of Suzano and Arujá deposit in Jambeiro and São Paulo.
*** The county of Piracicaba deposits in Rio das Pedras and Paulinia.
****The county of Marilia deposits in Quatá and Piratininga.
***** Four counties deposits in other States, Igarapava and Ituverava in the county of Uberaba, Minas Gerais and Arapeií and Bananal in the county of Barra Mansa, Rio de Janeiro.
Fonte: Elaborated by the authors from CETESB (2015).

Table 2. Electric generation plants, from MSW, in operation in the São Paulo state.

Power plant	Start operation	Technology	Power (MW)	City
Tecipar	30/10/2015	Landfill Biogas	4,278	Santana de Parnaíba
São João Biogás	27/03/2008	Landfill Biogas	24,640	São Paulo
Termoverde Caieiras	15/07/2016	Landfill Biogas	29,547	Caieiras
Guatapará	29/08/2014	Landfill Biogas	5,704	Guatapará
Bandeirantes	03/11/2014	Landfill Biogas	4,624	São Paulo
Sub-Total			68,793	
Power Plant	Authorization	Technology	Power (MW)	City
Paulinia	30/06/2015	Landfill Biogas	29,946	Paulínia
Barueri	12/11/2012	Incineration	20,000	Barueri
Sub-Total			49,946	
Total (operation + authorized)			118,739	

Source: BIG-ANEEL (2017).

The plants in operation in the State of São Paulo represent 59.33% of the potential of landfill biogas plants installed in Brazil, with the remaining ones being distributed in the states of Minas Gerais, Paraná, Santa Catarina, Rio Grande do Sul and Bahia. In the case of authorized plants, São Paulo's participation is 74.48% of the national total, with the others located in the states of Minas Gerais, Amapá and Paraíba. (BIG-ANEEL, 2017).

3 METHODOLOGY

The determination of the gravimetric composition of MSW is essential in the definition of the estimated energy potential contained in the wastes. Table 3 presents data on the gravimetric composition and mean LHV—Lower Heat Value of the landfill in the county of Santo André in 2015. These values were used as a reference for the calculation of the energy potential of the landfills.

From the gravimetric composition and LHV, estimates of the energy potential were calculated. A technology of landfill gas recovery was assumed for the landfills reported in Table 1.

In order to estimate the energy potential of the landfill gas, the theoretical model of estimation of the Intergovernmental Panel on Climate Change (IPCC, 1996) was used, which calculates the total methane emitted by the anaerobic waste degradation. Equation 1 of the model is reported below and refers to the total methane emission available in the landfill:

$$Q_{CH_4} = \frac{Pop_{urb}.RateSHW.USRf.L_0}{pCH_4} \qquad (1)$$

where: Q_{CH_4} = generated methane (m³CH$_4$/year); Pop_{urb} = urban population (inhabitants); $RateSHW$ = rate of solid household waste per inhabitant per year (kg de RSU/inhabitant year); $USRf$ = Fraction of urban solid residues deposited at solid waste disposal (%); L_0 = potential of methane generation from wastes (kg of CH$_4$/kg of USR); pCH4 = methane density (kg/m³).

The equation 2 is referred to the potential of methane generation from the USR:

$$L_0 = MCF.DOC.DOCf.F.\left(\frac{16}{12}\right) \qquad (2)$$

where: L_0 = Potential of methane generation from wastes (kg of CH$_4$/kg of USR); MCF = methane correction factor; DOC = degradable organic carbon (kg of C/kg of USR); $DOCf$ = fraction

Table 3. Gravimetric composition of the MSW of Santo André city.

Fraction	MSW (%)	LHV (MJ/kg)
Organic Matter	44.30	3.16
Sanitary wastes	11.90	7.67
Plastics	13.80	23.04
Paper/cardboard/tetra pack	9.90	9.16
Textiles	8.80	12.9
Inert materials (glass, metal and others)	11.30	0.00

Source: Gutierrez (2016).

Table 4. Values for the MCF for each type of landfill.

Type of local of final disposition	MCF
Dumps	0.4
Controlled dumps	0.8
Sanitary landfills	1
Locations without classification	0.6

Source: IPPC (1996).

Table 5. Content of degradable organic carbon.

Parameter	Type of residues	DOC (% mass)
A	Organic matter	15%
B	Sanitary residues	24%
C	Plastics	0%
D	Paper/cardboard/tetrapack	40%
E	Textiles	24%
F	Inert materials (glass, metal and others)	0%

Source: IPCC (2006); Gutierrez (2016).

of dissociated DOC (%); F = methane volumetric fraction in the landfill biogas (the pattern is 50%); $(16/12)$ = conversion factor of carbon in methane (kg de CH_4/kg de C).

The adopted value for the methane density was of 0.715 kg/Nm^3 (Fantozzi & Buratti, 2009), and the fraction of MSW considered was 100%.

For the Methane Correction Factor (MCF), the IPCC (1996) defines four categories of places for final disposition, according Table 4: i) inadequate landfills (dumps), ii) controlled dumps, iii) adequate landfills (sanitary landfills) and iv) landfills without classification. For the present study, a category of landfill with a value of MCF = 1.0 was adopted.

Another parameter of the IPCC model is related to the amount of degradable organic carbon (DOC), which was calculated by Equation 3 based on the composition of MSW and the amount of carbon in each mass fraction of the residue (IPCC, 2006) according to Table 5.

$$DOC = \sum (DOC_i \times W_i)$$ (3)

where: DOC = fraction of degradable organic carbon in wastes; DOC_i = degradable organic carbon in the residues type i; W_i = fraction of residue type i by waste category.

For the sanitary waste fraction, the DOC value of 24% for diapers, according to IPCC (2006), was adopted.

The fraction of DOC dissociated (DOCf), according to Birgemer & Crutzen (1987 apud IPCC, 1996), indicates the fraction of carbon that is available for the biochemical decomposition, and can be obtained by Equation 4, below.

4

$$DOCf = 0.014 \cdot T + 0,28 \tag{4}$$

where: DOCf = dissociated DOC fraction (%); T = temperature in the anaerobic zone (°C). The average temperature adopted for this study was 35°C, for the anaerobic zone in the final disposal site (Mendes, 2005).

With the data of methane content and other parameters, the calculation of electric energy generation using an internal combustion engine was carried out. Equations (5), (6) and (7) were used to calculate the available thermal power, the available electric power and the electric energy.

$$P_x = \left(\frac{Q_x \times P_{C(metano)}}{31,536,000} \right) \cdot E_C \tag{5}$$

$$P_{el} = P_x \cdot E_M \tag{6}$$

$$E_{available} = P_{el} \cdot 8000 \tag{7}$$

where: P_x = available power (kW); Q_x = methane flow rate per year (m³CH₄/year); P_c = methane lower heating value, (equivalent to 35,530 KJ/Nm³ of CH₄); Ec = efficiency of gas collection, the value of 65% was obtained from the Guatapará UTE Datasheet (2014); P_{el} = termal power, EM = engine efficiency of 39.6% (UTE Guatapará, 2014), $E_{available}$ = available energy (kWh); annual time availability of the generator = 8,000 hours, considering the maintenance stop times; 31,536,000 = amount of seconds per year for conversion from P_x to P_{el}.

4 RESULTS AND DISCUSSION

Table 6 presents the values of the calculated parameters related to the estimated methane generation from the landfills, relative to the quantity of USR available in each site.

According to Thompson et al. (2009), the IPCC method presents an uncertainty regarding the rate of methane production over the years, as well as the recovery of landfill gas, which may be different for landfills located in hot and cold climates requiring further research to calculate these parameters.

Thus, Table 7 presents the energy and power values for each landfill that receives its own municipal waste and also from others. The sum of the rest of the landfills is shown in the table as "others".

Due to the uncertainties in the IPCC model (Thompson et al, 2009), there is a difference between the theoretical energy potential and the actual potential of the waste in four landfills that have plants in operation. In this way, for the Caieiras landfill, a potential of 97 MW was calculated and its current power is 29.5 MW, corresponding to 30.4% of the theoretical potential. For the Guatapará landfill, there was a potential of 16.9 MW and its current power is 5.7 MW, corresponding to 33.7% of the theoretical potential. For the Santana de Parnaíba landfill, there was a calculated potential of 12.5 MW and its current power is of 4.3 MW, which corresponds to 34.1% of the theoretical potential. For the São João landfill in São Paulo, a potential of 67.5 MW was found and its current power is 26 MW, corresponding to 38.8% of the theoretical potential.

Table 6. Parameters to estimate methane generation.

Parameter	Result	Unit
DOCf =	0.77	
DOC =	0.156	
L0 =	0.112	m³CH₄/kgUSR
Methane density =	0.715	kg/Nm³

5

Table 7. Energy potential of São Paulo state.

Landfill	Energy (MWh/year)	Power (MW)	Landfill	Energy (MWh/year)	Power (MW)
Botucatu	12,960	1.62	Oscar Bressane	287	0.04
Cachoeira Paulista	31,168	3.90	Pariquera-Açu	1,595	0.20
Caieiras	776,373	97.05	Paulínia	283,227	35.40
Catanduva	24,197	3.02	Piacatú	631	0.08
Cesário Lange	10,786	1.35	Piratininga	28,037	3.50
Coroados	625	0.08	Quatá	22,697	2.84
Guará	11,045	1.38	Rio das Pedras	16,420	2.05
Guatapará	135,227	16.90	Sales Oliveira	1,472	0.18
Indaiatuba	24,332	3.04	Santa Fé do Sul	2,447	0.31
Iperó	84,920	10.61	Santa Salete	142	0.02
Itapevi	74,628	9.33	Santana de Parnaíba	100,503	12.56
Jambeiro	67,023	8.38	Santos	138,198	17.27
Jardinópolis	11,488	1.44	São Carlos	20,536	2.57
Júlio Mesquita	1,763	0.22	São Paulo	539,680	67.46
Leme	7,567	0.95	São Pedro	4,407	0.55
Mauá	192,025	24.00	Taquaral	518	0.06
Meridiano	15,022	1.88	Tremembé	41,478	5.18
Mesópolis	328	0.04	"Others"	978,758	122.34
Onda verde	50,027	6.25			
			Total	3,712,537	464

5 CONCLUSIONS

Considering the methodology of the IPCC (1996) and data from Gutierrez (2016), it was identified that the São Paulo State energy potential is of 464 MW, with an energy of 3,712,537 MWh/year. From this value, 341.7 MW (2,733,779 MWh/year) refers to landfills that receive MSW from two or more counties and 122.3 MW (978,758 MWh/year) refers to the others.

In 2015, the total amount of electricity consumed in the State of São Paulo was 130,598,789 MWh/year (SÃO PAULO STATE, 2016). Thus, with the technology considered, the energy value of the landfills would represent 2.84% of the total energy consumed in the State.

The state power plants have a power of 119 MW, from which 69 MW are related to plants in operation and 50 MW are referred to plants authorized by ANEEL.

It is suggested, for future studies, the economic-financial evaluation for the effective use of MSW as a source of renewable energy, also contributing to the progress of waste management in Brazil, which according to the National Solid Waste Policy provides for a final destination environmentally friendly .

REFERENCES

ANEEL – Agência Nacional de Energia Elétrica. 2017. *Banco de Informações de Geração (BIG)*. Available: <http://www2.aneel.gov.br/aplicacoes/capacidadebrasil/capacidadebrasil.cfm>. Accessed 03 Jan 2017.

Bingemer, H.Q. & Crutzen, P.J. 1987. Production of methane from solid waste. *Journal of Geophysical Research* 87 (D2): 2181–2187.

Brasil. Lei Nº 12.305, de 2 de agosto de 2010. 2010. Institui a Política Nacional de Resíduos Sólidos; altera a Lei Nº 9.605, de 12 de fevereiro de 1998; e dá outras providências. Diário Oficial [da] República Federativa do Brasil, Brasília, DF, 03 de ago. de 2010. Seção 1, p. 3.

CETESB. Companhia de Tecnologia de Saneamento Ambiental. 2016. *São Paulo. Inventário estadual de resíduos sólidos urbanos – 2015*. São Paulo: Cetesb. Available: <http://residuossolidos.cetesb.sp.gov.br/wp-content/uploads/sites/36/2013/11/inventario-RSD–2015.pdf>. Accessed 06 Set 2016.

Fantozzi, F. & Buratti, C. 2009. Biogas production from different substrates in an experimental continuously stirred tank reactor anaerobic digester. *Bioresource technology* (23): 5783–5789.

Gutierrez, A.C.G. 2016. *Caracterização da fração combustível de resíduos sólidos urbanos úmidos do município de Santo André visando seu aproveitamento energético por processos termoquímicos. Dissertação (Mestrado)*. Santo André: Universidade Federal do ABC.

Hoornweg, D. & Bhada-Tata, P. 2012. What a waste: a global review of solid waste management. *Urban Development Series Knowledge Paper* (World Bank, Washington, DC).

IBGE – Instituto Brasileiro de Geografia e Estatística. 2016. *Área Territorial Brasileira*. Available: <http://www.ibge.gov.br/home/geociencias/areaterritorial/principal.shtm>. Accessed 01 Feb 2017.

IPCC – International Panel on Climate Change. 2006. Chapter 2: Waste Generation, Composition and Management Data. *IPCC Guidelines for National Greenhouse Gas Inventories* 5.

IPCC – International Panel on Climate Change. 1996. Module 6 - WASTE. In *Guidelines for National Greenhouse Inventories: Reference Manual* 3.

Mendes, L.G.G. 2005. *Proposta de um sistema para aproveitamento energético de um aterro sanitário regional na cidade de Guaratinguetá. Dissertação (mestrado)*. UNESP – Guaratinguetá.

SÃO PAULO (ESTADO). SECRETARIA DE ENERGIA E MINERAÇÃO. 2016. *Anuário de Energéticos por Município no Estado de São Paulo – 2016 ano base 2015*. São Paulo.

SEADE – Fundação Sistema Estadual de Análise de Dados. 2016. *Produtos – Produto Interno Bruto*. Available: <http://www.seade.gov.br/produtos/pib-anual/>. Accessed 13 Feb. 2017.

Silva, T.N. & Freitas, F.S.N. & Candiani, G. 2013. Evaluation of surface emissions of gas from large landfills. *Revista de Engenharia Sanitária e Ambiental* 18(2): 95–104.

SNSA. Sistema Nacional de Informações sobre Saneamento. 2016. *Diagnóstico do manejo de resíduos sólidos urbanos – 2014*. Brasília: MCIDADES. Secretaria Nacional de Saneamento Ambiental.

Thompson, S. & Sawyer, J. & Bonam, R. & Valdivia, J.E. 2009. Building a better methane generation model: Validating models with methane recovery rates from 35 Canadian landfills. *Waste Management* 29(7): 2085–2091.

UTE Guatapará. 2014. *Ficha técnica da UTE Guatapará*. Available: < http://www.aneel.gov.br/consulta-processual>. Accessed 01 Feb. 2017.

WASTES – Solutions, Treatments and Opportunities II – Vilarinho, Castro & Lopes (Eds)
© 2018 Taylor & Francis Group, London, ISBN 978-1-138-19669-8

Recycled synthetic waste fibres for the reinforcement of concrete

K. Bendjillali
Department of Civil Engineering, Faculty of Technology, University Amar Telidji of Laghouat, Laghouat, Algeria

M. Chemrouk
Faculty of Civil Engineering, University of Sciences and Technology Houari Boumediene, Algiers, Algeria

B. Boulekbache
Department of Civil Engineering, University Hassiba Benbouali, Chlef, Algeria

ABSTRACT: In this paper the mechanical performances of concrete reinforced with synthetic waste fibres have been investigated. The tests were carried out to examine the effect of the length and the dosage of synthetic waste fibres on the workability, the flexural strength and the compressive strength of concrete. Several concretes were fabricated with three weight contents, 0.25, 0.5 and 1% of 30, 50 and 70 mm long fibres. On the basis of the tests performed, the addition of synthetic waste fibres to the different concretes has a negative effect on their workability but a positive effect on their mechanical behaviour. This effect of fibres is more significant with higher amount (1%) of longer fibres ($l_f = 70$ mm).

1 INTRODUCTION

Each year a large quantity of fibrous wastes is disposed in the nature; this not only causes economic and environmental problems for the society, but also represents a waste of resources (Wang 2010). The fibrous wastes are generally composed of natural and artificial material such as cotton, wool, steel, silk, polyester, nylon, polypropylene, etc. During these last years, many researchers have been focusing on the valorization of different wastes and residues in the field of construction (Benazzouk et al. 2006, Papakonstantinou & Tobolski 2006, Yazoghli-Marzouk et al. 2007, Debieb & Kenai 2008, Corinaldesi & Moriconi 2009, Evangelista & de Brito 2010, Saikia & de Brito 2012, Kunieda et al. 2014, Ahmadi et al. 2017). Such wastes' valorization can lead to cost reduction in the construction industry, enhance the structural performances and most of all protect the environment. Recycling waste fibres in the fabrication of cement based materials is gradually wide spreading throughout the world (Wang et al. 2000, Savastano Jr et al. 2005, Ochi et al. 2007, Sadrmomtazi & Haghi 2008, Aiello et al. 2009, Meddah & Bencheikh 2009, Pereira de Oliveira & Castro-Gomes 2011, Belferrag et al. 2013, Sebaibi et al. 2014, Fraternali et al. 2014, Ghernouti et al. 2015, Al-Tulaian et al. 2016, Borg et al. 2016). In this new recycling technology, synthetic fibres are largely used for reinforcing concrete and mortar compared to other fibres, because of their corrosion resistance and their high energy absorption capacity (Singh et al. 2004, García Santos et al. 2005, Sun & Xu 2009, Alamshahi et al. 2012, Kakooei et al. 2012, Soutsos et al. 2012, Zhang & Zhao 2012, Bendjillali et al. 2013, Cifuentes et al. 2013, Ramezanianpour et al. 2013, Yin et al. 2015). Polypropylene fibres affect less the workability of the fresh concrete (Dreux & Festa, 2002) and engender fewer difficulties during the mixing and placing of concretes than metallic fibres (Meddah & Bencheikh, 2009). They can significantly affect the lifespan of the structure by reducing the permeability, the amount of shrinkage and the expansion of concrete (Kakooei et al. 2012). Some researchers (Yin et al. 2015) have reported that the use of the macro plastic fibres in the construction of pavements, light precast elements and tunnel linings offer a significant cost and environmental benefits over traditional steel reinforcement. In low volume concentrations, the effect of fibres is like a secondary reinforcement for concrete since they control cracking but do not contribute to the load carrying capacity (Söylev & Özturan 2014). The fibres, when efficiently used, are able to increase the first cracking load and reduce the strain of the longitudinal reinforcement (Hamrat et al. 2016). Therefore, the objective of this research is to investigate the effectiveness of synthetic waste fibres in improving the mechanical behaviour of concrete and controlling its cracking.

2 EXPERIMENTAL PROGRAMME

2.1 *Materials*

The concrete used in this work is made from the following constituents:

- Portland cement with additives CEM II/B 42,5 N, fabricated by LAFARGE.
- Limestone crushed sand 0/3 mm with a fineness modulus of 2.73, a sand equivalent of 74% and an absorption of 3.8%. The rate of fine elements (<63 μm) is close to 11%.
- Limestone crushed gravels 3/8 and 8/15 mm with a specific gravity of 2.5, an absorption of 1.5% and 1.8% respectively.
- Superplasticizer of new generation (SIKA VISCOCRETE TEMPO12) in a liquid form, which is a high-range water-reducing admixture based on acrylic copolymer, used in a proportion of 0.65% by weight of cement.
- Synthetic polypropylene fibres coming from the waste of a nearby factory for the fabrication of domestic sweeps (PLAST BROS factory). The fibres have a diameter of the order of 0.80 mm, a specific gravity of 0.99 and a tensile strength of 160 MPa. Three lengths of fibres were used in this work 30, 50 and 70 mm (Fig. 1).

2.2 *Mixing and casting procedure*

This experimental investigation has been conducted on ten concrete mixes using plain concrete PC (without fibres) and fibre reinforced concrete FRC (Table 1). The fibre reinforced

Figure 1.   Synthetic waste fibres used in the concrete mix.

Table 1.   Compositions of concretes (kg/m³).

Mixture	Cement	Sand 0/3	Gravel 3/8	Gravel 8/15	Fibre	Water	Admixture	Slump (mm)
PC*	350	707	100	876	0.0	210	2.3	90
FRC 30–0.25**	350	707	100	876	5.6	210	2.3	75
FRC 30–0.5	350	707	100	876	11.2	210	2.3	55
FRC 30–1.0	350	707	100	876	22.4	210	2.3	30
FRC 50–0.25	350	707	100	876	5.6	210	2.3	53
FRC 50–0.5	350	707	100	876	11.2	210	2.3	30
FRC 50–1.0	350	707	100	876	22.4	210	2.3	00
FRC 70–0.25	350	707	100	876	5.6	210	2.3	35
FRC 70–0.5	350	707	100	876	11.2	210	2.3	10
FRC 70–1.0	350	707	100	876	22.4	210	2.3	00

*: No fibrous concrete;
**: Fibre reinforced concrete with a length of 30 mm and a weight content of 0.25%.

concretes were prepared with the same mix proportions as used in both plain concrete with three weight contents of fibres 0.25, 0.5 and 1%. Sand, gravels and cement were dry premixed first in a rotating drum-type concrete mixer; then, the superplasticizer and half of the total water content were added and mixed. Fibres were manually sprinkled into the mix and dispersed by hand to avoid the formation of fibres balls and achieve a uniform distribution of fibres. Then, the remaining water was added. For preventing any premature evaporation of the mixing water, the prepared specimens ($70 \times 70 \times 280$ mm) were put into plastic bags and cured at a temperature of $25 \pm 2°C$ for 28 days. To get a smooth surface for the application of load, the concrete surface of the specimens was levelled and finished properly.

2.3 *Tests*

The workability of concrete mixes was estimated by the measure of the concrete slump, according to French Standard NPF 18–451. The addition of synthetic waste fibres into the concrete affected negatively its workability, especially the longest fibres or those used with high dosage (Table 1). The flexural tests were carried out in three-point loading on prismatic specimens of 280 mm long and 70 mm square cross-section and the compressive tests were performed on cubes of 70 mm size.

3 RESULTS AND DISCUSSIONS

3.1 *Fresh concrete*

For evaluate the consistency of mixes, we have measured the slump of Abrams cone. The results are reported in the Figure 2. The incorporation of synthetic waste fibres into the concrete affected negatively its workability, especially the longest fibres or those used with high dosage. The use of the long fibres ($l_f = 70$ mm) leads to less workable mixes with a risk of formation of fibre clusters, if they are not perfectly dispersed by hand. On the other hand, short fibres can be easily dispersed in the concrete without the tendency to interlock. In general, the decrease of the workability in the cement composites is due to several factors, such as the nature, the geometry and the percentage of fibres and also to the adhesion of the interface matrix-fibre. The synthetic waste fibres used in this work do not absorb any water, but their incorporation in concrete decreases its workability by increasing the matrix cohesion, which results in the increase of the concrete stiffness. The fibre clusters formed with the use of the long fibres imprison a quantity of paste and consequently reduce the workability.

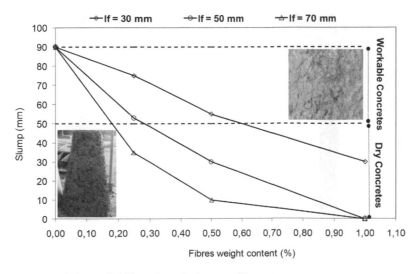

Figure 2.   Results of the workability of synthetic waste fibre concrete.

11

3.2 Flexural strength

The experimental results (Fig. 3) show a small increase in the flexural strength of the concrete at 28 days aging with the addition of synthetic waste fibres. The high flexural strengths are obtained in concrete FRC 70-1.0, with an increase of 11% compared to concrete without fibres. Such a flexural strength increase may not seem to be important, but the main benefit expected from the use of fibres is the improvement of the ductility and the cracking behaviour of the concrete material. It is reported that the toughness and the cracking resistance of concretes are enhanced by the use of the polypropylene fibres (Zhang & Zhao 2012). The quality of the interface matrix-fibre influences more the tensile properties of the composite, by increasing its energy absorption capacity during the cracking propagation. According to some literature (Singh et al. 2004), the interface matrix-fibre reaches its maximal resistance after only two curing days, while in the matrix, the development of the resistance requires more time. It was observed through this work that the flexural strengths of concrete were more improved when the content and the length of fibres were increased.

3.3 Compressive strength

The results of the compressive strength at 28 days of different concretes are presented in Figure 4. An improvement of the compressive behaviour of concretes was observed with the addition of synthetic waste fibres. The positive effect of fibres is more pronounced with higher proportions of longer fibres. The improvement exceeds 15, 18 and 20% in the concretes FRC 30-1, FRC 50-1 and FRC 70-1 respectively. Due to their flexibility, the used polypropylene fibres can easily bend and curve within the mixture without disturbing the aggregates arrangement. This will in turn improve the compactness of the synthetic fibrous concretes, leading to a relatively more important compressive strength of concrete. In this sense, it is worth noting, that some results reporting a decrease in the compressive strength have been published (Ramezanianpour et al. 2013, Boulekbache et al. 2015).

3.4 Failure modes

The concrete showed a significant softening, thanks to the presence of synthetic waste fibres. After the compressive test, the concrete samples without fibres are completely damaged, but with the addition of fibres, multiple thinner cracks are distributed on all the surface

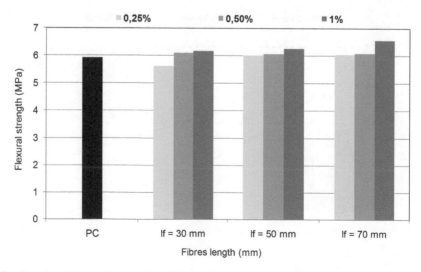

Figure 3. Results of flexural strength at 28 days of synthetic waste fibre concrete.

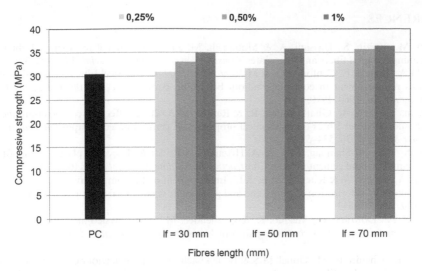

Figure 4.   Results of compressive strength at 28 days of synthetic waste fibre concrete.

a: PC   b: FRC 70-1.0.

Figure 5.   Failure of concrete.

of samples. Under the flexural test, the concrete samples without fibres failed suddenly (Fig. 5-a); in this case only one large crack appears on the concrete surface of the sample and divides it into two parts. The failure of the concrete sample is fragile and brittle. On the contrary, when synthetic waste fibres are added to the concrete, the fragile character of the material is softened and the brittle material becomes a ductile fibre reinforced concrete; the failure becomes more ductile and soft, with just a thinner crack appearing in the concrete sample (Fig. 5-b). Synthetic waste fibres bridge between the lips of cracks in a stitching manner and prevent them from opening. The stitching action of fibres depends on their distribution inside the cement matrix and hence on the workability of the fresh concrete (Chemrouk et al. 2013). Fibres transfer stresses from one side of a crack to the other and, in so doing; they increase the number of fine cracks that are not harmful to a concrete member (Fritih et al. 2013).

3.5 *Conclusions*

Based on the experimental research, the following conclusions can be drawn:

– The workability of synthetic waste fibre reinforced concrete decreases with the length and the weight content of fibres.
– Compared to concretes without fibre reinforcement, the synthetic waste fibre reinforced concretes have a better mechanical behaviour in flexure as well as in compression. The increase of both the weight content and the length of fibres leads to the best mechanical performance of concretes.
– With the presence of synthetic waste fibres, concrete had a better resistance to cracking, exhibited an enhanced ductility and showed a significant softening at failure.

REFERENCES

Ahmadi, M., Farzin, S., Hassani, A. & Motamedi, M. 2017. Mechanical properties of the concrete containing recycled fibers and aggregates. *Construction & Building Materials* 144: 392–398.

Aiello, M.A., Leuzzi, F., Centonze, G. & Maffezzoli, A. 2009. Use of steel fibres recovered from waste tyres as reinforcement in concrete: Pull-out behaviour, compressive and flexural strength. *Waste Management* 29: 1960–1970.

Alamshahi, V., Taeb, A., Ghaffarzadeh, R. & Rezaee, M.A. 2012. Effect of composition and length of PP and polyseter fibres on mechanical properties of cement based composites. *Construction & Building Materials* 36: 534–537.

Al-Tulaian, B.S., Al-Shannag, M.J. & Al-Hozaimy, A.R. 2016. Recycled plastic waste fibers for reinforcing Portland cement mortar. *Construction & Building Materials* 127: 102–110.

Belferrag, A., Kriker, A. & Khenfer, M.E. 2013. Improvement of the compressive strength of mortar in arid climates by valorization of dune sand and pneulmatic waste metal fiber. *Construction & Building Materials* 40: 847–853.

Benazzouk, A., Douzane, O., Mezreb, K. & Queneudec, M. 2006. Physico-mechanical properties of aerated cement composites containing shredded rubber waste. *Cement & Concrete Composites* 28: 650–657.

Bendjillali, K., Chemrouk, M., Goual, M.S. & Boulekbache, B. 2013. Behaviour of polypropylene fibre's mortars conserved in different environments. *European Journal of Environmental & Civil Engineering* 17(8): 687–699.

Borg, R.P., Baldacchino, O. & Ferrara, L. 2016. Early age performance and mechanical characteristics of recycled PET fibre reinforced concrete. *Construction & Building Materials* 108: 29–47.

Boulekbache, B., Hamrat, M., Chemrouk, M. & Amziane, S. 2015. Failure mechanism of fibre reinforced concrete under splitting test using digital image correlation. *Materials & Structures* 48: 2713–2726.

Chemrouk, M., Boulekbache, B., Hamrat, M. & Amziane, S. 2013. Improving the structural behaviour and the sustainability of high performances concrete with the addition of steel fibres. *Academic Journal of Science* 2(2): 359–377.

Cifuentes, H., Garcia, F., Maeso, O. & Medina, F. 2013. Influence of the properties of polypropylene fibres on the fracture behaviour of low, normal and high-strength FRC. *Construction & Building Materials* 45: 130–137.

Corinaldesi, V. & Moriconi, G. 2009. Behaviour of cementitious mortars containing different kinds of recycled aggregate. *Construction & Building Materials* 23(1): 289–294.

Debieb, F. & Kenai, S. 2008. The use of coarse and fine crushed bricks as aggregate in concrete. *Construction & Building Materials* 22: 886–893.

Dreux, G. & Festa, J. 2002. *Nouveau guide du béton et de ses constituants.* 8ème ed, Eyrolles.

Evangelista, L. & de Brito, J. 2010. Durability performance of concrete made with fine recycled concrete aggregates. *Cement & Concrete Composites* 32: 9–14.

French Standard NPF 18-451. 1981. *Bétons–Essai d'affaissement.*

Fraternali, F., Spadea, S. & Berardi, V.P. 2014. Effects of recycled PET fibres on the mechanical properties and seawater curing of Portland cement-based concretes. *Construction & Building Materials* 61: 293–302.

Fritih, Y., Vidal, T., Turatsinze, A. & Gérard, P. 2013. Flexural and shear behavior of steel fiber reinforced SCC beams. *KSCE Journal of Civil Engineering, Structural Engineering.* 17(6): 1383–1393.

García-Santos, A., Rincón, J. Ma., Romero, M. & Talero, R. 2005. Characterization of a polypropylene fibered cement composite using ESEM, FESEM and mechanical testing. *Construction & Building Materials* 19(5): 396–403.

Ghernouti, Y., Rabehi, B., Bouziani, T., Ghezraoui, H., Makhloufi, A. 2015. Fresh and hardened properties of self-compacting concrete containing plastic bag waste fibers (WFSCC). *Construction & Building Materials* 82: 98–100.

Hamrat, M., Boulekbache, B., Chemrouk, M. & Amziane, S. 2016. Flexural cracking behavior of normal strength, high strength and high strength fiber concrete beams, using Digital Image Correlation technique. *Construction & Building Materials* 106: 678–692.

Kakooei, S., Akil, H.M., Jamshidi, M. & Rouhi, J. 2012. The effects of polypropylene fibers on the properties of reinforced concrete structures. *Construction & Building Materials* 27: 73–77.

Kunieda, M., Ueda, N. & Nakamura, H. 2014. Ability of recycling on fiber reinforced concrete. *Construction & Building Materials* 67(C): 315–320.

Meddah, M.S. & Bencheikh, M. 2009. Properties of concrete reinforced with different kinds of industrial waste materials. *Construction & Building Materials* 23(10): 3196–3205.

Ochi, T., Okubo, S. & Fukui, K. 2007. Development of recycled PET fiber and its application as concrete-reinforcing fiber. *Cement & Concrete Composites* 29: 448–455.

Papakonstantinou, C.G. & Tobolski, M.J. 2006. Use of waste tire steel beads in Portland cement concrete. *Cement & Concrete Research* 36: 1686–1691.

Pereira de Oliveira, L.A. & Castro-Gomes, J.P. 2011. Physical and mechanical behaviour of recycled PET fibre reinforced mortar. *Construction & Building Materials* 25(4): 1712–1717.

Ramezamianpour, A.A., Esmaeili, M., Ghahari, S.A. & Najafi, M.H. 2013. Laboratory study on the effect of polypropylene fiber on durability and physical and mechanical characteristic of concrete for application in sleepers. *Construction & Building Materials* 44: 411–418.

Sadrmomtazi, A. & Haghi, A.K. 2008. Properties of cementitious composites containing polypropylene fiber waste. *Composite Interfaces* 15(7–9): 867–879.

Saikia, N. & de Brito, J. 2012. Use of plastic waste as aggregate in cement mortar and concrete preparation: A review. *Construction & Building Materials* 34: 385–401.

Savastano Jr, H., Warden, P.G. & Coutts, R.S.P. 2005. Microstructure and mechanical properties of waste fibre–cement composites. *Cement & Concrete Composites* 27: 583–592.

Sebaibi, N., Benzerzour, M. & Abriak, N.E. 2014. Influence of the distribution and orientation of fibres in a reinforced concrete with waste fibres and powders. *Construction & Building Materials* 65: 254–263.

Singh, S., Shukla, A. & Brown, R. 2004. Pullout behavior of polypropylene fibers from cementitious matrix. *Cement & Concrete Research* 34: 1919–1925.

Söylev, T.A. & Özturan, T. 2014. Durability, physical and mechanical properties of fiber-reinforced concretes at low-volume fraction. *Construction & Building Materials* 73: 67–75.

Soutsos, M.N., Le, T.T. & Lampropoulos, A.P. 2012. Flexural performance of fibre reinforced concrete made with steel and synthetic fibres. *Construction & Building Materials* 36: 704–710.

Sun, Z. & Xu, Q. 2009. Microscopic, physical and mechanical analysis of polypropylene fiber reinforced concrete. *Materials Science & Engineering A* 527: 198–204.

Wang, Y., Wu, H.C. & Li, V.C 2000. Concrete reinforcement with recycled fibers. *Journal of Materials in Civil Engineering* 12(4): 314–319.

Wang, Y. 2010. Fiber and Textile Waste Utilization. *Waste Biomass Valor*. Springer 1: 135–143.

Yazoghli-Marzouk, O., Dheilly, R.M. & Queneudec, M. 2007. Valorization of post-consumer waste plastic in cementitious concrete composites. *Waste Management* 27: 310–318.

Yin, S., Tuladhar, R., Shi, F., Combe, M., Collister, T. & Sivakugan, N. 2015. Use of macro plastic fibres in concrete: A review. *Construction & Building Materials* 93: 180–188.

Zhang, S. & Zhao, B. 2012. Influence of polypropylene fibre on the mechanical performance and durability of concrete materials. *European Journal of Environmental & Civil Engineering* 16(10): 1269–1277.

WASTES – Solutions, Treatments and Opportunities II – Vilarinho, Castro & Lopes (Eds)
© *2018 Taylor & Francis Group, London, ISBN 978-1-138-19669-8*

Leaching characteristics of co-bearing glasses obtained from spent Li-ion batteries

B.J. Forero, J.V. Díaz-Salaverría & P. Delvasto
Universidad Industrial de Santander, Bucaramanga, Santander, Colombia

C.X. Gouveia
Unidad de Gestión en Materiales y Procesos Sustentables, Fundación de Investigación y Desarrollo, Universidad Simón Bolívar, Caracas, Venezuela

ABSTRACT: In this work, the black powder cathode material of spent Li-ion batteries from mobile phones was extracted and then fused together with clear glass cullet and fluxing agents, to produce three formulations of Co-bearing glasses. The cathode powder material and the produced Co-bearing glasses were characterized by means of X-Ray Fluorescence (XRF) and X-Ray Diffraction (XRD) techniques; cathode material was also analyzed by Scanning Electron Microcopy (SEM) with Energy Dispersive Spectroscopy (EDS). In addition, the toxicity characteristics of the Co-bearing glasses obtained was assessed by standardized leaching procedures. The melting of this cathode powder with cullet and fluxes generated an amorphous structure, which confirmed its vitrification. Glasses containing 1.5% Co and 15% borax released the less heavy-metals to solution during leaching, but higher amounts of borax in glass formulation made the glass more leachable and, thus, less suitable for heavy-metal immobilization.

1 INTRODUCTION

Nowadays electronic waste, e-waste for short, is becoming a major problem for most waste managing systems around the world. E-waste fraction in municipal waste streams has been increasing rapidly, worsening the heavy-metal pollution problems associated with common landfilling procedures, particularly in emerging economies (Hilty 2005, Nnoroma & Osibanjob 2008, Robinson 2009, Amankwah-Amoah 2016). The spent batteries present in e-waste streams are particularly dangerous, because of its elevated concentration in toxic metallic elements, unstable chemistry and so on. For these reasons, most Latin American countries have been changing their environmental laws, in order to enforce the correct final disposition of spent household batteries (Blanco et al. 2017). For example, a novel legislation in Colombia (Department of Environment, Housing and Territorial Development 2010) states, as mandatory, the transformation of the metal-bearing substances contained in the spent batteries into valuable or useful materials and bans the final disposition of spent batteries into landfills. Metal-bearing active electrode materials contained in spent batteries can be recycled either by low temperature or by high temperature procedures, to recover metals or other valuable chemical compounds. Among the different available high temperature waste inertization/valorization techniques, vitrification is one of the most used. Vitrification consists in fusing the waste substances with silicate raw materials, to obtain a chemically stable glass matrix. Different waste substances have been treated in this way, including: fly ash (Park & Heo 2002), nuclear waste (Harrison 2014), metallurgical dusts (Pelino et al. 2002) and waste-water sludge (Kikuchi 1998). Nevertheless, vitrification seems to be the less explored valorization procedure for spent batteries. It has been reported that spent alkaline batteries were treated by vitrification techniques (Kuoa et al. 2009), but, to our knowledge, there are no reports on vitrification of the active electrode materials in other important types of batteries, such as Li-ion ones. These batteries contain high amounts of cobalt oxide related substances (Delvasto et al. 2015). Considering the above-mentioned ideas, in this work we extracted the cobalt-bearing fraction of spent Li-ion batteries from cell phones and fused it together with clear glass cullet and fluxing agents. The glasses obtained were chemically analyzed and its heavy-metal toxicity was assessed by standardized leaching procedures (US EPA 1992). The aim of the work is to report the leaching characteristics of some cobalt-bearing glasses obtained from spent Li-ion batteries, as cobalt and flux contents are varied in the formulation of the glass.

2 EXPERIMENT

A sample of 2.6 kg of prismatic Li-ion batteries associated with mobile phone devices was obtained from an e-waste collection system in the city of Bucaramanga, Colombia. The batteries were discharged and later disassembled using metallic pliers. Each of the components of these batteries, including anode and cathode materials, were separated by hand and weighted. The cathode powder material under study, was liberated from the aluminum foil (electron collector in the battery) using a cutting mill, before being separated by sieving (140 mesh). The cathode material powder was characterized by the following techniques: X-ray fluorescence (XRF), X-ray diffraction (XRD) and scanning electron microscopy (SEM-EDS).

The silicate base used to vitrify the cathodic material from Li-ion batteries was crushed glass from transparent beverage containers (hereafter cullet). Besides cathode material from Li-ion batteries and cullet, two types of fluxes were used in the formulation of three cobalt-bearing glasses, borax and lithium tetraborate. Such components were added in different proportions in each of the glasses, as shown in Table 1. The glass identified as "0" corresponds to the cullet used as silicate source.

The components of each glass were weighted, mixed and placed in fireclay crucibles and, subsequently, placed in a muffle. The materials were melted during 5 hours approximately, by heating it from 25°C to 1020°C at a constant rate of 200°C/h and holding the melt at 1020°C for 30 minutes. The molten glass was then poured on a metal tray with water, to promote the formation of a glass frit; excess water was removed from the trays by heat-drying at 100°C for two hours. Finally, the glass frits were grinded in a mortar, for the final production of cobalt-bearing glass powder, which was characterized by means of XRF and XRD techniques. In addition, the toxicity characteristics of the glass powders were determined, following the procedure given by US EPA Method 1311 (US EPA 1992). Heavy metals (Ba, Co, Cr, Mn, Ni, Pb and Zn) were measured in the leachate by means of atomic absorption spectroscopy (AAS) and expressed as ppm (mg/L).

Table 1. Cobalt-bearing glasses formulation.

Id.	Li-ion cathode black powder wt.%	Fluxes		Cullet wt.%
		Borax ($Na_2Ba_4O_7 \cdot 10H_2O$) wt.%	Lithium tetraborate ($Li_2Ba_4O_7$) wt.%	
0	0	0	0	100
1	3	15	1	81
2	10	15	1	74
3	10	30	1	59

3 RESULTS

The different components of the prismatic Li-ion batteries are shown in Figure 1. As it can be seen, most of the weight of the batteries is due to electrode active materials. These materials are two compound ribbons, the first one is made of carbon lithium powder adhered to a copper foil (anode) and the second one is a cobalt lithium oxide layer adhered to an aluminum foil (cathode) (Georgi-Maschlera et al. 2012). The cathode ribbon of the battery represents 37.5% of the total weight. Similar results have been reported in previous studies (Georgi-Maschlerà et al. 2012).

As our procedure showed, once the cathode ribbon is processed in the cutting mill, the 140-mesh passing black powder, containing most of the cobalt lithium oxide, renders 124 g per kilogram of battery. The X-ray fluorescence chemical elemental composition of the cathode black powder obtained is shown in Table 2. The analysis indicates that more than a half of the mass is cobalt, while some other metal elements, such as manganese and nickel are present in high proportions. The above-mentioned elements are usually added to modify the electrical properties of the cobalt lithium oxide (Nitta et al. 2015). It is important to point out that the analysis does not show the presence of lithium, because XRF techniques are unable to detect it. The elevated presence of aluminum is thought to be caused by contamination from the foil present in the cathode ribbon.

In the Figure 2, the morphology of the cathode black powder can be observed, as obtained by SEM in secondary electrons mode. The powder consists of homogeneous and disaggregated particles, with sizes ranging from 1 to 12 μm with a mean size of 6 μm. The energy dispersive x-ray spectroscopy analysis (EDS) performed on the image (embedded table), somehow agrees with the XRF results in Table 2. The X-ray diffraction analysis of the powder, Figure 3(a), indicates that it is mainly composed by a major phase of $LiCoO_2$, despite some peaks in the diffractogram could not be related to other phases within the cathode black powder.

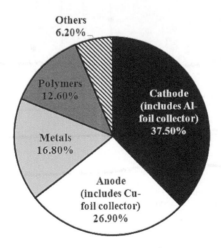

Figure 1.   Weight percentage of each component in the Li-ion batteries analyzed in this work.

Table 2.   Chemical composition of the cathode black powder obtained from the spent Li-ion batteries disassembled in this work.

Element, wt.%																
Al	Ca	Cl	Co	Cr	Cu	Fe	Mg	Mn	Na	Nb	Ni	P	Pb	S	Si	Ti
6.63	0.05	0.15	52.52	0.01	0.26	0.09	0.09	10.47	0.52	0.03	2.63	0.59	0.02	0.18	0.15	0.07

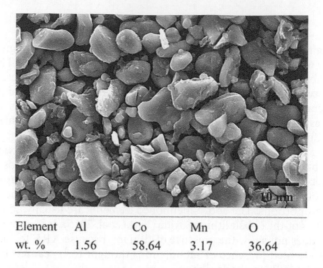

Element	Al	Co	Mn	O
wt. %	1.56	58.64	3.17	36.64

Figure 2. SEM image of the cathode black powder obtained from spent Li-ion batteries. EDS chemical analysis of the whole image is shown in the embedded table.

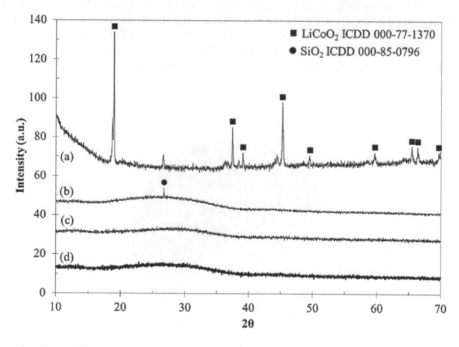

Figure 3. X-ray diffractograms of the black powder from the cathode of spent Li-ion batteries (a) and the three cobalt-bearing glasses prepared in this work: (b) glass 1 (3% cathode black powder, 15% borax); (c) glass 2 (10% cathode black powder, 15% borax) and (d) glass 3 (10% cathode black powder, 30% borax). All three glasses contained 1 % of lithium tetraborate.

The chemical composition of the glasses produced, as obtained by XRF, is shown in Table 3. It must be highlighted that the amount of Co, Al, Ni and Mn increases in the glass as the mass percentage of cathode black powder added increases in the mixture. Some other elements, such as Ca and Si have an opposite trend, because these latter elements are associated with the silicate matrix added. On the other hand, the Na present in the mixture mostly associated with the borax flux added, increases as this additive is incorporated into the

Table 3. XRF elemental chemical composition of the cobalt-bearing glasses prepared in this work.

Element, wt.%

Id.	Al	Ba	Ca	Cl	Co	Cr	Cu	Fe	K	Mg	Mn	Na	Ni	P	Pb	S	Si	Ti	Zr
0	0.84	–	8.62	0.02	–	–	–	0.09	0.44	0.26	–	11.08	–	–	0.01	0.07	32,72	0.04	0.01
1	0.88	–	8.50	0.02	1.53	0.01	–	0.17	0.46	0.24	0.36	12.07	0.08	0.05	0.01	0.07	30.84	0.04	0.01
2	1.03	–	7.83	0.04	5.24	0.01	0.03	0.12	0.43	0.25	1.21	11.75	0.28	0.04	0.01	0.08	28.56	0.04	0.01
3	1.22	0.02	7.41	0.04	6.11	–	0.03	0.13	0.44	0.23	1.41	13.50	0.32	0.05	0.01	0.08	26.87	0.05	0.02

Table 4. Toxicity test (TCLP) results indicating the amount of heavy metals leached out from the Co-bearing glasses prepared in this work.

Id.	Element, ppm						
	Ba	Co	Cr	Mn	Ni	Pb	Zn
0	0.26	0.00	0.00	0.00	0.00	0.00	0.19
1	0.58	7.10	0.01	0.11	0.12	0.00	0.21
2	0.43	107	0.00	0.85	1.18	0.00	0.12
3	8.73	302	0.00	6.07	14.1	0.00	0.01

mixture. The addition of flux in the mixture is intended to obtain a lower melting temperature in the glass. In fact, glasses "1", "2" and "3" were all blue colored glasses that were fluid at the pouring temperature of 1020°C used in this work. On the other hand, this temperature was not enough to melt completely the glass cullet (glass "0"), since beverage container transparent glass usually melts at temperatures above 1400°C (Shelby 2005). Values marked as "–" in Table 3, indicate that the measurement was below the detection limit of the technique (circa 10 mg/kg).

The diffraction pattern obtained for the glass samples "1", "2" and "3" can be seen in Figure 3(b), (c) and (d), respectively. These results show that, in all cases, the melting of the cathode black powder with cullet and fluxes at 1020°C generated an amorphous structure, typical of glassy materials, in which almost no diffraction maxima can be detected and only background signal is registered. Only for glass "1", Figure 3(b), a small peak at $2\theta \approx 27°$, which corresponds to SiO_2 phase, seems to indicate that a very small number of crystalline domains exist within the glassy matrix in this particular sample. This results confirms that the treatments employed contributed to vitrify the cobalt-bearing residue obtained from the cathode of the Li-ion spent batteries.

Nonetheless, the structural vitrification attained is not enough to judge if the cobalt and other heavy-metals contained in the residue are properly immobilized in the glass. To establish the chemical stability of the glasses, it is necessary to place it in contact with an aqueous medium that simulates an accelerated chemical attack of the vitrified solid waste. The TCLP test is useful to fulfill this objective. This test consists in placing the residue in contact with acetic acid containing reagents and measuring the amount of heavy-metals released by the residue to the aqueous solution (US EPA 1992). The results of the test are shown in Table 4.

Although the limits for TCLP results depend on the legislation or each country, if we consider Venezuelan standards (Minister Council 1998) as a Latin American reference, it declares as "non-toxic" any waste whose TCLP leachates do not overpass the following limits (in ppm): Ba: 30; Cr: 4; Ni: 5; Pb: 5. It is worthy to point out that this legislation establishes no specific limits for Zn, Co and Mn in solution. As per these limits, only glass "3" could be considered as toxic, because of its Ni leaching. This glass is also the one that dissolves more Co and Mn. When compared with glass "2" (15% borax by weight), this behavior can be caused by the addition of an excess of borax flux (30% by weight). A borax excess may have made the glass matrix hydrolysable (George 2015), which makes it less suitable for heavy-metal immobilization.

4 CONCLUDING REMARKS

Three formulations of Co-bearing glasses were prepared using the cathode black powder recovered from spent Li-ion batteries. The glasses contained from 1.5 to 6% of cobalt (by weight) and the amount of borax flux used in the formulation varied from 15% to 30% (by weight). It was found that glasses containing 1.5% Co and 15% borax released the least heavy-metals to solution in TCLP tests. Higher amounts of borax made the glass more leachable, which is not recommended if vitrification is chosen as an inertization procedure for this type of wastes.

REFERENCES

Amankwah-Amoah, J. 2016. Global business and emerging economies: Towards a new perspective on the effects of e-waste. *Technological Forecasting and Social Change* 105: 20–26.

Blanco, S., Orta, R., & Delvasto, P. 2017. Influence of Na^+, K^+, Mn^{2+}, Fe^{2+} and Zn^{2+} ions on the electrodeposition of Ni-Co alloys: Implications for the recycling of Ni-MH batteries. *Journal of Physics: Conference Series* 786(1).

Delvasto, P., Niño-Avendaño, C., & Moreno, I. 2015. Urban mining: spent batteries as a metalliferous resource. Paper presented at *VIII Congreso Internacional de Materiales, Paipa, 28–30 October.* Colombia: CIM. ISSN: 2500–6452.

Department of Environment, Housing and Territorial Development. 2010. *Resolution 1297.* (G.S. Office, Ed.) Bogotá, Colombia: Official Journal 47769.

George, J. 2015. Dissolution of borate glasses and precipitation of phosphate compounds. *Doctoral Dissertations.* 2382. USA: Missouri University of Science and Technology.

Georgi-Maschlera, T., Friedrich, B., Weyhe, R., Heegn, H., & Rutz, M. 2012. Development of a recycling process for Li-ion batteries. *Journal of Power Sources* 207: 173–182.

Harrison, M. 2014. Vitrification of High Level Waste in the UK. *Procedia Materials Science* 7: 10–15.

Hilty, L. 2005. Electronic waste—an emerging risk? *Environmental Impact Assessment Review* 25(5): 431–435.

Kikuchi, R. 1998. Vitrification process for treatment of sewage sludge and incineration ash. *Journal of the Air & Waste Management Association* 48(11): 1112–1115.

Kuoa, Y., Changb, J., Jinb, C., Linb, J., & Chang-Chiend, G. 2009. Vitrification for reclaiming spent alkaline batteries. *Waste Management* 29(7): 2132–2139.

Minister Council. 1998. Standards for the control of hazardous materials recovery and hazardous wastes management. *Decree 2635.* Caracas, Venezuela: Official Gazette 5245.

Nitta, N., Wu, F., Lee, J., & Yushin, G. 2015. Li-ion battery materials: present and future. *Materials Today* 18(5): 252–264.

Nnoroma, I., & Osibanjob, O. 2008. Overview of electronic waste (e-waste) management practices and legislations, and their poor applications in the developing countries. *Resources, Conservation and Recycling* 52(6): 843–858.

Park, Y., & Heo, J. 2002. Vitrification of fly ash from municipal solid waste incinerator. *Journal of Hazardous Materials* 91(1–3): 83–93.

Pelino, M., Karamanov, A., Pisciella, P., Crisucci, S., & Zonetti, D. 2002. Vitrification of electric arc furnace dusts. *Waste Management* 22(8): 945–949.

Robinson, B. 2009. E-waste: An assessment of global production and environmental impacts. *Science of The Total Environment* 408(2): 183–191.

Shelby, J. 2005. *Introduction to Glass Science and Technology* (2nd ed.). Cambridge, UK: Royal Society of Chemistry.

US EPA. 1992. *Method 1311: Toxicity Characteristic Leaching Procedure.* United States of America: Environmental Protection Agency.

Valorisation of different wastes: A sustainable approach in the design of new products

A. Teixeira, D. Monteiro, R. Ribeiro & V. Canavarro
University of Porto, Porto, Portugal

B. Rangel
Design Studio FEUP, Faculty of Engineering, University of Porto, Porto, Portugal

J.L. Alves
INEGI and Design Studio FEUP, Faculty of Engineering, University of Porto, Porto, Portugal

ABSTRACT: In 2014, 2,503 million of tons of waste were generated in EU-28 by all economic activities and households, and it is forecasted that this number is not going to be reduced. So, it is a society obligation to search for solutions for this problem, and the designers have an increased responsibility to change this reality. Different wastes were valued through design and engineering, seeking the creation of materials to be applied in the development of new products. The different proposals involved the use of fishing ropes and nets debris, almond by-product residues, coffee waste and Portuguese pine resin as an underexploited resource.

1   INTRODUCTION

According to Eurostat (Statistical Office of the European Union), Portugal generated in 2014, 14,586,917 tons of waste (Eurostat, 2017). On the other hand, Europe (28 Countries) ended the same year with 2,502,890,000 tons and, in 2012, in the United States, 624,700 tons of solid waste are daily generated and the forecast made by The World Bank Organization through the document "What a Waste" (Bhada-Tata, 2012) is that by 2025 it will be 701,709 tons.

Facing this situation, efforts should be made to develop more environmentally friendly solutions, intervening, for example, in using materials that can be reused or recycled. In 2015, under the Community Service Engineering European program, it was proposed a project WeWon'tWasteYOU for the students of the Master Program in Product and Industrial Design, in the scope of the Project Design course, in a partnership with the City Council of Matosinhos. The students were asked to develop products using wastes and/or residues of the industries of the city of Matosinhos that could be produced with low-tech tools, so that a small group of unemployed people were able to start a small production. After ending this unit course, the project motivated some students to continue exploring the use of wastes during their thesis. Below are presented four research works of four students which explored the use of fishing ropes and nets debris, almond by-products residues, coffee waste and Portuguese pine resin as an underexploited resource, that were applied as raw material for new products.

2   RESEARCH FOR USING WASTES IN NEW APPLICATIONS

2.1   *Design as a vehicle for using fishing waste to create new products*

The world's oceans are full of discarded debris that degrade and sink or drift ashore which have a very negative impact in the marine environment. Ghost nets, which are deliberately or accidentally lost in the oceans, usually by fishermen, are responsible of the entrapment and killing of many animals and they also may damage and destroy coral reefs through marine currents (World Ocean Review). According to United Nations Environment Programme, it is estimated that there are 640,000 tons of ghost nets worldwide (Shea, 2014).

Through this research, it was intended to study the potential of waste from fishing activities and the best way to transform them into raw material for new products. According to literature review, two companies deserve to be highlighted since they claim to use only recycled material in the creation of their products. Firstly, Aquafil Group, a global leader in the synthetic fibres industry, created the ECONYL®, a Nylon 6 100% regenerated yarn from fishing ropes and nets (FRN), carpets and cloths (Econyl). Secondly, Bureo conceives innovative solutions to the growing problem of plastic pollution in the oceans by developing products with sustainable design, as for example, skateboards and sunglasses (Bureo).

Followed by the literature review, it was proceeded to the collection of FRN debris at DocaPesca, in Matosinhos, to start the experimental work. Before starting with the experimentation of the transformation of FRN debris, it was important to identify their polymers. Bearing this in mind, a procedure (Monteiro, 2016) of identification of thermoplastics was used for four types of residues that consisted in immersion of the polymers in water and ethyl acetate, scratch them with the fingernail, exposure to the flame and smell of the smoke after being extinguished. The results showed that three of the samples correspond to a High-Density Polyethylene (HDPE), while the other is a Polyamide (PA).

The first transformation experiments were conducted in a vertical injection moulding machine using the debris. The materials were previously cut into small pieces to fill the cylinder of the machine and, then, injected into a small mould. With these experiments, it was verified that the recycled HDPE had a high viscosity, inhibiting the flow of the material through the sprue bushing, having never filled the mould. For last experiments, it was used a silicon mould of a cup where melted material was cast and manually pressed. These experiments showed that a good combination of parameters allows the filling of the mould. However, this moulding method has proved to be unsuitable due to the high viscosity of the material.

Considering the results obtained, a different approach was tried; the direct use of the recycled material. The product has the intention to keep the connection of the FRN debris to where they came from—the ocean and beach. Taking this into consideration, a "bag backboard", made of 100% recycled material from FRN, which allows the user to enjoy two functionalities in a single product was designed (Figure 1). This product will use the FRN in its natural state, fabric made of nylon and injected HDPE, so the bag component will be made using fishing net in its natural state allowing the sand that is always accumulated in towels and objects, such as toys, to fall off. Inside this component, there will be a small pocket made of nylon, so that the user can place smaller belongings with more value to ensure their safety.

Although the performed experiments with 100% recycled HDPE from FRN haven't been positive neither conclusive, the literature review indicated that using 100% recycled FRN and others is a reliable option, depending on the application. With this project, it can be concluded that using recycled materials, as for example, FRN, can be an excellent alternative in product design, creating more sustainable and still aesthetically pleasing solutions, avoiding the extraction of resources from nature.

Figure 1.   Product designed based on the raw material from FRN debris.

2.2 Application of almond nuts by-products in a biocomposite

This project focused on finding a solution through design for the problem of almond nuts industry residues which are by-products of the almond nuts, such as hulls, shells and skins. Through a literature review regarding the agro-industry, it was found that about one third of edible food produced for human consumption is wasted (Europen). In 2013, the almond industry produced 1,823,180 tons of residues (Faostat, International Nut and Dried Fruit Council Foundation, 2015). These, as well as other agricultural wastes, have not received enough attention, ending up being incinerated or used as biomass fuel to produce heat or generate electricity or dumped or used as animal feedstock. Besides these final uses, researchers have already studied the possibility of using the almond by-products in different applications with positive results such as, reinforcement in a polypropylene matrix and a wood based composite using urea-formaldehyde resin, absorbent of heavy metals and dyes, growing media for soilless culture and for preparing activated carbons.

According to the information from the previous paragraph, it was pretended with this project to study the possibility of employing the almond by-products in new products, by using them as a reinforcement of a polymer, thus creating a mouldable composite material. It is well known that oil-based plastic materials have unique and innumerable properties which have a big interest for manufacturers but it is also well known that they are one of the biggest pollutants on Earth. Concerning this, it was decided to use in the experimental work a PLA manufactured by NatureWorks® with the brand name of Ingeo™, which is a biodegradable biobased polymer, so a more environment friendly solution concerning its health.

Bearing this purpose in mind, by blending both materials, PLA and almond by-products, a biocomposite was obtained, which is a material 100% originated from renewable resources. For the experimental work, the almond by-products were grinded and milled in small particles. After this, the particles were sieved using a vibratory sieve shaker with the following sieves: 630, 400, 315, 125, 90 and 53 μm. Regarding the nature of these natural fillers, before proceeding to any blend, the particles were dried in an oven to remove the majority of its moisture. Plates were successfully created by blending 10% to 50% of reinforcement of different particle sizes. However, in all particle-sizes mixtures, it is notable the increasing of the roughness and brittleness with the amount of reinforcement. Besides, increasing the particle size changes the product aesthetics, since it is possible to clearly observe the particles in the PLA mixture. It is important to refer that there isn't an ideal blending composition since this will vary depending on the intended end use for the material.

After finishing the blending experimentation, a composite material mixture with 20% of reinforcement and particle-size range from 53 μm to 125 μm was manually pressed in a silicon mould of a cup. The result achieved was very interesting, with great aesthetics, but the material still needs to be characterized in order to know its properties.

After achieving the main goal of this research, the design project was the creation of a laptop stand, which was intended to be low weight, adjustable, small size, and, if possible, multifunctional or adaptable to tablets or smartphones. This product can be used as a vertical support for smartphones and tablets, with two adjustable positions in height and maximum dimensions of $5 \times 7 \times 2$ cm (length $\times$ width $\times$ height) (Figure 2). In order to be small and easy to pack, this product uses two separated laptop-stands. Considering the sizes of tablets, it may be necessary to use the two laptop stands together to hold the equipment.

This experimental work, using the almond by-products, presented positive results with room for improvement and more research. The composite can be produced with different

Figure 2.   Concept design of the laptop stand using the composite PLA + almond by-products.

reinforcement percentages and particle-size distribution, being also possible to combine bigger particles with very small particles. After characterization, it will be possible to know its potential applications.

2.3 *Coffee waste reused as a composite material*

Based on data provided by the International Coffee Organization, it is estimated that its overall consumption in 2014 was 149,8 million bags of 60 kg (Organization, 2015a). In the case of Portugal, in the same year, 823,000 bags of 60 kg were consumed, which represent a per capita consumption of around 4,7 kg (Organization, 2015b). According to the same source, the annual coffee demand from 2011 to 2014 increased by 2,4%. It is further noted that this increase is gradual, being registered every year. It is therefore plausible to deduce that this tendency will continue.

The research developed (Canavarro, 2016) aims to present a production method for reuse coffee ground leftovers. The main goal is to obtain a mouldable material, composed by coffee grounds and a suitable binder. It is also intended that the created material is durable and washable, with prospects of extending the realities in which it can be used. The composite material developed within this mixture can be used in a variety of products, from simple vases, coffee table tops, to light emitting objects with original effects. The options are vast and abundant.

The base material used in the experiments was coffee grounds, obtained domestically and the only treatment given was, in some cases, heating it in an oven at 100°C for 24 hours, in order to eliminate its humidity. In all other cases, it was used as is. So, in conclusion, it is obvious that the experimental work was fundamental in obtaining and developing this material. The epoxy resin experiment was valid in order to determine that coffee waste can, in fact, be mixed with other substances to achieve a composite material. It didn't work however, because it is not safe for food contact and is not biodegradable. Starch showed some promising results at the beginning, but with time it revealed great negative characteristics, with the appearance of mould on all samples. It also couldn't withstand contact with cold water, being therefore discarded. Pine resin displayed very weak results, because it couldn't be homogenized with coffee waste, not even with the addition of wax. PLA exceeded the initial expectations. It is biodegradable and the experiments showed very good results, with different formulas, permitting its use in different areas of interest. It is also FDA approved for food contact. The final material is not, however, perfect or completely developed. It does not withstand high temperatures like hot beverages (coffee). There is, for that reason, room for improvement.

The main design project is a tray with capacity for 6 cups. The cups are made out of our composite (preferably 40% coffee grounds waste), as well as the two side bases/handles. The tray platform is made out of light wood to introduce some contrast to the set. The cups may then be used for all kinds of cold beverages, desserts, condiments, spices, jams. The list is vast, as long as it's nothing hot (temperature wise) (Figure 3). The two side legs/handles are simple structural pieces with an orifice to facilitate handling. These were made to try to demonstrate other uses for the created material, with a higher thickness. The wooden tray has 6 holes for the cups and some geometric cuts at each end to make the joinery with the legs/handles.

Because the experimental work presented such positive results in terms of formulas between coffee waste and PLA, with so many different visual and physical effects, other

Figure 3.   Group of products created with the PLA+coffee waste mixture.

design possibilities (besides the main project) were thought of, trying to embrace the different potentials offered by the material. Some include the contact with food, others the good relation between coffee waste and plants, and even the effect that some samples have when placed against a light source.

2.4 *Bio-valorisation potential of the Portuguese pine resin as matrix of sustainable composites*

This project focused on the creation of proposals for revitalize the use of pine resin, a resource with high potential that is currently underexploited in Portugal. The idea is to take advantage of this natural resource as an opportunity to integrate matrices of sustainable composites of flax natural fibres, to apply in an ecological product line to promote the Portuguese Design. Due to the fragility characteristic of this resin, experiments were carried out with other materials, so that they can be used as additives to increase rosin plasticity. In this sense, materials from natural and synthetic waste were used, namely: cork powder, rubber granulate, wax, among others.

The traditional pine-tapping is the most common process for obtaining rosin, in which the resin appears naturally in the pine tree to protect and cover the wounds caused by the incisions, being collected and stored in plastic bags or vases (Abad Viñas et al., 2016). The potential for the production of rosin in Portugal is immense, since the pine tree (*Pinus pinaster*) represented in 2010 about 22% of the forest area in the national territory, corresponding to 714,000 hectares (ICNF, 2013). In Portugal, warm temperatures allow each pine to produce, on average, 3 to 4 kg of resin per year (Anastácio and Carvalho, 2008). However, the annual production of resin in Portugal in 2005 was about 5,000 tons, and there is currently installed capacity to increase this value considerably, to 20,000 tons (Anastácio and Carvalho, 2008). The applications of the pine resin derivatives are very broad and are mainly used by the processing industries as raw material present in industrial products, pharmaceuticals, and adhesives.

The research (Ruben, 2016) began with the preparation of several samples of rosin with each waste individually, varying the proportion of additive between 10% and 50%. Four different additives were considered: cork powder as a waste of natural origin; granulated rubber (particle size < 0.8 mm) and wax as wastes of synthetic origin; and EVA (ethylene vinyl acetate) in granules as a virgin material of synthetic origin. In terms of physical resistance, it was possible to see that the combination of pine resin and cork powder was brittle, since more voids were created that contribute to the fragilization of the samples. On the opposite side, the wax waste, when mixed with the pine resin, promoted an increase on samples ductility. However, the use of this feature modifies the original visual appearance of the resin, making it greenish. The processing of rubber granules led to the emission of fumes and odours, proving to be a dangerous and flammable additive. The results with this material were not very positive since the rubber waste granulate did not completely melt with the resin and the original appearance of the rosin changes completely, making the samples black and opaque. The virgin EVA presented itself as the most promising additive, showing complete compatibility with the pine resin, giving more flexibility, especially in the proportion of 30%, without significantly altering their original aspect.

After the research of the material for the composite matrix, pieces of furniture and lighting were designed (Figure 4). The concept consists of semi-structural composite pieces produced in a pine resin matrix with 30% EVA, reinforced by flax fibres. The composite pieces appear attached to complementary pieces produced in cork. The hollow composite part allows for some storage space for objects or a spotlight, so that the bench can be used as a lighting element.

With this work, it was possible to realize that the untapped potential of the Portuguese pine resin can also be considered a waste. The fragility of this natural resource can be improved with the use of natural and synthetic additives and the waste can have great potential as an alternative of less environmental impact. Moreover, since EVA was the only non-waste material, it is possible to carry out future experiments with EVA from shoe soles.

Figure 4.   Composition of the concept products idealized to apply the composite material.

3   CONCLUSIONS

From these experiences, it was concluded that waste in general is growing and there are no predictions for this growth to slow down. Although there are already waste treatment strategies in place, if it could be reused, this problem might not exist in a near future. Also, the use of most wastage materials allows for a low-cost approach, making this an unrivalled advantage. For this reason, it becomes apparent that waste really is a good choice for substances to be reused as a material on the creation of design products. It is also obvious that its abundance will continue to exist and designers should see this as an opportunity.

4   FUTURE DEVELOPMENTS OF THE PROJECT

The accomplishment of these research works with positive and promising results, raised a wave of motivation and interest for the use of wastes, entrepreneurially. Thus, in the near future, it is intended to continue the WeWon'tWasteYOU project by creating a social workshop—OFICINAdesign: School of Social Entrepreneurship—in which teams of designers and engineers will work together with groups of unemployed people. The necessary infrastructures are going to be prepared for this project for the creation of products based on the wastes generated by the local industry. Besides, training courses are going to be made for workers. With this initiative, it is expected to create products in a more sustainable way and appeal to the environment awareness of the society as well as creating more job opportunities for a vulnerable social group.

ACKNOWLEDGEMENTS

Funding of Project NORTE-01-0145-FEDER-000022 - SciTech—Science and Technology for Competitive and Sustainable Industries, cofinanced by Programa Operacional Regional do Norte (NORTE2020), through Fundo Europeu de Desenvolvimento Regional (FEDER), and Community Service Engineering European Project (LLP - 539642 - Community Service Engineering).

REFERENCES

Abad Viñas, R., Caudullo, G. & De Rigo, D. 2016. *Pinus pinaster* in Europe: distribution, habitat, usage and threats. *European Atlas of Forest Tree Species*. Luxembourg: European Commission.
Anastácio, D. & Carvalho, J. 2008. *Sector dos Resinosos em Portugal. Evolução e Análise*. Lisbon: Dgrf.
Bhada-Tata, Daniel Hoornweg e Perinaz. 2012. What A Waste—A Global Review Of Solid Waste Management. *Urban Development Series Knowledge Papers* 15:1–116.
BUREO. Bureo [Online]. Bureo. Available: http://bureo.co/ [Accessed 30/05/2016].
Canavarro, V. 2016. *Coffee Powder Reused as a Composite Material: a Step in the Right Direction*. Master Thesis MDIP. UPorto, Portugal.
ECONYL. *Econyl* [Online]. Econyl. Available: http://www.econyl.com/ [Accessed 30/05/2016].

EUROPEN. *Food Waste* [Online]. EUROPEN. Available: http://www.europen-packaging.eu/policy/7-food-waste.html [Accessed 07/08/2016].

EUROSTAT. 2017. *Generation of waste by waste category, hazardousness and NACE Rev. 2 activity* [Online]. Eurostat. Available: http://appsso.eurostat.ec.europa.eu/nui/show.do?dataset = env_ wasgen&lang = en [Accessed 15/06/2017].

FAOSTAT. FAOSTAT. Available: http://faostat3.fao.org/browse/Q/QC/E [Accessed 28/08/2016].

ICNF 2013. *IFN6-Áreas dos usos do solo e das espécies florestais de Portugal continental. Resultados preliminares*. Lisbon: Instituto da Conservação da Natureza e das Florestas.

International Nut and Dried Fruit Council Foundation 2015. Global Statistical Review 2014–2015. Inc.

Monteiro, D. 2016. *Design como veículo para o reaproveitamento dos resíduos de cordas e redes de pesca para a criação de produtos*. Master Thesis MDIP. UPorto, Portugal.

Organization, I.C. 2015a. *The Current State of the Global Coffee Trade* [Online]. Available: http://www.ico.org/monthly_coffee_trade_stats.asp [Accessed 10.12.2015 2015].

Organization, I.C. 2015b. *Portugal—Latest facts and figures about the global coffee trade from the International Coffee Organization* [Online]. Available: http://infogr.am/_/S268nTQ9nsOy58h5VJZH [Accessed 10.12.2015 2015].

Ruben, R. 2016. *Biocompósitos no Design: Estudo da Resina de Pinheiro Portuguesa e do Linho em Fibra*. Master Thesis MDIP. UPorto, Portugal.

Shea, J. 2014. Ghost Fishing Nets: Invisible Killers in the Oceans [Online]. *Earth Island J.* Available: http://www.earthisland.org/journal/index.php/elist/eListRead/ghost_fishing_nets_invisible_killers_in_the_oceans/ [Accessed 08/03/2016].

World Ocean Review. Litter—pervading the ocean [Online]. *World Ocean Review*. Available: http://worldoceanreview.com/en/wor-1/pollution/litter/ [Accessed 15/03/2016].

WASTES – Solutions, Treatments and Opportunities II – Vilarinho, Castro & Lopes (Eds)
© 2018 Taylor & Francis Group, London, ISBN 978-1-138-19669-8

Belgian and Portuguese apple tree bark and core: Comparison of antioxidant content

M.M. Moreira, M.F. Barroso, S. Morais & C. Delerue-Matos
REQUIMTE/LAQV, Instituto Superior de Engenharia do Instituto Politécnico do Porto, Porto, Portugal

A. Boeykens
School of Technology-Chemistry, Odisee University College, Ghent, Belgium
Cluster Bioengineering Technology (CBeT), Faculty of Engineering Technology, KU Leuven, Ghent, Belgium

H. Withouck
School of Technology-Chemistry, Odisee University College, Ghent, Belgium

ABSTRACT: This study investigated the total phenolic (TPC) and flavonoid (TFC) contents, as well as the antioxidant activity of extracts of bark and core from apple trees from Belgium and Portugal. Conventional and Microwave-Assisted Extraction (MAE) were tested and compared. Overall, the Portuguese apple trees extracts obtained by MAE exhibited the highest TPC (79.4 ± 3.7 mg GAE/g DW), 2,2-diphenyl-1-picrylhydrazyl radical scavenging activity (DPPH-RSA; 46.6 ± 4.8 mg TE/g DW) and ferric reducing antioxidant power (FRAP; 29.8 ± 2.4 mg AAE/g DW). HPLC analysis revealed that phloridzin was the predominant compound in all the obtained extracts of the apple tree bark and core from both countries investigated. This study highlighted the potential of using apple tree extracts as natural antioxidant substances for food and cosmetic applications.

1 INTRODUCTION

Malus domestica, which belongs to the Rosaceae family, is one of the most widely cultivated fruit tree. Different classes of phenolic compounds, such as hydroxycinnamic acids, dihydrochalcones, flavonols and catechins, have been identified and quantified in apple fruit, which health benefits are well recognized (Kalinowska et al., 2014). However, information concerning the phenolic composition and antioxidant activity from apple tree residues is scarce. Until now, only one report concerning the purification of phloretin from Fuji apple tree bark has been published (Xü et al., 2010). Therefore, the characterization of apple tree residues regarding their antioxidant activity is an important area of investigation.

Belgium and Portugal have a dominant apple-producing area, which in 2014 accounted for approximately 300 000 tons of apples produced (FAOSTAT, 2015), that generated large amounts of wood residues. These wastes are mainly used as firewood, which is not economically efficient and leads to environmental problems. Taking in consideration the antioxidant capacity associated with apple fruits (Kalinowska et al., 2014), the extraction of phenolic compounds from apple tree wood residues could represent an interesting way of valorization of these wastes. Thus, the aim of this study was to determine the phenolic composition and assess the *in vitro* antioxidant activity of the apple tree bark and core extracts from the *Malus domestica* 'King Jonagold' grown under Belgium and Portugal climatic conditions. A comparison of the antioxidant recoveries from trees of different origins will provide a basis for the use of apple tree wood residues as a functional food ingredient that can replace synthetic compounds. Moreover, some previous studies indicated that phenolic content and antioxidant activity are influenced by the differences in the agroclimatic conditions (Kalinowska et al., 2014, Yuri et al., 2009).

2 MATERIAL AND METHODS

2.1 *Chemicals and solvents*

The used solvents, reagents, and standards were of analytical grade. Ethanol absolute anhydrous (p.a.) was bought from Carlo Erba (Peypin, France), while methanol and formic acid (HPLC grade) were obtained from Merck (Darmstadt, Germany). Ascorbic acid (AA), gallic acid (GA), trolox (6-hydroxy-2,5,7,8-tetramethylchroman-2-carboxilic acid) and (-)-epicatechin were acquired from Sigma Aldrich (Madrid, Spain). DPPH (2,2-diphenyl-1-picrylhydrazyl), Folin–Ciocalteu reagent, sodium-carbonate, TPTZ (2,4,6-Tris(2-pyridyl)-s-triazine), aluminum chloride hexahydrate and the phenolic compound standards used in HPLC analysis were bought from Sigma-Aldrich (Steinheim, Germany).

2.2 *Sample*

Malus domestica 'King Jonagold' bark and core were separated from Belgium and Portuguese trees with 17 and 10-year-old, respectively. Samples were dried at 60°C until constant weight, grounded in a mill (Retsch ZM200, Germany) and sieved to particles smaller than 0.08 mm.

2.3 *Extraction techniques*

Microwave-assisted extraction (MAE) was carried out with a MARS-X 1500 W (Microwave Accelerated Reaction System for Extraction and Digestion, CEM, Mathews, NC, USA) as previously described by Moreira et al. (2017). 0.1 g of dried sample was extracted with 20 mL of ethanol:water (60:20, v/v) at 100°C, with medium stirring during 20 min.

The conventional extraction (CE) was performed as described by Moreira et al. (2017). Briefly, 0.5 g of dried bark or core were extracted in a water bath shaker (Shaker C type BSC127E, OVAN) at 100 rpm with 20 mL of ethanol:water (50:50, v/v) at 55°C during 2 h.

The attained extracts (from CE and MAE) were centrifuged at 4000 rpm during 10 min, and stored in the freezer at −20°C until further analysis. All the extractions were performed in duplicate.

2.4 *Total phenolic, flavonoid and antioxidant activity determination*

Total phenolic (TPC) and total flavonoid (TFC) contents were determined as described in detail by Paz et al. (2015). Antioxidant activity was evaluated by the DPPH-radical scavenging activity (DPPH-RSA) and ferric reducing antioxidant power (FRAP) assays (Paz et al., 2015). All measurements were performed in triplicate in a microplate reader (96-well plates, Nunc™ microwell, Denmark).

2.5 *High-performance liquid chromatography analysis*

Qualitative and quantitative analysis of phenolic compounds from apple wood extracts was performed by HPLC (Shimadzu, Japan) with diode array detector (DAD; SPD-M20 A, Shimadzu, Japan) according to Moreira et al. (2017). Phenolic compounds separation was carried out by a *Phenomenex* Gemini C_{18} column (250 mm × 4.6 mm, 5 μm), which was operated at a constant temperature of 25°C. The chromatographic conditions were set as follows: flow rate 1.0 mL/min, sample injection volume 20 μL, mobile phase A (methanol) and mobile phase B (water) both with 0.1% formic acid. The gradient program applied was: 0–13 min: 20–26.5% A; 13–18 min: 26.5% A; 18–25 min: 26.5–30% A; 25–50 min: 30–45% A; 50–60 min: 45–50% A; 60–70 min: 50–55% A; 70–90 min: 55–70% A; 90–100 min: 70–100% A, followed by 100% A for 5 min and back to 20% A in 10 min and 5 min of reconditioning before the next injection. Detection was carried out by scanning between 190 and 600 nm, and the quantification was made at 280, 320 or 360 nm depending on the maximum absorption of the phenolic compound.

3 RESULTS AND DISCUSSION

3.1 *Total phenolic and flavonoid contents*

In this study, the total amount of phenolic compounds in the extracts of apple tree core and bark ranged from 10.5 ± 0.6 to 27.0 ± 1.8 mg GAE/g DW and from 22.9 ± 0.3 to 79.4 ± 3.7 mg GAE/g DW, respectively (Table 1).

To the best of our knowledge, the comparison of phenolic compounds recovery from apple tree bark and core residues between different countries has not been reported yet. Despite of that, a comparison between our contents and the values reported in the literature for bark from other type of trees can be made (Alañón et al., 2011, Chupin et al., 2013, Deng et al., 2015, Naima et al., 2015). The results obtained in this work are comparable to the previous reported by Chupin et al. (2013) for bark of different maritime pine trees, which values ranged from 19.34 ± 4.90 to 96.81 ± 1.32 mg GAE/g DW. Also, the TPC found in the ethanol extracts of bark from *Solidago Canadensis* L. was at least 6-times lower (1.58 ± 0.02 mg GAE/g; Deng et al. (2015)), whereas Naima et al. (2015) reported higher amounts of phenolics in *Maroccan Acacia mollissima* bark, ranging from 131.5 ± 0.5 to 441.6 ± 0.3 mg GAE/g DW. Such differences could be related not only with the type of sample studied, but also with the extraction conditions and/or the applied technique. In fact, our results are in agreement with these observations, demonstrating that MAE was more efficient in polyphenols recovery from apple tree wood residues than CE technique (Table 1). These differences in TPC values between the extraction techniques were already expected as different authors have reported the same trend (Aspé and Fernández, 2011, Bouras et al., 2015, Naima et al., 2015). This may be explained by the principle behind the extraction technique applied. In MAE technique, the extraction solvent penetrates more easily inside the cell walls of the matrix and the linkages between the phenolics and cell walls are easily broken increasing the amount of phenolic compounds released (Drosou et al., 2015). Concerning the differences between the samples (Table 1), there is an ongoing trend with Portuguese apple tree samples presenting the highest phenolic and flavonoid contents. This pattern of variation may be explained by the different atmospheric conditions that the apple trees from both countries are subjected before the collection of the sample. Moreover, the soil composition, as well as the age of the studied apple tree can also justify these differences (Chupin et al., 2013, Dedrie et al., 2015, Naima et al.,

Table 1. Total phenolic (TPC) and total flavonoid (TFC) contents obtained for microwave-assisted extraction (MAE) and conventional extraction (CE) extracts of apple tree bark and core from Belgium and Portugal; values are mean ± standard deviation ($n = 3$).

Sample	Country	Extraction technique	TPC (mg GAE/g DW)*	TFC (mg EE/g DW)**
Bark	Belgium	MAE	39.9 ± 1.7	13.5 ± 0.3
		CE	22.9 ± 0.3	8.5 ± 0.6
	Portugal	MAE	79.4 ± 3.7	34.0 ± 2.4
		CE	51.9 ± 3.8	28.6 ± 2.4
Core	Belgium	MAE	19.6 ± 1.7	10.5 ± 0.7
		CE	10.5 ± 0.6	3.6 ± 0.3
	Portugal	MAE	27.0 ± 1.8	17.2 ± 0.9
		CE	17.6 ± 0.6	10.7 ± 0.9

*TPC values are expressed as mg gallic acid equivalents/g dry weight.
** TFC values are expressed as mg epicatechin equivalents/g dry weight.

2015, Yuri et al., 2009). Recently, Dedrie et al. (2015) evaluated the relationship between age and TPC of oak bark from Wallonia. These authors observed that the highest TPC were found in oak bark collected from 25-year-old trees (13.4 ± 0.2 mg GAE/g DW), which was 2.7 times higher than those reported for 140-year-old tree. A correlation between age of the tree and TPC values was also observed in our study, as the Portuguese apple tree was younger than the apple tree from Belgium.

Regarding the TFC assay, the same trend as previously described for TPC was found, with Portuguese apple tree extracts presenting at least 2 times more flavonoids than the Belgium apple tree extracts (34.0 ± 2.4 *versus* 13.5 ± 0.3 mg EE/g DW, respectively). As flavonoids are a group of phenolic compounds, a correlation between the results from TPC and TFC assays was already expected. In the present study, the correlation coefficient found was 0.89 advocating a good relationship for all the obtained results from both assays.

3.2 *Antioxidant activity*

It is well established that plants are a potential source of natural antioxidants, which act as reducing agents, hydrogen donators, oxidant and free radical scavengers. To evaluate the antioxidant activity of the obtained extracts, two antioxidant activity assays, namely DPPH-RSA and FRAP methods, were employed. As shown in Figure 1, there were wide variations in the DPPH and FRAP values of the apple tree bark and core extracts. For core extracts, the DPPH values ranged from 10.7 ± 0.7 (Belgium) to 32.3 ± 2.3 (Portugal) mg TE/g DW, and FRAP from 12.0 ± 0.8 (Belgium) to 22.2 ± 0.7 (Portugal) mg AAE/g DW. In the case of bark extracts, DPPH content varied from 21.3 ± 0.6 (Belgium) to 46.6 ± 4.8 (Portugal) mg TE/g DW, and FRAP assay from 11.3 ± 0.7 (Belgium) to 29.8 ± 2.4 (Portugal) mg AAE/g DW. The DPPH and FRAP values of bark were higher than those of core for both the extraction techniques tested. Deng et al. (2015) also reported that bark extracts from *Solidago Canadensis* L. prepared by advanced extraction techniques (ultrasound-assisted and high hydrostatic pressure-assisted extraction) exhibit higher DPPH values in comparison to ethanol extracts obtained by CE. In a previous work from Aspé & Fernández (2011), extracts from *Pinus radiata* bark obtained by MAE also revealed higher antiradical properties than the Soxhlet extracts. These authors demonstrated that MAE is more efficient in the preservation of the antioxidant activity from extracts than the traditional techniques, which is in accordance to the results found in our study.

As previously mentioned, the same behavior was observed for the TPC and TFC assays, with extracts from Portuguese apple tree presenting the highest antioxidant activity. In fact, there is a strong positive correlation between total phenolic as well as flavonoid contents and the antioxidant activity assessed by the DPPH-RSA and FRAP methods ($r = 0.82$ to 0.92). The strongest correlation was found between the antioxidant activity of the extracts of apple

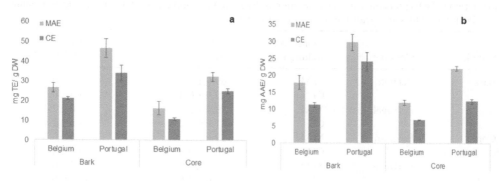

Figure 1. Antioxidant activity determined by a) DPPH-RSA and b) FRAP assays for microwave-assisted (MAE) and conventional extraction (CE) extracts of apple tree bark and core from Belgium and Portugal. Values are mean ± standard deviation ($n = 3$).

tree residues determined by the FRAP and TFC assays. Strong relationships between phenolic and flavonoid contents, as well as antioxidant activity of wood extracts has also been reported in other studies (Alañón et al., 2011, Bouras et al., 2015, Deng et al., 2015).

3.3   *Identification and quantification of phenolic compounds*

The HPLC phenolic profile from bark and core from Belgium and Portuguese apple trees was characterized. Figure 2 exhibits representative chromatograms of Belgium and Portuguese apple bark MAE extracts. HPLC chromatograms for CE extracts are not showed as their phenolic profile is identical to the MAE extracts, only with lower peak intensity. As can be seen in the Figure 2, no differences in the phenolic profile between the two MAE extracts were noticed. However, as expected the bark from Portuguese apple tree presented the highest content of phenolic compounds detected compared to the Belgium sample (sum of all identified phenolic compounds: 70.9 ± 3.5 (Portugal samples) *versus* 33.6 ± 1.7 mg/g DW (Belgium samples)). The major component identified in the bark extracts was the dihydrochalcone phloridzin, which accounted for at least 70% of all the phenolic compounds identified and quantified. These results are in close agreement with those found by other authors for apple tree leaves (Rana et al., 2016, Walia et al., 2016). The main differences in the phenolic profile between Belgium and Portuguese apple tree bark were found for the detected levels of ferulic acid (19.1 ± 1.0 *versus* 86.4 ± 4.3 mg/100 g DW, respectively) and cinnamic acid (56.4 ± 2.8 *versus* 191 ± 10 mg/100 g DW, respectively), and for the flavonoids catechin, quercetin-3-*O*-glucopyranoside and kaempferol-3-*O*-glucoside (18.6 ± 0.9 *versus* 187 ± 9, 161 ± 8 *versus* 230 ± 11 and 58.1 ± 2.9 *versus* 274 ± 14 mg/100 g DW, respectively). These quantitative differences might be associated with the geographical area, as well as with the climatic-meteorological and cultivation conditions.

In conclusion, the results of this study demonstrated that apple tree residues, and mainly bark, have potential to be a promising source of antioxidants and functional food ingredients. Their valorization may lead to economic gains and prevent or decrease environmental problems caused by accumulation of these wastes. Also, geographical origin of the trees may have a great impact on the levels of ferulic and cinnamic acids, and on the contents of flavonoids, such as catechin, quercetin-3-*O*-glucopyranoside and kaempferol-3-*O*-glucoside.

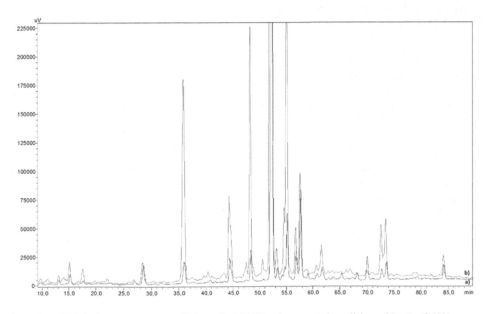

Figure 2.   HPLC chromatograms at 280 nm for MAE bark extracts (conditions: 20 mL of 60% aqueous ethanol, 100°C, 20 min, 0.1 g, and medium stirring speed) from a) Belgium and b) Portuguese apple tree.

ACKNOWLEDGMENTS

Manuela M. Moreira (SFRH/BPD/97049/2013) and M. F. Barroso (SFRH/BPD/78845/2011) are grateful for their postdoctoral fellowships financed by POPH-QREN—Tipologia 4.1—Formação Avançada, subsidized by Fundo Social Europeu and Ministério da Ciência, Tecnologia e Ensino Superior. Authors are also grateful for financial support from project NORTE-01-0145-FEDER-000011-311 Qualidade e Segurança Alimentar—uma abordagem (nano) tecnológica. The financial support from FCT/MEC through national funds and co-financed by FEDER, under the Partnership Agreement PT2020 through the project UID/QUI/50006/2013—POCI/01/0145/FERDER/007265, is also acknowledged; the project 6818 —Transnational Cooperation, Agreement between Portugal (FCT) and Serbia (MSTD) is also acknowledged. Annick Boeykens wishes to acknowledge Odisee for funding by means of a PWO grant (Flemish Government).

REFERENCES

Alañón, M.E., Castro-Vázquez, L., Díaz-Maroto, M.C., Hermosín-Gutiérrez, I., Gordon, M.H. & Pérez-Coello, M.S. 2011. Antioxidant capacity and phenolic composition of different woods used in cooperage. *Food Chem* 129: 1584–1590.

Aspé, E. & Fernández, K. 2011. The effect of different extraction techniques on extraction yield, total phenolic, and anti-radical capacity of extracts from Pinus radiata Bark. *Ind. Crops Prod.* 34: 838–844.

Bouras, M., Chadni, M., Barba, F.J., Grimi, N., Bals, O. & Vorobiev, E. 2015. Optimization of microwave-assisted extraction of polyphenols from Quercus bark. *Ind. Crops Prod.* 77: 590–601.

Chupin, L., Motillon, C., Charrier-El Bouhtoury, F., Pizzi, A. & Charrier, B. 2013. Characterisation of maritime pine (Pinus pinaster) bark tannins extracted under different conditions by spectroscopic methods, FTIR and HPLC. *Ind. Crops Prod.* 49: 897–903.

Dedrie, M., Jacquet, N., Bombeck, P.-L., Hébert, J. & Richel, A. 2015. Oak barks as raw materials for the extraction of polyphenols for the chemical and pharmaceutical sectors: A regional case study. *Ind. Crops Prod.* 70: 316–321.

Deng, Y., Zhao, Y., Padilla-Zakour, O. & Yang, G. 2015. Polyphenols, antioxidant and antimicrobial activities of leaf and bark extracts of Solidago canadensis L. *Ind. Crops Prod.* 74: 803–809.

Drosou, C., Kyriakopoulou, K., Bimpilas, A., Tsimogiannis, D. & Krokida, M. 2015. A comparative study on different extraction techniques to recover red grape pomace polyphenols from vinification byproducts. *Ind. Crops Prod.* 75, Part B: 141–149.

Faostat. 2015. *Food and Agriculture Organization of the United Nations. FAO Statistic Division* [Online]. Available: http://www.fao.org/faostat/en/#data [Accessed 28–01–2017 2017].

Kalinowska, M., Bielawska, A., Lewandowska-Siwkiewicz, H., Priebe, W. & Lewandowski, W. 2014. Apples: Content of phenolic compounds vs. variety, part of apple and cultivation model, extraction of phenolic compounds, biological properties. *Plant Physiol. Biochem.* 84: 169–188.

Moreira, M.M., Barroso, M.F., Boeykens, A., Withouck, H., Morais, S. & Delerue-Matos, C. 2017. Valorization of apple tree wood residues by polyphenols extraction: Comparison between conventional and microwave-assisted extraction. *Ind. Crops Prod.* 104: 210–220.

Naima, R., Oumam, M., Hannache, H., Sesbou, A., Charrier, B., Pizzi, A. & Charrier–El Bouhtoury, F. 2015. Comparison of the impact of different extraction methods on polyphenols yields and tannins extracted from Moroccan Acacia mollissima barks. *Ind. Crops Prod.* 70: 245–252.

Paz, M., Gúllon, P., Barroso, M.F., Carvalho, A.P., Domingues, V.F., Gomes, A.M., Becker, H., Longhinotti, E. & Delerue-Matos, C. 2015. Brazilian fruit pulps as functional foods and additives: Evaluation of bioactive compounds. *Food Chem.* 172: 462–468.

Rana, S., Kumar, S., Rana, A., Sharma, V., Katoch, P., Padwad, Y. & Bhushan, S. 2016. Phenolic constituents from apple tree leaves and their in vitro biological activity. *Ind. Crops Prod.* 90: 118–125.

Walia, M., Kumar, S. & Agnihotri, V.K. 2016. UPLC-PDA quantification of chemical constituents of two different varieties (golden and royal) of apple leaves and their antioxidant activity. *J Sci Food Agric.* 96: 1440–50.

Xü, K., Lü, H., Qü, B., Shan, H. & Song, J. 2010. High-speed counter-current chromatography preparative separation and purification of phloretin from apple tree bark. *Sep. Purif. Technol* 72: 406–409.

Yuri, J.A., Neira, A., Quilodran, A., Motomura, Y. & Palomo, I., 2009. Antioxidant activity and total phenolics concentration in apple peel and flesh is determined by cultivar and agroclimatic growing regions in Chile. *J. Food Agric. Environ* 7: 513–517.

Process development for a combined treatment of EAFD and jarosite

S. Wegscheider & S. Steinlechner
Nonferrous Metallurgy, Montanuniversitaet Leoben, Leoben, Austria

G. Hanke
Geology and Economic Geology, Montanuniversitaet Leoben, Leoben, Austria

J. Antrekowitsch
Christian Doppler Laboratory for Optimization and Biomass Utilization in Heavy Metal Recycling, Montanuniversitaet Leoben, Leoben, Austria

ABSTRACT: Investigation of the chemical and mineralogical properties of electric arc furnace dust and jarosite indicates a high potential for recycling, due to contained valuable metals such as indium and silver beside zinc, lead and iron. The development of new appropriate metallurgical processes establishes new possibilities to reduce the amount of hazardous wastes and to open up new chances for secondary resources. Based on successful volatilization trials with a mix of jarosite and Electric Arc Furnace Dust (EAFD), a metal bath reduction process was developed due to some advantages in further usage of the products. The purpose of the process is to obtain an iron alloy, a metal-bearing off-gas and a heavy metal free slag. Results from characterization and first metal bath reduction trials are presented in this paper.

1 INTRODUCTION

In view of the ever-growing demand of raw materials and the difficult search for new primary deposits, by-products and wastes of the metal producing industry become more and more interesting due to their high amount of valuable metals. The exploitation of existing dumps and continuously formed materials does not only have advantage in supply of metals, it also reduces the residues forming which are usually treated as hazardous waste (Rathore et al., 2008). However, electric arc furnace dust and especially jarosite dumps are very versatile in their composition thus making it difficult to develop a matching method of recovering. Therefore, tribute must be payed to the characterization of the starting product. Furthermore, detailed thermodynamic calculations and process development are necessary to recover the minor as well as the base metals from these residues.

2 CHARACTERIZATION

Detailed mineralogical and chemical characterization of the material is of prime importance for developing appropriate methods to recover the metals of interest. Scanning electron microscopy and different methods of chemical analysis allow a characterization of the electric arc furnace dust and the jarosite. Mineralogical and chemical analysis is of course limited to grains larger than 1 μm and the detection limit of about 0.5 wt-% due to the energy-dispersive X-ray method used.

2.1 Electric arc furnace dust

Electric arc furnace dust consists mainly of particles of very small grain size of about 1–2 μm (Muchado et al., 2006). About 39 wt-% of the analyzed material is made up of zinc and nearly all of it occurs in the smallest fraction of the electric arc furnace dust. The extremely small size does not allow single grain analysis by using scanning electron microscopy. However, a bulk-like analysis indicates that the grains are composed of zinc and iron, probably zinc-ferrite. Larger particles reach sizes of nearly 100 μm and consist very often of adhesions of different iron rich phases. Sometimes they also bear up to 20 wt-% of calcium, silicium and/or aluminum and few percent of manganese and chromium. Lead occurs in zinc and iron containing particles as well as in higher amounts with silicium and zinc. Together, calcium and silicon make up about 5.5 wt-% of the whole sample and most likely are derived from the degradation of the refractories (Rizescu et al., 2010).

The very fine-grained particles in Figure 1(a) are iron-zinc oxides with small amounts of other elements like lead and manganese. The large, often spherical grains are mainly iron-oxides and differ in their additional amount of zinc, which is generally lower than in the small grains. Very bright colors indicate higher lead content. The chemical analysis of the material analyzed is listed in Table 1.

2.2 Jarosite

Similar to the EAFD, the jarosite is also characterized by very small grain sizes, but it is more variable in its mineralogical and chemical composition (Pappu et al., 2006). X-ray diffractometric analysis identifies gypsum and jarosite as main mineral phases. However, many

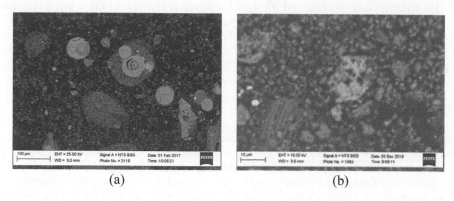

(a) (b)

Figure 1. Characteristic SEM image of an electric arc furnace dust showing globules of iron oxides in a fine-grained matrix composed of iron-zinc oxide (a) and SEM image of fine grained jarosite particles and a silver containing grain in the center of the picture (b).

Table 1. Chemical analysis of the EAFD.

	Fe [%]	Zn [%]	Pb [%]	S [%]	Cl [%]	F [%]	Al [%]	Ca [%]	Si [%]	Mg [%]
EAFD	19.0	39.3	3.2	0.56	3.85	0.06	0.7	3.9	1.6	1.2

Table 2. Chemical analysis of the jarosite material.

	Ag [ppm]	In [ppm]	Fe [%]	Zn [%]	Pb [%]	S [%]	Al [%]	Ca [%]	Si [%]	Mg [%]
Jarosite	180	230	27.1	6.5	6.2	8.4	0.9	0.57	3.3	0.06

other interesting metal-bearing phases have been identified. The very fine-grained fraction of smaller than 1 μm is not quantifiable with the methods used. As the mineral jarosite forms as a precipitation product during the hydrometallurgical processing of zinc, it is the main component of this fraction. Similar to the electric arc furnace dust, the larger particles reach sizes of up to 100 μm. These larger particles often include quartz, barite, zinc-ferrite and different zinc, iron or lead sulfides and/or sulfates. However, other phases such as feldspar, olivine, wustite and spinel are also present. As already mentioned, gypsum is the second major phase besides jarosite. It forms relatively large crystals, which can easily be seen using a stereo microscope. According to the chemical analysis, the jarosite usually bears, besides the already mentioned metals, considerable amounts of silver, indium, germanium and gallium. These metals are in the focus of the research. Currently indium, germanium and gallium have not been found in single grain analysis. However, it was possible to identify silver containing phases. The silver bearing particles (see Figure 1 (b)) contain up to 2 wt-% silver, are about 10 μm in size and consist of more than 20 wt-% copper. The chemical analysis of the jarosite is shown in Table 2.

The jarosite as well as the electric arc furnace dust are constantly produced in large amounts all over the world. Whilst about 50% of the electric arc furnace dust is recycled, the major part of the jarosite is dumped despite its high potential for valuable metals. As both materials are quite similar to each other, they could be processed together. This would bring two main benefits: on the one hand, it helps to decrease the tonnage of material that must be dumped, on the other hand it brings a new secondary resource for many valuable metals.

3 EXPERIMENT

3.1 *Process proposal*

Volatilization trials (with a solid phase) in a small muffle furnace at 1100°C with a mix of both residues showed an evaporation of valuable metals such as silver and indium as compounds. In Table 3 the recovery rates of one volatilization trial for selected elements are listed. Elements or rather compounds of these elements like silver, lead, sulfur, chlorine and fluorine are highly volatile. In case of indium a recovery rate of more than 50% could be achieved, iron remained in the mixture. Only 13% of the zinc evaporates, but at higher temperatures, higher recovery rates are achievable (nearly 100% at the metal bath trial, see Table 7).

Based on the results of the test series a concept for a pyrometallurgical treatment of a mix of jarosite and EAFD was developed (see Figure 2).

Table 3. Recovery rates of selected elements.

Sample	Ag [%]	In [%]	Fe [%]	Zn [%]	Pb [%]	S [%]	Cl [%]	F [%]
Mix with C	>89.91	52.61	3.09	12.81	95.04	97.79	99.75	96.04

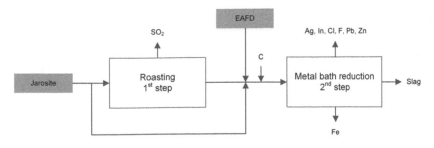

Figure 2. Flow sheet of the pyrometallurgical concept.

A possible first step of the concept is a roasting of the jarosite at about 650°C to get rid of the SO$_2$ and to decompose the jarosite. Afterwards the jarosite (natural or roasted) is mixed with EAFD and enters a metal bath process. In this second step silver, indium, lead and zinc can be recovered by volatilization, moreover two other products arise, an iron alloy and a slag.

3.2 *Thermodynamic calculation*

After characterization thermodynamic calculations were performed to obtain an appropriate mixture of jarosite and electric arc furnace dust (EAFD). On the one hand the jarosite is a carrier of valuable metals like indium and silver, on the other hand EAFD supplies the necessary chlorine to build volatile compounds with indium and silver. Both residues are similar regarding their composition in terms of base metals and complement each other. The stoichiometric amount of the reducing agent to reduce the desired compounds was calculated by mass balances.

Moreover quaternary systems were used to calculate the slag behavior during the pyrometallurgical process.

3.3 *Metal bath reduction*

At the laboratory scale experiment only material without a pretreatment (roasting step) was used yet. The ratio of the amount of jarosite to EAFD always remained the same. Findings from thermodynamic calculations showed that a temperature above 900°C can be sufficient for a selective volatilization, nevertheless to obtain a slag with low viscosity temperatures of 1500°C were necessary. The residence time of the residue in the furnace depends on the reaction rate to reduce and volatilize the desired compounds. Trials showed an average residence time of 60 to 90 min.

A preparation of the input materials was the first experimental part. Due to its high moisture content, the precipitation residue jarosite was dried at 120°C for 48 h, where a mass loss of about 25% can be detected. The jarosite and the EAFD were mixed well in the defined ratio.

The pyrometallurgical investigations were carried out in an induction furnace with a graphite crucible. After charging the material into the furnace, the surface was covered with Desulco (a high puritiy carbon additive) to ensure a reducing atmosphere. Additionally, in some trials finely ground Desulco was mixed together with the jarosite and the EAFD. After the reduction and vaporization of the desired elements the crucible was removed from the furnace and cooled down by air. Subsequently the generated alloy, the slag and the Desulco that was not consumed were separated for mass balance and analysis.

4   EXPERIMENTAL RESULTS

The experimental results are divided into two parts—the mass balance results from selected trials and the analysis of one trial.

4.1 *Mass balance*

Three different trials concerning the mass balance are listed below in Table 4.

At trial V1 and V3 the reductant is twice stoichiometric, in trial V2 only once stoichiometric. The Desulco reducing agent covers the surface and hence, reducing conditions are provided. Beside the consumption of Desulco also carbon from the graphite crucible takes part at the reduction. The difference between trial V1 and V3 is that half of the amount of the reducing agent (grinded) in V3 is mixed together with the input material EAFD and jarosite. The exhaust gas amount is calculated and correlates with the input amount of the mixture and Desulco (V3) minimized by the amount of iron alloy and slag.

4.2 *Chemical analysis and recovery rates of the slag and the iron alloy*

In Table 5 the chemical analysis of the iron alloy is listed.

Beside iron, also sulfur and carbon are present in appreciable amounts in the alloy. Zinc and lead are more or less completely removed from the alloy. Also, 80 ppm of silver can be detected in the alloy. Unfortunately, indium was not detectable in the input material and therefore no evidence regarding an indium volatilization is provided.

Figure 4 (a) shows three different phases. The iron matrix (medium grey) hosts iron-sulfide impurities (dark grey) which contain few percent of chromium, manganese, and copper. The latter of which is also the main element of the white colored phases. They also contain several percent of iron, antimony, silver, sulfur and calcium.

The chemical analysis of the slag is shown in Table 6.

The slag is low in lead and zinc and shows small amounts of silver with 60 ppm. Regarding the final composition of the slag, a relatively low melting point (~1350°C) could be achieved. The melting point and the composition are illustrated in the quaternary system in Figure 3.

Table 4. Mass balance of selected trials.

		V1	V2	V3
	Mixture of jarosite and EAFD [g]	200	600	440
Input	Desulco for covering the surface [g]	40	60	40
	Desulco [g]	–	–	40
	Iron alloy [g]	38.65	90.01	81.50
Output	Slag [g]	32.75	131.60	65.95
	Exhaust gas, dust [g]*	128.60	378.40	292.55

*Calculated, without Desulco consumption from surface covering.

Table 5. Chemical analysis of the iron alloy.

	Ag [ppm]	In [ppm]	Fe [%]	Zn [%]	Pb [%]	S [%]	C [%]
Iron alloy	80	<50	88.0	0.05	<0.02	6.3	0.6

Table 6. Chemical analysis of the slag.

	Ag [ppm]	In [ppm]	FeO [%]	Zn [%]	Pb [%]	S [%]	Al_2O_3 [%]	CaO [%]	SiO_2 [%]	MgO [%]
Slag	60	<50	7.59	<0.02	<0.02	3.52	14.17	16.09	39.14	7.13

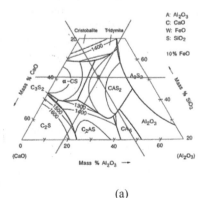

(a)

	FeO [%]	Al_2O_3 [%]	CaO [%]	SiO_2 [%]
Slag	9.9	18.4	20.9	50.8

(b)

Figure 3.   Quaternary slag diagram (a) (Allibert et al., 2008) and components (b).

41

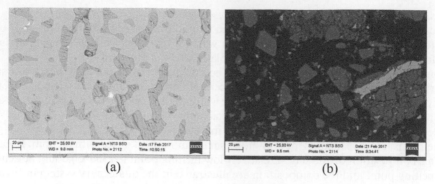

(a)   (b)

Figure 4. SEM picture of the iron alloy (a) and the slag (b).

Table 7. Recovery rates of selected elements to iron alloy, slag and off-gas/dust.

	Ag [%]	In [%]	Fe [%]	Zn [%]	Pb [%]	S [%]	Cl [%]	F [%]	Al [%]	Ca [%]	Si [%]	Mg [%]
Iron alloy	13,60	–	81.5	0.04	0.09	30.4						
Slag	8,25	–	4.4	0.01	0.07	13.8			100*	100*	100*	100*
Off-gas/dust**	78,15	–	14.1	99.95	99.84	55.8	100	100	–*	–*	–*	–*

*Without carry over.
**Calculated.

In Figure 4 (b) the dark grey particles are very variable in their chemistry and consists mainly of aluminum, calcium and silicon. The light grey colored phases, including the large, asymmetrical grains as well as the small spherical ones, are mainly composed of iron including several percent of chromium and manganese.

A chemical analysis of the off-gas/dust was not possible due to the low amount of input material.

The recovery rates of different elements into the iron alloy, the slag as well as the off-gas/dust are displayed in Table 7.

Nearly 82% of the iron is in the alloy, minor parts of iron stay oxidic in the slag and about 14% of iron can be found in the off-gas system. Regarding zinc and lead the metal bath process worked quite successful, as nearly 100% of both elements are volatilized to the off-gas system. Neither in the alloy, nor in the slag, appreciable amounts of lead and zinc could be detected.

Unfortunately, indium is below the detection limits in the input material and hence no recovery rates are given in Table 7. However, the recovery rate for silver to the offgas system is more than 78%, which is quite successful. Minor parts of silver stay in the iron alloy and the slag. Moreover, the distribution rate of sulfur is improvable, which may be realized by an input material consisting of roasted jarosite and EAFD and hence, further trials are necessary.

5 CONCLUSION

Scanning electron microscopy allows characterization of jarosite as well as electric arc furnace dust as long as grains are more than a few microns in size. Bulk-like, semi quantitative analysis is also possible on smaller particles and allows a coarse estimation of the chemical composition and the mineralogy. As silver is one of the main metals of interest its positive detection is an intermediate success. It was found in very similar grains in jarosite of two different locations. Quantification is not possible because it is still not sure how much of the total silver is hosted by these copper particles. Moreover, the silver occurred in particles with more than 20 wt-% copper, but it is not known yet if particles with lower copper percentage do not contain any silver, or if it is just below the detection limit of the scanning electron microscope. The next steps will include microprobe analysis with a detection limit of about 300 ppm, which should bring clearance to this question. Host particles for indium, gallium and germanium have not been identified so far.

As shown in previous trials, a volatilization of valuable metals such as silver and indium from a mix of residues (jarosite and electric arc furnace dust) is possible and also realizable in a metal bath process for silver yet. Furthermore, it was possible to obtain an iron alloy, a slag as well as a Zn- and Pb-bearing off-gas. To improve the existing concept and to recover indium too, more experiments are required.

ACKNOWLEDGEMENT

The authors want to thank the Austrian Research Promotion Agency (FFG) and the Federal Ministry of Science, Research and Economy (BMWFW) for the financial support.

REFERENCES

Allibert M. et al. 2008. *Slag Atlas*. Düsseldorf: Stahleisen.
Muchado J.G. et al. 2006. Chemical, physical, structural and morphological characterization of the electric arc furnace dust. *Journal of hazardous materials* 136.3: 953–960.
Pappu A et al. 2006. Jarosite characteristics and its utilization potentials. *Science of the Total Environment* 359.1: 232–243.
Rathore N. et al. 2008. Utilization of Jarosite generated from Lead-Zinc Smelter for various Applications: a Review. *International Journal of Civil Engineering and Technology* 5.11: 192–200.
Rizescu, C.Z. et al. 2010. Characterization of steel mill electric-arc furnace dust. *Advances in Waste. Management*: 139–143.

WASTES – Solutions, Treatments and Opportunities II – Vilarinho, Castro & Lopes (Eds)
© 2018 Taylor & Francis Group, London, ISBN 978-1-138-19669-8

Mechanical damage of a nonwoven geotextile induced by recycled aggregates

J.R. Carneiro & M.L. Lopes
Construct-Geo, Faculty of Engineering, University of Porto, Porto, Portugal

A. da Silva
Faculty of Engineering, University of Porto, Porto, Portugal

ABSTRACT: The installation process (which many times induces mechanical damage) may provoke unwanted changes in the mechanical behaviour of the geosynthetics. This work evaluates the effect of the aggregate on the mechanical damage under repeated loading suffered by a geotextile. For that purpose, the geotextile was damaged in laboratory (procedure based on EN ISO 10722) with recycled aggregates (construction and demolition waste and incineration bottom ashes) and *corundum* (standard aggregate used in EN ISO 10722). The damage occurred in the geotextile (in the mechanical damage tests) was evaluated by visual inspection and by monitoring its tensile and static puncture properties. Results showed that the damage caused by the recycled aggregates in the short-term behaviour of the geotextile was lower than the damage induced by *corundum*. Thus, and concerning mechanical damage, there are good perspectives for the use of the recycled aggregates as backfills in contact with geotextiles.

1 INTRODUCTION

Nowadays, and for achieving environmental sustainability, it is important to reduce the exploitation of natural resources. Many materials used in construction are natural and, when possible, their replacement by recycled materials is being encouraged (for example, the use of recycled aggregates in geotechnical works). In many situations, the recycled materials may be in contact with other construction materials (like geosynthetics) and, in order to be applied, they cannot have a negative impact on the other materials (or at least a more negative impact than the natural materials they are intending to replace).

Geosynthetics are polymeric products used in the construction of many infrastructures (such as waste landfills, roads, embankments, reservoirs or coastal engineering structures) due to their excellent performance and relatively low cost. These materials can be typically divided in geotextiles, geomembranes and related-products (like geogrids, geonets or geocomposites).

The installation process can induce mechanical damage on the geosynthetics, provoking unwanted changes in their physical, mechanical and hydraulic properties. The damage that occurs during installation (like cuts in fibres, formation of holes, abrasion or reductions in mechanical resistance) is mainly caused by handling the materials and by the placement and compaction of backfills over them (Pinho-Lopes & Lopes 2010). The backfills most often have natural origin and may eventually be replaced by recycled aggregates in some applications. The existing studies about the use of recycled aggregates in contact with geosynthetics are relatively few and are mainly about interface properties (for instance, Vieira & Pereira 2016 or Vieira et al. 2016), being necessary the development of additional research. For example, there is a lack of studies on the damage that recycled aggregates may cause to geosynthetics during the installation process (a key issue for the use of aggregates as backfills).

The damage that occurs during the installation of geosynthetics can be evaluated by (1) laboratory tests that simulate the damaging actions or by (2) field damage tests (installation under real conditions and immediate recovery of the materials for analysis). The European normative EN ISO 10722 describes a laboratory procedure to induce mechanical damage on geosynthetics. Many authors used this procedure to evaluate the damage occurred in geosynthetics during their installation process, while others tried to establish correlations with field tests.

In this work, a geotextile was submitted to laboratory mechanical damage tests (with exception for the use of recycled aggregates, tests according to EN ISO 10722), being monitored the changes occurred in its short-term behaviour. The geotextile was damaged with *corundum* (aggregate used in EN ISO 10722) and with four recycled aggregates: three different types of construction and demolition waste (C&DW) and incineration bottom ashes (IBA). The main goals of the work included: (1) evaluation of the effect of different recycled aggregates on the damage suffered by the geotextile, (2) comparison of the damage provoked by *corundum* (standard aggregate) with the damage caused by the recycled aggregates and (3) evaluation if the recycled aggregates (in terms of mechanical damage induced to the geotextile) can be used as viable alternatives for backfills.

2 EXPERIMENTAL DESCRIPTION

2.1 Geosynthetics

This work studied a nonwoven polypropylene geotextile with a mass per unit area of 400 g.m^{-2} and a thickness of 3.0 mm. The sampling process was carried out according to EN ISO 9862.

2.2 Mechanical damage under repeated loading tests

With exception for the granular materials, the mechanical damage under repeated loading tests (for simplification, hereinafter called by mechanical damage tests) were carried out according to EN ISO 10722. These tests were performed in a prototype equipment developed at the Faculty of Engineering of Porto University. The equipment is basically formed by a container (rigid metal box where the geotextile and granular materials were placed), a loading plate and a compression machine.

The geotextile (specimens with a width of 250 mm and a length of 500 mm) was placed between two layers of a granular material and subjected to dynamic loading between (5 ± 0.5) kPa and (500 ± 10) kPa at the frequency of 1 Hz for 200 cycles. The granular materials used in this work included *corundum* (synthetic aggregate used in EN ISO 10722) and four recycled aggregates: three C&DW and IBA (Table 1) (Fig. 1). The recycled C&DW were collected from a Portuguese recycling plant (RCD, SA—Figueira da Foz) (coming mainly from the demolition of buildings and cleaning of lands with illegal deposition of C&DW) and the IBA came from a Portuguese incineration plant (Lipor II, Maia) (incineration of municipal waste).

The grain size distributions of the recycled aggregates were determined by sieving (tests carried out according to EN 933–1). Some parameters for the characterization of the grain size distributions (D_X –x% grain size passing and D_{max} – maximum grain size) can be found in Table 2.

The standardized aggregate (*corundum*) had a grain size between 5 mm and 10 mm. Following the indications of EN ISO 10722, *corundum* was sieved (5 mm aperture sieve) after every 3 uses (the particles retained in the sieve were reused) and totally discarded after 20 uses.

Table 1. Recycled aggregates used in the mechanical damage tests.

Designation	Residue type	Main components
C&DW I	Recycled C&DW (fine grain)	Soil, masonry, concrete, aggregates
C&DW II	Recycled ceramic aggregate	Tiles, bricks, mosaics, masonry units
C&DW III	Recycled concrete aggregate	Cement, mortars, aggregates
IBA	Recycled incineration bottom ashes	Minerals, glass, metals

2.3 *Evaluation of the damage suffered by the geotextile*

The damage occurred in the geotextile during the mechanical damage tests was evaluated qualitatively (by visual inspection) and quantitatively (using reference tests). The reference tests included tensile tests (according to EN ISO 10319) and static puncture tests (according to EN ISO 12236). The experimental conditions of the tests are summarized in Table 3.

The tensile and puncture tests were performed in a tensile/compression machine from *Lloyd Instruments* (model LR 50 K). The mechanical parameters determined in the tensile tests included tensile strength (TS, in $kN.m^{-1}$) and elongation at maximum load (E_{ML}, in%), while static puncture resistance (F_P, in kN) and push-trough displacement (displacement at maximum force) (h_P, in mm) were the parameters determined in the puncture tests. The tensile tests were carried out in the machine direction of production of the geotextile.

The tensile (TS and E_{ML}) and puncture (F_P and h_P) properties are expressed with 95% confidence intervals determined according to Montgomery & Runger (2010). Some results are presented in terms of retained strength (RS, in%). The RS was obtained by the quotient

Figure 1.   Granular materials: (a) C&DW I; (b) C&DW II; (c) C&DW III; (d) IBA; (e) *corundum*.

Table 2.   Characterization of the grain size distribution of the recycled aggregates.

Recycled aggregate	D_{10} (mm)	D_{30} (mm)	D_{50} (mm)	D_{60} (mm)	D_{max} (mm)
C&DW I	–	0.3	0.5	0.8	6.3
C&DW II	0.7	8.4	12.8	14.6	20.0
C&DW III	0.9	6.6	14.8	18.7	31.5
IBA	0.2	1.3	3.8	5.5	12.5

Table 3.   Experimental conditions of the characterization tests.

Test	Standard	Specimens type	Specimens size	N	Test speed
Tensile	EN ISO 10319	Rectangular	200 mm × 100 mm*	5	20 mm.min^{-1}
Puncture	EN ISO 12236	Circular	150 mm**	5	50 mm.min^{-1}

(N – number of specimens; *length between grips; **diameter between grips).

between the resistances (tensile or static puncture) of the damaged samples and the respective resistances of the reference sample (undamaged).

3 MAIN RESULTS

3.1 *Visual analysis of damage*

The mechanical damage tests did not cause very pronounced signs of degradation in the geotextile (Fig. 2). In all cases, it was possible to find some fine particles (constituent particles of the granular materials or particles resulting from their fragmentation) imprisoned in the nonwoven structure. With exception for C&DW I, the recycled aggregates and *corundum* provoked small punctures on the geotextile (no holes were found). Although on a small scale, all granular materials induced some abrasion on the surface of the geotextile.

3.2 *Tensile properties*

The mechanical damage tests with the C&DW did not provoke relevant changes in the tensile strength of the geotextile (Table 4). The minor variations occurred in tensile strength (retained tensile strengths between 97.5% and 101.6%) can be explained by the heterogeneity of the geotextile. Indeed, it is well known that nonwoven geotextiles typically present some heterogeneity arising from their manufacturing process (this heterogeneity can be responsible for small differences between properties determined in consecutive repetitions of the same test).

Like the tests with C&DW, the test with IBA did not cause a very pronounced change in the tensile strength of the geotextile (retained tensile strength of 93.8%). Contrary to the tests with recycled aggregates, the test with the standardized aggregate (*corundum*) induced a relevant decrease in the tensile strength of the geotextile (retained tensile strength of 83.9%).

The elongation at maximum load of the geotextile suffered a decrease after the mechanical damage tests (Table 4). The reductions provoked by the tests with the recycled aggregates were very identical (decreases from 81.3% to 67.1%–68.6%) and lower than that caused by the test with *corundum* (decrease from 81.3% to 59.3%). The reductions observed in elongation at maximum load were accompanied by increases in the stiffness of the geotextile.

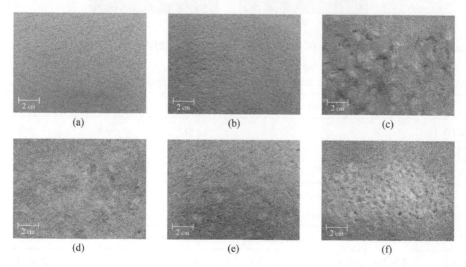

Figure 2.   Geotextile before (a) and after (b-f) the mechanical damage tests: (b) damage with C&DW I; (c) damage with C&DW II; (d) damage with C&DW III; (e) damage with IBA; (f) damage with *corundum*.

Table 4. Tensile properties of the geotextile before and after the mechanical damage tests.

Mechanical damage test	T (kN.m^{-1})	E$_{ML}$ (%)
Undamaged	24.60 (± 1.76)	81.3 (± 3.6)
Mechanical damage with C&DW I	24.99 (± 1.01)	68.2 (± 7.4)
Mechanical damage with C&DW II	23.99 (± 1.25)	68.6 (± 9.6)
Mechanical damage with C&DW I	24.99 (± 1.01)	68.2 (± 7.4)
Mechanical damage with C&DW III	24.57 (± 2.31)	67.6 (± 5.5)
Mechanical damage with IBA	23.08 (± 1.69)	67.1 (± 6.8)
Mechanical damage with *corundum*	20.65 (± 1.48)	59.3 (± 6.6)

(in brackets are the 95% confidence intervals).

Table 5. Puncture properties of the geotextile before and after the mechanical damage tests.

Mechanical damage test	F$_p$ (kN)	h$_p$ (mm)
Undamaged	4.93 (± 0.32)	64.6 (± 2.0)
Mechanical damage with C&DW I	5.18 (± 0.45)	63.1 (± 1.0)
Mechanical damage with C&DW II	4.86 (± 0.49)	58.5 (± 3.6)
Mechanical damage with C&DW III	4.68 (± 0.51)	56.5 (± 3.6)
Mechanical damage with IBA	4.86 (± 0.32)	56.0 (± 4.0)
Mechanical damage with *corundum*	3.99 (± 0.28)	52.6 (± 1.5)

(in brackets are the 95% confidence intervals).

3.3 *Static puncture properties*

The changes occurred in the puncture properties after the mechanical damage tests were similar to those occurred in the tensile properties. Like tensile strength, the puncture resistance also did not suffer relevant modifications after the tests with the recycled aggregates (retained puncture resistances between 94.9% and 105.1%) (Table 5). By contrast, and similarly to what happened for tensile strength, the test with *corundum* provoked a decrease in puncture resistance (retained puncture resistance of 80.9%). The push-trough displacement remained practically unchanged after the test with C&DW I and decreased after the other tests (highest reduction observed after the test with *corundum*) (Table 5).

4 CONCLUSIONS

The mechanical damage tests with the recycled aggregates did not cause much relevant changes in the tensile and puncture properties of the geotextile (only relatively small decreases in elongation at maximum load and push-trough displacement). By contrast, the tests with *corundum* (aggregate used in the standard method for inducing mechanical damage on geosynthetics) led to reductions in the tensile and puncture properties of the geotextile. Figure 3 compares the retained strengths (tensile and puncture) of the geotextile after the mechanical damage tests.

The changes occurred in tensile strength after the mechanical damage tests were identical to the changes occurred in puncture resistance. *Corundum* was the most damaging aggregate, provoking some reductions in the mechanical resistance of the geotextile (retained tensile strength of 83.9% and retained puncture resistance of 80.9%). *Corundum* is a uniform well-graded synthetic aggregate and, despite not having so large particles as the C&DW II, C&DW III and IBA, it has angular particles capable of inducing damage to the geotextile.

Contrarily to *corundum*, the three different C&DW and IBA did not provoke relevant damage (decrease of mechanical resistance) in the geotextile. Regarding the possibility of inserting these recycled aggregates as backfill materials in contact with geotextiles (and/or other geosynthetics), this study does not allow definitive conclusions, since it only evaluated the mechanical damage induced by the aggregates to the geotextile. Still, the damage provoked by the recycled

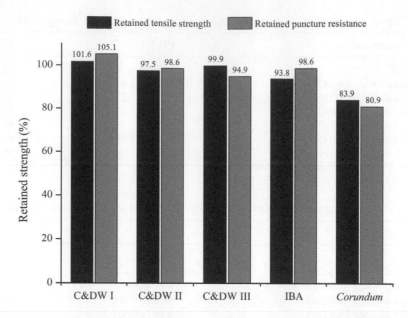

Figure 3. Comparison of the retained strengths of the geotextile after the mechanical damage tests.

aggregates in the short-term behaviour of the geotextile was lower than the damage caused by the standard aggregate used in EN ISO 10722, which opens good perspectives for the use of these type of aggregates in the construction of Civil Engineering infrastructures, allowing their recovery and reuse. However, and before such application, further studies are needed.

ACKNOWLEDGEMENTS

This work was financially supported by Project POCI-01-0145-FEDER-007457, funded by FEDER funds through "COMPETE 2020 – Programa Operacional Competitividade e Internacionalização" (POCI) and by national funds through "FCT—Fundação para a Ciência e a Tecnologia". J.R. Carneiro would also like to thank "FCT" for the research grant SFRH/BPD/88730/2012 (grant supported by POPH/POCH/FSE funding).

REFERENCES

EN 933–1. 2012. *Tests for geometrical properties of aggregates. Part 1: Determination of particle size distribution—Sieving method.*

EN ISO 9862. 2005. *Geosynthetics—Sample and preparation of test specimens.*

EN ISO 10319. 2008. *Geosynthetics—Wide-width tensile test.*

EN ISO 10722. 2007. *Geosynthetics—Index test procedure for the evaluation of mechanical damage under repeated loading. Damage caused by granular material.*

EN ISO 12236. 2006. *Geosynthetics—Static puncture test (CBR test).*

Montgomery, D.C. & Runger, G.C. 2010. *Applied Statistics and Probability for Engineers.* 5th ed. New York: John Wiley & Sons.

Pinho-Lopes, M.J.F. & Lopes, M.L. 2010. *Durability of Geosynthetics.* Porto: FEUP (in Portuguese).

Vieira, C.S. & Pereira, P.M. 2016. Interface shear properties of geosynthetics and construction and demolition waste from large-scale direct shear tests. *Geosynthetics International* 23 (1): 62–70.

Vieira, C.S., Pereira, P.M. & Lopes, M.L. 2016. Recycled construction and demolition wastes as filling material for geosynthetics reinforced structures. Interface properties. *Journal of Cleaner Production* 124: 299–311.

Smart biofilms produced from fish filleting wastes

R.S. Brito, C.A. Araújo & L.F.H. Lourenço
Faculty of Food Engineering, Federal University of Pará, Pará, Brazil

E.J.G. Pino-Hernádez & G.S. Souza Matos
Centre for Biological Engineering, University of Minho, Minho, Portugal

M.R.S. Peixoto Joele
Federal Institute of Education, Science and Technology of Pará, Pará, Brazil

ABSTRACT: The myofibrillar proteins utilization from fish processing wastes to produce films has an importance to decrease the environmental impact due to usual inappropriate disposal of these wastes. The aim of this work was to make and characterize biodegradable films developed from myofibrillar proteins of gó whitefish (Macrodon ancylodon) wastes, with anthocyanin addition as a pH change indicator. The material proteic extracted was chacterized presenting high concentration of total proteins (95.01%), important for the development of filmogenic solution. According to the results, there was a significant difference ($p < 0.05$) between the control and the added with anthocyanin biofilms. The film with lower concentration of anthocyanin presented better barrier property, with water vapor permeability of 3.8×10^{-11} g.m^{-1}.s^{-1}.Pa^{-1}. There is a great potential for fish processing wastes utilization to obtain biodegradable films even as the anthocyanin pigment powder is promising to act as pH indicator.

1 INTRODUCTION

The fishing industry during the processing generates a volume of waste higher than 50% (Gonçalves, 2011; Balbinot, 2015), which by-products of fish filleting (flour, surimi, for example) are compound by high levels of proteins, requiring a better use of this raw material. There are three types of proteins in fish: sarcoplasmic, stromal and myofibrillar (Dangaran et al. 2009) and among them myofibrillar proteins extracted from the fish muscles have been used as film-forming material by several researchers (Iwata et al. 2000; Paschoalick et al. 2003; Shiku et al. 2003, 2004; Artharn et al. 2007).

The large environmental impact generate from waste disposal of non-biodegradable plastic materials has become a global concern. Therefore, to partially replace this type of material, researches are being done to find or to develop ecological and renewable polymer materials (Khwaldia et al. 2010). In this context, the development of biofilms with characteristics of biodegradable packaging has become a priority because it contributes to the reduction of environmental problems caused by the constant use of petroleum-based polymers (Davanço et al. 2007; Arrieta et al. 2014).

Biodegradable films obtained from biological materials act as a barrier to external elements, protecting packaged products from physical and biological damage, preventing the volatilization of compounds and moisture loss, increasing the shelf life of the product. Recent studies have demonstrated that myofibrillar proteins have excellent ability to form biofilms, specifically those of fish (Arfat et al. 2014) and beef (Souza et al. 2012).

There are different types of smart packaging, among them those that indicate the pH change in the food for the consumer, by color change of the pigments present in the packaging material that are sensitive to pH alterations (Rebello, 2009). According to Bridle and Timberlake (1997), the largest group of water-soluble pigments in the Plant Kingdom are anthocyanins, which are studied all over the world as natural coloring agents in foods, responsible for the tones ranging from red to blue. The properties of anthocyanins, including their color, are strongly influenced by their structure and pH (Lopes et al. 2007).

Therefore, the aim of this work was to develop a smart biodegradable film from fish proteins added with anthocyanins and evaluate the film change color according to pH of the packaged product.

2 MATERIAL AND METHODS

2.1 Material

The gó whitefish (*Macrodon ancylodon*) wastes were obtained from fish filleting processing of fishing industry Ecomar, located in Vigia City (Pará State—Brazil). The natural powdered anthocyanin was obtained from the Centro de Valorização Agroalimentar dos Compostos Bioativos da Amazônia (CVACBA).

2.2 Extraction of myofibrillar proteins

The methodology proposed by Limpan et al. (2010) was used with some modifications, according to the Figure 1. The proteic material extracted was characterized by analysis of protein, ash, moisture and lipids performed according to AOAC method (2000).

2.3 Development of smart biodegradable films

The biofilms were developed according to Limpan et al. (2010) and Arenas (2012), with modifications. To obtain the filmogenic solution was mixed 1% of lyophilized myofibrillar proteins with distilled water and 40% of glycerol as plasticizer. The pH was adjusted to 3 with hydrochloric acid 2M and the solution was homogenized under 10000 rpm for 5 minutes using homogenizer Turratec (Tecnal, TE-102). Then, the solution was heated in a water bath (Tecnal, TE-057) for 30 minutes at 60°C, under manual stirring with glass rod for better heat distribution. After cooling, 7 (seven) portions of the solution were separated, in 6 (six) was added natural anthocyanin at concentrations ranging from 0.05 to 0.1 g anthocyanin/ 100 g of filmogenic solution, and 1 (one) was the control solution (without addition of anthocyanin).

According to *casting* method González et al. (2017) with some modifications, 130 mL of filmogenic solution was added in silicone support (22 cm in diameter × 3 cm in height) and dried in incubator stove (Quimis, Q315M) for 16 hours at 25°C. After drying, the biofilms were stored in polyethylene packaging at room temperature (22°C).

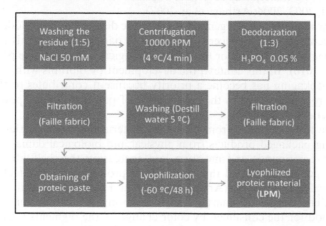

Figure 1.   Flowchart of lyophilized myofibrillar proteins obtention from gó whitefish residues.

2.4   Characterization of smart biodegradable films

– *Thickness*: use of digital micrometer with 0.001 mm resolution;
– *Tensile strength and percent elongation*: ASTM D882-91 methodology (ASTM, 1996) using a texturometer (QTS, Brookfield);
– *Water vapor permeability*: ASTM D882-95 (ASTM, 1995) modified method described by Arfat et al. (2014);
– *Solubility*: according to Gontard et al. (1994) methodology;
– *Color*: according to Hernández et al. (2017) methodology. Using MINOLTA colorimeter model CR 310, obtaining L*, a* and b* parameters.

2.5   Evaluation of films pH indicator activity

30 g portions of gó whitefish fillet *in natura* were placed in glass beakers covered with control (without anthocyanin on formulation) and smart biofilms, and stored under refrigeration (8°C). Monitoring was done by instrumental color and pH analysis within 24 hours.

2.6   Statistical analysis

The results obtained were evaluated by Analysis of Variance (ANOVA) and Tukey test at $p < 0.05$ significance level, on STATISTICA 7 software.

3   RESULTS AND DISCUSSION

3.1   Characterization of proteic material

The physico-chemical analysis results of lyophilized myofibrillar protein indicates the efficiency of the process performed, with low values of moisture and lipids and high protein contain (Table 1). The amount of protein obtained is considered relevant because from them will be produced the biofilms.

3.2   Characterization of biodegradable films

The films thickness evaluation demonstrates that films with higher anthocyanin content presented higher thickness values (Table 2), as expected, due to total solids content increase by addition of powdered anthocyanin.

All the analyzed films had a significant difference ($p < 0.05$) in relation to the water vapor permeability (WVP), being control film with the highest value and the film with 0.05 g of anthocyanin the lowest (Table 2). These results indicate that the presence of anthocyanin turned the films less permeable, because the WVP can be affected by several factor of the material, including the additives incorporated in the polymeric matrix. The films also presented a significant difference ($p < 0.05$) in relation to the solubility, being the control film with the higher value. However, all films presented satisfactory values regarding to this characteristic evaluated. In addition, the films with lower anthocyanin concentrations presented higher values to tensile strength (TS) and percent elongation (E%), as can be seen in Table 2.

Table 1.   Centesimal composition of lyophilized proteic material.

Compounds[1] (%)	LPM[2]
Moisture	3.86 ± 0.1
Ash	0.90 ± 0.4
Lipids	2.44 ± 0.03
Proteins	95.01 ± 0.1

[1]Results on a dry basis; [2]Lyophilized proteic material.

Table 2. Results of thickness, water vapor permeability (WVP), solubility, tensile strength (TS) and percent elongation (E%).

Films	Thickness (mm)	WVP ($\times 10^{-11}$ g.m^{-1}.s^{-1}.Pa^{-1})	Solubility (%)	TS (MPa)	E (%)
Control	0.035 ± 0.02^d	5.23 ± 0.53^a	36.55 ± 1.20^a	3.78 ± 0.63^d	10.4 ± 1.1^e
0.05*	0.038 ± 0.02^d	3.18 ± 0.38^g	12.65 ± 0.05^e	11.93 ± 0.40	252.65 ± 0.4^a
0.06*	0.041 ± 0.01^a	3.51 ± 0.40^e	16.34 ± 0.07^d	7.07 ± 0.30^b	111.27 ± 0.8^{bc}
0.07*	0.042 ± 0.04^c	3.48 ± 0.58^f	18.28 ± 0.04^d	7.84 ± 0.73^b	111.86 ± 0.2^b
0.08*	0.047 ± 0.02^b	4.48 ± 0.30^c	19.04 ± 0.05^d	4.69 ± 0.50^{cd}	109.02 ± 0.7^c
0.09*	0.048 ± 0.07^b	4.20 ± 0.36^d	22.76 ± 0.05^c	4.85 ± 0.61^{cd}	105.14 ± 0.2^d
0.1*	0.052 ± 0.08^a	4.89 ± 0.42^b	29.55 ± 0.06^b	8.17 ± 0.56^b	109.3 ± 0.7^{bc}

*Anthocyanin concentration (g/100 g of filmogenic solution); Same letters in the same column did not differ significantly at the level 5% ($p < 0.05$).

Table 3. Results of L*, a* and b* color parameters of biodegradable films.

Films	L*	a*	b*
Control	95.54 ± 0.02^a	-5.46 ± 0.01^e	8.16 ± 0.01^a
0.05**	56.94 ± 0.02^b	25.60 ± 0.02^d	-6.52 ± 0.01^b
0.06**	54.95 ± 0.02^b	27.76 ± 0.01^{cd}	-6.83 ± 0.01^b
0.07**	50.71 ± 0.01^c	34.4 ± 0.02^a	-6.94 ± 0.01^b
0.08**	50.02 ± 0.03^c	28.43 ± 0.02^c	-7.03 ± 0.02^b
0.09**	50.70 ± 0.02^c	29.36 ± 0.01^{bc}	-9.95 ± 0.02^c
0.1**	46.26 ± 0.01^d	31.68 ± 0.02^{ab}	-8.02 ± 0.02^{bc}

*Color parameters; ** Anthocyanin concentration (g/100 g of filmogenic solution); Same letters in the same column did not differ significantly at the level 5% ($p < 0.05$).

The color evaluation indicates that the anthocyanin addition left the films significantly more opaque, as can be observed by the parameter L* decrease (Table 3), as expected. Evaluating the results of coordinates a* and b* can be seen the change from greenish to red with the mix of blue tones, due the anthocyanin presence.

According to the found results on biofilms characterization, higher tensile strength, lower WVP and less opaque color, the formulation with lower anthocyanin concentration (0.05 g/100 g of filmogenic solution) was selected to evaluate the film application on gó whitefish. It is important to emphasize that all formulations produced could be used to many purposes.

3.3 Evaluation of smart film as pH change indicator

On Table 4, it is possible to see that the biodegradable film added with anthocyanin (0.05 g/100 g of filmogenic solution), during the fish storage time at 8°C, presented significant changes on L* and a* parameters, according to fish pH alterations. This result offers a greater safety to the consumer, which can evaluate the food quality and freshness through the package, in this case the smart biofilm, and at the same time avoid the food waste (Pacquit et al. 2006).

The b* parameter (blue—yellow axis) and the film pH did not present significant difference over the storage days. Although this parameter did not present a significant difference, the results obtained showed negative values, indicating the presence of the blue component in the film with anthocyanin, that decreased throughout the storage time, due to the volatile bases generated by the fish deterioration, making the medium more alkali.

Table 4. Color parameters of biodegradable films and fish pH. stored at 8°C.

Storage days	L*	a*	b*	pH
	Film added with anthocyanin			
0	55.41 ± 0.02[c]	32.01 ± 0.01[a]	−6.14 ± 0.01[a]	6.24 ± 0.02[a]
1	64.27 ± 0.02[a]	29.74 ± 0.03[a]	−5.81 ± 0.02[a]	6.64 ± 0.03[a]
2	59.27 ± 0.01[b]	31.11 ± 0.02[a]	−6.84 ± 0.01[a]	6.90 ± 0.01[a]
3	56.86 ± 0.03[bc]	19.56 ± 0.01[b]	−4.79 ± 0.02[a]	7.33 ± 0.02[a]
	Control film			
0	90.16 ± 0.02[a]	−4.54 ± 0.02[a]	8.95 ± 0.02[bc]	6.24 ± 0.01[a]
1	91.30 ± 0.01[a]	−4.80 ± 0.02[a]	7.41 ± 0.01[c]	6.73 ± 0.02[a]
2	90.82 ± 0.03[a]	−4.27 ± 0.03[a]	10.71 ± 0.03[ab]	7.08 ± 0.01[a]
3	91.45 ± 0.2[a]	−4.30 ± 0.02[a]	11.52 ± 0.01[a]	7.50 ± 0.01[a]

Same letters in the same column did not differ significantly at the level 5% ($p < 0.05$).

The biodegradable film without anthocyanin, used as control film, stored under the same conditions that the film with anthocyanin did not presented difference on L* and a* color parameters, indicating that the biodegradable film keep stable the color parameters independent of the packed product pH.

According to Yoshida et al. (2014); Luchese et al. (2017) the colorimetric difference in anthocyanin-added films is visibly perceptible to the human eye, this nonconformity is primarily due to the anthocyanin source and the pH, so further research is needed to adequately correlate the color change with the shelf life of products.

4 CONCLUSIONS

The results obtained in this research showed the great potential of wastes from fishing industry to the development of a biodegradable film as food package, and the powdered anthocyanin also has a great potential as indicator to pH change suggesting the change in food freshness. However, for industrial scale production it is necessary to improve this material, making it more stable under conditions of high humidity.

REFERENCES

American Society for Testing and Materials—ASTM 1996. *D882-91: Standard Test Methods for Tensile Properties of Thin Plastic Sheeting*. Philadelphia (Annual Book of ASTM Standards).

American Society for Testing and Materials—ASTM, 1995. *E96-95: Designation Standard Method for Water Vapor Transmission of Materials*. Philadelphia (Annual Book of ASTM Standards).

Arenas, A.M.Z. 2012. *Filme biodegradável à base de fécula de mandioca. Como potencial indicador de mudança de pH*. Dissertação (Mestrado)—Escola Politécnica de São Paulo, São Paulo.

Arfat, Y.A., Benjakul, S., Prodpran, T., Osako, K. 2014. Development and characterisation of blen films on fish protein isolate and fish gelatin. *Food Hydrocolloids* 39: 58–67.

Arrieta, M.P., Lopez, J., Hernadez, A., Rayón, E. 2014. Ternary PLA-PHB-Limonene blends intended for biodegradable food packaging appications. *European Polymer Journal* 50: 255–270.

Artharn, A., Benjakul, S., Prodpran, T., Tanaka, M. 2007. Properties of a protein-based film from round scad (Decapterus maruadsi) as affected by muscle types and washing. *Food Chemistry, London* 103 (3): 867–874.

Association of Official Analytical Chemists—AOAC 2000. *Official Methods of Analysis of AOAC International*. 17th ed. Washington DC. v. 1 e v.2.

Balbinot, E. 2015. *Eletrocoagulação no tratamento de efluentes da filetagem do pescado*. Universidade Federal de Santa Catarina (UFSC) – Florianópolis, SC.

Bridle, P., Timberlake, C.F. 1997. Anthocyanins as natural food colours – selected aspects. *Food Chemistry* 58 (1–2): 103–109.

Dangaran, K., Tomasula, P.M., QI, P. 2009. Structure and function of protein-based edible films and coatings. In: Embuscado, M.E., Kerry C., Huber, K.C. (Eds.). Edible Films and Coatings for Food Applications. New York: Springer.

Davanço, T., Tanada-Palmu, P., Grosso, C. 2007. Filmes compostos de gelatina, triacetina, ácido esteárico ou capróico: efeito do pH e da adição de surfactantes sobre a funcionalidade dos filmes. *Ciência e Tecnologia de Alimentos*. Campinas. 27 (2): 408–416.

Gonçalves, A.A. 2011. *Tecnologia do Pescado: Ciência, Tecnologia, Inovação e Legislação*. São Paulo: Atheneu.

Gontard, N., Duchez, C., Cuq, J., Guilbert, S. 1994. Edible composite films of wheat gluten and lipids: Water vapor permeability and other physical properties. *International Journal Food Science Technology*. Oxford. 29 (1): 39–50.

González, K., Martin, L., González, A., Retegi, A., Eceiza, A., Gabilondo, N. 2017. D-isosorbide and 1,3-propanediol as plasticizers for starch-based films: Characterization and aging study. *Journal of Applied Polymer Science* 134 (20): 1–10.

Hernández, E.J.G.P., de Carvalho, R.N., Joele, M.R.S.P., da Silva Araújo, C., & Lourenço, L.D.F.H. (2017). Effects of modified atmosphere packing over the shelf life of sous vide from captive pirarucu (Arapaima gigas). *Innovative Food Science & Emerging Technologies* 39: 94–100.

Iwata, K., Ishizaki, S., Handa, A., Tanaka, M. 2000. Preparation and characterization of edible films from fish water-soluble proteins. Fisheries *Science*. Tokyo. 66 (2): 372–378.

Khwaldia, K., Arab-Tehrany, E., Desobry, S. 2010. Biopolymer coatings on paper packaging materials. *Comprehensive Reviews in Food Science and Food Safety*. Chicago. 9 (1): 82–91.

Limpan, N., Prodpran, T., Benjakul, S., Prasarpran, S. 2010. Properties of biodegradable blend films based on fish myofibrillar protein and polyvinyl alcohol as influenced by blend composition and pH level. *Journal Food Engineering*. Essex. 100 (1): 85–92.

Lopes, T.J., Xavier, M.F., QUadri, M.G.N., QuadrI, M.B. 2007. *Anthocyanins: A Brief Review of Structural Characteristics and Stability*. Departamento de Engenharia Química e Engenharia de Alimentos - CPGENQ—UFSC.

Luchese, C.L., Sperotto, N., Spada, J.C., Tessaro, I.C. 2017. Effect of blueberry agro-industrial waste addition to corn starch-based films for the production of a pH-indicator film. *International journal of biological macromolecules* 104: 11–18.

Pacquit A., Frisby J., Diamond D., Lau K., Farrell A., Quilty B., Diamond D. 2006. Development of a smart packaging for the monitoring of fish spoilage. *Food Chemistry* 102: 466–470.

Paschoalick, T.M., Garcia, F.T., Sobral, P.J.A., Habitante, A.M.Q.B. 2003. Characterization of some functional properties of edible films based on muscle proteins of Nile tilapia. *Food Hydrocolloids*. Oxford. 17 (4): 419–427.

Rebello, F.F.P. 2009. Novas tecnologias aplicadas às embalagens de alimentos—Revisão. *Revista Agroambiental*. Instituto Federal Sul de Minas.

Shiku, Y., Hamaguchi, P., Tanaka, M. 2003. Effect of pH on the preparation of edible films based on fish myofibrillar proteins. *Fisheries Science*. Tokyo. 69 (5): 1026–1032.

Shiku, Y., Hamaguchi, P., Benjakul, S., Visessanguan, W., Tanaka M. 2004. Effect of surimi quality on properties of edible films based on Alaska pollack. *Food Chem* 86 (4): 493–499.

Souza, S.M.A., Sobral, P.J.A., Menegalli, F.C. 2012. Physical properties of edible films based on bovine myofibril proteins. *Seminário: Ciências Agrárias*. 33: 283–296.

Yoshida, C.M.P., Maciel, V.B.V., Mendonça, M.E.D., Franco, T.T. 2014. Chitosan biobased and intelligent films: Monitoring pH variations. *LWT—Food Science and Technology* 55: 83–89.

WASTES – Solutions, Treatments and Opportunities II – Vilarinho, Castro & Lopes (Eds)
© 2018 Taylor & Francis Group, London, ISBN 978-1-138-19669-8

PAC and tannin as coagulants in swine slaughterhouse wastewater treatment

M.C. Bongiovani, T. Werberich, R.M. Schneider & A.G. do Amaral
Mato Grosso Federal University, Sinop, MT, Brazil

ABSTRACT: The objective of the present investigation is to evaluate the influence of two coagulants, an inorganic and natural (PAC and tannin), in the treatment of effluent from pig slaughtering. Both coagulants were prepared in aqueous solution, 1% (v/v) to the PAC and 0.5% (w/v) to tannin. To evaluate the effect of coagulants in the effluent, coagulation/flocculation assays were performed using the jar-test apparatus. The dosages evaluated for both coagulants were 50, 100, 150, 200, 250 and 300 mg L^{-1}. The parameters analyzed were color and turbidity. According to the results of the statistics analysis, the optimal dosage obtained for both coagulants was 250 mg L^{-1}. With these dosages, it was observed turbidity and color removals higher than 94% and 85%, respectively, with residuals less than 40 NTU for turbidity and 400 uH for color, indicating with this optimal dosage best results mainly when tannin is used as coagulant.

1 INTRODUCTION

The modern agroindustry, characterized by large numbers of animals slaughtered, generates large volumes of wastewater. When this effluent is disposed in the environment, without treatment, can cause very serious environmental problems, which can affect soil and water quality (Drogui et al., 2008).

Approximately 40% of the water used in the agroindustrial process of a slaughterhouse is disposed as effluent. This effluent consists of a large amount of organic matter, as well as a high concentration of suspended solids and high biochemical oxygen demand (BOD) (Amuda & Alade, 2006; Bazrafshan et al., 2012).

The treatment of these effluents has been extensively studied in conventional biological treatments (Escobar et al., 2005; Chan et al., 2009; López-López et al., 2010). However, due to the high organic load with low efficiency removal (Mittal, 2006; Tariq et al., 2012; Bugallo et al., 2014), long hydraulic retention times and greater requirements of area, a pre-treatment, as coagulation/flocculation, can be an alternative (Mittal, 2006; Amuda & Alade, 2006; Drogui et al., 2008; Tariq et al., 2012). This physic-chemical treatment is increasingly being used for the preliminary treatment due to the fact that this treatment is able to remove toxic and non-biodegradable substances (high organic load and fat content) that biological treatment alone cannot remove (Oller & Sanchez-Perez, 2011). Furthermore, the pre-separation of organic matter means that less will have to be treated by any subsequent biological step and reduce area required (Aguilar et al, 2005).

Inorganic coagulants as ferric sulphate, $Al_2(SO_4)_3$ and polyaluminium chloride were used to treat abattoir effluent (Aguilar et al., 2003; Amuda & Alade, 2006; Bazrafshan et al., 2012; Zueva et al., 2013). However, recent studies have discussed several serious drawbacks of using the alum salts, such as its inefficiency at low temperature and the pH of the treated effluent, a great production of sludge with high concentrations of aluminum, which makes difficult the disposal in the soil, due to the contamination and the accumulation of this metal, which can cause harmful effects to human health (Vijayaraghavan et al., 2011).

In this sense, the search for natural coagulants has been the target of many researches in the treatment of wastewaters (Sánchez-Martín et al., 2010; Lagasi et al., 2014; Del Real-Olvera et al., 2015). These coagulants/flocculants have shown advantages in relation to chemical, specifically with regard to biodegradability, low toxicity and low residual sludge production rate (Beltrán-Heredia et al., 2010; Chun-Yang et al., 2010; Saranya et al., 2013).

Because of the great need for sustainable and economical alternative wastewater treatment, the purpose of this research was to evaluate the performance of two coagulants, inorganic and natural (Polyaluminium chloride – PAC and tannin) in the treatment of effluent from swine slaughterhouse.

2 MATERIAL AND METHODS

The raw effluents for coagulation/flocculation tests were collected from a swine slaughterhouse plant at Sinop, MT, i.e. any preliminary treatment was realized before. The collected sample was packaged in a container and kept cooled until initial characterization by laboratory analysis. The analyses were color, turbidity, pH, $UV_{254\,nm}$, COD, BOD and TSS, performed by Standard Methods for the Examination of Water and Wastewater (APHA, 2012).

The agroindustry treatment system is only for stabilization lagoons. The objective of coagulation/flocculation treatment before biological treatment is remove the organic matter and improve the efficiency of the treatments or substitute the biological treatment.

2.1 Coagulants used in the tests

The coagulants used for the tests were a tannin-based product called Tanfloc SG®, a trademark that belongs to TANAC (Montenegro, Brazil) and PAC (Aluminium polychloride). The tannin, supplied as a solid, is modified by a physic-chemical process from the leaching of the bark of the Acacia mearnsii de Wild. This commercial tannin containing both amine and phenolic groups is cationic in nature (Beltrán-Heredia et al., 2011). This tree is very common in Brazil and it has a high concentration of tannins with high flocculant power. Standard solutions of both coagulants were prepared with distilled water at the time of the test.

The coagulant PAC was prepared in solution 1% (v/v) (1 mL PAC for 99 mL of distilled water) and the tannin coagulant was prepared in solution 0.5% (w/v) (0.5 g of tannin to 100 mL of distilled water). Subsequently both were shaken for 30 minutes.

2.2 Coagulation/flocculation tests

The experimental tests were performed in the jar-test apparatus with six samples of the effluent simultaneously (250 mL). The pH of the samples was kept constant during the experiment. The conditions used in each jar-test with PAC and tannin as coagulants are shown in Table 1.

The dosages evaluated for both coagulants were 50, 100, 150, 200, 250 and 300 mg.L^{-1}. The optimal dosage value was the one with the most significant reduction of turbidity and color in terms of percentage.

Table 1. Operational conditions of the jar-test.

Conditions	PAC[1]	Tannin[2]
Rapid mixing velocity (rpm)	200	120
Rapid mixing time (min)	5	2.5
Slow mixing velocity (rpm)	20	20
Slow mixing time (min)	20	20
Settling time (min)	60	20

[1]Amuda & Alade (2006); [2]Bongiovani et al. (2010).

58

The optimal coagulant dosage is defined as the value above which there is no significant difference in the increase in the removal efficiency with increasing coagulant dosage (Aguilar, et al., 2005).

2.3   *Statistical analysis*

To evaluate the best dosages range of agents used after the coagulation/flocculation process, design of experiments was carried out using a completely randomized design (CRD) in factorial arrangement 2×6, which factors: coagulants (PAC and tannin) and dosages (six dosages for both coagulants), with three replications, comprising a total of 18 experiments for each coagulant.

The results were submitted to analysis of variance (ANOVA) by Sisvar® software, and when significant difference between the treatments was verified, the results were submitted to Tukey's test at a level of 5% of significance (p-value < 0.05), to verify the significance of differences among the means.

In these tests, the evaluated parameters were turbidity and color. The turbidity was measured in a turbidimeter and the color in a colorimeter.

3   RESULTS AND DISCUSSIONS

The characterization of the effluent is presented in Table 2.

High values of BOD, COD and suspended solids were observed, showing that the raw effluent from the experiments had a high organic load constituted, basically, of blood and organic material, which are responsible for the red color and most of the turbidity. Both of these components inhibit the coagulation process due to their complexity in relation to their removal from water.

In the blank tests, a decline in color (3433.33 uH) and turbidity (404 NTU) levels were observed when compared to the raw effluent.

The averages of color and turbidity residuals of the effluent treated with PAC and tannin are presented in Table 3.

Table 2.   Effluent characterization.

Parameters	Mean values
Turbidity (NTU)	870
Color (uH)	5700
pH	6.58
$UV_{254\,nm}$ (cm^{-1})	2.653
COD $(mg.L^{-1})$	1611
BOD $(mg.L^{-1})$	928
TSS $(mg.L^{-1})$	1080

Table 3.   Averages of turbidity and color residuals in determination of coagulants optimum dosage.

Dosage $(mg\ L^{-1})$	Residual turbidity (NTU)		Residual color (uH)	
	PAC	Tannin	PAC	Tannin
50	169.7 ± 5.0	206.7 ± 10.1	2440.0 ± 20.0	2373.3 ± 50.3
100	101.3 ± 1.6	112.3 ± 10.1	1526.7 ± 80.8	1386.7 ± 83.3
150	79.9 ± 0.5	112.7 ± 4.0	1120.0 ± 20.0	1233.3 ± 41.6
200	54.5 ± 0.6	33.6 ± 5.0	543.7 ± 33.5	480.0 ± 72.1
250	27.2 ± 3.5	20.1 ± 1.3	406.0 ± 69.2	277.0 ± 57.0
300	14.3 ± 3.6	4.4 ± 0.5	302.7 ± 39.1	151.0 ± 10.6

In general, analyzing each parameter, lower residuals (when compared with blank tests) of turbidity and color were observed with dosage in the range of 250–300 mg.L^{-1} for the coagulant PAC and 200–300 mg.L^{-1} for the coagulant tannin. In both cases, the turbidity results obtained were in accordance with established by CONAMA Resolution 430/2011 (Brazil, 2011), which establishes the value of 40 NTU as the maximum residual turbidity for the effluent discharge in a water body, i.e. it isn't necessary a post treatment, only the coagulation/flocculation treatment is enough to achieve the legislation. In addition, lower residuals of color and turbidity were observed, mainly if the tannin is used as coagulant in the dosage range of 200 a 300 mg.L^{-1}.

Figures 1a and 1b present the graphs of the variables involved according to the factorial design, the removal efficiencies for turbidity and color parameters were evaluated.

The Tukey test for multiple comparison was used to determine the optimal coagulant dosage and which is the best coagulant to be used.

Regarding the removal parameters, the statistical analysis indicates that the different coagulants are statistically different between some individual dosages (lowercase letters) only for turbidity parameter. When evaluating each coagulant in different dosages (uppercase letters), there is a statistical difference between the dosages, and for both coagulants, the best dosage range was 250–300 mg.L^{-1}, with recommended use of the lowest dosage (250 mg.L^{-1}), since it minimizes costs in the treatment of the effluent. With these dosages, it was observed high removals of the analyzed parameters (> 94% for turbidity and > 85% for color) with residuals (< 40 NTU for turbidity and < 400 uH for color), highlighting at this optimal dosage, better results when tannin is used as a coagulant.

Aside from the evident sustainable and environmentally friendly aspects, the high efficiency of this natural coagulant can be explained by the higher resistance of flocs formed to shear forces in a turbulent flow due to the bridging effect. In the other words, bridging linkages are more resistant to breakage at high shear levels when compared to non-polymeric coagulants such as alum (Kim, 1995; Beltrán-Heredia et al., 2010). Another advantage is that the use of natural coagulant produces less quantity of sludge (Aboulhassan et al., 2016) and doesn't have metals from the coagulant with possibility of soil disposal on comparing with chemical ones (Jain et al., 2015). However, it is necessary to reduce the amount of sludge that need to be disposed. In this context, the development of anaerobic reactors based on such natural anaerobic systems could produce eco-technologies for the effective management of a wide variety of solid wastes and industrial wastewater (Gijzen, 2002). In this sludge treatment, the residence time of the natural sludge could be less than sludge obtained using chemical coagulants, because this can contain high levels of easily biodegradable materials (Chen et al., 2008). These aspects can increase the demand of natural coagulants production.

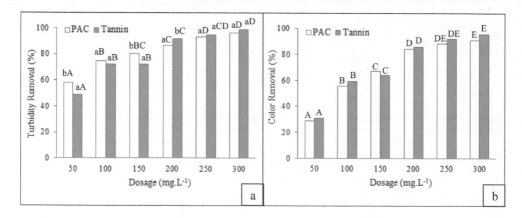

Figure 1. Analysis of the parameters (a) turbidity removal and (b) color removal using the coagulation/flocculation process with the PAC and Tannin coagulants. Lowercase letters (coagulant evaluation at each dosage) and uppercase letters (assessment of dosage for each coagulant) in the graphics identify different statistically significant groups (Tukey's test, p<0.05).

There are few studies in the literature related to the treatment of slaughterhouse effluents using chemical or natural coagulants in the coagulation/flocculation process.

Al-Mutairi et al. (2004) also studied coagulation/flocculation of slaughterhouse wastewater, however using alum as coagulant in the dosage range between 0 and 1000 mg.L^{-1}. The optimal dosages efficiency for suspended solids and turbidity (>99%) were 400 and 200 mg. L^{-1}, respectively, almost the same than obtained in this research (250 mg.L^{-1}) for both coagulants. In addition, the same authors observed that the alum caused inhibitory effects to the system, i.e. the solids collected were much more toxic than those from the effluents (Al-Mutairi et al., 2006).

As well as in this research, De Sena et al. (2008) also compared the tannin Tanfloc SG® with inorganic coagulants (alum and ferric salts) and four coagulation aids. The results show that the tannin had high efficiency removal of organic matter, mainly BOD parameter (>80%) when it was used 25 and 3 mg.L^{-1} of the coagulant and coagulant aid, respectively. A possible alternative to reduce the optimal dosage of this tannin obtained in this research would be add a coagulant aid to enhance process efficiencies by the increase of the overall removal of organic matter.

4 CONCLUSIONS

Optimal dosage of 250 mg.L^{-1} was obtained for both PAC and tannin coagulants, while both in this dosage were in accordance with the current legislation regarding residual turbidity.

In view of the lowest residuals of color and turbidity obtained at this optimal dosage, the tannin can be recommended as a promising coagulant agent in the treatment slaughterhouse wastewater, when compared to the chemical coagulant PAC, since it becomes environmentally correct because it is natural.

REFERENCES

Aboulhassan, M.A., Souabi, S., Yaacoubi, A., Baudu, M. 2016. Coagulation efficacy of a tannin coagulant agent compared to metal salts for paint manufacturing wastewater treatment. *Desalination and Water Treatment* 57(41): 19199–19205.

Aguilar, M.I., Sáez, J., Lloréns, M., Soler, A., Ortuño, J.F. 2003. Microscopic observation of particle reduction in slaughterhouse wastewater by of coagulation–flocculation using anionic ferric sulphate as coagulant and different coagulant aids. *Water Research* 37(9): 2233–2241.

Aguilar, M.I., Sáez, J., Lloréns, M., Soler, A., Ortuño, J.F., Meseguer, V., Fuentes, A. 2005. Improvement of coagulation–flocculation process using anionic polyacrylamide as coagulant aid. *Chemosphere* 58(1): 47–56.

Al-Mutairi, N.Z., Hamoda, M.F., Al-Ghusain, I. 2004. Coagulant selection and sludge conditioning in a slaughterhouse wastewater treatment plant. *Bioresource Technology* 95(2): 115–119.

Al-Mutairi, N.Z. 2006. Coagulant toxicity and effectiveness in a slaughterhouse wastewater treatment plant. *Ecotoxicology and Environmental Safety* 65(1): 74–83.

Amuda, O.S., Alade, A. 2006. Coagulation/ flocculation process in the treatment of abattoir wastewater. *Desalination* 196(1–3): 22–31.

APHA American Public Health Association. 2012. *Standard Methods for the Examination of Water and Wastewater.* 22 st, Centennial Edition: Washington.

Bazrafshan, E., Mostafapour, F.K., Farzadkia, M., Ownagh, K.A., Mahvi, A.H. 2012. Slaughterhouse wastewater treatment by combined chemical coagulation and electrocoagulation process. *PLoS One* 7(6): 1–8.

Beltrán-Heredia, J., Sánchez-Martín, J., Martín-Sánchez, C. 2011. Remediation of Dye-Polluted Solutions by a New Tannin-Based Coagulant. *Industrial Engineering Chemical Research* 50(2): 686–693.

Beltrán-Heredia, J., Sánchez-Martín, J., Gómez-Muñoz, M.C. 2010. New coagulant agents from tannin extracts: Preliminary optimisation studies. *Chemical Engineering Journal* 162(3): 1019–1025.

Bongiovani, M.C., Konradt-Moraes, L.C., Begamasco, R., Lourenço, B.S.S., Tavares, C.R.G. 2010. Os benefícios da utilização de coagulantes naturais para a obtenção de água potável. *Acta Scientiarum Technology* 32(2): 167–170.

Brazil. 2011. CONAMA Resolution n° 430/2011 of May 13, 2011. Provides about conditions and standards for discharging of liquid effluent, complements and amends Resolution 357/05. *Official Journal of the Union n° 92 of May 16, 2011*. Brasília: DF.

Bugallo, P.M.B., Andrade, L.C., De la Torre, M.A., López, R.T. 2014. Analysis of the slaughterhouses in Galicia (NW Spain). *Science of the Total Environment* 481(1): 656–661.

Chan, Y.J., Chong, M.F., Law, C.L., Hassel, D. 2009. A review on anaerobic-aerobic treatment of industrial and municipal wastewater. *Chemical Engineering Journal* 155(1–2): 1–18.

Chun-Yang Y. 2010. Emerging usage of plant-based coagulants for water and wastewater treatment. *Process Biochemistry* 45(9): 1437–1444.

Chen, Y., Cheng, J.J., Creamer, K.S. 2008. Inhibition of anaerobic digestion process: a review. *Bioresourse Technology* 99(10): 4044–4064.

De Sena, R.F., Claudino, A., Moretti, K., Bonfanti, I.C.P., Moreira, R.F.P.M., José, H.J. 2008. Biofuel application of biomass obtained from a meat industry wastewater plant through the flotation process—A case study. *Resources Conservation & Recycling* 52(3): 557–569.

Del Real-Olvera, J., Rustrian-Portilla, E., Houbron, E., Landa-Huerta, F.J. 2016. Adsorption of organic pollutants from slaughterhouse wastewater using powder of *Moringa oleifera* seeds as a natural coagulant. *Desalination and Water Treatment* 57(21): 9971–9981.

Drogui, P., Asselin, M., Brar, S.K., Benmoussa, H., Blais, J.F. 2008. Electrochemical removal of pollutants from agro-industry wastewaters. *Separation and Purification Technology* 61(3): 301–310.

Escobar, F.C., Marín, J.P., Mateos, P.A., Guzman, F.R., Barrantes, M.M.D. 2005. Aerobic purification of dairy wastewater in continuous regime: Part II: Kinetic study of the organic matter removal in two reactor configurations. *Biochemical Engineering Journal* 22(2): 117–124.

Gijzen, H.J. 2002. Anaerobic digestion for sustainable development: A natural approach. *Water Science & Technology* 45(10): 321–328.

Jain, R.K., Dange, P.S., Lad, R.K. 2015. A treatment of domestic sewage and generation of bio sludge using natural coagulants. *International Journal of Research in Engineering and Technology (IJRET)* 4(7): 152–156.

Kim, Y.H. 1995. *Coagulants and Flocculants. Theory and Practice*. Tall Oak Publishing: Littleton.

Lagasi, J.F., Agunwamba, J.C., Aho, M. 2014. Comparative studies on the use of ordinary and de-oiled *Moringa oleifera* in the treatment of abattoir wastewater. *The International Journal of Engineering and Science* 3(2): 1–7. *Water Science & Technology* 45(10): 321–328.

López-López, A., Vallejo-Rodriguez, R., Méndez-Romero, D.C. 2010. Evaluation of a combined anaerobic and aerobic system for the treatment of slaughterhouse wastewater. *Environmental Technology* 31(3): 319–326.

Oller, I., Malato, S., Sanchez-Perez, J. 2011. Combination of advanced oxidation processes and biological treatments for wastewater decontamination—a review. *Science of Total Environment* 409(20): 4141–4166.

Sánchez-Martín, J., Beltrán-Heredia, J., Solera-Hernández, C. 2010. Surface water and wastewater treatment using a new tannin-based coagulant. Pilot plant trials. *Journal of Environmental Management* 91(10): 2051–2058.

Saranya, P., Ramesh, S.T., Gandhimathi, R. 2014. Effectiveness of natural coagulants from non-plant-based sources for water and wastewater treatment—a review. *Desalination and Water Treatment* 52(31–33): 6030–6039.

Tariq, M., Ahmad, M., Siddique, S., Waheed, A., Shafiq, T., Khan, M.H. 2012. Optimization of coagulation process for the treatment of the characterized slaughterhouse wastewater. *Pakistan Journal of Science and Industrial Research* 55(1): 42–48.

Vijayaraghavan, G., Sivakumar, T., Kumar, A.V. 2011. Application of plant based coagulants for waste water treatment. *International Journal of Advanced Engineering Research and Studies* 1 (1): 88–92.

Zueva, S.B., Ostrikov, A.N., Ilyina, N.M., De Michelis, I., Vegliò, F. 2013. Coagulation Processes for Treatment of Waste Water from Meat Industry. *International Journal of Waste Resources* 3(2): 1–4.

WASTES – Solutions, Treatments and Opportunities II – Vilarinho, Castro & Lopes (Eds)
© 2018 Taylor & Francis Group, London, ISBN 978-1-138-19669-8

Waste to energy as a complementary energy source in Abuja, Nigeria

O.M. Aderoju & G.A. Dias
Faculty of Sciences, University of Porto, Porto, Portugal

ABSTRACT: The Nigerian national power output is less than 4000 MW from fluctuating sources such as hydro power (seasonal) and gas-fired plants (prone to sabotage). Municipal Solid Waste (MSW) is a burden to the environment in terms of its health risk and its management but also an underrated energy source in Nigeria. Hence, this study aims at investigating the MSW in Abuja as a supplementary resource for power generation. The study focused on Waste to Energy by incineration as an instant solution for both MSW reduction as well as power generation. Also, the study employs proximate and ultimate analysis to evaluate the Lower calorific Values (LHV) for Porto & Abuja to determine the Power Generation Potential (P_{gp}). Furthermore, a comparison approach is considered to establish a relationship between $P_{gp\,(Porto)}$ & $P_{gp\,(Abuja)}$ from MSW to estimate, project and predict energy output from WTE incineration plant in Abuja, Nigeria.

1 INTRODUCTION

The growing concern on an eco-friendly environment to sustain the globe for the generations to come is of significant interest. Current paces of urbanization, consumerist societies and waste generation have challenged the global sustainability in many ways (Visvanathan *et al.*, 2007). Most cities in Africa are confronted with the intricacy of Municipal Solid Waste (MSW) management, its collection and disposal. The primary sources of MSW are from residences, institutions, commercial establishments, inert waste from industries and hospitals, and agricultural activities. MSW is obviously becoming a major challenge in major cities of developing nations due to rapid urbanization and increasing population growth (Aderoju *et al.*, 2015). The World Bank (2012) reported that the estimate of the global MSW generated is about 1.3 billion tonnes yearly and it increases by 1% yearly. Recovery is one aspect of sustainable waste management that is based on the well-known hierarchy of "prevention" "reuse", "recycling", "recovery" and "disposal". Waste-to-energy (WTE) refers to the recovery of heat and power from waste, and in particular non-recyclable waste (Ryu and Shin, 2013). The increasing clamour for energy and satisfying it with a combination of conventional and renewable resources is a huge challenge (Jain *et al.*, 2014). Non-conventional energy exploitation through the useful harnessing of biomass energy locked up in the urban solid waste stream into grid energy seems to be a more probable option that has gained both political and public discussions on alternative energy sources (Akuffo, 1998). The WTE industry has proven itself to be an environmentally friendly solution to the disposal of MSW and the production of energy (Kumar *et al.*, 2010). The decomposition of organic waste in dumpsites continuously emits greenhouse gases, which contribute enormously to the depletion of the ozone layer and global warming as a whole. The proper reuse of MSW improves sanitation in urban centres, decreases emissions of greenhouse gases due to its decomposition and in turn helps to reduce the consumption of fossil fuels (Possoli *et al.*, 2013). The sporadic supply of electricity has brought a damaging setback to the Nigerian economy such that a complete-halt of operation by some major companies making them flee the country to nearby countries where constant electric power supply is assured. With haphazard dumpsites defacing the Nigerian cities, little or no effort has been made to utilize MSW as a resource towards a waste-to-energy (WTE) solution. The over-dependence on both hydro (which is seasonal), and natural gas (which is prone to sabotage) for power generation has led to the production of less than 4000 MW for a population of over 183 million people. In general, the concept of WTE in Nigeria is a welcoming development and a great positivity to the economy by harnessing MSW stream as a resource for electrical power generation. WTE technology significantly reduces the menace of excessive and continuous MSW generated in our major cities, promotes a sustainable aesthetic environment, electrical power generation and distribution, job creation, revenue generation and transition to a clean energy economy in Nigeria. On this note, the study aims to utilize MSW stream as a supplementary solution for power generation and distribution Nigeria. The study limits its investigation of the two most populated local government area in Abuja which are AMAC (Abuja Municipal Area Council) and Bwari LGA such that their MSW can be utilized for power generation.

2 AN OVERVIEW OF WTE (INCINERATION) CONCEPTS IN ENERGY RECOVERY

The continued concerns over energy prices, increasing population and climate change issues has led the drive towards a need for alternative and new energy sources (Amoo and Fagbenle, 2013). However, renewable energy resource remains that which is boundless and it is replenished naturally such as hydro, wind, solar, and biomass. MSW has its major constituent as organic waste and others like plastics which are products of fossil fuel are continuously generated by humans in the environment. Therefore, WTE is recognized as a promising alternative for waste management to overcome the waste generation problem and as a potential renewable energy source (S.T Tan *et al.*, 2015). WTE technologies are able to convert the energy content of different types of waste in various forms of valuable energy, hence electric power is produced and distributed through local and national grid systems (WEC, 2013). About 130 million tons of MSW are worldwide burned annually in 600 plants based on WTE that generate electricity, steam for heating purpose and recovered metals for recycling (Themelis, 2003). Waste combustion according to Voelker (2000) provides integrated solutions to the problems of modern era "by recovering otherwise lost energy, thereby reducing our use of precious natural resources, cutting down our emissions of greenhouse gases, saving valuable land that would otherwise be destined to become landfill and recovering land once scarified to product of consumerism. Kagkelidou in 2005 estimated that the net electrical energy provided to utilities for each ton of solid waste corresponds to saving 170 liters of fuel oil. Recent technological advances and tighter pollution regulations ensure that modern WTE facilities are cleaner than almost all major manufacturing industries (Hazardous Waste Resource Center, 2000). According to Suberu *et al.*, (2012), thermo-chemical conversion is preferably used due to its ability to ensure that the contribution of both biodegradable and non-biodegradable components of the waste are used for the energy output. Studies have shown that, the lower operational cost, higher energy output and less complexity in technology, and its instant operation makes WTE by incineration the most suitable. MSW incineration to energy appears to be most suitable for Nigeria due to the massive reduction of MSW in the environment and also as an immediate additional solution to the current electric power crisis.

3 METHODOLOGY

This study is based on utilizing MSW in Abuja Municipal Area Council (AMAC) and Bwari area council for electricity. This study intends to demonstrate a possible concept to evaluate, predict and project the power generation potential of the MSW in Abuja through a comparative approach using two cities (Porto and Abuja). Also, estimating of the calorific values of the MSW samples and the energy recovery potential for (Porto & Abuja) is essential using this approach.

3.1 *Study area*

Abuja is the capital city of Nigeria, located in the center of Nigeria in the Federal Capital Territory (FCT) within latitude 7° 25¢N and 9° 20° N of the equator and longitude 5° 45¢E and 7° 39¢E of the meridian. The Federal Capital Territory has a land area of 7,753.85 km² square kilometers. It has a population of about 2,238,800 persons (NPC, 2011). It experiences majorly wet and dry seasons. Abuja experiences an average daily minimum and maximum temperature of 20.5°C and 30.8°C respectively. It has a mean rainfall and humidity of about 119.2 mm and 58.4% respectively with the highest rainfall in August and lowest between November and March respectively.

3.2 Basic characterization and sample preparation of MSW in AMAC and Bwari area council

The physical characteristics to be measured in waste are basically the classification of the waste stream components, quantity in terms of mass or volume, and the degree of wetness. In the study area, the four major dumpsites identified and visited were Kubwa, Dutse, Gossa and Karshi dumpsites. Furthermore, 10 kg of freshly mixed MSW samples were collected at each dumpsite for characterization 3 times a week. The mixing, manual sorting and weighing was done to determine the MSW composition. The physical analysis was based on weight ratio of different components in the MSW stream. The percentage of samples composition was categorized into 8 different components which are: food waste/organic; textile; glass/Ceramics; metal; paper; plastics; rubber: and other waste. For the purpose of power generation from MSW, incombustible components like metal, glass, ceramic wastes are separated for recycling.

3.3 Proximate analysis and ultimate analysis

In General, moisture content affects the self-sustained combustibility and the calorific value of the waste stream (Komislis et al., 2014). According to Dong et al., in 2002, proximate analysis determines the moisture, volatile matter, fixed carbon and ash content of the waste sample. With the assumption that calorific value is proportional to the carbon content and hydrogen in the sample, then Higher Heat Value (HHV) is assumed to be a function of the quantity of fixed carbon in percentage present in which volatile matter percentage has its main components as carbon and hydrogen (Gunnamatta, 2016). The combustible components in the waste sample are weighed and heated separately in the oven for 1 hr to determine the moisture content and dry mass of each component in the MSW sample. An additional loss of weight on ignition at 950°C for 30 min is used to determine the % volatile matter (dry basis) and % ash content (dry basis).

The ultimate analysis of waste analyzes the percent of carbon, hydrogen, oxygen, nitrogen, sulphur and ash present in the MSW sample. Provided the dry mass of each component in the MSW sample is determined, the elemental contents like carbon, hydrogen, oxygen, nitrogen, sulphur and the ash content for waste sample can be determined using the standard table of ultimate analysis of combustible waste in Tchobanoglaus et al., (1977); Wess et al., (2004); Othman (2008) and it is expressed as;

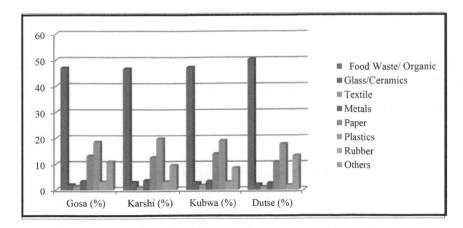

Figure 1. Dumpsites composition in AMAC and Bwari area council. Source: Author.

Elemental Content (kg) = (Dry mass*Standard element (C/H/O/N/S) value)/100   (1)

where;

The Standard element (C/H/O/N/S) value in the MSW components is obtained from the standard table of ultimate analysis of combustible waste. This is estimated for total components present in the waste sample and cumulated to arrive at a (C, H, O, N, S) value of the sample which is substituted into the Dulong equation. The HHV value can be determined using the modified Dulong equation which considers nitrogen and it is expressed as;

$$\text{HHV (kJ/kg)} = 337C + 1419(H_2O - 0.1250_2) + 93S + 23\ N \qquad (2)$$

where; C = Carbon (%), H = Hydrogen (%), O = Oxygen (%), S = Sulphur (%), N = Nitrogen (%)

3.4 *Calorific value estimation for MSW energy potential*

Calorific measurement can be described as an assessment of the energy output and the degree of combustion of the waste stream that is used in evaluating the economic viability of a WTE project. The calorific value is classified into the Higher Heat Value (HHV) and Lower Heat Value (LHV). HHV is described as the quantity of heat generated by a complete combustion of a unit mass of sample in air or oxygen, such that the product of combustion cooled down to the room temperature and still remains in liquid form (Franjo *et al.*, 1992). However, LHV is the net heat produced when a unit mass of the sample is completely burnt in air or oxygen, such that the product of combustion is allowed to escape as steam. LHV of the MSW sample is calculated in Prasada Rao *et al.*, (2010) as;

$$\text{LHV} = \text{HHV} - 9\ (H\%) * \text{LHS} \qquad (3)$$

where; LHV = Lower Heat Value of each component (kcal/kg), HHV = Higher Calorific Value of each component (kcal/kg), LHS = Latent Heat of Steam which is 587 (kcal/kg), (H%) = hydrogen percentage of the sample.

3.5 *Energy recovery potential and power generation potential*

The possible amount of energy recovered from MSW based on different conversion methods is a function of its calorific value and organic content (Tsunatu *et al.*, 2015). Thus, thermo-chemical conversion utilizes all its organic matter, biodegradable as well as non-biodegradable for energy output. The Energy Recovery Potential for the thermo-chemical conversion of MSW is expressed as;

$$E_{rp}\ \text{(kWh)} = \text{LHV} * Wt * (1000/860) * \eta \qquad (4)$$

where; E_{rp} (kWh) = energy recovery potential of MSW sample in kWh, LHV = Lower Heat Value (kcal/kg), Wt = Weight of waste (tons), η = Conversion Efficiency which ranges between 22–28% (IEA, 2007).

Furthermore, the Power Generation Potential (P_{gp}) is the amount of energy that can possibly be generated daily. It is expressed as;

$$P_{gp}\ \text{(kW)} = E_{rp}/24 \qquad (5)$$

3.6 *Comparative power generation potential (Porto and Abuja)*

The comparison has the tendency of establishing a relationship between the $P_{gp\ (Porto)}$ and $P_{gp\ (Abuja)}$. This approach considers the LHV of the MSW sample and the MSW (Wt) to obtain the $E_{rp\ (Porto)}$ & $E_{rp\ (Abuja)}$ respectively. Furthermore, the $P_{gp\ (Porto)}$ & $P_{gp\ (Abuja)}$ having known the E_{rp} of both locations is expressed as;

$$P_{gp\,(Abuja)} = E_{rp\,(Abuja)}/24 \tag{6}$$

$$P_{gp\,(Porto)} = E_{rp\,(Porto)}/24 \tag{7}$$

By substituting Eq. (4) into Eq. (6) & Eq. (7);

$$\text{Hence; } P_{gp\,(Abuja)} = LHV_{(Abuja)} * Wt * 0.04845 * \eta \tag{8}$$

$$P_{gp\,(Porto)} = LHV_{(Porto)} * Wt * 0.04845 * \eta \tag{9}$$

The relationship between the $P_{gp\,(Porto)}$ & $P_{gp\,(Abuja)}$ can be established by evaluating the $P_{gp\,(Porto)}$ & $P_{gp\,(Abuja)}$ using the Wt repeatedly at varying and regular intervals in Eq. (8) & Eq. (9). Also, the evaluated results are plotted against the Wt such that a possible mathematical correlation is expected between $P_{gp\,(Porto)}$ & $P_{gp\,(Abuja)}$ respectively.

4 DISCUSSION

Generally, the P_{gp} using MSW in a WTE plant is dependent on the LHV and the quantity of MSW (tons). Also, the LHV basically requires that the condition of the waste stream in terms of chemical and physical characteristics is satisfactory before it is used as a resource for power generation. Using proximate and ultimate analysis, the HHV is obtained with the aim to evaluate the LHV. Recall that there are no similarities in terms of per capital daily generation, LHV, and the population of both locations (Porto & Abuja). However, for comparison approach the same weight of MSW is used at varying interval like (100 tons, 200 tons, ... 1000 tons) alongside the $LHV_{(Porto)}$ & $LHV_{(Abuja)}$ in Eq. (8) & Eq. (9) to arrive at P_{gp} for both locations respectively. The $P_{gp\,(Porto)}$ & $P_{gp\,(Abuja)}$ will be plotted against the MSW weights (Wt) to establish a relationship if any. In case there is a relationship, it can be used to estimate and make projections for the possible amount of power in kW for every ton of MSW in Wt.

5 CONCLUSION

The practice of generating energy from waste for domestic and other uses has been demonstrated in commercial quantity by most developed countries in Europe, America and Asia. Incineration with energy recovery indeed has been proven environmentally sound due to the advancement in technology over the years. The problem of exponential generation of MSW in Nigeria has become an overburden on both the environment and authorities. The study employed a comparative approach for Power Generation Potential (P_{gp}) of Porto (Portugal) and Abuja (Nigeria) to give an idea on how to arrive at a standard relation in the form of a mathematical expression for the prediction and projection of the possible amount of electric energy (kW) from a certain quantity of MSW (tons) for Nigeria, provided all the necessary criteria are fulfilled. The merit of this study tends to reduce the volume of MSW streams in major Nigerian cities and simultaneously serve a resource for electric power generation. This study provides a new scope for government and other stakeholders to invest towards providing basic sustainable infrastructure for better quality of life and concurrently create an eco-friendly environment for the generations to come.

REFERENCES

Aderoju O.M., A Guerner Dias and Guimaraes R. 2015. Building Capacity an Integrated Perception and Attitude towards Municipal Solid Waste Management in Nigeria. *WASTES: Solutions, Treatments and Opportunities*, pp. 7–12. London: CRC Press, Taylors and Francis Group.

Amoo O.M. and Fagbenle R.L. 2013. Renewable Municipal Solid Waste pathways for Energy Generation and Sustainable Development in the Nigerian context. *International Journal of Energy and Environmental Engineering* 4/42.

Akuffo F.O. 1988. Options for Meeting Ghana's Future Power needs; Renewable Energy Sources. *High Street Journal, Week Ending* 24.

Dong C., Jin B. and Li D. 2003. Predicting the Heating Value of MSW with a Feed Forward Neutral Network. *Waste Management* 23: 103–106.

Franjo C.F., Ledo J.P., Rodriguez Anon J.S. and Nunez L. 1992. Calorific Value of Municipal Solid Waste. *Environmental Technology* 13: 11, 1085–1089.

Gunamantha M. 2016. Prediction of Higher Heating Value Bioorganic Fraction of Municipal Solid Waste from Proximate Analysis data. *International Journal of Engineering Research & Technology* 5 (2): 442–447.

Hazardous Waste Resource Center 2000. *Hazardous Waste Incineration: Advanced Technology to Protect the Environment. Environmental Technology Center*, http://www.etc.org/technologicalandenvironmentalissues/treatmenttechnologies/incineration/ assessed on 5/01/17.

IEA 2007. *Biomass for Power Generation and CHP*. IEA Energy Technology Essential ETE03, pp 1–4.

Jain P., Handa K. and Paul A. 2014. Studies on Waste-to-Energy Technologies in India & a detailed study of Waste-to-Energy Plants in Delhi. *International Journal of Advance Research 2/1*: 109.

Kagkelidou M. 2005. Burning Waste for Energy. In Athens News 15/07/2005:A06.

Komilis D., Kissas K. and Symeonidis A. 2014. Effect of Organic Matter and Moisture on the Calorific Value of Solid Waste: An update of the Tunner diagram. *Waste Management* 34: 249–255.

Kumar J.S., Subbaiah K.V. and Prasada Rao P.V.V. 2010. Waste to Energy—A Case Study of Eluru City, Andhra Pradesh, India. *International Journal of Environmental Science* 1(2): 151–162.

National Population Commission (NPC) 2011. *Nigeria*. Federal Capital territory: City Population; Retrieved from *www.citypopulation.de* on the 20th of Nov, 2016.

Othman N, Basri N.E.A., Yunus N.M. and Sidek L.M. 2008. Determination of physical and chemical characteristics of electronic plastic waste (Ep-Waste) resin using proximate and ultimate analysis method. *International conference on construction and building technology*, pp 169–180.

Possoli L., Coelho V.L., Ando Junior O.H., Neto J.M., Spacek A.D., Olivera M.O., Schaeffer L. and Bretas A.S. 2013. Electricity Generation by Use of Urban Solid Waste. *International Conference on Renewable Energies and Power Quality (ICREPQ '13), (RE and PQJ)*. ISSN 2172–038 X, No 11.

Prasada Rao P.V., Venkata K.S. and Sudhir J.K. 2010. Waste to energy: A case study of Eluru, A.P, India. *International Journal of Environmental Science and Development* 1(2): 151–162.

Ryu, C. and Shin, D. 2013. Combined Heat and Power from Municipal Solid Waste: Current Status and Issues in South Korea. *Energies* 6: 45–57.

Suberu M.Y., Mokhtar A.S. and Bashir N. 2012. Renewable Power Generation Opportunity from Municipal Solid Waste: A Case Study of Lagos Metropolis (Nigeria). *Journal of Energy Technologies and Policy* 2(2).

Tan S.T., Ho W.S., Hashim H., Lee C.T., Taib M.R. and Ho C.S. 2015. Energy, Economic and Environment (3E) Analysis of Waste–to Energy (WTE) Strategies for Municipal Solid Waste (MSW) Management in Malaysia. *Journal for Energy Conversion and Management* 102: 111–120.

Tchobanoglaus G., Theisen H. and Eliassen R. 1977. Solid wastes: engineering principles and management issues. New York: McGraw-Hill.

Themelis, N.J. 2003. An Overview of the Global Waste—to-Energy Industry, WTERT (2003).

Tsunatu D.Y., Tickson T.S, San K.D. and Namo J.M. 2015. *Municipal Solid Waste as Alternative Source of Energy Generation: A Case Study of Jalingo Metropolis, Taraba State, Nigeria*. 5(3): 185–193.

Visvanathan C., Adhikari R. and Ananth A.P. 2007. 3R Practices for Municipal Solid Waste Management in Asia: *The Second Baltic Symposium on Environmental Chemistry, KALMAR, SWEDEN*. November 26–28.

Voelker B.M. 2000. Waste–to-Energy, Solution for solid Waste Problem of 21st Century. http//www/p2pays.org/ref/09/08624.pdf.14/03/08.

WEC 2013. *World Energy Resources; Waste to Energy* 7b: 2–11.

Wess J.A., Olsen L.D. and Sweeney M.H. 2004. *Asphalt (bitumen): Concise International Chemical Assessment*. Document 59 (2012). World health organization.

World Bank 2012. *What a waste: a global review of solid waste management, urban development*. Series knowledge papers.

Study on properties related to energy recovery from waste streams in Finland

E. Sermyagina, M. Nikku, E. Vakkilainen & T. Hyppänen
Lappeenranta University of Technology, Lappeenranta, Finland

ABSTRACT: Current paper presents the results of proximate analysis and heating value measurements for several different municipal solid waste (paper, cardboard, waste wood, textile, juice box carton and sewage sludge) and industrial waste fractions (bark, sawmill residue and industrial sludge). The values are compared with the fuel samples (forest residue, eucalyptus wood chips, peat and coal). Municipal solid waste fractions exhibit quite similar composition, comparable on dry basis with the industrial wood wastes, woody fuels and peat.

1 INTRODUCTION

1.1 *Waste management in Finland*

Reduction of waste and improvement of waste management are among the main routes in achieving resource efficient and low-waste society. The legislative framework of five-step waste hierarchy was introduced within Waste Framework Directive (EU, 2008) by the European Parliament: the waste prevention is the preferable option, followed by material re-use, recycling and other forms of recovery, with disposal (e.g. landfill) as the last choice. The main objective of the hierarchy is to extract the maximum practical benefit from the materials and to minimize amount of landfilled waste.

The waste management policies in Finland are based on the European Union waste strategy. The National Waste Plan was successfully providing the guidance for the waste management within the waste hierarchy principles until the year 2016 (the newer version of the national waste plan is currently developing). Among its objectives were to decrease of the total volume of generated municipal solid waste (MSW), to recycle 50% of municipal waste streams, generate energy from 30% and dispose less than 20% at the landfills (Ministry of the Environment, 2009). The waste disposal at landfills in Finland has significantly reduced during the recent years, as the energy recovery has increased: almost 90% of the generated MSW was recovered in 2015.

The municipal waste consists largely of waste produced by households, though it also includes such sources as offices, public institutions and commerce. The waste streams include different organic and inorganic fractions like food waste, paper, wood, plastics, glass, metal and other inert materials. Generally, the waste is collected by the municipality; and the composition of the waste may vary from municipality to municipality. With regard to industrial waste, it is generally subjected to material or energy recovery on site when possible, as frequently done for example in pulp and paper industry. At the same time, certain waste fractions have to be send to the municipal waste treatment and should be subsequently treated together with the other MSW streams.

1.2 *MSW composition and analysis*

Several researchers have evaluated the chemical composition of mixed and/or separated fractions of MSW with respect to various thermochemical treatments. With regard to Northern European countries, an extensive work to assess the chemical content (heavy metals, ash content and the heating value) of the Danish MSW streams has been presented by Riber et al. (2009) and Götze et al. (2016). The notable differences in the studied properties of the residual and source-segregated waste materials were found for the parameters related to organic matter and for the elements of environmental concern. Pyrolytic degradation characteristics by thermogravimetric analysis (TGA) together with proximate and ultimate analyses were evaluated for the different waste components of MSW in Norway (Sørum et al. 2001). In an another study, the household waste in the Netherlands has been assessed with respect to elemental (heavy metals and additionally heating value) and quantitative composition (Cornelissen & Otte, 1995).

Both quantity and chemical content of the waste materials can differ significantly with region and the season. As far as the waste treatment in Finland is concerned, the knowledge about chemical composition of MSW streams is relatively scarce. In the paper by Horttanainen et al. (2013), the qualitative composition of the mixed MSW was studied. The heating value and renewable share of energy content of the mixed waste stream was determined but the detailed evaluation of chemical composition for the separate waste streams is limited. Additionally, some data on combustible properties of the municipal solid waste in Finland is presented by Technical Research Centre of Finland (Nashrullah 2015; Ajanko et al. 2005; Alakangas, 2000). Wilén et al. (2004) discussed the possible solutions for energy generation from the waste streams in Finland with a focus on co-firing in combined heat and power plants, gasification perspectives and advanced gas cleaning.

Even though the mentioned studies cover certain amount of data on waste from different regions worldwide, this information has limited application to the specific region due to variations in the waste components and chemical composition of the waste. A comprehensive characterization of MSW is scarce, possibly due to the highly heterogeneous nature of MSW and expensive procedures for this process. The objective of this paper is to improve the knowledge concerning the fuel properties of the main waste components in Finland. Plastics, paper, cardboard, textile, sewage sludge and demolition wood are the typical fractions of municipal waste in Finland. (Nasrullah 2015; Wilén et al. 2004) These waste streams were analyzed and compared along with the most common industrial wastes from a pulp and paper plant and several fuel samples (both renewable and fossil). This information is crucially important for quantifying the possibilities of energy generation from MSW.

2 MATERIALS AND METHODS

2.1 Samples

Municipal solid waste samples were collected from a landfill site in South Karelia, Finland, serving some 130.000 habitants in the region. Horttanainen et al (2013) studied the same landfill site, comparing and finding the waste composition to be similar with other Finnish landfills. Their study also indicated limited variation in the heating value of combustible waste over a longer period of time. The waste was source separated in to energy waste that could be combusted for energy production. Different fractions of common household waste, such as paper, cardboard, plastic, textile and juice box carton were sampled by hand from the waste piles. Waste (demolition) wood was also gathered from the landfill site. Sewage sludge was obtained from wastewater treatment in city of Lappeenranta after the mechanical water removal after which it is currently sent for landfilling.

Industrial waste samples were collected from a pulp and paper plant operating in the same region: samples contained bark, residue from a sawmill and industrial sludge from the plant.

Biomass fuel samples of forest residue and peat were obtained from the local power plant. Additionally, the samples of Polish (bituminous) coal and Brazilian eucalyptus (mixture of *Eucalyptus Grandis* and *Eucalyptus Urophylla* from Minas Gerais state, Brazil) were tested.

2.2 Chemical characterization

Standard procedures were used to characterize the waste components and fuel samples. Each sample was analyzed at least twice, and the average value was then calculated. Additionally, the samples were pelletized for determination of the heating value in a bomb calorimeter.

The total moisture content was determined with simplified oven dry method (CEN/TS, 2010; SFS, 2010a): samples were dried in a laboratory oven at a temperature of 105°C in air atmosphere until the constant mass was reached. Dried samples were then ground to under 1 mm particles for further analysis. The ash content of the samples was determined with the standardized procedures of EN 15403 (SFS, 2011c) and EN 14775 (SFS, 2009) by heating the sample to 550°C and maintaining it in a constant temperature for at least 2 h. The procedure for determination of the volatile matter is described in the standards EN 15402 (SFS, 2011b) and EN 15148 (SFS, 2010c). The volatile mass of the sample is given by the mass loss at the temperature of 900 ± 10°C in 7 min without contact with air. Fixed carbon was determined by reducing the mass of ash and volatiles from the initial mass of the dry sample. The higher heating value was determined with the Parr 6400 calorimeter, following the standard EN 15400 (SFS, 2011a) for waste components and EN 14918 (SFS, 2010b) for fuel samples.

3 RESULTS AND DISCUSSION

3.1 *Proximate analysis*

The results of the proximate analysis and higher heating value (HHV) measurements on dry basis are presented in Table 1. The deviation of the resulting values between the experimental runs was below 0.3% on average for the volatile matter and the heating value and around 2.5% on average for the ash content measurements. Plastic had more than 90% of volatile matter and insignificant amount of fixed carbon (0.2%). The ash content for all investigated MSW materials was moderate (6.6% on average) with the only exception of paper with ash content of 36%. Similar results have been published by Sørum et al. (2001) for recycled and glossy paper samples. Along with plastic, sewage sludge and juice box carton had the highest share of volatiles (82%). The studied industrial waste materials showed comparable results: all the samples have on average 2% of ash, 78% of volatiles and 20% of fixed carbon.

Eucalyptus wood chips showed the lowest ash content among all the studied materials streams in this study, additionally they had a high volatile content. The characteristics of forest residue were naturally quite similar to the values of industrial wood waste samples due to their similar origin. Peat and coal had the highest content of fixed carbon and the lowest volatile content among all the samples, while the coal had quite high ash content (16.3%).

It can be noted that the composition of cardboard, waste wood, textile and juice box carton are comparable to forest residue fuel and industrial wastes. At the same time, the elemental composition should be accurately assessed prior the further treatment, since these materials can contain elements that are less favorable for combustion equipment (e.g. alkalis or chlorine) or the environment (e.g. heavy metals). Especially the heavy metals are enriched in the sewage sludge.

In addition, the initial moisture content of the waste streams affect significantly on their potential utilization. Figure 1 illustrates the results of the proximate analysis for the studied materials on wet basis. The data indicates that the industrial waste and biomass samples as well as sludges had the highest moisture contents, in order of 50% of the fuel mass for biomasses and over 70% for sludges. The high moisture content of sewage sludge is due to generally applied mechanical dewatering techniques: as the sludge is usually transported for landfilling, the dewatering allows to reduce the transportation and landfilling expenses. The lower moisture content could be obtained with, for example, thermal drying, but this would require additional investment on equipment and would increase the operating costs significantly.

The results obtained in the current study for different fractions of municipal and industrial wastes are compared with the ones available in the literature in Figure 2. Green waste consists of grass, leaves and kitchen garbage. The comparison of the volatile matter against the fixed carbon content for each sample presents a simple classification of different waste streams. The presented data illustrates the significant heterogeneity between the typical wastes: origin, treatment and utilization methods of the initial material can affect considerably on the chemical properties of the generated waste. It can be noted that even for materials with the same origin (for instance paper materials and woody wastes) there are certain differences in the chemical composition among them due the processing.

Table 1. The proximate analysis and higher heating value of municipal and industrial wastes and fuel samples wastes on dry basis. AC: ash content; VM: volatile matter; FC: fixed carbon; HHV: higher heating value.

	Sample	Proximate analysis [wt%]			HHV [MJ/kg]
		AC	VM	FC	
Municipal waste	Paper	36.02	56.35	7.63	12.17
	Cardboard	6.18	78.65	15.17	17.91
	Sewage sludge	11.20	82.20	6.60	19.56
	Waste wood	1.28	79.04	19.67	20.26
	Textile	6.51	76.31	17.18	23.67
	Juice box	7.18	82.84	9.97	23.95
	Plastic	7.36	92.43	0.21	40.72
Industrial waste	Industrial sludge	1.90	81.03	17.07	19.89
	Bark, paper line	2.15	80.52	17.33	20.18
	Sawmill residue	2.02	74.00	23.98	20.72
	Bark, pulp line	2.74	75.19	22.07	21.06
Fuel	Eucalyptus	0.16	91.77	8.06	19.11
	Forest residue	4.34	76.45	19.20	20.67
	Peat	7.84	63.55	28.61	21.21
	Coal	16.26	24.28	59.47	29.53

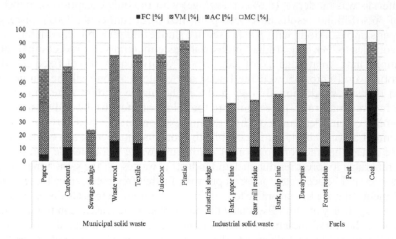

Figure 1. Proximate analysis of municipal and industrial wastes and fuel samples on wet basis. FC: fixed carbon; VM: volatile matter; AC: ash content; MC: moisture content.

3.2 *Heating value*

The higher heating values for all the studied samples are presented in Figure 3. It can be seen that the values for different waste fractions are of the same order as those of widely utilized fuels. The lowest heating value was found for paper, which can be explained by high share of ash (Table 1). The high heating value of plastic waste can be explained by the lower oxygen content and higher content of hydrocarbons in this oil based product compared to the cellulosic materials (Sørum et al. 2001). Since the juice box carton consists of both cellulosic fraction and some plastic with aluminum, it had higher heating value than the other studied cellulosic waste materials. It should be noted that the moisture content lowers the effective, lower heating value of the material during combustion. This results to sewage sludge with the high moisture not being as potential fuel as woody biomass.

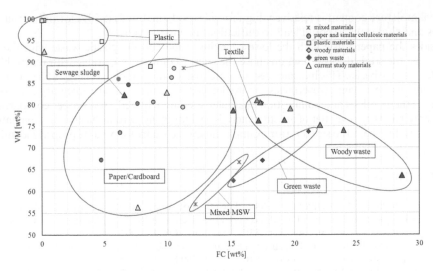

Figure 2. Volatile matter (VM) against fixed carbon (FC) of waste samples (dry basis) of the current study (triangular markers) and the available literature (Alakangas, 2000; Buah et al. 2007; Gunasee et al. 2016; Horttanainen et al. 2013; Liu et al. 2016; Sørum et al. 2001; Zhu et al. 2016).

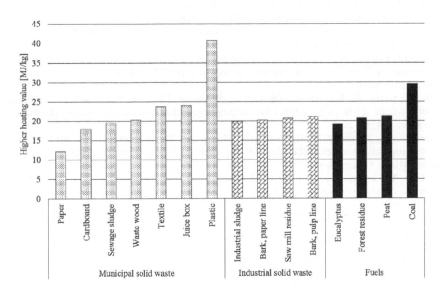

Figure 3. Higher heating values of municipal and industrial wastes and fuel samples on dry basis.

4 CONCLUSIONS

The fuel characteristics of common Finnish municipal solid waste fractions were determined and compared with the industrial wood wastes and four different fuels (eucalyptus chips, forest residues, peat and bituminous coal). The results of proximate analysis indicate a large heterogeneity concerning the chemical composition for the studied waste samples. Paper indicated the highest ash content (36%) and the lowest higher heating value (12.2 MJ/kg) among all samples. Plastic had the highest volatile content (92.4%) and the lowest fixed carbon content (0.2%). On dry basis, MSW fractions along with sewage sludge are quite similar by characteristics with industrial wood wastes, woody fuels and peat. On the other hand, the original moisture content is relatively high for all woody materials (>50% for industrial wood wastes and sludges) and could lead to difficulties in energy recovery.

Besides the paper waste with the lowest heating value and the plastic waste with the highest one, all the studied waste fractions have the energy content comparable with renewable fuel samples (on dry basis). This indicates certain potential for the combined energy utilization of the considered waste streams. Nevertheless, more detailed evaluation of the elementary composition as well as the pyrolysis and combustion characteristics are required for a proper comparison of the energy recovery potential of waste materials. Additionally, certain hazardous elements that can be found in the waste streams (for example, in sewage sludge) should be properly identified.

REFERENCES

Ajanko, S., Moilanen, A., & Juvonen, J. 2005. The effect of wastes' source separation system and handling technique on the quality of solid recovered fuel. Research Notes 2317. Espoo: VTT. 83 p. + app. 21 p. [In Finnish, abstract in English.]. ISBN 951-38-6753-6; 951-38-6754-4.

Alakangas, E. 2000. Properties of fuels used in Finland. Research Notes 2045. Espoo: VTT. 172 p. + app. 17 p. [In Finnish, abstract in English.]. ISBN 951-38-5699-2; 951-38-5740-9.

Buah, W.K., Cunliffe, A.M., & Williams, P.T. 2007. Characterization of Products from the Pyrolysis of Municipal Solid Waste. *Process Safety and Environmental Protection* 85(5): 450–457.

CEN/TS. 2010. CEN/TS 15414-2:en Solid recovered fuels. Determination of moisture content using the oven dry method. Part 2: Determination of total moisture content by a simplified method.

Cornelissen, A.A.J., & Otte, P.F. 1995. *Physical investigation of the composition of household waste in the Netherlands*. National Institute of public health and environmental protection. Report 776201011.

EU 2008. *Directive 2008/98/EC of the European Parliament and of the Council of 19 November 2008 on waste and repealing certain Directives*.

Gunasee, S.D., Carrier, M., Gorgens, J.F., & Mohee, R. 2016. Pyrolysis and combustion of municipal solid wastes: Evaluation of synergistic effects using TGA-MS. *Journal of Analytical and Applied Pyrolysis* 121: 50–61.

Götze, R., Pivnenko, K., Boldrin, A., Scheutz, C., & Astrup, T.F. 2016. Physico-chemical characterisation of material fractions in residual and source-segregated household waste in Denmark. *Waste Management* 54: 13–26.

Horttanainen, M., Teirasvuo, N., Kapustina, V., Hupponen, M., & Luoranen, M. 2013. The composition, heating value and renewable share of the energy content of mixed municipal solid waste in Finland. *Waste Management* 33(12): 2680–2686.

Liu, G., Liao, Y., Guo, S., Ma, X., Zeng, C., & Wu, J. 2016. Thermal behavior and kinetics of municipal solid waste during pyrolysis and combustion process. *Applied Thermal Engineering* 98: 400–408.

Ministry of the Environment 2009. *Towards a recycling society*. The National Waste Plan for 2016. Retrieved from https://helda.helsinki.fi/bitstream/handle/10138/38022/FE_14_2009.pdf?sequence = 1.

Nasrullah, M. 2015. *Material and energy balance of solid recovered fuel production*. Doctoral dissertation, VTT Science 115. 147 p.

Riber, C., Petersen, C., & Christensen, T.H. 2009. Chemical composition of material fractions in Danish household waste. *Waste Management* 29(4): 1251–1257.

SFS. 2009. SFS-EN 14775:en Solid biofuels. Determination of ash content.

SFS. 2010a. SFS-EN 14774-2:en Solid biofuels. Determination of moisture content using the oven dry method. Part 2: Determination of total moisture content by a simplified method.

SFS. 2010b. SFS-EN 14918:en Solid biofuels. Determination of calorific value.

SFS. 2010c. SFS-EN 15148:en Solid biofuels. Determination of the content of volatile matter.

SFS. 2011a. SFS-EN 15400:en Solid recovered fuels. Determination of calorific value.

SFS. 2011b. SFS-EN 15402:en Solid recovered fuels. Determination of the content of volatile matter.

SFS. 2011c. SFS-EN 15403:en Solid recovered fuels. Determination of ash content.

Sørum, L., Gronli, M.G., & Hustad, J.E. 2001. Pyrolysis characteristics and kinetics of municipal solid wastes. *Fuel* 80(9): 1217–1227.

Wilén, C., Salokoski, P., Kurkela, E., & Sipilä, K. 2004. *Finnish expert report on best available techniques in energy production from solid recovered fuels*. Helsinki, Finland. Retrieved from https://helda.helsinki.fi/bitstream/handle/10138/40639/FE_688.pdf?sequence=1.

Zhu, L., Zhang, L., Fan, J., Jiang, P., & Li, L. 2016. MSW to synthetic natural gas: System modeling and thermodynamics assessment. *Waste Management* 48: 257–264.

Chars from co-gasification of rice wastes as Cr(III) removal agents

D. Dias, W. Ribeiro & N. Lapa
LAQV, REQUIMTE, Departamento de Ciências e Tecnologia da Biomassa, Faculdade de Ciências e Tecnologia, Universidade Nova de Lisboa, Lisboa, Portugal

M. Bernardo, I. Matos & I. Fonseca
LAQV, REQUIMTE, Departamento de Química, Faculdade de Ciências e Tecnologia, Universidade Nova de Lisboa, Lisboa, Portugal

F. Pinto
Laboratório Nacional de Energia e Geologia, Unidade de Bioenergia, Lisboa, Portugal

ABSTRACT: Rice husk (80% w/w) and polyethylene (20% w/w) were submitted for co-gasification. The resulting char (GC) was characterized and used in Cr(III) liquid-phase removal assays. A Commercial Activated Carbon (CAC) was also used for comparison purposes. GC was mainly composed of ashes (68.3% w/w), with Si being the major mineral element and K the most soluble one. GC presented an alkaline character, which was responsible for an ecotoxic level in the char. In the Cr(III) removal assays, three Solid/Liquid ratios (S/L) were used: 2.5, 5.0 and 10.0 g L^{-1}. Generally, GC performed better than CAC: in the S/L of 10.0 g L^{-1}, GC removed 98.9% of Cr(III) (mainly by precipitation), while in the S/L of 5.0 g L^{-1}, GC removed 42.3% of Cr(III) by adsorption. The ionic exchanges of K$^+$ from GC and Cr(III) in the liquid medium were the predominant removal mechanism.

1 INTRODUCTION

Rice is one of the most highly produced cereals in the world. In Europe, Portugal is one of the major producers and consumers of rice. High rice production leads to a high generation of different waste types, such as rice husk (RH), rice straw and plastics from seeds and fertilizers packaging, for example polyethylene (PE). Frequently, the final destination of most of these forms of waste is not environmentally adequate. This has led to several studies being conducted regarding the different valorization routes for these various forms of waste, such as pyrolysis and gasification (Chakma et al., 2016; Lim et al., 2012; Quispe et al., 2017).

Thermal processes, such as pyrolysis and gasification, are interesting technologies that are being used to convert wastes into energy and materials. In the case of pyrolysis, three products are formed—namely gases, liquids (bio-oils) and solids (chars) (Bridgwater, 2012) – while in the case of gasification, mainly syngas and some chars are produced, due to the higher temperatures used (Kumar et al., 2009). Some tars are also formed, though in minor amounts.

Different valorization pathways for chars have been demonstrated, with the adsorption of pollutants from aqueous effluents being one of the most interesting (Bernardo et al., 2013; Galhetas et al., 2014). Chars obtained from gasification and pyrolysis may present porous structures, similarly to commercial activated carbons, which are the most common adsorbent materials used in the adsorption of contaminants. Nevertheless, the presence of mineral species or functional groups on chars' surfaces makes them interesting adsorbent materials. Additionally, chars are characterized by lower costs and environmental benefits, especially if the raw materials used in thermal processes are bio-wastes with no (or low) commercial value.

Although there are some studies on the adsorption of pollutants by using pyrolysis chars from rice wastes (Liu et al., 2011; Tong and Xu, 2013), the use of gasification chars from rice wastes as adsorbent materials has not yet been studied.

On the other hand, Cr is one of Europe's critical raw materials due to its high economic importance to the European industry (European Commission, 2014). Cr is present in several industrial wastewaters as a toxic contaminant. Therefore, its recovery from liquid effluents is a priority for the economic and environmental sustainability of the European industry.

The main goal of this work was to assemble these two issues and to study the removal capacity of Cr(III) from aqueous effluents by using a char produced in the co-gasification of rice wastes. A commercial activated carbon was also used for comparison purposes.

2 MATERIALS AND METHODS

2.1 Co-gasification assays

RH (80% w/w) and PE (20% w/w) were blended and submitted for co-gasification assays at 850°C under a feedstock flow of 5 g daf min^{-1} (daf: dry ash-free), an equivalent ratio of 0.2 (O_2) and a steam flow of 5 g min^{-1}. Several mixtures of RH and PE have been gasified. In the present study, only the char produced with 80% RH and 20% PE (w/w) was used because it produced the highest amount of chars and the lowest amount of tars. On the other hand, the amount of rice wastes produced was much bigger than that of plastic wastes. Although it was not the scope of the work, the results obtained showed that co-gasification was a viable option for rice and plastic wastes valorisation, as the syngas produced has suitable composition and calorific value for different applications, though some of them required some syngas upgrading (Pinto et al., 2016). The gasification reactor (0.08 m diameter × 1.5 m height) had a bubbling fluidized bed constituted of fine sand. More details about the experimental installation are available in previous works (Pinto et al., 2016). At the end of the co-gasification assays, the char and sand were separated through sieving.

2.2 Characterizations of the materials

The co-gasification char (GC) and the commercial activated carbon (CAC) (Norit GAC 1240) were milled, sieved to 100 μm and submitted for the following analyses:

i. Elemental analysis and ash quantification—CHNS were quantified in an Elemental Thermo Finnigan analyzer (ASTM D 5373 for C, H and N, and ASTM D4239 for S); ashes were quantified by the gravimetric method EN 14775;

ii. Mineral content—mineral content was determined by acidic digestion (EN 15290) and comprised the quantification of Ca, Fe, Mg, Al, K, Na, Si and Ti in the acidic eluates by ICP-AES;

iii. Mineral mobility and ecotoxic level—mineral mobility was determined through a leaching test with deionized water (EN 12457-2) and comprised the characterization of eluates for: (a) the quantification of Ca, Fe, Mg, Al, K, Na, Si and Ti by ICP-AES, and (b) an ecotoxic test, with and without pH correction, through a bioluminescence inhibition assessment of the bacterium *Vibrio fischeri* (Microtox assay) (ISO 11348-3). The ecotoxic results were expressed as EC_{50}–30 min (% v/v), which shows that the effective concentration of the eluate promoted a 50% decrease in bioluminescence after 30 minutes of exposure;

iv. Textural analysis—this analysis included the determination of surface area, pore volume, and pore size distribution by N_2 adsorption-desorption isotherm at 77 K, after sample degasification under vacuum conditions at 150°C;

v. pH_{pzc}—0.1 M NaCl solutions with an initial pH between 2 and 12 were prepared. 0.1 g of char was added to 20 ml of each 0.1 M NaCl solution; the mixtures were agitated in a roller-table device at 150 rpm for 24 h; at the end of the agitation time, the final pH (pH_f) was measured. The pH_{pzc} value corresponds to the point where $pH_i = pH_f$.

2.3 Cr(III) removal assays

Batch studies were performed with GC and a CAC (Norit GAC 1240) under the following conditions: Cr(III) initial concentration = 70 mg L^{-1}; initial pH = 4.5; S/L = 2.5, 5.0 and 10.0 g L^{-1}; sample volume = 25 ml; agitation time = 24 h. A pH of 4.5 was used because Cr(III) starts to precipitate at pH above 5.0. All the removal assays were conducted in a roller-table shaker, under constant mixing of 150 rpm. A standard $Cr(NO_3)_3$ solution of 1000 mg Cr(III) L^{-1} in 0.5 M of HNO_3 (Merck) was used for preparing a 70 mg L^{-1} Cr(III) solution. After the batch experiments, the samples were filtered through cellulose nitrate membranes with a porosity of 0.45 μm. The pH was measured in the filtrates and the Cr concentration was quantified by ICP-AES.

The removal efficiency, η (%), and adsorbent uptake capacity, q_{exp} (mg g^{-1}), were calculated using equations 1 and 2, respectively:

$$\eta(\%) = \frac{(C_0 - C_f)}{C_0} \times 100 \tag{1}$$

$$q_{exp} = \frac{(C_0 - C_f)}{m} \times V \tag{2}$$

where C_0 and C_f are Cr(III) concentrations (mg L^{-1}) before and after the batch assays, respectively, m is the adsorbent mass (g) and V the solution volume (L).

3 RESULTS AND DISCUSSION

3.1 Characterizations of the materials

3.1.1 Elemental analysis and ashes

GC was mainly composed of ashes (68.3% w/w) (Table 1), probably due to both the ash content in the raw materials and the relatively high temperature used in the co-gasification assays (850°C) that converted most of the organic matter into syngas, concentrating the mineral matter in the char. The ash content of CAC was significantly lower (6.25% w/w), but its C content was much higher (86.3% w/w) than for GC (25.9% w/w).

3.1.2 Mineral content and mobility

Figure 1 shows the mineral content and mobility of GC. Si was the major element present in GC, as the main raw material used in the co-gasification assay was RH, which is known to have high concentrations of Si (Lim et al., 2012). K, Ca, Ti, and Mg also registered high concentrations in the char, although at lower levels than Si.

In general, the mobility of minerals was low. K was the exception, with a concentration of 1405 mg kg^{-1} db, which corresponded to a solubility of 18.1%. The mobility of K may be important for ion exchanges with Cr(III).

The mineral content and mobility of CAC were not quantified because this char did not use any ion exchanges in the Cr(III) removal processes.

Table 1. Elemental analysis and ash content of GC and CAC (results in% w/w as received).

Parameter	GC	CAC
C	25.9	86.3
H	2.88	0.47
N	<0.2	<0.2
S	<0.03	0.57
Ash content	68.3	6.25

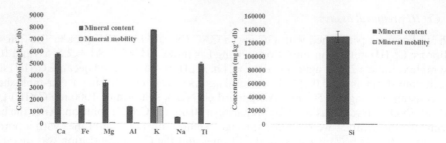

Figure 1.   Mineral content and mobility of GC.

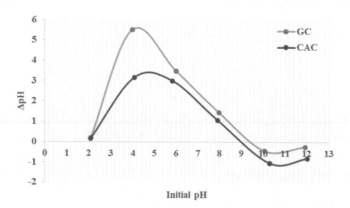

Figure 2.   pH$_{pzc}$ of GC and CAC.

Table 2.   Ecotoxic levels of GC eluate for the bacterium *Vibrio fischeri*, before and after pH correction.

Ecotoxic assay	Before pH correction	After pH correction
Luminescence inhibition of the bacterium *V. fischeri* EC$_{50}$–30 min (% v/v)	20.5	> 99.0

3.1.3   *pH$_{pzc}$*

Both adsorbents had similar values of pH$_{pzc}$ (9.6 for GC and 9.1 for CAC) (Figure 2), with both being characterized as alkaline materials. The basic nature of GC was related to its higher mineral content.

3.1.4   *Ecotoxic level*

Table 2 shows the ecotoxic levels of GC eluate before and after pH correction.

The alkaline nature of GC generated an ecotoxic level of its eluate to the bacterium *V. fischeri*. This is a very sensitive bacterium to pH variations. After the pH correction to the optimum pH interval (7.5–8.5) for *V. fischeri*, no ecotoxic level was detected. The ecotoxicity can be associated to alkali and/or alkali-earth metals that attribute the alkaline character to the char eluate. A prior leaching process could remove these alkali and alkali-earth metals responsible for the eluate ecotoxicity, but it could also compromise the Cr(III) adsorption capacity as these metals participate in ion exchange mechanisms involved in Cr(III) removal (data not shown in the present study).

3.1.5   *Textural analysis*

Table 3 shows the textural properties of GC and CAC.

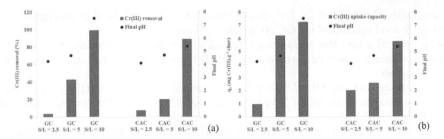

Figure 3.   Cr(III) removal (a) and uptake capacity (b) for GC and CAC (S/L expressed in g L^{-1}). The final pH are indicated as dark dots.

Table 3.   Textural properties of GC and CAC.

Textural parameter	GC	CAC
S_{BET} (m^2 g^{-1})	3.5	1034
V_{micro} (cm^3 g^{-1})	n.q.	0.30
V_{meso} (cm^3 g^{-1})	0.01	0.17
V_{total} (cm^3 g^{-1})	0.01	0.56

n.q.: not quantifiable.

CAC presented a high surface area and total pore volume. In contrast, GC presented a very low surface area, being a non-porous material. This was expected, given the high ash content of GC and because this is a non-activated adsorbent material.

3.2   *Cr(III) removal assays*

Figure 3 shows the results achieved in the Cr(III) removal assays by GC and CAC, and the final pH obtained in each treated solution.

In general, Cr(III) removal was more efficient for GC then for CAC (Figure 3a).

For the S/L of 10.0 g L^{-1}, GC and CAC removed almost all the Cr(III) in the solution (98.9% and 88.8%, respectively). For this S/L, both adsorbents promoted a final pH above 5.0, and under this condition Cr(III) precipitates.

For the S/L of 2.5 g L^{-1}, CAC presented a slightly higher Cr(III) removal than GC; however, both adsorbents had a relatively weak performance ($\eta < 7\%$).

For the S/L of 5.0 g L^{-1}, GC removed 42.3% of Cr(III) from the solution, while CAC removed only 19.7%. The final pH for both chars was quite similar and below 5; thus, it is likely that no precipitation occurred.

The Cr(III) uptake capacity (Figure 3b) was in agreement with removal efficiencies. Regarding the S/L of 5.0 and 10.0 g L^{-1}, the Cr(III) uptake capacity for GC was higher than for CAC. The highest Cr(III) uptake capacity was found for GC at the L/S of 10.0 g L^{-1} (7.23 mg g^{-1}), due to precipitation. Regarding the S/L of 5.0 g L^{-1} where is it probable that no precipitation occurred, GC and CAC presented uptake capacities of around 6 mg g^{-1} and 2.5 mg g^{-1}, respectively.

Although CAC showed better textural properties, it did not present the best performance on Cr(III) removal, indicating that the mechanism behind the uptake is not exclusively controlled by pore filling. The higher removal of Cr(III) achieved by the mineral-rich GC sample is associated with ion exchange reactions on the char's external surface, probably due to K$^+$ exchange, as this was identified in the mineral mobility assay.

4   CONCLUSIONS

GC was mainly composed of ashes (68.3% w/w), which in turn were mainly composed of Si, followed by K, Ca, Ti, and Mg. K was the element with the highest mobility rate in water (18.1%).

The alkaline character of the GC sample caused some ecotoxic level on its eluate; however, no ecotoxic level was found for GC eluate after pH correction.

GC presented a very low surface area, typical of non-porous materials. Nevertheless, this sample presented the best performance on Cr(III) removal in comparison to the commercial activated carbon.

For the assay with the S/L of 10.0 g L^{-1}, GC removed 98.9% of Cr(III) by precipitation, while in the assay with the S/L of 5.0 g L^{-1}, it removed 42.3% of Cr(III) through ion exchange, probably due to K^+.

The Cr(III) uptake capacity of GC for the S/L of 5.0 and 10.0 g L^{-1} (6.20 mg g^{-1} and 7.23 mg g^{-1}, respectively) were also higher than for CAC for the same S/L values.

Overall, GC performed better than CAC on the removal of Cr(III) from liquid effluents. It can be concluded that GC may be an economic and sustainable alternative to CAC in the removal of Cr(III) from liquid effluents.

ACKNOWLEDGEMENTS

This research was funded by FEDER through the Operational Program for Competitive Factors of COMPETE and by Portuguese funds through FCT (Foundation for Science and Technology) through the project PTDC/AAG-REC/3477/2012 – RICEVALOR "Energetic valorisation of wastes obtained during rice production in Portugal", FCOMP-01-0124-FEDER-027827, a project sponsored by FCT/MTCES, QREN, COMPETE and FEDER.

The authors also acknowledge the Foundation for Science and Technology for funding Maria Bernardo's post-doc fellowship (SFRH/BPD/93407/2013), Diogo Dias's PhD fellowship (SFRH/BD/101751/2014), and LAQV/REQUIMTE through Portuguese funds (UID/QUI/50006/2013) and co-funds by the ERDF under the PT2020 Partnership Agreement (POCI-01-0145-FEDER-007265).

REFERENCES

Bernardo, M., Mendes, S., Lapa, N., Gonçalves, M., Mendes, B., Pinto, F., Lopes, H., Fonseca, I. 2013. Removal of lead (Pb2+) from aqueous medium by using chars from co-pyrolysis. *J. Colloid Interface Sci.* 409, 158–65. doi:10.1016/j.jcis.2013.07.050.

Bridgwater, A.V. 2012. Review of fast pyrolysis of biomass and product upgrading. *Biomass and Bioenergy* 38: 68–94. doi:10.1016/j.biombioe.2011.01.048.

Chakma, S., Ranjan, A., Choudhury, H.A., Dikshit, P.K., Moholkar, V.S. 2016. Bioenergy from rice crop residues: Role in developing economies. *Clean Technol. Environ.* Policy 18, 373–394. doi:10.1007/s10098-015-1051-5.

European Commission 2014. *On the review of the list of critical raw materials for the EU and the implementation of the Raw Materials Initiative*. Brussels.

Galhetas, M., Mestre, A.S., Pinto, M.L., Gulyurtlu, I., Lopes, H., Carvalho, A.P. 2014. Chars from gasification of coal and pine activated with K2CO3: acetaminophen and caffeine adsorption from aqueous solutions. *J. Colloid Interface Sci.* 433: 94–103. doi:10.1016/j.jcis.2014.06.043.

Kumar, A., Jones, D.D., Hanna, M.A. 2009. Thermochemical biomass gasification: A review of the current status of the technology. *Energies* 2: 556–581. doi:10.3390/en20300556.

Lim, J.S., Abdul Manan, Z., Wan Alwi, S.R., Hashim, H. 2012. A review on utilisation of biomass from rice industry as a source of renewable energy. *Renew. Sustain. Energy Rev.* 16: 3084–3094. doi:10.1016/j.rser.2012.02.051.

Liu, W.-J., Zeng, F.-X., Jiang, H., Zhang, X.-S. 2011. Preparation of high adsorption capacity bio-chars from waste biomass. *Bioresour. Technol.* 102: 8247–52. doi:10.1016/j.biortech.2011.06.014.

Pinto, F., André, R., Miranda, M., Neves, D., Varela, F., Santos, J. 2016. Effect of gasification agent on co-gasification of rice production wastes mixtures. *Fuel* 180: 407–416. doi:10.1016/j.fuel.2016.04.048.

Quispe, I., Navia, R., Kahhat, R. 2017. Energy potential from rice husk through direct combustion and fast pyrolysis: A review. *Waste Manag.* 59: 200–210. doi:10.1016/j.wasman.2016.10.001.

Tong, X., Xu, R. 2013. Removal of Cu(II) from acidic electroplating effluent by biochars generated from crop straws. *J. Environ. Sci.* 25: 652–658. doi:10.1016/S1001-0742(12)60118-1.

Anaerobic digestion sludge composting—assessment of the star-up process

M.E. Silva
Instituto Politécnico de Viseu, Viseu, Portugal
LEPABE, Porto, Portugal

S. Araújo & I. Brás
Instituto Politécnico de Viseu, Viseu, Portugal

G. Lobo, A. Cordeiro & M. Faria
Ferrovial Serviços S.A., Barreiro de Besteiros, Portugal

A.C. Cunha-Queda
Universidade de Lisboa, Lisboa, Portugal

O.C. Nunes
LEPABE, Faculdade de Engenharia, Universidade do Porto, Porto, Portugal

ABSTRACT: The aim of this work was to follow the start-up of the composting process of sludge from the anaerobic digestion of municipal solid waste and to assess the quality of the final product. The temperature profile and dry matter content was registered over 10 weeks. The composting process was monitored by standard physical and chemical parameters and the quality of the final compost was assessed by phytotoxicity and stability assays. The low average temperature values registered during composting suggest over aeration. In addition, all the standard physical and chemical parameters analysed showed slight variations over composting, suggesting that the anaerobic digestion sludge was already a stable raw material. The final compost was not phytotoxic and was stable, but contained some inert materials. Overall, the final compost had quality compatible to be used as soil amendment. Nevertheless, the operational conditions optimization should be carried out to improve its quality.

1 INTRODUCTION

Organic wastes valorisation is a main objective of European waste management policy, with a valorisation goal of 65% until 2020. To reach this goal, the majority of the municipal solid waste (MSW) management associations had implemented the Mechanical Biological Treatment (MBT) for recycling the MSW organic fraction. The MBT relies upon the mechanical separation of non-biodegradable wastes from biodegradable, enclosing physical unit operations to recover recyclable non-biodegradable fractions like plastics such as PET, metals or even tetrapack containers. The biodegradable organic fraction valorisation occurs through anaerobic digestion followed by a composting process of the digested sludge.

Composting is a biological treatment in which aerobic thermophilic and mesophilic microorganisms use organic matter (OM) as a substrate. The main products of this process are mineralised materials (CO_2, H_2O, NH_4^+) and stabilised OM (mostly humic substances) (Hernanez-Apaolaza et al., 2000). The final composting product is defined as a stabilised material that can be used as amendment in agricultural soils, if reaching high quality. Alternatively, the MBT refuse fraction can be used to produce Refuse Derived Fuels (RDFs), which is a mix of materials with a homogenous calorific value to burn in incinerators or in some cement kilns.

The aim of this work was to follow the star-up of the composting process of a municipal MBT unit. Additionally, the quality and the calorific value of the final compost were determined in order to evaluate its agronomic value and/or its energetic potential.

2 MATERIALS AND METHODS

2.1 The composting unit of the MBT process

After the anaerobic digestion process of the MSW organic fraction, the resulting anaerobic digestion sludge (DS) is submitted to a mechanical process to promote its dehydration, followed by a composting process. The DS physicochemical characterization is shown in Table 1.

After dehydration, the sludge batches were placed in the composting tunnels, and mixed with wood chips, which were used as bulking agent, in a ratio of 60:40, respectively. The composting tunnels were 30 m length and 2.8 m high. Inside the tunnels, the composting pile was mechanically turned at regular intervals, in order to be homogenized and aerated. Aeration was also achieved with a mechanical device located at the bottom of the tunnels. The air flow ranged between 30% and 100% (Table 2). During two weeks the pile was not aerated. The bulking agent addition, the periodically turning and the mechanical aeration was implemented to minimize the formation of anaerobic spots and to prevent excessively high temperature within the pile. Temperature was measured daily with a probe (HI 762PWL; Hanna Instruments) at 5 points along the pile. The weekly temperature average and the dry matter content profile over 10 weeks (T1 – T10) is shown in Table 2. Dry matter content was evaluated from the weight loss after sample drying at 105°C for 24 h (EN 13040, 1999).

After 10 weeks, the final compost was sieved in order to attain particles with size under 2 mm, and to separate the bulking agent and some physical contaminants, such as contaminant

Table 1. Anaerobic digestion sludge characterization.

Sample	pH	EC mS/cm	OM % dm	Moisture %	TN % dm	TC %	C/N
DS	7.8 ± 0.1	2.24 ± 0.15	33.4 ± 3.2	38.4 ± 1.6	1.04 ± 0.09	16.7 ± 1.5	16.1

EC – Electric Conductivity; OM – Organic Matter; TN – Total Nitrogen; TC – Total Carbon; dm – dry matter.

Table 2. Weekly temperature of the five sampling points, dry matter content and aeration output by mechanical device.

Time (weeks)	Temperature (°C) 1	2	3	4	5	Average	Dry Mater (%)	Aeration (%)
T1	30.3 ± 2.9	30.9 ± 0.8	26.1 ± 2.6	30.4 ± 2.3	33.5 ± 0.9	29.4 ± 0.8	59.3 ± 2.4	100
T2	37.2 ± 1.9	19.6 ± 0.9	17.3 ± 0.9	21.7 ± 0.9	17.4 ± 0.9	25.7 ± 0.1	58.3 ± 1.8	100
T3	51.8 ± 1.3	17.6 ± 1.4	14.9 ± 0.3	24.3 ± 1.6	16.0 ± 0.4	24.9 ± 0.7	63.1 ± 1.7	100
T4	34.1 ± 0.6	16.8 ± 0.8	15.8 ± 0.7	16.5 ± 0.4	14.9 ± 0.3	19.6 ± 0.3	63.0 ± 0.5	WA
T5	52.1 ± 0.8	21.3 ± 1.6	18.7 ± 0.5	20.0 ± 1.7	16.0 ± 0.3	25.6 ± 0.6	64.0 ± 0.3	WA
T6	50.9 ± 2.3	25.6 ± 0.9	20.8 ± 0.9	20.9 ± 1.1	16.9 ± 0.4	27.0 ± 0.2	66.6 ± 1.8	30
T7	50.1 ± 2.3	40.0 ± 1.1	25.3 ± 0.4	30.5 ± 1.2	18.9 ± 0.3	33.0 ± 0.1	66.1 ± 1.3	30
T8	39.1 ± 2.3	31.2 ± 0.9	28.7 ± 0.4	38.7 ± 1.2	24.3 ± 0.3	32.4 ± 0.1	58.4 ± 1.0	30
T9	23.4 ± 1.1	28.9 ± 0.9	29.2 ± 0.4	24.2 ± 1.2	19.9 ± 0.3	25.1 ± 0.1	65.5 ± 0.8	30
T10	18.7 ± 1.3	22.3 ± 2.3	23.1 ± 2.8	20.1 ± 0.4	19.3 ± 0.1	20.8 ± 0.1	65.6 ± 0.2	30

WA – Without aeration.

inert materials not removed during the initial mechanical separation of the MSW organic fraction.

2.2 *Sampling and characterization of the compost*

Sampling was made at the beginning of the composting process (T1), after four weeks (T4) and at the end of the process (T10), corresponding to the final compost. Samples of about 3 kg were collected from the top to the bottom of the pile with a sampling pipe, and homogenized by mixing. Approximately 2 kg of the sample was returned to the pile and the remaining part was stored at 4°C for further analysis. The pile was sampled in five different points. Electrical conductivity (EC) and pH measurements were performed by electrometric determination in aqueous extracts (1:5, weight/volume) of the samples. The organic matter and ash contents were determined by loss-on-ignition, at 550°C for 8 h (EN 13039, 1999). Total carbon was estimated as the product of the organic matter content and the empirical coefficient 0.5 (Zucconi & Bertoldi, 1987). The Kjeldahl digestion method was followed to determine the total nitrogen content (EN 13654-1). All analyses were performed in triplicate. Stability was assessed through the self-heating test using about 1 kg of compost sample, with a standard moisture content of 35% in a Dewar vessel with 2 L capacity. The higher heating value (HHV) was carried out following the CN/TS 15400 (2015).

The phytotoxicity was assessed through the germination index (GI) with an aqueous extract of the compost and was calculated based on the following formula:

$$GI = \frac{\overline{GS_s} * \overline{RL_s}}{\overline{GS_c} * \overline{RL_c}} * 100 \tag{1}$$

where $\overline{GS}$ represents the average number of germinated seeds and $\overline{RL}$ the average root length (mm) of seeds in samples – s, and the control – c tests.

3 RESULTS AND DISCUSSION

3.1 *Anaerobic digestion sludge characterization*

The physicochemical properties of the DS used as raw material of the composting are shown in Table 1. Overall, the DS was a non-acidic waste, and had low values for almost all parameters analysed. The C/N ratio was lower than the range recommended for composting processes (25–30) (Vallini, 1995).

3.2 *Anaerobic digestion sludge composting star-up*

Briefly, the MBT process used to valorise the MSW organic fraction is represented in Figure 1.

In this MBT unit, DS was composted with wood chips because they behave as bulking agents, improving the pile structure and allowing air circulation. As described before, the composting pile in the tunnel was periodically aerated and turned. The temperature and the dry matter content were chosen as indicators of the composting process developing (Table 2).

In general it was observed that the average temperatures were always below 35°C, suggesting that the conditions were not ideal for the development of microorganisms. This may arise from the fact that the organic matter of the sludge was already stabilized after the anaerobic digestion process. The aeration conditions were promoted using three methods: addition of bulking agent (40%), periodical pile turning and mechanical aeration. However, when analysing the temperature at the five points, it was observed that was observed that the first sampling point showed always the highest temperature value. This sampling point is located at the beginning of the tunnel without influence of the mechanical aeration, which may explain the highest temperature value. Additionally, it may promote the understanding of the effect of this aeration method in the composting process. Moreover, the aeration output variation

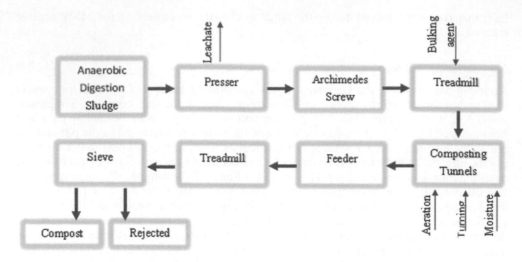

Figure 1.   Flowchart of the processes used in the organic waste valorisation.

from 100%, without aeration and 30%, indicated that this operational parameter influences the temperature profile. Lower temperature values were registered with a 100% aeration output, suggesting that high aeration output may cause a temperature decrease. However, this operational parameter did not have effect in the dry matter content. According with Cukjati et al. (2012) the aeration at the beginning of the process must be of 76% and must be reduced to 30% at the final phase. Another operational parameter important for composting process is the moisture content, crucial to the microbiological metabolism. This value must be between 45 and 60% to favor growth and activity of the microbial populations (Abdullah & Chin, 2010). Over the composting process, the moisture content was lower than the optimum value, ranging from 40% to 35%. These low moisture values may have contributed to hamper the microorganisms' development and thus, to observe low temperature values. In further studies, the initial DS dewatering should be carefully followed in order to achieve the optimum water content values in the raw material. Moreover, the moisture content must be controlled over the composting process to attain optimal values.

The evolution of the physicochemical parameters during the composting process are shown in Table 3.

Results showed almost no variation over time for the parameters analysed. pH values were stable and presented a slight alkaline character. EC values at the end of process were approximately 2 mS/cm, which is lower than 4 mS/cm, the higher limit considered tolerable by plants of medium sensitivity (Lasaridi et al., 2006). The OM evolution confirmed the low microbial activity during the composting time, as already suggested by the temperature profile. At the beginning of composting, the C/N ratio was lower than the C/N ratio optimum for composting, which may contribute for the low microbiological activity herein observed. The C/N ratio is an essential factor for microorganisms to perform the decomposition of organic wastes during the composting process. Carbon is needed by the microorganisms to obtain energy and nitrogen permits the microorganisms to thrive.

The final compost, obtained after a ten weeks period (T10), was sieved as represented in Figure 1. The screening was not efficient in the separation of the physical contaminants, since the final compost had 35% of total inerts (range 20 mm –1 mm). To avoid this situation it is recommended an additional mechanical treatment, such as a ballistic separator and/or an electro iman.

Nevertheless, the final compost registered a C/N ratio lower than the initial value. Regarding the toxicity characteristics, composts should have high germination index (GI) values if they will be further used as organic amendment of agricultural soils. GI reflects the degree of salinity and the presence of substances with phytotoxic effects such as heavy metals,

Table 3. Physicochemical parameters evolution of the composting material.

Sample	pH	EC mS/cm	OM % dm	TN % dm	TC %	C/N	GI %	Stability class	HHV MJ/kg
T1	8.0 ± 0.1	2.5 ± 0.1	30.1 ± 3.5	0.93 ± 0.17	15.0 ± 1.8	16.2	ND	ND	ND
T4	8.0 ± 0.1	2.1 ± 0.1	29.8 ± 0.3	0.90 ± 0.90	16.9 ± 0.7	18.8	ND	ND	5
T10	8.0 ± 0.0	1.7 ± 0.1	31.7 ± 1.3	1.08 ± 1.36	15.8 ± 0.2	14.6	133 ± 4	V	7

EC – Electric Conductivity; OM – Organic Matter; TN – Total Nitrogen; TC – Total Carbon; GI – Germination Index; HHV – Higher Heating Value; ND – not determined; dm – dry matter.

indicating its possible phytotoxicity if applied to soil. The high GI value of the final compost herein obtained indicates that it is not phytotoxic.

Once submitted to the self-heating test in Dewar vessels, compost reached the highest stability class (V) according to the Laga-Merkblatt M10 (1995). Composts of class I reach maximum temperature values above 60°C, whereas those of classes II and III show maximum temperature values between 50 and 60°C and 40 and 50°C, respectively. Only the composts that reach maximum temperature values lower than 40°C are considered stable. These include classes IV and V, which reach maximum temperature values between 30 and 40°C and below 30°C, respectively (Laga-Merkblatt M10, 1995). The final compost herein obtained reached a maximum temperature of 25°C, suggesting that the composting process was sufficient to obtain stable compost.

The overall parameters analysed for the final compost indicate that it could be used as organic amendment for agricultural soils. To access the feasibility of using the final compost as fuel, its heating value was determined. A value of 7 MJ/kg was obtained, which is low when compared to the conventional fuels, namely with coal (28.2 MJ/kg), which is used as a reference. Nevertheless, in order to improve the heating value it may be mixed with other wastes.

4 CONCLUSIONS

Biological treatments of the studied MBT unit seem to be effective methods for producing stabilized organic end-products, ensuring their maximum benefit for agriculture. The properties of final compost indicate that it had standard quality for the parameters analysed. Nevertheless, the process can be further improved through operating conditions optimization, namely, aeration.

ACKNOWLEDGMENT

The authors would like to express their gratitude to the Instituto Politécnico de Viseu, the Center for Studies in Education, Technologies and Health (CI&DETS) and the Portuguese Foundation for Science and Technology (FCT).

REFERENCES

Abdullah, N. & Chin, N.L. 2010. Simplex-centroid mixture formulation for optimized composting of kitchen waste. *Bioresource Technology* 101 (21): 8205–8210.
CEN, E.C. for S. 2005. *Solid recovered fuels—Methods for the determination of calorific value*, CEN/TS 15400.
Cukjati, N., Zupančič, G.D., Roš, M. & Grilc, V. 2012. Composting of anaerobic sludge: An economically feasible element of a sustainable sewage sludge management. *Journal of Environmental Management* 106: 48–55.

EN 13039 1999. *Soil Improvers and Growing Media—Determination of organic matter content and ash.* European Committee for Standardization. Technical Committee CEN/TC 223.

EN 13040 1999. *Soil Improvers and Growing Media—Sample preparation for chemical and physical tests, determination of dry matter content, moisture content and laboratory compacted bulk density.* European Committee for Standardization, Technical Committee CEN/TC 223.

EN 13654-1 2001. *Soil Improvers and Growing Media—Determination of Nitrogen—Part 1: Modified Kjeldahl Method.* European Committee for Standardization, Technical Committee CEN/TC 223.

Hernanez-Apaolaza, L., Gasco, J.M., Guerrero, F. 2000. Initial organic matter transformation of soil amended with composted sewage sludge. *Biol. Fertil. Soils* 32: 421–426.

Laga-Merkblatt M10, 1995. *Qualitätskriterien und Anwendungsempfehlungen für Kompost.* Müll-Handbuch Lfg. 5/95, Kennziffer 6856, Verlag, Berlin.

Lasaridi, K., Protopapa, I., Kotsou, M., Pilidis, G. Manios, T., Kyriacou, A. 2006. Quality assessment of composts in the Greek market: The need for standards and quality assurance. *J. of Environ. Manage* 80: 58–65.

Vallini, G. (1995). Il Compostaggio. In: Bertini, I., Cipollini, R. & Tundo, P. (Eds): *La protezione dell'ambiente in Italia*, pp. 83–134. Consiglio Nazionale delle Richerche, Società Chimica Italiana e Consorzio Interuniversitario Chimica per l'Ambiente. Bologna.

Zucconi, F. & De Bertoldi, M. 1987. Compost specifications for the production and characterization of compost from municipal solid waste. In M. De Bertoldi, M.P. Ferranti, P.L'Hermite, F. Zucconi (Eds.), *Compost: production, quality and use*, Elsevier Applied Science Publishers Ltd, Essex, pp. 30–50.

WASTES – Solutions, Treatments and Opportunities II – Vilarinho, Castro & Lopes (Eds)
© 2018 Taylor & Francis Group, London, ISBN 978-1-138-19669-8

Diagnosis and assessment of the management of sanitary landfill leachates in Portugal

A. Fernandes, L. Ciríaco, M.J. Pacheco & A. Lopes
FibEnTech-UBI and Department of Chemistry, Universidade da Beira Interior, Covilhã, Portugal

A. Albuquerque
FibEnTech-UBI and Department of Civil Engineering and Architecture, Universidade da Beira Interior, Covilhã, Portugal

ABSTRACT: Municipal Solid Wastes (MSW) management involves serious environmental impacts, namely the production of complex effluents—the leachates. These effluents, resulting mainly from the wastes deposition in sanitary landfills, possess a complex and recalcitrant composition and represent a significant source of pollution, presenting an accumulative, threatening and detrimental effect to the survival of aquatic life forms and ecological balances. This paper presents the current situation regarding the effluents generated from municipal solid wastes management in Portugal. The study presented, which began in 2015, refers to the situation in Portugal in 2014. The results and conclusions obtained, based on the available information, allowed the characterization of the effluents produced at the Portuguese MSW management sites, the identification of the treatments currently applied and the destination of the treated effluents.

1 INTRODUCTION

The growing production of municipal solid wastes (MSW) is an inevitable consequence of population growth and nowadays patterns of consumption. Over the last decades MSW management suffered an outstanding evolution of legislative and political character. Despite all the efforts and actions taken to prevent wastes generation and promote its reuse and recycling, a significant amount of wastes continues to be deposited in sanitary landfills. According to Eurostat statistics, in 2014, 66 million tons of municipal wastes were landfilled in the EU-28.

As a result of the MSW deposition in sanitary landfills, a very complex effluent, usually known as sanitary landfill leachate (SLL), is produced. SLL is one of the major environmental problems concerning water pollution, since it is a very complex wastewater containing different heavy metals, organic and inorganic compounds, some of them refractory and toxic, which possesses color and odor (Eggen et al. 2010). Due to its complex, recalcitrant and variable composition, SLL present an accumulative, threatening and detrimental effect to the survival of aquatic life forms and ecological balances (Foo & Hameed 2009, Mavakala et al. 2016). Thus, an adequate collection and treatment of these effluents is required, to prevent contamination of water resources, at the surface and groundwater, and soils.

The type of technology to be used in SLL treatment is conditioned, among other issues, by the quality of the leachate, which is influenced by wastes composition, climatic conditions and filling process regarding waste compaction, landfill cover and height of landfill layers, and landfill age (Koc-Jurczyk & Jurczyk 2011, Renou et al. 2008). Furthermore, the leachate treatment plants should have sufficient capacity to absorb leachate inflow associated with exceptional rainfall conditions and should allow the interruption of its operation in case of need.

The aim of this work is to present the current situation of the management of SLL generated in Portugal, regarding characterization of the sanitary landfill leachates produced in Portugal, identification of leachates treatments currently applied, destination of the treated effluents and of the resulting sludge and concentrates, when membranes processes are applied, as well as identification of the main difficulties observed by the MSW management entities.

2 MATERIALS AND METHODS

This study began in 2015 and refers to the situation in Portugal in 2014, excluding Azores and Madeira islands. The methodology adopted was based on a national survey of the 23 MSW management systems (SGRU) in continental territory of Portugal and visits to some SGRU facilities. The purpose of the inquiry was the acquisition of precise data regarding the MSW generated, the characterization and destination of the effluents produced and the existing infrastructures in the SGRUs. Standard statistical procedures, available in Microsoft Excel and IBM SPSS Statistics, were used to treat the acquired data.

3 RESULTS AND DISCUSSION

3.1 *MSW management in Portugal*

Table 1 describes the Portuguese SGRUs in 2014, their main infrastructures and the amount of MSW managed. The size of the 23 SGRUs vary significantly, having the smallest system only 0.26% of the population and the largest 16%. This disparity is due to the population distribution in Portugal, being northern and coastal population densities significantly higher, i.e., in Lisbon and Porto the population density exceeds 600 inhabitants/km^2, whereas in some inland regions this figure is about 3 inhabitants/km^2. Thus, there is a great heterogeneity among the 23 SGRUs concerning the number of municipalities covered, geographic dispersion, demography and socio-economic conditions, which is reflected in the options adopted to collect and treat the MSW, as well as in the equipment and infrastructures used for the MSW management.

Table 1. Characterization of the SGRUs existing in mainland Portugal and MSW managed, in 2014.

SGRU	Region	Inhabitants	Sorting station	Organic valorization	Incineration unit	Landfill	MSW managed /10^3 tons per year
1	North	337,644	2	–	–	2	128
2	interior	275,139	1	–	–	1	100
3		143,564	–	1	–	1	78
4		947,916	4	1	–	5	343
5	North	1,000,000	1	1	1	1	500
6	coast	321,776	1	–	–	1	130
7		442,866	1	1	–	1	184
8		79,631	1	–	–	1	36
9	Centre	342,371	1	–	–	1	134
10	interior	126,658	–	–	–	1	59
11		194,954	1	1	–	1	77
12		209,587	1	–	–	1	164
13		272,000	2	1	–	2	131
14	Centre	938,367	2	2	–	2	388
15	coast	841,073	1	1	–	–	390
16		304,000	1	1	–	1	107
17		1,600,000	2	1	1	2	894
18	South	115,437	1	1	–	1	63
19	interior	155,268	1	1	–	1	81
20		95,866	–	1	–	1	49
21	South	442,358	2	4	–	2	360
22	coast	781,044	1	2	–	2	439
23		25,485	1	–	–	1	14

The 23 SGRUs serve a population of approximately 10 million inhabitants and operate 28 sorting stations, 20 organic valorization units, 2 incineration units and 32 sanitary landfills. SGRUs that serve a larger population, and consequently manage higher amount of MSW, possess a larger number of infrastructures: the two incineration units that exist in Portugal are located in the two SGRU that handle higher amount of MSW, in Lisbon and Porto regions.

The conventional Portuguese MSW management organization involves 3 steps: i) Collection (unselected and selected solid wastes); ii) Treatment (organic valorization or incineration for the unselected materials, and sorting station for the selected ones); iii) Disposal of the refused materials from the treatment step in the sanitary landfill.

According to data collected in this study, about 4.8 million tons of MSW were generated in Portugal during 2014. When data of wastes managed vs. population served are plotted in a graph (Fig. 1a), it can be seen that the amount of wastes managed is proportional to the population served, and that the SGRU that managed higher MSW are located in the coast area of Portugal, where the main cities are located.

Figure 1b presents the distribution, in terms of final destination, of the MSW managed in Portugal in 2014. About 55% of the MSW had landfilling as final destination. Despite being a huge amount, the efforts and investments in valorization infrastructures led to a significant reduction in the ratio of MSW landfilled over the past 12 years, when compared to 2002, in which 76% of MSW were sent to landfills (Magrinho et al. 2006). Although the most part of the MSW landfilled were directly deposited by the municipalities (72%), around 22% came from the sorting stations and the organic valorization units. Less than half of the MSW forwarded to organic valorization units were submitted to organic valorization processes. This was due to the fact that some organic valorization units started its operation recently and, therefore, are still at a refining stage, and others are functioning only as mechanical treatment units. Nevertheless, around 440 thousand tons of recyclable wastes were recovered and 63 thousand tons of compost were produced in 2014 by the 23 Portuguese SGRU.

Approximately 1 million tons of MSW were sent for incineration units. This operation is recommended for the fraction of MSW that cannot be reused or recycled, since it produces electrical energy and slags of ferrous and nonferrous metals that can be recycled. Also, it provides a significant reduction of the volume of MSW to be deposited in landfills.

Being sanitary landfilling the main method for MSW treatment, recovery of landfill biogas must be considered, since this gas has a high calorific value. In contrast with what was reported by Magrinho et al. (2006), which states that in 2002 landfill biogas energy recovery was not conducted in Portugal, in 2014 landfill biogas energy recovery is performed by most of the SGRU, being the energy produced in 2014 higher than 6×10^9 kWh.

According to the objectives of the strategic plan for MSW (PERSU 2020), the amount of MSW landfilled should significantly reduce in the next years by increasing the fraction sent to valorization units. In fact, MSW valorization, organic and/or energetic, is the way for a sustainable waste management, as it brings economic and environmental advantages.

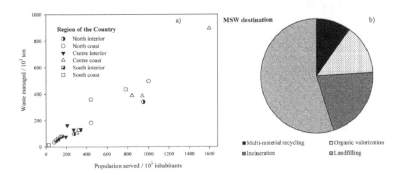

Figure 1.   (a) Relation between the amount of MSW managed by the different SGRU and the population served and (b) final destination of the MSW managed in Portugal in 2014.

3.2 Sanitary landfill leachates characterization

SLL comprise the liquids produced during wastes decomposition and the liquids that come into the landfill from external sources, such as rainwater. Data collected showed that, in 2014, the mean value of the average influent flow of leachate at the Portuguese SGRU was 160, 86 and 38 m^3 per day, for the north, center and south regions of Portugal, respectively. These results compare with the ratio between the amounts of leachates generated and rainwater, since rainfall is more intense in the north region of the country and less significant in the south.

Table 2 presents a summary of the main characteristics of the sanitary landfill leachates produced in Portugal in 2014. As it can be inferred from data showed in Table 2, leachates composition varies significantly between the different SGRU. This variation depends mainly on the type of wastes landfilled, the seasonal weather variation, the filling process and the age of the sanitary landfill (Koc-Jurczyk & Jurczyk 2011, Renou et al. 2008).

In Figure 2a, the average chemical oxygen demand (COD) values in leachates, according to the location of the sanitary landfill, is displayed. It can be seen that the highest organic loads are associated to the regions of the country where the population density is higher and, consequently, the amount of MSW produced too, showing a relation between the amount of wastes landfilled and the organic load of the leachates. The relative low value presented by the north coast, despite its high population density, is probably due to the existence of the incineration unit that provides a significant reduction in the volume of MSW landfilled. Also, the high COD value presented by the south coast, despite the low number of inhabitants, can be associated to the high number of tourists who frequent this region of the country.

Besides SLL, a new type of effluent is being increasingly produced at the SGRU, coming from the valorization units. It presents, generally, very high organic loads and is often sent for treatment together with SLL. According to data collected, in 2014, in Portugal, 83% of the effluents generated at the SGRU were from sanitary landfills and 17% were from valorization units. It is expected that, in the next years, the percentage of effluents generated at valorization units increases, since the national strategic plan for urban waste (PERSU 2020) establishes actions to reduce the amount of MSW landfilled, including the use of the biodegradable MSW for the production of compost and waste fuels. Besides the impact that these measures will have on the amount of effluents produced by the valorization units, they will also interfere in the quality of the leachates produced. Reducing the amount of biodegradable MSW in landfills will decrease the biodegradability of the leachates formed, which will be more difficult to treat by the conventional biological technologies.

Figure 2b shows the trend of average COD values according to the effluent origin. When leachates from sealed and active sanitary landfills are compared, it can be seen that the organic load is much lower in leachates from sealed landfills, which is in agreement with data reported in the literature (Foo & Hameed 2009).

Analyzing the COD values of the effluents produced at valorization units, it can be seen that they are higher than those presented by leachates. In fact, despite the volume of effluent produced in valorization units being much lower than the one generated in landfills, these effluents from valorization units present higher organic load and suspended solids content,

Table 2. Main characteristics of the sanitary landfill leachates produced in Portugal in 2014.

Parameter	Mean value	Minimum value	Maximum value
COD/mg L^{-1}	6622	190	41,000
BOD$_5$*/mg L^{-1}	2056	140	19,000
TOC**/mg L^{-1}	2459	69	11,400
Ammonium nitrogen/mg L^{-1}	2167	16	6150
Suspended solids/mg L^{-1}	1468	23	2200
pH	8.6	9.5	7.4

*Biochemical oxygen demand.
**Total organic carbon.

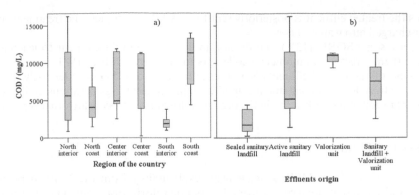

Figure 2. Boxplot of the average COD values in leachates produced in Portugal in 2014 according to (a) the location of the sanitary landfill and (b) the effluent origin. The box represents the interquartile (IQ) range which contains the middle 50% of the records. The line across the box indicates the median. The whiskers present the highest and lowest values that are no greater than 1.5 times the IQ range.

being more difficult to treat by conventional processes. This is a real problem that has been the focus of concern among the SGRU, who seek for a solution.

3.3 Sanitary landfill leachates treatment and destination

Generally, the type of treatment applied depends on the available municipal drainage infrastructures or existent watercourses. Whenever possible, leachates should be discharged, after pretreatment, into municipal collectors to be treated in the municipal wastewater treatment plants (MWTP), minimizing the costs associated to a tertiary treatment. When it is not possible, the leachates should be fully treated in situ, and the quality of the treated leachate must ensure compliance with the discharge standards set by the national authority. According to data collected, most of the conventional biological and physicochemical technologies that were initially implemented at the SGRU are not capable to pretreat leachates efficiently and their discharge into municipal wastewater treatment plants is compromising the MWTP efficiency. Thus, SGRU are not allowed to discharge leachates into the MWTP and have to find a solution for the leachate management. In several SGRU, the conventional biological and physicochemical technologies that were initially implemented were disabled and substituted by/or complemented with membrane technologies, as an effort to improve the quality of the treated leachate.

According to data collected in this study, currently in Portugal, 39% of the leachates treatment plants are equipped with reverse osmosis (RO) technology, 35% with biological/physico-chemical treatments, 18% with biological/physico-chemical + reverse osmosis, 4% with biological/physico-chemical + ultrafiltration, and 4% with other treatment options. This means that 57% of the leachates treatment plants are equipped RO technology. In 2004, the RO technology was used in 35% of the existing leachates treatment plants in Portugal, as a secondary treatment. The number of RO infrastructures to treat SLL has increased during the last 10 years and, according to the information collected in the present study, it should increase in the next years, since, despite their expensive investment and maintenance costs, RO is, for the moment, the only technology available in the market capable of providing a treated effluent that can be discharged into MWTP. Besides costs, RO presents as disadvantages the regular fouling of the semipermeable membrane and the problems associated with the discharge of RO concentrates. These concentrates are liquid wastes, around 40% of the leachate, and, as a common practice, are being deposited in the sanitary landfills. The average leachate treatment cost applying a single reverse osmosis treatment was estimated as 10.00 €/m³ of treated effluent.

Regarding the destination of the treated effluents, it was found that when a reverse osmosis treatment is used, the treated effluent is discharged directly into watercourses. When different treatment processes are applied, the quality parameters for discharge are not always fulfilled and thus the treated effluent is posteriorly sent to MWTP to complete the treatment required to be discharged into watercourses.

There are some SGRU (13%) that do not apply any kind of treatment to leachates and discharge them directly into municipal collectors to be treated in the MWTP.

The reuse of the treated effluents is a practice used in some SGRU. Generally the reused effluents are from RO treatment and are applied mainly in cleaning operations. Irrigation, reagents preparation and reutilization in the organic valorization process are other reported uses.

4 CONCLUSIONS

Effluents generated from MSW management result mainly from the MSW deposition in sanitary landfills and, more recently, from the organic valorization units. They are characterized by a complex composition that varies from place to place and over time. In Portugal, conventional biological and physicochemical treatments are not enough to reach the level of purification needed to eliminate their negative impact on the environment. Thus, they are being replaced by or complemented with RO units, which allow the treated effluent to be discharged into watercourses. There are still 50% of the MSW management sites that discharge their effluents into municipal collectors to be treated, partially or totally, in MWTP. However, due to inefficiency of the pretreatment technologies used, many SGRU are having problems in discharge their effluents into municipal collectors, since the admission criteria are not being accomplished. This problem will be aggravated in the coming years, since, with the increase in the organic valorization infrastructures, the biodegradability index of the leachates will become lower. Furthermore, the amount of effluents generated from the valorization infrastructures will raise, meaning effluents much more polluted and with high content of suspended solids. This is a real problem to SGRU that desperately seek for a solution. RO technology, despite the good results in terms of the quality of the treated effluent, presents the problem of membrane fouling, reducing the efficiency of the process and requiring frequent and costly maintenance, and moreover, the problem presented by the concentrates.

ACKNOWLEDGMENTS

We thank Fundação para a Ciência e a Tecnologia, FCT, for funding the FibEnTech Research Unit, project UID/Multi/00195/2013, and for the grant awarded to A. Fernandes, SFRH/BPD/103615/2014, and also to the MSW management systems and Empresa Geral do Fomento (EGF) that provided us the information used in this study, and to Professor Doutor Mário Russo, Eng.º João Pedro Rodrigues and Eng.º Artur Cabeças, for the technical support.

REFERENCES

Eggen, T., Moeder, M. & Arukwe, A. 2010. Municipal landfill leachates: A significant source for new and emerging pollutants. *Science of the Total Environment* 408: 5147–5157.

Foo, K.Y. & Hameed, B.H. 2009. An overview of landfill leachate treatment via activated carbon adsorption process. *Journal of Hazardous Materials* 171: 54–60.

Koc-Jurczyk, J. & Jurczyk, L. 2011. The influence of waste landfills on ground and water environment, in: *Contemporary Problems of Management and Environmental Protection*, No. 9, Some Aspects of Environmental Impact of Waste Dumps, Chapter 2, pp. 29–40.

Magrinho, A., Didelet, F., Semião, V. 2006. Municipal solid waste disposal in Portugal. *Waste Management* 26: 1477–1489.

Mavakala, B.K., Le Faucheur, S., Mulaji, C.K., Laffite, A., Devarajan, N., Biey, E.M., Giuliani, G., Otamonga, J.-P., Kabatusuila, P., Mpiana, P.T. & Poté, J. 2016. Leachates draining from controlled municipal solid waste landfill: Detailed geochemical characterization and toxicity tests. *Waste Management* 55: 238–248.

Renou, S., Givaudan, J.G., Poulain, S., Dirassouyan, F. & Moulin, P. 2008. Landfill leachate treatment: Review and opportunity. *Journal of Hazardous Materials* 150: 468–493.

WASTES – Solutions, Treatments and Opportunities II – Vilarinho, Castro & Lopes (Eds)
© 2018 Taylor & Francis Group, London, ISBN 978-1-138-19669-8

Truck tire pyrolysis optimization using the self-produced carbon black as a catalyst

N. Akkouche, M. Balistrou, M. Hachemi & N. Himrane
Laboratoire d'Energétique, Mécanique et Ingénieries, Université M'hamed Bougara, Boumerdes, Algeria

K. Loubar & M. Tazerout
Laboratoire GEPEA, UMR 6144 CNRS, Ecole des Mines de Nantes, Nantes, France

ABSTRACT: Powder of scrap truck tire was pyrolysed and both quantitative and qualitative analyses of pyrolytic vapors were performed using a sampling system, GC and GC-MS-FID analyzes. The process optimization consists in optimizing the thermal cracking process of the pyrolytic vapors by reusing the produced carbon black (CB) as a catalyst. It is obtained that reusing 33 wt% of CB increases the gas yield from 10 to 17 wt% at the expense of the oil yield. As regard the gaseous phase, the catalyst significantly increases the yields of the H_2 and CO species, and reduces the yields of the CO_2, C_3H_6 and C_3H_8 species. It improves its LHV from 41.2 to 43.4 MJ.kg^{-1}. As regard the liquid phase, the catalyst results in the decreasing of C_5 species, so that almost half converts to gas. The C_{10} species have also decreased in favor of C_6 and C_8 species, producing light gases.

1 INTRODUCTION

In recent years, tire waste are a major environmental problem. The year 2010 was marked by a global production of 17 million tons. It is estimated that 3.3 million tons per year are generated in the European Community, 2.5 million tons per year in North America and 1 million tons per year in Japan (Akkouche et al. 2017, Williams 2013, Rodriguez et al. 2001). China itself generates 5.2 million tons of used tires each year (Oyedun et al. 2012). In 2006, Algeria generated a deposit of more than 25000 tons of waste tires (Trouzine et al. 2011). The latter are eligible to increase due to the increasing number of vehicles in recent years. Indeed, the number of vehicles in Algeria increased by 29% between 2006 and 2012.

Energy recovery from tire waste is limited mainly by sulfur oxides (SOx) and polycyclic aromatic hydrocarbons (PAHs), which form during their submission at high temperatures. In oxygen-rich flames (incineration), SO, SO_2 and SO_3 dominate the products of combustion, whereas in oxygen-poor flames (pyro-gasification), the undesirable species that dominate the products are H_2S, S_2 and SH (Martinez et al. 2013, Giere et al. 2006, Conesa et al. 2008).

During pyrolysis, waste tire, consisting mainly of about 12% of metal, 10% of mineral and 77.5% of organic material, turns into syngas, pyrolytic oil, carbon black and metal. In this context, the energy recovery from waste tire using pyrolysis appears as an attractive alternative to fossil fuels. It offers the possibility to optimize its products and to use them separately (Barbooti et al. 2004, Raj et al. 2013, Lombardi, et al. 2015).

In the present study, the waste truck tire powder, recovered from the waste tire treatment industry in Sidi Aiche–Bejaia–Algeria, is pyrolyzed in a lab-scale fixed bed batch reactor up to 700°C under heating rate of 10°C.min^{-1}. In order to optimize the amount and quality of pyrolytic vapors, the pyrolysis products from various tests, in particular gas and oil are fractionated during the time depending on the reactor temperature. Chromatographic analyses (GC analysis for gas and GC-MS-FID analysis for oil) are performed on the sampled and fractionated products to determine the distribution of the hydrocarbon species. In order to improve the thermal cracking process, tests have been carried out by reusing the produced carbon black (CB) as a catalyst. Since the pyrolysis of a tire filler results in the production of 38 wt% of CB, the series of analyzes concerns the reuse of 25 and 33% of the produced CB as catalyst.

2 EXPERIMENTAL PROCEDURES

2.1 Materials and methods

The used raw material consist of the powder of scrap truck tires, recovered from a Waste Tire Processing Company in Algeria. The powder, with a particle size of about 1 mm, consists of 0.64 wt% of moisture, 63.26 wt% of volatile matter, 30.72 wt% of fixed carbon and 5.39 wt% of ash.

In order to investigate the effect of the reuse of the pyrolysis solid residue (called carbon black) as a catalyst on the thermochemical behavior and consequently on the yields of the pyrolysis, various tests were carried out heating a feed of 300 g up to about 700°C, under constant heating rate of 10°C min⁻¹.

The test series consists of using 25 and 33% of the CB amount, produced under the same operating conditions of the pyrolysis. This amount of CB, produced by each pyrolysis test of 300 g of waste tire, was about 113 g. The catalytic cracking column containing the reused carbon black is heated so that the pyrolysis vapors will be passed through the catalyst bed at 700°C.

2.2 Experimental device

Pyrolysis experiments were carried out in a lab-scale fixed bed reactor, as described in Figure 1. The batch reactor (1) consists of a 190 mm diameter and 240 mm height cylindrical, stainless steel, external electrical wall heated. It opens into a 42 mm internal diameter and 180 mm height cylindrical steel column (2), which is also emerged in a heating zone. To evaluate the temperature of the pyrolysis reaction, three thermocouples were placed in different locations (internal wall of reactor, internal wall of column, atmosphere of reaction). In the range of progress of the pyrolysis reaction (devolatilization and cracking phenomena), the difference between the three recorded temperatures is negligible. Therefore, the temperature of pyrolysis is considered as that of atmosphere, measured using thermocouple placed on the axis at 120 mm above the reactor bottom. A nitrogen flow is used to purge reactor before the beginning of each experiment and to sweep the last pyrolysis vapors at the end of each experiment.

The produced pyrolysis vapors are directed towards a water-cooled (5°C) condenser (3), where pyrolysis oil is then decanted in a separatory funnel (4) and the synthesis gas is collected in a sampling bag (5). The system is used for sampling, identifying and quantifying

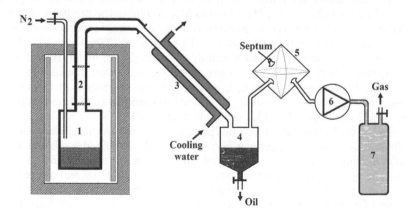

Figure 1.   Experimental setup of the pyrolysis of waste tires.

the total pyrolysis vapors (oil and gas). Each sample is identified by the reactor temperature which corresponds to the time of its sampling.

Using the valve of the funnel, each quantity of decanted oil (5–10 g) is stored in a vial and then weighed and identified using GC-MS-FID analysis. In order to characterize the overall produced oil, tests have been made without sampling the oil, and the analyses (density, viscosity, HHV and GC-MS-FID) were performed.

The system (Sampling Bag/Compressor/Tank) is set up for sampling the produced synthesis gas. After collecting each sample in the bag (about 4 liters), a sample collection (25 ml) was taken using a syringe through the septum of the sampling bag, and then the bag content is discharged through the compressor (6) to tank (7) where its temperature and pressure are measured. The syringe content is intended for chromatographic analysis (micro-GC) while the tank content is ejected to the atmosphere. The pyrolysis products identification and quantification methods have been clearly presented elsewhere (Akkouche, et al. 2017).

3   RESULTS AND DISCUSSION

It should be recalled that the mass of the produced CB during the pyrolysis of 300 g of tire is about 113 g. It is insensitive to the amount of recycled CB, which aims to crack pyrolytic vapors. The reused amount of the CB as catalyst are 28.25 and 37.67 g. They represent respectively 25 and 33% of the mass of the produced CB by the pyrolysis itself.

Figure 2 shows the yields of pyrolysis products. Since cracking only concerns pyrolytic vapors, the CB yield is insensitive to the amount of the reused CB as a catalyst. It is 38% and represents the mass of CB and metals used in the manufacturing and vulcanization processes of tires.

The results show that the reuse of the CB as a catalyst has a significant advantage over the cracking of pyrolytic oils by converting them into syngas. By reusing one third of the produced amount to crack the pyrolytic vapors, it makes it possible to increase the gas yield by 10 to 17% at the expense of the pyrolytic oil yield. The catalytic effect of the CB intervenes not only in the cracking reactions of the heavy vapors, but also in the gasification of the CB itself and in the soot and char deposit during cracking of the pyrolytic vapors. Indeed, the amount of the catalyst remains constant, while its composition and the size of its particles have slightly changed.

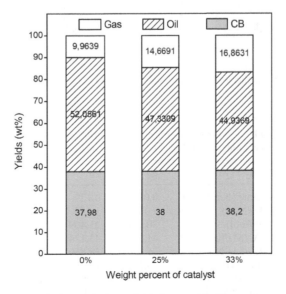

Figure 2.   Yield (wt%) of catalytic pyrolysis products from truck tire waste.

3.1 Syngas characteristics

In Figure 3, the produced gas composition is given both as a weight percent (wt%) and as a volume percent (vol%). It makes it possible to visualize the effect of the catalyst on the concentration of each chemical species.

It is noticed that the reuse of CB as catalyst has several advantages. It significantly increases the yields (concentrations) of the H_2 and CO species, and reduces the yields of the CO_2, C_3H_6 and C_3H_8 species. The yield lowering of the CO_2 species decreases the acidity of the gas and raises its calorific value, whereas the yield lowering of the C_3H_6 and C_3H_8 species raises the methane number of syngas, which is too low (30) compared to the required limits by engine manufacturers. The catalyst also has the advantage of converting more sulfur, contained in the oils, into H_2S. The main physicochemical characteristics of the gases are summarized in Table 1.

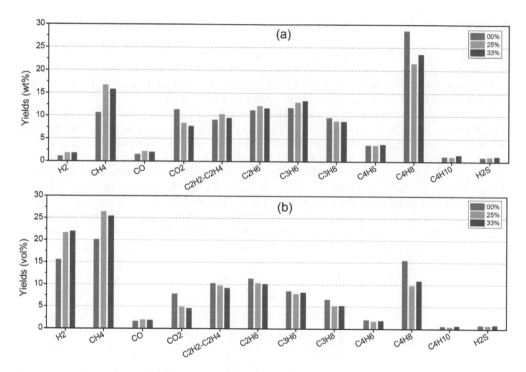

Figure 3. Distribution of the mass (a) and volume (b) gas composition according to the catalyst amount.

Table 1. Physicochemical characteristics of syngas from catalytic pyrolysis of truck tire waste.

Catalyst (wt%)	00	25	33
Volume (Nm³/kg_tire)	0,074	0,129	0,146
ρ (kg/Nm³)	1,351	1,133	1,153
M (g/mol)	30,281	25,409	25,850
LHV (MJ/kg)	41,23	43,124	43,390
LHV (MJ/Nm³)	55,696	48,880	50,037
HHV (MJ/kg)	44,556	46,786	47,056
HHV(MJ/Nm³)	60,189	53,031	54,264
Methane number	~38	~41	~43

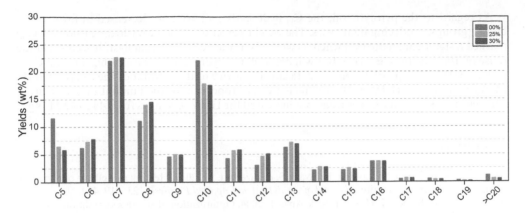

Figure 4. Carbon groups distribution of the pyrolytic oils from catalytic pyrolysis of truck tire waste.

3.2 *Pyrolytic oil characteristics*

Thanks to the performed CG-MS-FID analysis and to the developed calibration method (Akkouche, et al. 2017), the yields (wt%) of the oil hydrocarbons species were determined and the results were classified according to the carbon groups number. Figure 4 shows the results, carried out on pyrolytic oils from catalytic pyrolysis, classified according to carbon number groups.

The cracking reactions of the heavy vapors have been promoted by passing the pyrolytic vapors through a CB bed, heated to 600–700°C. The obtained pyrolytic oil consists mainly of light hydrocarbons (up to C_{13}). Indeed, the increase in the contact surface between the vapors and the catalyst has favored the cracking reactions of heavy the pyrolytic vapors.

Increasing amount of the recycled CB results in the decreasing of C_5 species, so that almost half (6%) converts to gas. It is already noted that the gas yield almost doubled using the CB as catalyst (Figure 2). The C_{10} species have also decreased in favor of C_6 and C_8 species, producing light gases. The insensitivity to the catalyst of the weight percent of the C_7, C_9 species and groups greater than C_{13} does not imply that they are not concerned by the catalytic cracking reactions. They are certainly both cracked to give other less heavy compounds, and are produced as a result of the cracking of the heavier compounds. Indeed, the weight percents of the heavy hydrocarbons ($> C_{20}$) marked a significant reduction. It seems that the reuse of 100% of the CB as catalyst will crack all the hydrocarbons whose carbon groups are higher than C_{16}. The HHV of the pyrolytic oil is practically insensitive to the amount of the catalyst (40–41 MJ.kg^{-1}).

4 CONCLUSION

The pyrolysis of the truck tire powder showed significant yields of pyrolytic vapors (62 wt%). The carbonization efficiency (CB yield) is practically insensitive to operating conditions (38 wt%). It represents the mass of the carbon black and metals used in the tire manufacturing process.

The reuse of the CB as a catalyst has a significant advantage over the cracking of pyrolytic oils by converting them into syngas. The reuse of one third of the produced CB makes it possible to increase the efficiency of the gas by 10 to 17% at the expense of the pyrolytic oil yield. Indeed, this catalyst considerably increases the yields of the H_2 and CO species and reduces the yields of the CO_2, C_3H_6 and C_3H_8 species. The lowering of the CO_2 yield decreases the acidity of syngas and raises its calorific value, whereas the lowering of the C_3H_6 and C_3H_8 components raises the methane number of syngas, which is too low compared to the required limits by engine manufacturers. The catalyst also has the advantage of converting more sulfur, contained in the oils, into H_2S.

As regard to the pyrolytic oil, the C_{10} hydrocarbons have also decreased in favor of C_6 and C_8 groups, producing light gases. The weight percents of the heavy hydrocarbons ($> C_{20}$) marked a significant reduction when 33 wt% of CB was reused as a catalyst. It seems that the reuse of 100% of the CB as catalyst will crack all the hydrocarbons groups higher than C_{16}.

The HHV of the pyrolytic oil is practically insensitive to the amount of the catalyst (40–41 MJ.kg^{-1}), while the HHV of the gas increases in proportion to the amount of the catalyst. The reuse of 33% of the CB as a catalyst increased the HHV of syngas from 44 to 47 MJ.kg^{-1}.

REFERENCES

Akkouche, N. Balistrou, M. Loubar, K. Awad, S. & Tazerout, M. 2017. Heating rate effects on pyrolytic vapors from scrap truck tires. *Journal of Analytical and Applied Pyrolysis* 123: 419–429.

Barbooti, M. Mohamed, T. Hussain, A. & Abas, F. 2004. Optimization of pyrolysis conditions of scrap tires under inert gas atmosphere. *Analytical and Applied Pyrolysis* 72: 165–170.

Conesa, J. Galvez, A. Mateos, F. Martin-Gullon, I. & Font, R. 2008. Organic and inorganic pollutants from cement kiln stack feeding alternative fuels. *Journal of Hazardous Materials* 158: 585–592.

Giere, R. Smith, K. & Blackford, M. 2006. Chemical composition of fuels and emissions from a coal and tire combustion experiment in a power station. *Fuel* 85: 2278–2285.

Lombardi, L. Carnevale, E. & Corti, A. 2015. A review of technologies and performances of thermal treatment systems for energy recovery from waste. *Waste Management* 37: 26–44.

Martinez, J.D. Puy, N. Murillo, R. Garcia, T. Navarro, M.V. & Mastral, A.M. 2013. Waste tyre pyrolysis—A review. *Renewable and Sustainable Energy Reviews* 23: 179–213.

Oyedun, A. Lam, K.-L. Fittkau, M. & Hui, C.-W. 2012. Optimisation of particle size in waste tyre pyrolysis. *Fuel* 95: 417–424.

Raj, R.E. Kennedy, Z.R. & Pillai, B. 2013. Optimization of process parameters in flash pyrolysis of waste tyres to liquid and gaseous fuel in a fluidized bed reactor. *Energy Conversion and Management* 67: 145–151.

Rodriguez, I.D. Laresgoiti, M. Cabrero, M. Torres, A. & Chomon, M. 2001. Pyrolysis of scrap tyres. *Fuel Processing Technology* 72: 9–22.

Trouzine, H. Asroun, A. Asroun, N. Belabdelouhab, F. & Long, N.T. 2011. Problématique des pneumatiques usagés en Algérie. *Nature & Technologie* (5): 28–35.

Williams, P.T. 2013. Pyrolysis of waste tyres—A review. *Waste Management* 33: 1714–1728.

Waste cooking oils: Low-cost substrate for co-production of lipase and microbial lipids

S.M. Miranda, A.S. Pereira, I. Belo & M. Lopes
Centre of Biological Engineering, University of Minho, Braga, Portugal

ABSTRACT: Waste Cooking Oils (WCO), generated from vegetable oils used at high temperatures in food frying, represent a highly dangerous threat to the environment. The ability of *Y. lipolytica* to grow and simultaneously produce lipase and accumulate lipids on this cheap and renewable carbon source was evaluated. A statistical experimental design based on Taguchi method was employed to assess the individual and combined effect of pH, WCO concentration and arabic gum concentration on lipase and microbial lipids production. pH was found to be the most significant parameter and the interaction between WCO and arabic gum concentration had the higher influence for both lipase and microbial lipids production. The scale-up of the process in a 2-L bioreactor demonstrate that oxygen concentration was also a crucial parameter to consider for lipase production. Increasing oxygen availability, a 38-fold improvement in lipase activity was achieved.

1 INTRODUCTION

Current domestic and industrial practices have led to an excessive generation of various low-value byproducts, which could have an adverse impact to environment and human health due to the presence of undesired substances. One example are waste cooking oils (WCO), generated from vegetable oils used at high temperatures in food frying.

It is estimated that approximately 700,000–100,000,000 ton of WCO per year are produced in European Union (Hanisah et al., 2013) and according to Portuguese Decree-Law no. 267/2009, approximately 65,000 ton of WCO are generated in Portugal. In 2011 only 9436 ton of WCO were sent for recycling or valorization (Portal de gestão de resíduos—http://www.netresiduos.com/homepage.aspx?menuid = 31. Accessed in June, 2016). WCO are often discharged through public sewerage system, which can reduce the sewers diameter and consequently block pipes and cause flooding (Ortner et al., 2016). Moreover, these wastes, in which the lipids are often the main and most problematic components can hinder the sewage treatment at wastewater treatment plants (Appels et al., 2011).

Due to the environmental problems associated with WCO, it becomes urgent the recycling and valorization of these pollutants residues. There are some applications for WCO that have been explored (Panadare & Rathod, 2015;), but most of the research is focused on the biodiesel generation from WCO (Fu et al., 2016). However, use of WCO as a feedstock for biodiesel production has some disadvantages: the presence of polar compounds and impurities, and high free fatty acids and water content might interfere in the biodiesel production and final quality (Panadare & Rathod, 2015).

Other alternative for WCO valorization is its utilization as bioprocesses media component for the cost efficient production of added-value compounds by microbial strains. *Yarrowia lipolytica* is well known by its capacity to assimilate fatty substrates (triglycerides, fats and oils) (Fickers et al., 2005) and to use efficiently carbon sources from agro-industrial wastes such as olive mill wastewater and crude glycerol (Lopes et al., 2009; Ferreira et al., 2016), producing lipase and microbial lipids.

Although some works have reported the production of lipase or microbial lipids by *Y. lipolytica* strains from WCO (Domínguez et al., 2010; Liu et al., 2015), no studies on the simultaneous effect of pH, WCO concentration and the surfactant arabic gum concentration are available in the literature. Thus, in the present work, the effect of such parameters on maximum production of lipase and microbial lipids from WCO was assessed by an experimental design based on Taguchi method. Additionally, the effect of oxygenation was also evaluated in bioreactor experiments.

2 MATERIALS AND METHODS

2.1 *Yeast strain and WCO characterization*

Y. lipolytica W29 (ATCC 20460) was maintained on YPDA medium (yeast extract 10 g·L^{-1}, peptone 20 g·L^{-1}, glucose 20 g·L^{-1} and agar 20 g·L^{-1}) at 4°C to a maximum of two weeks.

WCO were collected from a public school canteen and were composed by stearic acid 3%, palmitic acid 8%, linoleic acid 64%, and oleic acid 25%.

2.2 *Experimental design: Optimization of microbial lipids and lipase production*

A statistical technique of designing experiments based on Taguchi method was used to reach the highest lipase and microbial lipids production by optimizing the factors pH, WCO and arabic gum concentration. In the literature, carbon source limiting nutrients and pH are cited as factors that affect lipase and microbial lipids production. Arabic gum, used as emulsifier, could increase the bioavailability of the oils to the yeast cells (Bellou et al., 2016).

These factors were varied in three levels and combined in a total of nine experiments (Table 1), which were carried out in 500 mL baffled Erlenmeyer flasks with 200 mL of production medium at 27°C and at 170 rpm. Yeast cells, which grown overnight in YPD medium (yeast extract 10 g·L^{-1}, peptone 20 g·L^{-1} and glucose 20 g·L^{-1}), were centrifuged and

Table 1. Factors and levels used in the experimental design and maximum lipase activity, microbial lipids content and fatty acids composition obtained in batch cultures of *Y. lipolytica* W29 in the experiments designed by Taguchi L9 orthogonal array. Data are average ± standard deviation of two independent replicates.

Assay	pH	WCO	Gum	Lipase activity (U·L^{-1})	Microbial lipids (%, w/w)	Long chain fatty acids (%)			
						Stearic	Palmitic	Oleic	Linoleic
1	1	1	1	230.5 ± 79.2	44 ± 5	3.4 ± 0.3	28.1 ± 0.5	39.8 ± 0.4	28.7 ± 0.3
2	1	2	2	79.1 ± 47.2	53 ± 1	3.6 ± 0.2	33.1 ± 0.5	38.2 ± 0.3	25.1 ± 0.2
3	1	3	3	64.4 ± 33.8	46 ± 3	3.6 ± 0.2	31.8 ± 0.3	35.5 ± 0.3	29.0 ± 0.4
4	2	1	2	285.4 ± 10.8	32 ± 2	3.1 ± 0.3	26.4 ± 0.4	31.2 ± 0.4	39.3 ± 0.2
5	2	2	3	371.9 ± 7.9	35 ± 6	2.8 ± 0.1	28.8 ± 0.2	30.6 ± 0.2	37.8 ± 0.3
6	2	3	1	313.2 ± 10.2	22 ± 1	2.7 ± 0.2	24.9 ± 0.5	28.5 ± 0.4	43.9 ± 0.3
7	3	1	3	488.6 ± 53.5	24 ± 1	5.1 ± 0.3	27.4 ± 0.5	30.8 ± 0.5	36.7 ± 0.4
8	3	2	1	145.7 ± 29.4	21 ± 1	2.3 ± 0.2	26.8 ± 0.4	26.1 ± 0.4	44.8 ± 0.5
9	3	3	2	523.1 ± 15.9	27 ± 1	2.5 ± 0.3	30.0 ± 0.4	27.0 ± 0.4	40.5 ± 0.3

Level	pH	WCO (g·L^{-1})	Arabic gum (g·L^{-1})
1	5.6	10	5
2	6.5	30	10
3	7.2	50	20

resuspended in production medium (WCO, yeast nitrogen base without aminoacids (YNB) 6.7 g·L⁻¹ and arabic gum dissolved in Tris-HCl buffer 0.4 M). The responses of interest were lipase activity (U·L⁻¹) and microbial lipids content (expressed as the ratio of lipids mass and cellular mass, %). Both responses were processed in the Qualitek-4 software applying "bigger is better" quality characteristics to assess the optimal culture conditions for maximum lipase and microbial lipids production. One-way analysis of variance (ANOVA) was employed for statistical analysis of the results.

2.3 Bioreactor experiments

Batch cultivations in a 2-L bioreactor (BIOLAB, B. Braun, Germany) were performed with 1.6-L of production medium (WCO 10 g·L⁻¹ and YNB 6.7 g·L⁻¹ dissolved in Tris-HCl buffer 0.4 M, pH 7.2) at 27°C. Yeast cells were pre-grown overnight in 500 mL Erlenmeyer flasks with 200 mL of YPD medium at 27°C and 170 rpm, centrifuged and resuspended in production medium (0.5 g·L⁻¹ of initial cellular density). To evaluate the effect of oxygenation on lipase and microbial lipids production, experiments were performed at an air flow rate of 0.5 L · min⁻¹ (0.31 vvm) and varying the stirring speed (200 rpm and 400 rpm). pH was controlled with the addition of KOH 2M. Dissolved oxygen during cell cultivation was measured with an optical probe (Inpro 6000, Metler Toledo, USA).

2.4 Analytical methods

Extracellular lipase activity was quantified in the sample supernatant using p-nitrophenyl butyrate 0.42 mM and acetone 4% (v/v) in phosphate buffer 50 mM, pH 7.3 as substrate (adapted from Gomes et al., 2011). One unit of activity was expressed as the quantity of enzyme that produces 1 µmol of p-nitrophenol per minute in the assay conditions. Microbial lipids were extracted from lyophilized cells with methanol and chloroform (1:1, v/v) and quantified by phospho-vanillin colorimetric method (Inouye and Lotufo, 2006). A mixture of methanol acidified with sulfuric acid 10% (v/v) was used to convert the microbial lipids into their corresponding methyl esters, which are then extracted with n-hexane (1:2, v/v). The fatty acid methyl esters in organic phase (FAME) was quantified by gas chromatography (CP-3800 gas chromatograph (Varian Inc., USA) fitted with FID detector and Bruker BRU-MS nonpolar column), using methyl heptadecanoate (C17:0) as internal standard. The separation of FAME was attained using a temperature gradient: from 40°C to 150°C at a flow of 30°C·min⁻¹ and from 150°C to 250°C at a flow of 5°C·min⁻¹. The injector and detector temperatures were 250°C and 280°C, respectively, and helium was used as carrier gas.

3 RESULTS AND DISCUSSION

3.1 Experimental design

Lipase activity (U·L⁻¹) and microbial lipids content obtained from the nine experiments are shown in Table 1. Both lipase and microbial lipids production were dependent on the combination of the various parameters studied and ranged, respectively, from 64.4 U·L⁻¹ to 523.1 U·L⁻¹ and from 21% to 53%.

The results obtained for lipase production shows the potential of WCO to successfully induce lipase in *Y. lipolytica* W29 cultures. In all assays, cells of *Y. lipolytica* W29 accumulated more than 20% of their cell dry weight as intracellular lipids, which is in accordance with the premise of oleaginous yeasts. Particularly for the assay 2, this value exceeded 50%, revealing that the combination of this strain with WCO in the same bioprocess is advantageous for microbial lipids production.

The percentage of long chain fatty acids varied between the nine experiments and was dependent on the combination of the factors studied. Negligible amounts of stearic acid were detected (2%–5%), while considerable quantities of linoleic (25%–45%), oleic (26%–40%), and palmitic (25%–33%) acids were accumulated. This shows that fatty acid profile of lipids accumulated by *Y. lipolytica* W29 could be modulated through the manipulation of medium composition. Due to their composition rich in oleic and linoleic acids (mono and polyunsaturated fatty acids, respectively), these intracellular fatty acids can be used as an excellent food supplement.

The effect of individual parameters at different levels on lipase and microbial lipids production by *Y. lipolytica* W29 is shown in Figure 1. The increase of pH from level 1 (5.6) to level 3 (7.2) had a remarkable positive effect on lipase synthesis, but an opposite effect was observed for microbial lipids production. Change WCO and arabic gum concentration had a less relevant effect than pH variation, since the difference in both responses due to changes in the parameters levels were small. It should be stressed out that only 10 g·L^{-1} of WCO is enough to induce the maximum lipase production and nevertheless no inhibition effect was observed by increasing WCO concentration up to 50 g·L^{-1}. Concerning microbial lipids content, the intermediate level of WCO (30 g·L^{-1}) and arabic gum (5 g·L^{-1}) was found as the optimum for production maximization.

According to the ANOVA results for the L9 orthogonal array, pH was by far the most significant parameter, with 84.8% and 80.6% contribution toward variation in lipase and microbial lipids production, respectively. WCO and arabic gum concentration had no influence on lipase production by *Y. lipolytica* W29 in the range of the present study. Regarding microbial lipids production also WCO concentration had a negligible effect, as opposed to

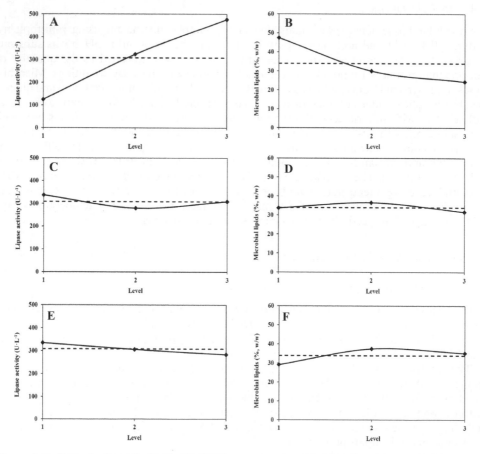

Figure 1. Individual effect of pH (A, B), WCO concentration (C, D) and arabic gum concentration (E, F) on lipase (left column) and microbial lipids (right column) production by *Y. lipolytica* W29 obtained from Qualitek-4 software. pH levels: 1–5.6, 2–6.5, 3–7.2; WCO concentration (g·L^{-1}) levels: 1–10, 2–30, 3–50; arabic gum concentration (g·L^{-1}) levels: 1–5, 2–10, 3–20.

arabic gum concentration which had a significant effect. The importance of medium pH on lipase production was previously reported and, apparently, lipase secretion depends on yeast strain (Lopes et al., 2008; Gonçalves et al., 2013). Papanikolaou et al. (2002) also identified the medium pH as a crucial factor for lipids production by *Y. lipolytica*.

Taguchi method established the optimum conditions for maximization of lipase (pH 7.2; WCO 10 $g \cdot L^{-1}$; arabic gum 0 $g \cdot L^{-1}$) and microbial lipids (pH 5.6; WCO 30 $g \cdot L^{-1}$; arabic gum 5 $g \cdot L^{-1}$) production considering the experimental data, and predicted a theoretical value in such conditions (531.4 $U \cdot L^{-1}$ for lipase and 53% for microbial lipids).

To attest the validity of the experimental design, two verification assays with three replicates were performed at optimal conditions predicted by the method, one in the optimal conditions for lipase secretion and other for microbial lipids production. Additionally, an assay with olive oil instead of WCO was carried out, since this vegetable oil is traditionally used for lipase induction (Lopes et al., 2008). Extracellular lipase activity profiles were similar in WCO and olive oil-based media, but higher values of enzyme activity were obtained when WCO was used as both carbon and inducer sources. The maximum lipase activity attained in WCO medium (531.7 $U \cdot L^{-1}$) was close to the value predicted by Taguchi method and the double of that obtained with olive oil (273.9 $U \cdot L^{-1}$). This result confirms the suitability of WCO as an inducer of lipase biosynthesis and is particularly promising since olive oil is much more expensive than WCO. The content of microbial lipids attained in the optimal conditions for lipids accumulation (51%) was close to the predicted one and demonstrates the ability of *Y. lipolytica* W29 to accumulate large amounts of lipids from WCO. This yeast oil content is one of the higher reported for a wild *Y. lipolytica* strain (El Bialy et al., 2011; Saygün et al., 2014).

3.2 *Bioreactor experiments*

Experiments in a 2-L BIOLAB bioreactor under the optimum conditions optimized at flask scale were performed. To study the effect of oxygenation on the process, two different stirring rates were used (Figure 2).

The increase of oxygen mass transfer by varying the stirring rate from 200 rpm to 400 rpm led to a considerable enhancement of lipase production. A 38- and 56-fold improvement in maximum enzyme activity and productivity, respectively, was achieved in the condition with higher stirring rate. In the experiments carried out at lower stirring rate, the process was limited by oxygen concentration, since dissolved oxygen dropped to 0% in the first hour and remained in this value until 32 h. This result proves that an adequate oxygen concentration is critical to maximal production of lipase by *Y. lipolytica* W29 from WCO.

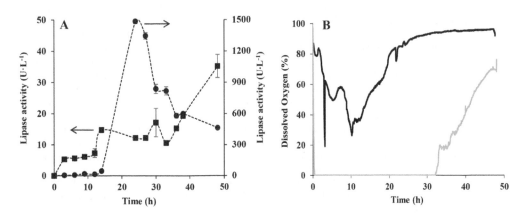

Figure 2.   Lipase activity (A) and dissolved oxygen (B) profiles obtained in the experiments carried out in 2-L BIOLAB reactor at different stirring rates: (■) 200 rpm, primary axis; (●) 400 rpm, secondly axis. The air flow rate in both experiments was 0.5 $L \cdot min^{-1}$ (0.31 vvm).

ACKNOWLEDGEMENTS

This study was supported by the Portuguese Foundation for Science and Technology (FCT) under the scope of the strategic funding of UID/BIO/04469/2013 unit and COMPETE 2020 (POCI-01–0145-FEDER-006684), Post-Doctoral grant (SFRH/BPD/101034/2014) attributed to Marlene Lopes, Project TUBITAK/0009/2014 and BioTecNorte operation (NORTE-01–0145-FEDER-000004) funded by the European Regional Development Fund under the scope of Norte2020—Programa Operacional Regional do Norte.

REFERENCES

Appels, L., Lauwers, J., Degrève, J., Helsen, L., Lievens, B., Willems, K., Impe, J.V., Dewil, R. 2011. Anaerobic digestion in global bio-energy production: Potential and research challenges. *Renewable and Sustainable Energy Reviews* 15: 4295–4301.

Bellou, S., Triantaphyllidou, I.-E., Aggeli, D., Elazzazy, A.M., Baeshen, M.N., Aggelis, G. 2016. Microbial oils as food additives: recent approaches for improving microbial oil production and its polyunsaturated fatty acid content. *Current Opinion in Biotechnology* 37: 24–35.

Brígida, A.I.S., Amaral, P.F.F., Coelho, M.A.Z., Gonçalves, L.R.B. 2014. Lipase from *Yarrowia lipolytica*: Production, characterization and application as an industrial biocatalyst. *Journal of Molecular Catalysis - B Enzymatic* 101: 148–158.

Domínguez, A., Deive, F.J., Sanromán, M.A., Longo, M.A. 2010. Biodegradation and utilization of waste cooking oil by Yarrowia lipolytica CECT 1240. *European Journal of Lipid Science and Technology* 112: 1200–1208.

Donot, F., Fontana, A., Baccou, J.C., Strub, C., Schorr-Galindo, S. 2014. Single cell oils (SCOs) from oleaginous yeasts and moulds: Production and genetics. *Biomass and Bioenergy* 68: 135–150.

Ferreira, P., Lopes, M., Mota, M., Belo, I. 2016. Oxygen mass transfer impact on citric acid production by *Yarrowia lipolytica* from crude glycerol. *Biochemical Engineering Journal* 110: 35–42.

Fickers, P., Benetti, P.-H., Waché, Y., Marty, A., Mauersberger, S., Smit, M.S., Nicaud, J.-M. 2005. Hydrophobic substrate utilisation by the yeast *Yarrowia lipolytica*, and its potential applications. *FEMS Yeast Research* 5: 527–543.

Fu, J., Turn, S.Q., Takushi, B.M., Kawamata, C.L. 2016. Storage and oxidation stabilities of biodiesel derived from waste cooking oil. *Fuel* 167: 89–97.

Gomes, N., Gonçalves, C., García-Román, M., Teixeira, J.A., Belo, I. 2011. Optimization of a colometric assay for yeast lipase activity in complex systems. *Analytical Methods* 3:1008–1013.

Gonçalves, F.A.G., Colen, G., Takahashi, J.A. 2013. Optimization of cultivation conditions for extracellular lipase production by *Yarrowia lipolytica* using response surface method. *African Journal of Biotechnology* 12(17): 2270–2278.

Hanisah, K., Kumar, S., Tajul, A.Y. 2013. The Management of waste cooking oil: a preliminary survey. *Health and the Environment Journal* 4(1): 76–81.

Inouye, L.S. and Lotufo, G.R. 2006. Comparison of macro-gravimetric and micro-colorimetric lipid determination methods. *Talanta* 70(3):584–587.

Liu, X., Lv, J., Xu, J., Zhang, T., Deng, Y., He, J. 2015. Citric acid production in *Yarrowia lipolytica* SWJ-1b yeast when grown on waste cooking oil. *Applied Biochemistry and Biotechnology* 175(5): 2347–2356.

Lopes, M., Gomes, N., Gonçalves, C., Coelho, M.A.Z., Mota, M., Belo, I. 2008. *Yarrowia lipolytica* lipase production enhanced by increased air pressure. *Letters in Applied Microbiology* 46: 255–260.

Lopes, M., Araújo, C., Gomes, N., Gonçalves, C., Aguedo, M., Belo, I. 2009. Use of olive mill wastewater by wild type *Yarrowia lipolytica* strains: effect of medium supplementation and surfactant presence. *Journal of Chemical Technology and Biotechnology* 84(4): 533–537.

Ortner, M.E., Müller, M., Schneider, I., Bockreis, A. 2016. Environmental assessment of three different utilization paths of waste cooking oil from households. *Resources, Conservation and Recylcing* 106: 59–67.

Panadare, D.C., Rathod, V.K. 2015. Applications of waste cooking oil other than biodiesel: a review. *Iranian Journal of Chemistry and Chemical Engineering* 12(3): 55–76.

Papanikolaou S., Chevalot I., Komaitis., Marc I., Aggelis G. 2002. Single cell oil production by Yarrowia lipolytica growing on an industrial derivate of animal fat in batch cultures. *Applied Microbiology and Biotechnology* 58: 308–312.

Saygün A., Şahin-Yeşilçubuk, Aran N. 2014. Effects of different oil sources and residues on biomass and metabolite production by *Yarrowia lipolytica* YB 423–12. *Journal of the American Oil Chemists' Society* 91:1521–1530.

WASTES – Solutions, Treatments and Opportunities II – Vilarinho, Castro & Lopes (Eds)
© 2018 Taylor & Francis Group, London, ISBN 978-1-138-19669-8

Proposal for MSW management facilities location in a state of Brazil

D.A. Colvero
University of Aveiro, Aveiro, Portugal
Brazilian National Council for Scientific and Technological Development (CNPq), Brazil

A.P.D. Gomes, L.A.C. Tarelho & M.A.A. Matos
University of Aveiro, Aveiro, Portugal

K.A. Santos
University of Goiás, Goiás, Brazil

ABSTRACT: The municipalities of the Goiás State, Brazil, have difficulties in the management of their Municipal Solid Waste (MSW). For instance of 20 municipalities in the micro-region of the Federal District Surroundings (FDS) only two dispose their MSW in licensed landfills. Considering this situation, the goal of this study was to propose host municipalities for the ins-tallation of shared MSW Management Facilities (MSWMF) in the micro-region of the FDS. In order to identify restriction-free areas for MSW treatment facilities location, a geographic information system was used. Subsequently, through the Center of Mass (CM) methodology, host municipalities for MSWMF were obtained. Results show that the FDS micro-region about 40% of its area which is suitable for MSW treatment facilities installation. From the CM, five MSWMF are proposed for the FDS: two municipalities with separate facilities, and the other three to be shared.

1 INTRODUCTION

The difficulty in properly managing municipal solid waste (MSW) has raised concern of muni-cipalities in Brazil. In order to evaluate such difficulty it was considered the Goiás State, as a case study. According to the Environment Secretariat, Water Resources, Infrastructure, Cities and Metropolitan Affairs—SECIMA/GO, of the 246 municipalities in Goiás, only 15 have licensed landfills that receive MSW from 16 municipalities in Goiás, as Cidade Ociden-tal shares its MSW disposal system with the municipality of Valparaíso de Goiás (SECIMA/GO, 2015; Colvero *et al.*, 2015).

In order to change this scenario, the State of Goiás must define management strategies for MSW that assess technical, economic and legislative aspects, to avoid health risks and environmental damage (Guerrero, *et al.* 2013; Soltani *et al.*, 2015). Therefore, the study has identified appropriate areas for MSW treatment facilities such as landfills. It has also taken into account that these facilities need a waste disposal process that cannot be reused or con-verted into a value-added product (Cherubini *et al.*, 2009).

In addition, the identification of available areas for landfills will help the definition of all municipal solid waste management facilities (MSWMF) of a given region of Goiás. These MSWMF, which according to the National Policy on Solid Waste should preferably serve two or more municipalities (shared management), need to be optimized in location terms, so that MSW transportation costs can be minimized (Brasil, 2010; Chen e Lo, 2016). To do that,

MSWMF should be centralized among municipalities, as a result of the principles of self suf-ficiency and proximity as referred in Directive 2008/98/EC (EC, 2008). Moreover, since MSW treatment facilities will be next to the generating sources, the emissions associated with the transportation of these wastes will be lower (Silva *et al.* 2012).

Thus, in view of the need for MSW treatment and disposal technologies in Brazil, this study aims to propose host municipalities (HM) of future MSWMF for Goiás micro-region of the Federal District Surroundings (FDS). This micro-region was chosen because it is the second largest population of the State. Moreover, it has only one landfill site licensed by SECIMA/GO (IBGE, 2016; SECIMA/GO, 2015).

2 METHODOLOGY

This study established the HM for the management facilities of the FDS micro-region. Municipalities of neighboring micro-regions may also be part of one of the FDS facilities. The proxy-mity among municipalities will define this issue.

2.1 *Study coverage area*

The Goiás State is located in the Central-West region of Brazil. In 2015, Goiás had 6 610 681 inhabitants (IBGE, 2016), distributed in 246 municipalities. The State is divided into 18 micro-regions. These micro-regions are groupings of neighboring municipalities whose goal is to make decisions about the location of economic and social activities, as well as validate stu-dies on infrastructure planning and location to be installed in urban and rural areas, such as the MSWMF (IMB, 2014). One of the 18 micro-regions is the FDS, which includes 20 municipalities and that in 2015 had 1 185 247 inhabitants (IBGE, 2016; IMB, 2014). This micro-region is the second largest MSW generator in the Goiás State, behind the Goiânia Metropolitan micro-region, where the State capital is located. (Table 1).

2.2 *Identification of suitable areas for landfill installation in the FDS micro-region*

Firstly it was essential to learn about availability, subject to approval and restric-ted areas for MSW treatment facilities installation in the FDS micro-region. In order to make a preliminary selection of free or restricted areas for the installation of landfills in the FDS micro-region, five legal documents were used as reference to define which geographic and environmental aspects should be considered (Table 2), as criteria for MSW treatment facili-ties location, since these aspects must be subject to approval by competent environmental authorities.

From these documents, the restrictive aspects were considered (as well as the ones subject to approval by the competent environmental control agency—ECA) for the installation of MSW treatment facilities in the FDS micro-region. As Ferreira and Ferreira (2014) indi-cated, the mapping of available areas for construction of facilities to MSW should consider environmental aspects and distance criteria.

Table 1. Number of municipalities, area, population and MSW production in the Goiás State and in the FDS micro-region.

Location	No. of municipalities	Area (km²)	Population in 2015 IBGE estimate– (inhabitants)	MSW production in 2015 (t/day)	Population density (inhab/km²)
FDS	20	33 164	1 185 247	800	35.7
Goiás State	246	340 111	6 610 681	5 829	19.4

Source: Adapted from IMB, 2014; IBGE, 2016.

Table 2. Restrictive and subject to approval criteria for landfill installation.

Criteria	Restrictive values for the installation of landfills	Legal documents
Distance from urban perimeter	3 km	CEMAm Resolution No. 05/2014
Land Slope	Bigger than 1% and shorter than 20%	NBR ABNT No. 13 896/1997 and CEMAm Resolution No. 05/2014
Distance from surface water bodies	0.3 km from any water bodies 0.5 km from water bodies for supply 2.5 km from the collection spot for public supply	CEMAm Resolution No. 05/2014
Distance from Conservation Units	3 km from the Conservation Unit limit (subject to approval distance)	CONAMA Resolution No. 428/2010 CEMAm Resolution No. 05/2014
Distance from aerodromes	20 km (subject to approval distance)	Federal Law No. 12 725/2012
Existence of remnant native vegetation	It should be located outside legal reserves and also where, preferably, there is no need for deforestation	CEMAm Resolution No. 05/2014
Quilombola lands and Indigenous lands	8 km (subject to approval distance)	Interministerial Ordinance No. 60/2015 of the Ministry of Environment

Source: Adapted from ABNT, 1997; Brasil, 2012; CONAMA, 2010; MMA, 2015, SEMARH/GO, 2014.

Thus, based on the restrictions established in legal documents, a geographic information system (GIS) application, (ESRI, version 10.3.1), was used to locate and identify suitable areas for landfill construction, as in the studies of Gbanie *et al.* (2013), Gorsevski *et al.* (2012), Ferreira and Ferreira (2014) and NURSOL/UFG (2015). GIS is one of the most sophisticated spatial analy—sis technologies in which geospatial or geographically referenced data are collected, stored, managed, integrated, manipulated and analyzed (Hannan *et al.*, 2015).

2.3 Geographically centralized Goiás municipalities

Once available areas for landfill installation were established, the second criteria of the HM was established: geographic location. As the municipalities which will integrate each MSWMF were defined according to proximity, the CM of the FDS and also eight neighboring micro-regions were calculated. This means that a municipality of the FDS can be part of another micro-region system, just as the municipality of another micro-region may integrate some proposed MSWMF to the FDS micro-region. The established HM were those geographically centralized within each of the nine assessed micro-regions and that have interconnection with paved roads (NURSOL/UFG, 2015; Russo, 2003). The quality of roads as well as the availability of MSW management facilities reduce MSW collection and transport time and, consequently, management costs (Guerrero *et al.*, 2013).

Given the large road distances among municipalities, in some micro-regions it was necessary to group the municipalities, in order to obtain associations of municipalities by proximity. To achieve this, the mass geometry methodology was chosen. It uses the x and y coordinates (in decimal values) and the MSW production (t/day) of each municipality that is part of the region in which finding the center of mass (CM) is desired. Thus, the geographic coordinates allied to the quantity of wastes of each location were used, in order to obtain the CM of a given region. The CM calculations were performed by applying the expressions presented in Equations 1 and 2 (Pereira *et al.*, 2013; Russo, 2003).

$$x = \frac{\sum x_i * P_i}{\sum P_i} \qquad (1)$$

$$y = \frac{\sum y_i * P_i}{\sum P_i} \qquad (2)$$

where: x = longitude; y = latitude; x_i and y_i = geographical coordinates of waste production centers (geographical center of the municipalities' urban area); P_i = quantity of municipal waste from each municipality. This way, by knowing the values of x and y, it is possible to obtain the geographical coordinate in which the CM of a region of interest is located.

The CM definition enables cost reductions with MSW transportation, since the MSWMF will be closer to the generators of these wastes (Pereira *et al.*, 2013).

2.4 *Distance from host municipalities and other municipalities*

Once the HM were selected, the other municipalities had to define where to send their MSW, having more than one HM to choose from. Thus, the chosen HM for the other municipalities would be the closest one (by paved roads).

The maximum distance between the HM and the other municipalities (that will be sending their MSW to the HM) must be of 25 km (Chen and Lo, 2016; FEAM and Engebio, 2010). This distance was set to limit the time spent by MSW transportation to the MSWMF by one hour, as these trucks move at an average speed of 50 km/h and must make the round trip to the MSW source generator (Akiko and Suzuki, 2009).

Municipalities which are over 25 km away from a HM must send their MSW to a transfer station (TS). Then, the MSW will be forwarded to the HM management facilities. The maximum distance from the municipalities which are not hosts and the TS must be 25 km. As for the distance between the TS and the HM, it must be 100 km maximum (FEAM and Engebio, 2010). The distance from TS starts at 25 km, as this is considered the turning point, i.e., starting at 25 km, it is economically viable to have a TS. (US EPA, 2002).

3 RESULTS

3.1 *Restricted, subject to approval or available for landfill installation*

The FDS micro-region has 60% of all its territorial extension restricted and 16% of available area for landfill installation (Table 3). Another 24% depend on ECA authorization for landfills to be built. In the Águas Lindas de Goiás, Novo Gama and Valparaíso de Goiás municipalities there is no free area for landfill installation. In the FDS micro-region, 17 unlicensed landfills and one licensed landfill by SECIMA/GO were identified. The only licensed landfill, in an area subject to approval, is located in Cidade Ocidental, and also receives MSW from the neighboring municipality of Valparaiso de Goiás. As for the other landfills (all unlicensed), 11 are in restricted areas, four are in areas subject to approval and two are in free areas for landfill installation.

Table 3. MSW production (t/day) and areas (km² and %) with restrictions, subject to approval and availability for landfill installation, of the Federal District Surrounding micro-region.

Micro-region	Total area	Restricted area for landfilling		Subject to approval area for landfilling		Available area for landfilling	
	km²	km²	%	km²	%	km²	%
FDS	38 108	22 879	60	6 133	16	9096	24

3.2 Proposal for host municipalities of Goiás MSW management facilities

There were five MSWMF proposed to serve the municipalities of the FDS micro-region (Figure 1). This definition was made by taking the following into account: CM of the FDS micro-region, the eight neighboring micro-regions and the available or subject to approval areas. These facilities, which will cover the FDS main treatment and disposal technologies, will serve 16 of the FDS 20 municipalities. Four other municipalities will be part of three other neighboring systems to FDS micro-region. In addition, to ensure that the largest number of municipalities will share the same system benefiting from lower transportation costs, eight TS were proposed for the FDS MSWMF.

The total population served, the amount produced and the number of municipalities in each of the five MSWMF proposed for the FDS micro-region are presented in Table 4. Cocalzinho de Goiás will have a MSWMF that will serve six municipalities, while the Cidade Ocidental MSWMF will have five municipalities, and Planaltina, three municipalities. Cabeceiras and Mimoso de Goiás will have individual MSWMF.

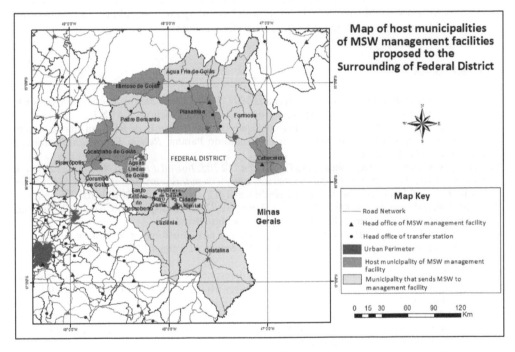

Figure 1. Municipal solid waste management facilities proposed to the Surroundings of the Federal District.

Table 4. Host municipality, served population, current MSW production and number of municipalities that make part of each MSW management facility of Federal District Surroundings micro-region.

Host municipality of the MSW management facility	Served population (inhabitants)	MSW production in 2015 (t/day)	Quantitative of served municipalities
Cabeceiras	7 829	11	1
Cocalzinho de Goiás	342 709	199	6
Cidade Ocidental	571 500	363	5
Mimoso de Goiás	2 715	1	1
Planaltina	205 217	167	3

4 CONCLUSION

The FDS micro-region has about 40% of its area either free or subject to restrictions for location of municipal waste treatment facilities. This means that the locations for the FDS MSWMF should be well planned. Especially in municipalities which do not have free areas, such as Águas Lindas de Goiás, Novo Gama and Valparaíso de Goiás.

The delimitation of free or subject to approval areas, combined with the waste production mass center methodology, were important to define the approximate location for MSWMF facilities, considering a set of chosen host municipalities of the FDS micro-region.

Thus, five host municipalities of MSWMF were established for the FDS micro-region. Considering maximum displacements of 25 km for those that directly send their MSW to the MSWMF and up to 100 km for those that have a transfer station.

It is also worth noting that each proposed HM for the MSWMF of the FDS micro-region will need an environmental impact study and report in order to establish the exact location of each facility. Concerning this work, proposed methodology results can be of great relevance to the municipalities of the FDS micro-region, as well as to other municipalities in Goiás, as it points out how to identify appropriate areas for the installation of MSWMF. This work shows how to apply established criteria under which municipalities should be unified in order to share and locate waste treatment facilities transfer stations.

REFERENCES

ABNT—Associação Brasileira de Normas Técnicas 1997. *NBR 13.896: aterros de resíduos sólidos não perigosos—critérios para projeto, implantação e operação.* Rio de Janeiro/RJ: ABNT.

Akiko, J., & Suzuki, N. 2009. Consórcios intermunicipais para a destinação de RSU em aterros regionais : estudo prospectivo para os municípios no Estado do Paraná. *Revista Engenharia Sanitária e Ambiental—RESA* 14(2): 155–158.

Brasil 2010. Presidência da República. Casa Civil. Lei n.° 12.305. *Institui a Política Nacional de Resíduos Sólidos; altera a Lei no 9.605, de 12 de fevereiro de 1998; e dá outras providências,* Pub. L. No. 12.305. Brasil. Retrieved from http://www.planalto.gov.br/ccivil_03/_ato2007–2010/2010/lei/l12305.htm.

Brasil 2012. Presidência da República. Casa Civil. Lei n.° 12.725/2012: dispõe sobre o controle da fauna nas imediações de aeródromos. Brasília, DF, 16 de outubro. Brasil.

Chen, Y.-C., & Lo, S.-L. 2016. Evaluation of greenhouse gas emissions for several municipal solid waste management strategies. *Journal of Cleaner Production* 113: 606–612. https://doi.org/10.1016/j.jclepro.2015.11.058.

Cherubini, F., Bargigli, S., Ulgiati, S. 2009. Life cycle assessment (LCA) of waste management strategies: Landfilling, sorting plant and incineration. *Energy* 34(12): 2116–2123. http://doi.org/10.1016/j.energy.2008.08.023.

Colvero, D.A., Gomes, A.P.D., & Pfeiffer, S.C. 2015. Análise dos custos das rotas tecnológicas dos resíduos sólidos urbanos de Cidade Ocidental, Goiás. *Sodebrás* 10(117): 196–204. Retrieved from http://www.sodebras.com.br/edicoes/N117.pdf.

CONAMA—Conselho Nacional do Meio Ambiente 2010. *Resolução CONAMA n° 428/2010: dispõe, no âmbito do licenciamento ambiental sobre a autorização do órgão responsável pela administração da Unidade de Conservação (UC).* Retrieved from http://www.mma.gov.br/port/conama/legiabre.cfm?codlegi = 641.

EC—European Commission 2008. *Directive 2008/98/EC of the European Parliament and of the Council, of 19 november 2008: on waste and repealing certain Directives,* 312 Jornal Oficial da União Europeia 3–30.

FEAM—Fundação Estadual do Meio Ambiente & Engebio 2010. *Estudo do estado da arte e análise de viabilidade técnica, econômica e ambiental da implantação de uma usina de tratamento térmico de resíduos sólidos urbanos com geração de energia elétrica no estado de Minas Gerais—Relatório 1.* Belo Horizonte, MG, Brazil.

Ferreira, W.A. de A., Ferreira, N.C. 2014. Seleção Preliminar de áreas para instalação de aterros sanitários na região metropolitana de Goiânia. *XI Congresso Nacional de Meio Ambiente de Poços de Caldas, Minas Gerais.* 21 a 23 de maio de 2014. 10 p.

Gbanie, S.P., Tengbe, P.B., Momoh, J.S., Medo, J., & Kabba, V.T. S. 2013. Modelling landfill location using Geographic Information Systems (GIS) and Multi-Criteria Decision Analysis (MCDA):

Case study Bo, Southern Sierra Leone. *Applied Geography* 36: 3–12. https://doi.org/10.1016/j. apgeog.2012.06.013.

Gorsevski, P.V., Donevska, K.R., Mitrovski, C.D., Frizado, J.P. 2012. Integrating multi-criteria evaluation techniques with geographic information systems for landfill site selection: A case study using ordered weighted average. *Waste Management* 32(2): 287–296. http://doi.org/10.1016/j. wasman.2011.09.023.

Guerrero, L.A., Maas, G., & Hogland, W. 2013. Solid waste management challenges for cities in developing countries. *Waste Management* 33(1): 220–32. https://doi.org/10.1016/j.wasman.2012.09.008.

Hannan, M.A., Abdulla Al Mamun, M., Hussain, A., Basri, H., Begum, R.A. 2015. A review on technologies and their usage in solid waste monitoring and management systems: Issues and challenges. *Waste Management* 43: 509–523. http://doi.org/10.1016/j.wasman.2015.05.033.

IBGE—Instituto Brasileiro de Geografia e Estatística 2016. *Cidades@Goiás*. Retrieved from: http:// cod.ibge.gov.br/1V4.

IMB—Instituto Mauro Borges de Estatísticas e Estudos Socioeconômicos 2014. *Estatísticas das Meso e Microrregiões do Estado de Goiás – 2013*. Goiânia/GO.

MMA—Ministério do Meio Ambiente 2015. *Portaria Interministerial nº 60, de 24 de março de 2015. Estabelece procedimentos administrativos em processos de licenciamento ambiental de competência do Instituto Brasileiro do Meio Ambiente e dos Recursos Naturais Renováveis—IBAMA*.

NURSOL/UFG—Núcleo de Resíduos Sólidos e Líquidos 2015. *Plano de resíduos sólidos do Estado de Goiás: produto final (produto 10)*. Goiânia/GO.

Pereira, C.D., Franco, D., & Castilhos Jr., A. B. de 2013. Implantação de Estação de Transferência de Resíduos Sólidos Urbanos utilizando Tecnologia SIG. *Revista Brasileira de Ciências Ambientais*, (ISSN Impresso 1808–4524/ISSN Eletrônico: 2176–9478), 71–84.

Russo, M. A. T. 2003. *Tratamento de resíduos sólidos*. Universidade de Coimbra, 196 p.

SECIMA/GO—Secretaria de Meio Ambiente, Recursos Hídricos, Infraestrutura, Cidades e Assuntos Metropolitanos 2015. *Nota técnica – aterros sanitários*. Goiânia, GO, Brasil.

SEMARH/GO—Secretaria do Meio Ambiente e dos Recursos Hídricos do Estado de Goiás 2014. *Resolução nº 005/2014 – CEMAm, de 26 de fevereiro de 2014: dispõe sobre os procedimentos de licenciamento ambiental dos projetos de disposição final dos resíduos sólidos urbanos, na modalidade aterro sanitário, nos municípios do Estado de Goiás*. Goiânia/GO.

Silva C., L., Costa R., B., & Rathmann, R. 2012. Gestão de resíduos sólidos urbanos na cidade do Porto (Portugal): um exemplo de prática sustentável? *Revista de Gestão Social e Ambiental* 6(2): 60–78. https://doi.org/10.5773/rgsa.v6i2.372.

Soltani, A., Hewage, K., Reza, B., Sadiq, R. 2015. Multiple stakeholders in multi criteria decision-making in the context of municipal solid waste management: A review. *Waste Management* 35: 318–328. http://doi.org/10.1016/j.wasman.2014.09.010.

US EPA—United States Environmental Protection Agency 2002. *Solid waste and emergency response. A waste transfer ttation: a manual for decision-making*. Washington, United States.

WASTES – Solutions, Treatments and Opportunities II – Vilarinho, Castro & Lopes (Eds)
© 2018 Taylor & Francis Group, London, ISBN 978-1-138-19669-8

Production of tannin-based adsorbents and their use for arsenic uptake from water

H.A.M. Bacelo, C.M.S. Botelho & S.C.R. Santos
Laboratory of Separation and Reaction Engineering-Laboratory of Catalysis and Materials (LSRE-LCM), Chemical Engineering Department, Faculdade de Engenharia da Universidade do Porto, Porto, Portugal

ABSTRACT: Tannins have been shown to be excellent candidates to produce biosorbents. These ubiquitous and inexpensive natural biopolymers are of easy extraction and conversion into insoluble matrices. Furthermore, it is somewhat easy to chemically modify a TBA (Tannin Based Adsorbent), in order to enhance its adsorption ability towards specific substances. In this work, a pine-tannin based adsorbent was modified with iron loading in order to assess its ability to uptake arsenic. Oxidation of the tannin gel prior to iron treatment was proven to be essential for the iron uptake. Moreover, the iron content of the tannin gel was shown to improve its affinity towards arsenic. It was achieved maximum adsorption capacities of 1.92 and 2.3 mg g^{-1} of As(III) and As(V), respectively, at 20°C and pH 3.5. Hence, the adsorbent here synthesized presents itself as a potential alternative for arsenic removal from aqueous solutions through adsorption.

1 INTRODUCTION

Adsorption processes are viewed as relatively simple methods, effective for the removal of various contaminants from aqueous solution. Depending on the water characteristics, adsorption can be applied as an alternative to conventional treatment processes (coagulation/flocculation, biological treatment) or as a final step, at a post-treatment level, in water or industrial wastewater treatment. In order to reduce the environmental impact of adsorption processes, adsorbents should be ideally effective, renewable and abundant materials, and should require minimal processing before use.

Tannins are inexpensive and ubiquitous natural polymers, polyphenolic secondary metabolites of higher plants, mainly present in soft tissues. After cellulose, hemicellulose, and lignin, tannins are the most abundant compounds extracted from biomass (Arbenz and Averous, 2015), leaves, roots, bark, seeds, wood and fruits. Regarding the woodland conservation, particularly to avoid the fell of trees, alternative tannin sources, such as locally available residues or by-products, should be selected. Maritime pine (*Pinus pinaster*) is one of the most common trees in Portugal, making the bark of these trees one of the preferable tannin sources for the national production of tannin-based adsorbents (TBAs). The extraction of tannins from vegetable residues constitutes then an important contribute for their reuse and valorisation, and for tannins sustainable production.

The presence of phenolic groups in tannins clearly indicates its anionic nature (Yin, 2010). The anionic nature of tannins surface groups opens the possibility to use them as adsorbents for cationic species. However, by chemical surface modifications, adsorption of soluble anionic species, which arsenic is an example of, can also be possible. Arsenic (As) is a toxic, non-degradable metalloid which is ubiquitous in the environment and highly toxic to all forms of life. Arsenic leaches to ground water through natural weathering reactions, biological activity, geochemical reactions and volcanic emissions. Moreover, anthropogenic sources as mining activity, combustion of fossil fuels, smelting of metals, use of arsenical compounds in agriculture (insecticides, pesticides and herbicides) and in livestock feed are additional pathways for environmental arsenic problems (Ungureanu et al., 2015). It is known that arsenite is more toxic than arsenate, and inorganic As more toxic than organic As.

Thus, taking all the above information into account, it is of extreme importance the existence of cost-effective and environmentally-benign remediation techniques specific for arsenic-contaminated water incidents. Although adsorption is a relatively simple method, the adsorptive As removal is not easy using many low-cost adsorbents, due to the neutral and anionic nature of the contaminant and its difficult complexation with many functional groups. Iron-based adsorbents (iron oxides, iron impregnated materials and bimetal oxides) have been identified as promising adsorbents for arsenic. An interesting study concerning the recovery of phosphates in an iron-treated TBA and its further direct reuse as a fertilizer is reported by Ogata et al. (2011). This kind of iron-loaded biosorbent seemed to be also an interesting option to arsenic, which has some similarities with phosphate in aqueous solution. Based on that, this work aims to assess the arsenic removal from aqueous solutions using a lab-produced pine-tannin-based adsorbent.

2 MATERIALS AND METHODS

2.1 Adsorbent preparation and characterization

2.1.1 Tannin extraction
Pinus pinaster bark was collected and broken up (pieces roughly ranging from 2 mm to 2 cm). The tannin extraction procedure was adapted from Sanchéz-Martín et al. (2011). The bark was immersed in distilled water (600 mL per 100 g of bark). Then, sodium hydroxide (analytical grade, pellets, *Absolve*) was added (5 g per 100 g of bark) and the mixture was stirred and heated at 85°C for 90 min. Solids were separated from the liquid fraction by filtration (qualitative filter paper, *Whatman*). Finally, the liquid fraction was evaporated in a heating plate until the sample contained almost no water. The humid precipitant was dried in an oven at 65°C and considered the tannin extract, denoted as TE.

2.1.2 Tannin gelification
The tannin gelification procedure was adapted from Xie et al. (2013) and Nakano and Ogata. (2005). The tannin extract was dissolved in 0.25 mol L^{-1} NaOH solution (4 mL per g of tannin extract) at room temperature, which was followed by addition of 0.4 mL of formaldehyde (37 wt%, analytical grade, *Labsolve*) per g of tannin extract, as a crosslinking reagent. After gelification at 80°C for 12 h, the product was dried in an oven at 65°C, following by washing successively with HNO_3 solution (0.05 mol L^{-1}) and distilled water to remove unreacted substances. Finally, the obtained adsorbent was once more dried at 65°C. The resultant product was considered the tannin-based adsorbent gel and denoted in the present work as TG.

2.1.3 Preparation of iron-loaded adsorbent
Based on the results previously obtained (Bacelo, 2014), an unmodified tannin gel does not present significant arsenic uptake ability (6% of As(III) removal was found from a 24 mg L^{-1}, at pH 4, 20°C and using a TG dosage of 10 g L^{-1}). However, the literature allowed to infer the possibility of using chemically modified tannin gels (iron-loaded TG) to treat arsenic-contaminated waters.

Therefore, TG was subjected to an oxidation and a subsequent iron-loading processing step prior to its application in arsenic solutions. The procedure of oxidation was optimized for the pine tannins used in this work. The following variables were studied in order to maximize the amount of iron impregnated in the tannin gel: nitric acid concentration (1 and 2 mol L^{-1}), temperature (40 and 60°C) and time of oxidation (30–120 min). One gram of TG was stirred with 100 mL of 1 mol L^{-1} nitric acid solution at 60 °C for 30, 50 or 120 min, or 2 mol L^{-1}, at 40 or 60°C for 90 min. Next, the adsorbent was washed with water and, finally, dried in an oven at 65°C. The resultant products were considered the oxidized tannin-based adsorbent gels.

One gram of the oxidized tannin-based adsorbent gel particles was added to 50 mL of a saturated solution of $FeCl_3$ at pH 2 (adjusted with HNO_3). The mixture was kept under stirring at room temperature for 48 h, followed by washing with water to remove unreacted species. Finally, the adsorbent was dried in an oven at 65°C. The resultant product was considered the iron-loaded tannin-based adsorbent gel. Also, non-oxidized gel was also subjected to same iron treatment. The resultant products were named as presented in Table 1.

2.1.4 Determination of iron content

Iron content present in the iron-loaded tannin-based adsorbents and in TG was determined by digesting 0.5 g of adsorbent in glass tubes at 150°C for 2 h with 5.0 mL of distilled water, 12.0 mL of HCl and 4.0 mL of HNO_3 (concentrated acids, 37% and 65% w/w respectively, analytical grade, *Merck*), using *DK6 Heating Digester Velp Scientifica*. Digested solutions were filtered through cellulose acetate membrane filters (0.45 µm porosity) and iron concentrations were determined by flame atomic absorption spectrometry (FAAS) – equipment: *GBC 932 Plus*—using a single-element cathode lamp. The amounts of iron present in the gel matrix per g of adsorbent were calculated.

2.2 Adsorption assays

2.2.1 Effect of pH

Adsorption assays were executed through the application of TG-Ox1-50-Fe in As(III) or As(V) solutions (10 mg L^{-1}) with a dosage of 5 g L^{-1}, which was kept under stirring in constant temperature (20°C) and pH for 24 h (more than enough to reach the equilibrium). The samples were made in duplicate and several pH values were evaluated: 2.0, 3.5 and 4.5. There was a preference towards pH in the acid region since the arsenic-contaminated waters generally present an acidic feature. After equilibrium was reached, cellulose acetate filters were used to filter every sample. The liquid samples were then analyzed for arsenic concentration (flame or graphite furnace AAS) and the amount of metalloid removed per mass unit of adsorbent (q) was calculated. Total dissolved iron leached from the adsorbent to the solution was also analyzed (FAAS).

2.2.2 Kinetic assay

An arsenic (III or V) solution (500 mL, 10 mg L^{-1}) was stirred with 5.0 g of adsorbent (TG-Ox1-50-Fe) at pH 3.5, 20°C. Samples were retrieved at 5, 15, 30, 60, 120 and 240 min after the start of the assay and were immediately filtrated. Then, the samples were analysed in terms of arsenic and iron content.

2.2.3 Adsorption equilibrium isotherms

In order to determine adsorption isotherms and establish the maximum adsorption capacities of TG-Ox1-50-Fe, for As(III) and As(V), equilibrium experiments were performed. 0.3 g of adsorbent was applied to 30 mL of arsenic (III or V) solutions of different concentrations (1–40 mg L^{-1}) at pH 3.5 and 20°C. After the equilibrium was reached (stirring time: 24 h), the samples were filtrated, arsenic and iron content in the liquid phase analyzed and the amounts of As removed (q_{eq}, mg/g), per gram of adsorbent calculated.

Table 1. Iron content of several TBAs (value ± deviation between average and duplicate measurements).

Adsorbent	Oxidation conditions	Iron-treatment	Iron content (mg g^{-1})
TG	–	–	0.20 ± 0.01
TG-Fe	–	+	2.1 ± 0.4
TG-Ox2-40-Fe	2 M, 90 min, 40°C	+	5.9 ± 0.1
TG-Ox2-60-Fe	2 M, 90 min, 60°C	+	13 ± 1
TG-Ox1-30-Fe	1 M, 30 min, 60°C	+	10 ± 1
TG-Ox1-50-Fe	1 M, 50 min, 60°C	+	24.4 ± 0.7
TG-Ox1-120-Fe	1 M, 120 min, 60°C	+	7.8 ± 0.2

3 RESULTS AND DISCUSSION

3.1 *Preparation of iron-loaded tannin-based adsorbents*

Under the experimental conditions used (alkaline extraction, at 85 °C, for 90 min), the average extractable tannin content was 59 mg per g of bark. During tannins gelification an average global mass loss (evaluated comparing the TE mass used and the final TG mass obtained) of 58% was obtained.

A few different sets of conditions for the oxidation procedure were tested in order to maximize the iron uptake by the adsorbent. The conditions stated in the original procedure presented by Ogata et al. (2011) (nitric acid 2 mol L^{-1}, 120 min, 60°C) proved in this work to be too aggressive for the adsorbent. Using this high concentration, and oxidation time of 90 min, a great portion of adsorbent was dissolved and an excessive oxidation occurred, causing a 53% loss in mass (from the comparison of the amounts of TG used and TG-Ox2-60 obtained). The use of a lower acid concentration (1 mol L^{-1}) at 60°C and for an oxidation time of 50 min leads to a reduced mass loss (8%), being the best result achieved. The optimization of the procedure to prepare iron-loaded TBAs should take in account not specifically this mass loss, but the amount of iron impregnated in the adsorbent.

3.2 *Iron content*

To optimize the oxidation conditions (nitric acid concentration and time), a correlation between these parameters and the iron content of the adsorbents was evaluated (Table 1). The TG sample was used as a control, presenting, as expected, negligible iron content. Comparing the iron content of the non-oxidized iron-loaded adsorbent (TG-Fe) with the content of any of the oxidized adsorbents (TG-Ox-Fe samples), it is possible to conclude the importance of the oxidation step to maximize the iron uptake. In the different conditions studied, the oxidation step was responsible for 3–12 times greater Fe levels in the adsorbents. Furthermore, it can be seen that too little or too much oxidation decreases the iron uptake by the TBA. Therefore, the optimal oxidation conditions for posterior iron treatment was found to be 50 min in 1 mol L^{-1} HNO$_3$ at 60°C, where a content of 24.4 ± 0.7 mg of iron per g of adsorbent was achieved (Table 1).

3.3 *Adsorption assays*

3.3.1 *Effect of pH*

After optimizing the tannin preparation, a tannin gel with high iron content was obtained: TG-Ox1-50-Fe. The adsorbent was tested on arsenic uptake at pH 2, 3.5 and 4.5 and the results can be observed in Fig. 1. The maximum adsorbed amount was achieved at pH 3.5 (close to the pH$_{PZC}$ = 2.93 ± 0.02) for the pentavalent arsenic, which was 1.8 mg g^{-1}. The trivalent form of arsenic was better removed at pH 4.5 (q = 1.7 mg g^{-1}). Some results present almost no variation while other present a considerable variation. However, it does

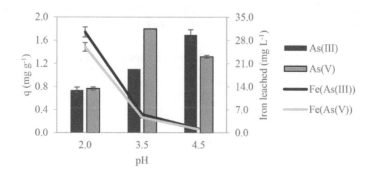

Figure 1. The effect of pH on the amount of As adsorbed by TG-0 × 1-50-Fe and on the iron leached to the solution: C$_{in}$ = 10 mg L^{-1}; adsorbent dosage = 5 g L^{-1}; T = 20°C; contact time = 24 h.

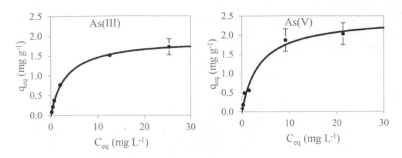

Figure 2. Equilibrium adsorption isotherms, experimental and Langmuir modelling: adsorbed amount (q_{eq}) as a function of the concentration of arsenic in aqueous solution (C_{eq}). Adsorbent dosage = 10 g L^{-1}; T = 20°C; contact time = 24 h.

not seem to be for a specific reason other than randomness. Of note, due to time limitations the assays were not replicated as much as it is desirable, thus the results here present are merely suggestive and not yet conclusive.

Such results indicate the optimal pH for adsorption as pH ≈ 4. Considering the impossibility to establish electrostatic interactions between As(III) and As(V) species and the iron-loaded adsorbents, the present results suggest that the iron provides sites for specific As adsorption. As(V) has been reported to adsorb onto iron sites by the formation of surface complexes and As(III) follows the same mechanism, after previous oxidation to the pentavalent form (Vieira et al., 2017). The results obtained and the ease to control pH during the experiment were used as criteria to choose pH = 3.5 for the kinetic and equilibrium assays, performed with the TG-O × 1-50-Fe for both As(III) and As(V).

3.3.2 *Adsorption kinetics*
Results obtained (not illustrated) showed that arsenic adsorption by TG-O × 1-50-Fe is rapid and, for both arsenic species, equilibrium was observed after two hours, achieving similar final values with a slightly better performance for arsenate uptake. Additionally, it is clear that As(V) is adsorbed faster that As(III). In fact, the concentration of As(V) is halved after 30 min of contact with the adsorbent, while that only happens with As(III) after 60 min.

3.3.3 *Adsorption equilibrium*
Equilibrium isotherms obtained for the adsorption of arsenite and arsenate by TG-O × 1-50-Fe, at 20°C and pH 3.5, were present in Fig. 2. For the experimental conditions studied, the maximum adsorbed amounts of As(III) and As(V) were 1.7 ± 0.2 mg g^{-1} and 2.0 ± 0.3 mg g^{-1} at equilibrium concentrations of 25.4 mg L^{-1} and 21.4 mg L^{-1}, respectively. Removal ranged from 40 to 83% and from 49 to 90% for As(III) and As(V), respectively. It is also observable a higher absolute uncertainty for results at higher equilibrium concentrations, although similar relative errors were obtained for all the concentration range.

Both Langmuir and Freundlich models were assessed and the former was found to better adjust to experimental data. According to the Langmuir model fitting (Fig. 2), the adsorbent maximum adsorption capacities (Q_m) were 1.92 ± 0.09 and 2.3 ± 0.2 mg g^{-1}, respectively for As(III) and As(V). These values fit somewhat reasonably when compared with other values reported in literature for unconventional adsorbents (Kuriakose et al., 2004, Gu et al., 2005, Gupta et al., 2009, Vieira et al., 2017).

4 CONCLUDING REMARKS

In this work, the ability of an iron-loaded tannin gel to uptake arsenic from aqueous solutions was assessed. It was concluded that an oxidation step prior to the treatment with iron is of great importance to the optimization of adsorbent preparation. The optimal conditions for the oxidation step were found to be 60°C, 1 mol L^{-1} of HNO_3 and 50 min of contact time.

Under these conditions, a Fe-loaded tannin gel adsorbent (TG-Ox1–50-Fe) was produced, with an iron content of 24.4 ± 0.9 mg g^{-1}.

Slightly acidic conditions showed to favour the adsorption of As(III) and As(V), with optimum pH of 3.5 (for arsenite) and 4.5 (for arsenate). Since the pH$_{PZC}$ was found to be 2.93 ± 0.02, such behaviour suggests that adsorption must be happening mostly through chemical interaction with iron and not through electrostatic attraction. The adsorbent seems to have affinity towards both oxidation states of arsenic, making it advantageous for real scenario applications. Since some arsenic-contaminated waters, like those leached from mines, contain both arsenic species.

Furthermore, equilibrium isotherms followed the Langmuir model, with predicted maximum adsorption capacities of 1.92 mg g^{-1} (As(III)) and 2.3 mg g^{-1} (As(V)), at 20°C and pH 3.5. For both As oxidation states, the adsorption equilibrium was reached in two hours. In almost every case, the As(III) adsorption leaches more iron than that of As(V). The evidences found in the present work are in general agreement with an adsorption mechanism proposed in literature for other iron-based adsorbents (stating that oxidation of As(III) into As(V) occurs prior to adsorption). However, future work needs to be done to prove the occurrence of As(III) conversion to As(V) and Fe(III) to Fe(II).

ACKNOWLEDGEMENTS

This work is a result of project "AIProcMat@N2020 – Advanced Industrial Processes and Materials for a Sustainable Northern Region of Portugal 2020", with the reference NORTE-01-0145-FEDER-000006, supported by Norte Portugal Regional Operational Programme (NORTE 2020), under the Portugal 2020 Partnership Agreement, through the European Regional Development Fund (ERDF) and of Project POCI-01-0145-FEDER-006984 – Associate Laboratory LSRE-LCM funded by ERDF through COMPETE2020 – Programa Operacional Competitividade e Internacionalização (POCI) – and by national funds through FCT—Fundação para a Ciência e a Tecnologia.

REFERENCES

Arbenz, A. & Averous, L. 2015. Chemical modification of tannins to elaborate aromatic biobased macromolecular architectures. *Green Chemistry* 17: 2626–2646.

Bacelo, H. 2014. *Natural-based adsorbents for wastewater treatment applications*, Universidade do Porto.

Gu, Z.M., Fang, J. & Deng, B.L. 2005. Preparation and evaluation of GAC-based ironcontaining adsorbents for arsenic removal. *Environ. Sci. Technol.* 39: 3833–3843.

Gupta, A., Chauhan, V.S. & Sankararamakrishnan, N. 2009. Preparation and evaluation of iron-chitosan composites for removal of As(III) and As(V) from arsenic contaminated real life groundwater. *Water Research* 43: 3862–3870.

Kuriakose, S., Singh, T.S. & Pant, K.K. 2004. Adsorption of As(III) from aqueous solution onto iron oxide impregnated activated alumina. *Water Qual. Res. J. Can.* 39: 258–266.

Nakano, Y. & Ogata., T. 2005. Mechanisms of gold recovery from aqueous solutions using a novel tannin gel adsorbent synthesized from natural condensed tannin. *Water Research* 39: 4281–4286.

Ogata, T., Morisada, S., Oinuma, Y., Seida, Y. & Nakano, Y. 2011. Preparation of adsorbent for phosphate recovery from aqueous solutions based on condensed tannin gel. *Journal of Hazardous Materials* 192: 698–703.

Sanchéz-Martín, J., Beltrán-Heredia, J. & Gibello-Pérez, P. 2011. Adsorbent biopolymers from tannin extracts for water treatment. *Chemical Engineering Journal* 168: 1241–1247.

Ungureanu, G., Santos, S., Boaventura, R. & Botelho., C. 2015. Arsenic and antimony in water and wastewater: Overview of removal techniques with special reference to latest advances in adsorption. *Journal of Environmental Management* 152: 326–342.

Vieira, B., Pintor, A., Boaventura, R., Botelho, C. & Santos, S. 2017. Arsenic removal from water using iron-coated seaweeds. *Journal of Environmental Management* 192: 224–233.

Xie, F., Fan, Z.J., Zhang, Q.L. & Luo, Z.R. 2013. Selective Adsorption of Au3+ from Aqueous Solutions Using Persimmon Powder-Formaldehyde Resin. *Journal of Applied Polymer Science* 130: 3937–3946.

Yin, C.Y. 2010. Emerging usage of plant-based coagulants for water and wastewater treatment. *Process Biochemistry* 45: 1437–1444.

WASTES – Solutions, Treatments and Opportunities II – Vilarinho, Castro & Lopes (Eds)
© 2018 Taylor & Francis Group, London, ISBN 978-1-138-19669-8

Damage induced by recycled C&D wastes on the short-term tensile behaviour of a geogrid

P.M. Pereira, C.S. Vieira & M.L. Lopes
CONSTRUCT-Geo, Faculty of Engineering, University of Porto, Porto, Portugal

ABSTRACT: The valorisation of Construction and Demolition Wastes (C&DW) is nowadays an imperative since it reduces the use of natural resources and avoids congesting landfills with these inert materials. This paper presents the mechanical, chemical and environmental degradation induced by fine grain recycled C&DW on the short-term tensile behaviour of an extruded uniaxial High Density Polyethylene (HDPE) geogrid. In order to study the chemical and environmental degradation a damage trial embankment was constructed using C&DW as filling material. The damage caused by the mechanical actions during installation was also simulated by installation damage laboratory tests. Wide width tensile tests were carried out on geogrid samples exhumed from the trial embankment after 12 months of exposure, on laboratory damaged samples and on intact samples. Their short-term tensile behaviour is compared. Scanning electron microscope (SEM) images of intact and exhumed specimens are also presented.

1 INTRODUCTION

Construction and Demolition Wastes (C&DW) are wastes derived from construction, reconstruction, cleaning of the work site and earthworks, demolition and collapse of buildings, maintenance and rehabilitation of existing constructions. Recycled C&DW have been considered as alternative materials for use as aggregates in concrete and pavement layers of transport infrastructures (Vieira & Pereira, 2015a).

Considering the need to find new ways of avoiding landfilling of inert waste and preserving natural resources, recent studies have been carried out on the reuse of recycled C&DW in geosynthetic reinforced structures (Arulrajah et al., 2014, Vieira & Pereira, 2015a, b, Pereira et al., 2015, Vieira & Pereira, 2016, Vieira et al., 2016). However, one of the main issues regarding the use of geosynthetics in contact with alternative materials is their durability.

The damage caused by mechanical actions during installation and the chemical and biological degradation are important issues to be considered in geosynthetics behaviour. The changes in their physical, mechanical and hydraulic properties, induced by the above-mentioned degradation processes, can control the performance of the structures where these materials are used.

Within the framework of a research project damage trial embankments have been constructed to study degradation induced by recycled C&DW on different geosynthetics. The exhumation of geosynthetic samples from these embankments was done after 6, 12 and 24 months of exposure. The results herein presented are related to HDPE geogrid samples exhumed after 12 months of installation.

The mechanical damage induced by this recycled material on the geogrid was simulated by laboratory installation damage tests.

2 MATERIALS AND METHODS

One of the geosynthetics tested was an extruded uniaxial high density polyethylene (HDPE) geogrid (Figure 1) with aperture dimensions of 16 mm × 219 mm. To minimize the influence of external factors intact (as provided by the manufacturers), damaged in the laboratory and exhumed samples were taken from the same roll of material and tested using the same methods and equipment.

The damage trial embankments were constructed using fine grain recycled C&DW coming mainly from maintenance works or demolitions of small buildings and cleaning of lands with illegal deposition of C&DW. At the lateral slopes of the embankment, coarse aggregates from recycled C&DW were placed to prevent erosion by rain water (Figure 2a). To minimize the installation mechanical damage on the geosynthetics, a lightweight compaction process was adopted (forward compaction plate with weight of 94 kg).

The tensile behaviour of exhumed specimens presented in this paper is related to geogrid samples exhumed after 12 months of installation (Figure 2b). The samples were carefully exhumed to prevent additional damage, being the material just above the geosynthetics removed carefully with the hands.

The predominant materials of the fine recycled C&DW used in the embankments construction are concrete, masonry, unbound aggregates, natural stones, as well as, a significant portion of soil. These recycled materials were provided by a Portuguese Recycling plant located in Centre region. Details on embankment construction, characterization of recycled C&DW and exhumation are available in Vieira & Pereira (2015b).

Scanning Electron Microscope (SEM) analyses were performed using a high resolution Environmental Scanning Electron Microscope with X-Ray Microanalysis and Electron Backscattered Diffraction analysis (Quanta 400 FEG ESEM/EDAX Genesis X4M) from the Materials Centre of University of Porto.

Laboratory installation damage tests were also carried out, using recycled C&DW similar to the one used in the embankments construction (coming from the same batch), to study the mechanical damage induced by these recycled aggregates on the geogrid tensile behaviour.

The mechanical damage tests were performed using a laboratory prototype developed at the University of Porto (Lopes & Lopes, 2003). The apparatus is composed by a rigid

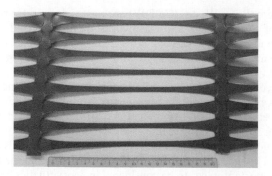

Figure 1. Visual aspect of the geogrid (ruler in centimetres).

Figure 2. Trial embankments: (a) one of the embankments at the end of construction; (b) geogrid during exhumation.

container (300 mm × 300 mm × 150 mm) divided in two boxes (where the geogrid and the C&DW were placed), a loading plate and a hydraulic compression system.

The geogrid specimens were cut with a width of 200 mm (9 longitudinal bars) and length of 500 mm. Each specimen was placed between two layers of C&DW and submitted to repeated loading. The layer placed under the specimen consisted in two sublayers (each 37.5 mm high) compacted by a flat plate loaded to a pressure of 200 ± 2 kPa, during 60 s, over the whole area of the test container. The layer placed over the specimen consisted in loose C&DW with 75 mm high. Each specimen was subjected to dynamic loading (ranging between 5 ± 0.5 and 500 ± 10 kPa) at a frequency of 1 Hz and for 200 cycles. Finished the loading, the specimen was removed carefully from the test container, avoiding additional damage.

Tensile tests carried out on intact (as provided by the manufacturers), exhumed and damaged specimens were performed in accordance with the European Standard EN ISO 10319 (2008). Five specimens (for each condition) and a strain rate of 20%/min were used.

3   RESULTS AND DISCUSSION

3.1   *Specimens exhumed from the embankment*

Although the preliminary visual inspections of the exhumed samples have not revealed significant damages. SEM analyses were carried out to evaluate potential damages in more detail.

Figure 3 illustrates SEM images (at 500 × magnification) of intact (Figure 3a) and exhumed specimens of geogrid (Figure 3b). The exhumed sample shows very small cavities and grooves appearing to be provoked by hard particles of the recycled C&DW. The intact sample also shows some irregularities.

Load-strain curves of exhumed geogrid specimens resulting from tensile tests carried out according to EN ISO 10319 (2008) are illustrated in Figure 4. The mean curve is also represented. The maximum tensile strength (T_{max}), the geogrid strain for T_{max} (ε_{Tmax}), the secant stiffness modulus at strain of 2% ($J_{2\%}$) and the secant stiffness modulus at ε_{Tmax} (J_{Tmax}) for the five specimens are summarized in Table 1. The mean values of these parameters and the 95% confidence intervals assuming a Student's t-distribution were also included in Table 1. Analysing Figure 4 and Table 1 it is clear the low variability of the results.

The tensile behaviour of exhumed and intact specimens will be compared and discussed in section 3.3.

3.2   *Specimens damaged in laboratory*

After the laboratory mechanical damage tests previously mentioned, the specimens were subjected to tensile load tests following similar procedures to those used for intact or exhumed specimens. The load-strain curves of damaged geogrid specimens, as well as the mean curve corresponding to the 5 samples are represented in Figure 5. Table 2 summarizes the values

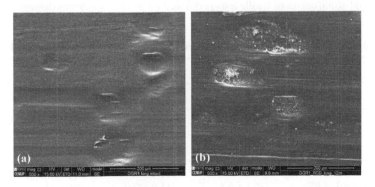

Figure 3.   SEM images of geogrid specimens (× 500): (a) intact; (b) exhumed after 12 months.

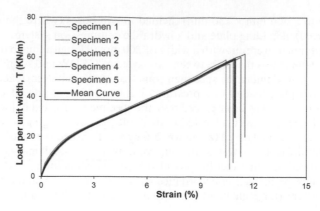

Figure 4. Load-strain curves of tensile tests performed on exhumed geogrid specimens.

Table 1. Summary of results of tensile tests carried out on exhumed geogrid specimens.

	T_{max} (kN/m)	ε_{Tmax} (%)	$J_{2\%}$ (kN/m)	J_{Tmax} (kN/m)
Specimen 1	58.8	10.4	1111	565
Specimen 2	56.8	10.6	1069	534
Specimen 3	61.8	11.5	1080	538
Specimen 4	60.2	11.3	1071	535
Specimen 5	59.0	10.8	1049	545
Mean value	59.3	10.9	1076	544
Confidence interval of 95%	59.3 ± 2.3	10.9 ± 0.5	1076 ± 28	544 ± 16

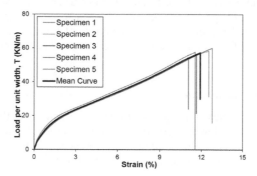

Figure 5. Load-strain curves of tensile tests performed on geogrid specimens damaged in laboratory.

Table 2. Summary of results of tensile tests carried out on geogrid specimens damaged in laboratory.

	T_{max} (kN/m)	ε_{Tmax} (%)	$J_{2\%}$ (kN/m)	J_{Tmax} (kN/m)
Specimen 1	57.5	11.5	1040	499
Specimen 2	58.2	12.5	931	464
Specimen 3	55.8	11.6	975	479
Specimen 4	54.1	11.1	965	489
Specimen 5	59.9	12.8	941	468
Mean value	57.1	11.9	970	480
Confidence interval of 95%	57.1 ± 2.8	11.9 ± 0.9	970 ± 53	480 ± 18

of maximum tensile strength (T_{max}), geogrid strain for T_{max} (ε_{Tmax}), secant stiffness modulus at strain of 2% ($J_{2\%}$) and at ε_{Tmax} (J_{Tmax}). The mean values of these parameters and the 95% confidence intervals assuming a Student's t-distribution were also tabulated.

3.3 *Comparison and discussion of results*

The tensile behaviour of intact specimens was reported in a previous publication (Vieira and Pereira, 2015b). Table 3 summarizes the main results.

The comparison of these results with those presented in Tables 1 and 2 points out that of the loss of strength caused either by the exposure to the recycled C&DW for 12 months or by the laboratory installation damage tests is very small (loss of 2% and 6% on average, respectively). A slight decrease on the geogrid tensile stiffness is also noticed.

It would be expected that the variability of the results for exhumed and damaged specimens would be greater than that of intact specimens, due to the different mechanisms that might contribute to the geogrid damage however, tensile tests carried out with intact specimens have shown greater variability. This variability is usual and it might be attributed to the production process.

Figure 6 compares the mean curves for intact, exhumed and damaged specimens. The shape of curves is quite similar but the coordinates at failure were shifted. The geogrid tensile stiffness for very small strains (initial stiffness) did not change significantly while a decrease on the secant stiffness modulus at ε_{Tmax} (J_{Tmax}) is noticeable. The reduction on the secant stiffness was greater in samples damaged in laboratory than in samples exposed 12 months to the recycled C&DW.

The damage on geosynthetics used as reinforcement is currently quantified by the retained values of relevant parameters, such as, the tensile strength, the strain at maximum load or the secant stiffness modulus. The retained value of the parameter X can be defined as the ratio between the mean value of the parameter X for damaged or exhumed specimens and the corresponding mean value for intact specimens.

The mean values of the retained tensile strength, R_T, retained peak strain, $R\varepsilon$, and retained secant modulus at 2% of strain, $R_{J2\%}$ are presented in Table 4. As previously mentioned, the mechanical damage induced in laboratory was slightly more pronounced.

Table 3. Summary of results of tensile tests carried out on intact geogrid specimens (Vieira & Pereira, 2015b).

	T_{max} (kN/m)	ε_{Tmax} (%)	$J_{2\%}$ (kN/m)	J_{Tmax} (kN/m)
Specimen 1	63.6	10.4	1144	612
Specimen 2	57.8	9.8	1074	593
Specimen 3	58.8	10.4	1063	564
Specimen 4	62.2	9.8	1149	638
Specimen 5	58.9	10.2	996	578
Mean value	60.3	10.1	1085	597
Confidence interval of 95%	60.3 ± 3.1	10.1 ± 0.4	1085 ± 79	597 ± 36

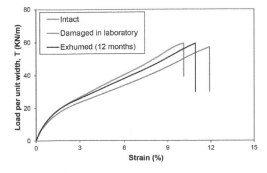

Figure 6. Comparison of mean load-strain curves of intact, damaged and exhumed specimens.

123

Table 4. Mean values of retained tensile strength, R_T, retained peak strain, $R\varepsilon$, and retained modulus, $R_{J2\%}$.

Exhumed after 12 months			Mechanical damaged		
R_T (%)	$R\varepsilon$ (%)	$R_{J2\%}$ (%)	R_T (%)	$R\varepsilon$ (%)	$R_{J2\%}$ (%)
98.41	108.1	99.2	94.7	117.9	89.4

4 CONCLUSIONS

The mechanical, chemical and environmental degradation induced by fine grain recycled C&DW on the short-term tensile behaviour of an extruded uniaxial HDPE geogrid was presented. The research herein reported has shown that the mechanical damage induced in the laboratory is more aggressive than that caused during trial embankments construction. That means that, it is possible to consider that the installation damage during the construction of the trial embankments was, as expected, insignificant and therefore, the damages recorded on exhumed specimens could be attributed to the chemical and environmental degradation.

Results of tensile tests carried out on exhumed specimens have shown that the exposure of this geogrid to fine grain C&DW under real atmospheric conditions for 12 months did not induce geogrid degradation.

The results and conclusions presented in this paper are partial results from a broaden research project. Results related to exhumed specimens submitted to longer periods of exposure to C&DW and the comparison with the damage caused by the exposure to a natural aggregate will be presented in the near future.

ACKNOWLEDGEMENTS

This work was financially supported by: Project POCI-01-0145-FEDER-007457 – CONSTRUCT—Institute of R&D In Structures and Construction funded by FEDER funds through COMPETE2020 – Programa Operacional Competitividade e Internacionalização (POCI) – and by national funds through FCT; Research Project: FCOMP-01-0124-FEDER-028842, RCD-VALOR – (PTDC/ECM-GEO/0622/2012).

REFERENCES

Arulrajah, A., Rahman, M., Piratheepan, J., Bo, M. & Imteaz, M. 2014. Evaluation of Interface Shear Strength Properties of Geogrid-Reinforced Construction and Demolition Materials Using a Modified Large-Scale Direct Shear Testing Apparatus. *Journal of Materials in Civil Engineering* 26 (5): 974–982.

EN ISO 10319 2008. *Geosynthetics—Wide width tensile test*, ISO, TC 221.

Lopes, M.P. & Lopes, M.L. 2003. Equipment to carry out laboratory damage during installation tests on geosynthetics. *Journal of the Portuguese Geotechnical Society* 98: 7–24 (in Portuguese).

Pereira, P.M., Vieira, C.S. & Lopes, M.L. 2015 Characterization of construction and demolition wastes (C&DW)/geogrid interfaces. In *Wastes: Solutions, Treatments and Opportunities—Selected Papers from the 3rd Edition of the International Conference on Wastes: Solutions, Treatments and Opportunities, 2015.* Viana do Castelo, Portugal.

Vieira, C.S. & Pereira, P.M. 2015a. Use of recycled construction and demolition materials in geotechnical applications: A review. *Resources, Conservation and Recycling* 103: 192–204.

Vieira, C.S. & Pereira, P.M. 2015b. Damage induced by recycled Construction and Demolition Wastes on the short-term tensile behaviour of two geosynthetics. *Transportation Geotechnics* 4: 64–75.

Vieira, C.S. & Pereira, P.M. 2016. Interface shear properties of geosynthetics and construction and demolition waste from large-scale direct shear tests. *Geosynthetics International* 23 (1): 62–70.

Vieira, C.S., Pereira, P.M. & Lopes, M.L. 2016. Recycled Construction and Demolition Wastes as filling material for geosynthetic reinforced structures. Interface properties. *Journal of Cleaner Production* 124: 299–311.

Lean-green synergy awareness: A Portuguese survey

M.F. Abreu, A.C. Alves & F. Moreira
*Department of Production and Systems, Centro ALGORITMI, University of Minho,
Guimarães, Portugal*

ABSTRACT: Although Lean has been around for some decades, the thorough contribution towards a greener industrial activity is still not completely known, therefore its use and application is justified mainly for the purpose of improvements on productivity, product quality and consumer satisfaction. The authors reviewed the nature of this synergetic link, aiming at a dual approach towards further effectiveness on both the production activities and the environmental impacts. This paper presents a survey on industrial companies, located in the north of Portugal, aiming for an enhanced understanding on the contemporary awareness on the Lean-Green link. Results of the study show that a great majority of the respondents do effectively know and apply Lean strategies, but a rather small fraction only know the virtues of considering the cleaner production lemma for justifying the application of Lean. The positive multi-dimensional outcomes, seem therefore rather unintentional, and accordingly, should be encouraged and improved.

1 INTRODUCTION

Nowadays, to cope with the ever-increasing needs of a growing consumer society, production has assumed a prime relevance since this activity provides the products demanded by the marketplace. This activity pressures the environment, given that it requires the exploitation of natural resources and releases pollutants. The governments, societies and the companies have to work together to mitigate these pressures so as balance this activity with desired environment boundaries. To do this, companies have been adopting some organizational methodologies and strategies that endorse ideas of "doing more with less" as promoted by Lean Production and of "creating more with less" as encouraged by the eco-efficiency concept. The synergies between these two strategies are undeniable and resulted in the known Lean-Green approach (Alves *et al.* 2016).

Lean production, as an organizational methodology, has been spreading through all economic activity sectors. It has roots on a new production approach conceived by the Toyota Motor Company, after the Second World War, called Toyota Production System (Monden 1983, Ohno 1988). This new production system arose because it was a time of financial restrain and resource scarcity in Japan, and Toyota was looking for a solution that accomplished what mass production did best, that was, to do things spending the minimum resources, while retaining its ability to adapt to changing circumstances. This new paradigm was coined by the MIT researchers as "Lean Production" and became internationally known after the publication of the book "The Machine that changed the world" from Womack (Womack et al. 1990).

Toyota developed a solution where it spent less of everything, i.e. less resources, less human effort, less space and fewer inventories by the elimination of all wastes. Wastes are all the activities that do not add value to the products. Ohno (1988) classified them in seven categories: 1) overproduction; 2) over processing; 3) transports; 4) defects; 5) motion; 6) inventory and 7) waiting. Later, untapped human potential was considered the eighth waste (Liker 2004). In 1996, and in order to systematically eliminate these wastes, Womack and Jones developed the five Lean thinking principles: value; value stream; flow; pull production and pursuit of perfection (Womack & Jones 1996). Pursuit of perfection implies searching for continuous improvement (*Kaizen*) and people play an important role in this aim, because a real Lean culture environment promotes people involvement and creativity (Alves et al. 2012). A Lean company is always, and continuously, concerned about waste reduction, it is a never-ending process.

In 1989, the UNEP/UNIDO (UN Environment Programme and UN Industrial Development Organization, respectively) launched the Green Cleaner Production Programme (CP), aiming "the continuous application of an integrated preventive environmental strategy applied to processes, products and services to reduce risks to humans and the environment" (UNEP 1996). The UNEP and the WBCSD (World Business Council for Sustainable Development) recognized that the main contributions of the Cleaner Production and of the Eco-efficiency consists in encouraging and strength the sustainability trend (WBCSD/UNEP 1998).

The link between Lean Production and Green (Production), called the Lean-Green link was investigated from the 1990 onwards (Maxwell et al. 1993, Maxwell et al. 1998, Larson & Greenwood 2004, Pojasek 2008). Although Lean was not mainly designed to address sustainability issues (EPA 2003, Moreira et al. 2010), some authors have revealed that their principles and practices brought several benefits that could be placed under the umbrella of Green (Klassen 2000; Rothenberg et al. 2001; Found 2009). The Lean focus is on reducing waste, while the Green focus is on reducing the environmental impact. As the Lean approach highlights—adding value to the operations grounded on the continuous elimination of all kinds of waste—seems to suit well a principle of nature, that is: "*within nature, nothing is lost, all is transformed*", and consequently, the waste concept does not seem to apply within nature. Therefore, the Lean-Green approach might offer a framework for delivering cleaner and valuable products. The U.S. EPA (Environmental Protection Agency) has been working to relate Lean with the environment. One of the designed toolkits is "The Lean and Environment Toolkit" (US-EPA 2007), where the environmental waste concept is defined as "any unnecessary use of resources or a substance released into the air, water, or land that could harm human health or the environment." and demonstrate that Lean tools can be applied to reduce environmental wastes. The environmental wastes can happen when companies use resources to deliver products or services to customers, and/or when customers use and dispose products (Maia et al. 2013).

In order to answer to the research question: "Is a Lean company more sustainable?" a survey was conducted. The purpose of this paper is mainly to present the results of the survey while attempting to bring some light on the underlying research question. After the introduction, which presents a brief background on Lean and the Lean-Green link, four more sections follow. Section two describes the research methodology. The third section presents the survey results, followed by the analysis and discussion in the fourth. Finally, in the last section, conclusions and future research lines are drawn.

2 METHODOLOGY

The research methodology employed was a survey grounded on a questionnaire entitled "Lean Production contributions for company's sustainability". The purpose of the questionnaire was to investigate the awareness on Lean and Green production methodologies, to know if they were implemented in companies and the extent to which its implementation contributes to the company's productivity and sustainability.

The survey was developed at University of Minho, Portugal, and addressed a database of national and international companies', with activity on the northern region of Portugal. The questionnaire had three main parts, holding 20 questions in total: Part I was dedicated to the characterization of the respondent, and was divided into four questions; Part II was intended to identify and characterize the company, by identifying, among others, the main sector of activity, the main product/service, number of workers, years in operation and main market (seven questions); and Part III related to the company's management system and characterization of the production model (nine questions), with the aim of knowing the management system and production models implemented; if the production model implemented is considered to promote sustainability; the benefits achieved by the company; the tools embraced for the improvement of the environmental performance; the sustainability indicators in use; if the company considers to implement the Lean Production.

After design and validation of the questionnaire (Saunders et al. 2009) it was sent via email with a cover letter, revealing the purpose of the study and the survey link. A first distribution was made in March 2016 and a second one in April 2016. The questionnaire was available on-line four months, between March and June 2016, and included open and closed questions.

From the 447 questionnaires sent, 357 were effectively delivered. The number of valid questionnaires was 42, which were subsequently analyzed. This corresponds to a response rate of about 11.8% (42 out of 357) and attending to the survey instrument used, was considered an acceptable response rate. The data was analysed using Microsoft Excel® and SPSS® (V.24).

3 SURVEY RESULTS

3.1 *Respondents and companies' characterization*

The open question on the respondents profile had a great variety of responses, ranging from CEO to process engineer, product design director, etc. The profile most represented was "department manager". Three respondents specifically replied "Lean managers" and three more were focused on quality and/or environment management. About 45% were at the company for less than 10 years, 41% between 11 and 20 years and about 14% were more than 21 and less than 30 years. There was no one working for more than 30 years at the company.

Regarding the characterization of the companies, most respondents pertained to large size companies (69%), followed by medium companies with 21.4% of the total, and only 9.5% were small companies. There were no micro-companies. Concerning the activity sector, it was used the Portuguese ranking of economic activities (CAE), defined by the National Statistics Institute (INE). The most representative sector was Manufacturing, representing 64.3% of the total, followed by the Wholesale and Retail Trade and Motor Vehicle Repair (7.1%); Construction (4.8%) and Water Distribution (2.4%). The "Other sectors" represented 21.4% of the total. The majority of the companies had prevalence on the marketplace of over 30 years (62%), followed by companies in the range of 10 to 30 years (36%) in operation. Only 2% were less than 10 years old. Respondents were essentially international companies (69%).

3.2 *Production models, main concerns and benefits*

Regarding the production system, respondents were asked to identify the production model(s) nearest to the one adopted by the company, from a list of 10 possible options (they can choose all the options, if they wanted). The Lean Production was the most popular (74%) followed by Kaizen System (55%), Kanban Systems (43%) and JIT (36%).

Respondents were asked to select from a list of 16 possible benefits the ones that the company achieved. As shown in Figure 1, the mostly reported benefit was the increase in productivity (88%), followed by waste reduction (79%) and customer satisfaction improvement (79%).

Concerning the opinion on the contribution of the production model to focus on the main concerns of the company, respondents had to select among a list of: 1) environment; 2) people; 3) social responsibility; 4) wastes; 5) costs; 6) productivity. The most reported concern was productivity (almost 86%), followed by the costs (81%), people (about 71%), wastes (69%) and environment (almost 55%).

Twelve companies were identified to select eight benefits (the ones with higher percentage). Two thirds of the companies (67%) belong to the Manufacturing activity sector and the other third (33%) belong to the "Others" activity sector.

3.3 *Sustainability awareness and approaches for improvement environmental performance*

Overall, 98% of the respondents agreed that the adopted production model promoted the sustainability of the company. One company answered that they did not know. Concerning the awareness of sustainability indicators, as shown in Figure 2, the most reported indicator (among nine in total) was energy consumption (88%), followed by the solid and liquid waste production (74%), the water consumption (71%), materials consumption, (67%), treatment and waste disposal cost (55%).

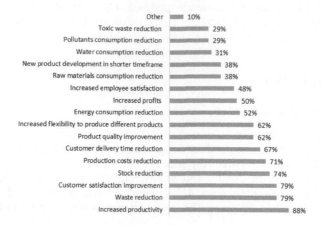

Figure 1.   Benefits achieved by the company.

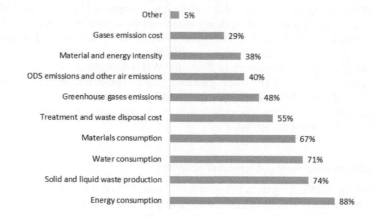

Figure 2.   Sustainability indicators.

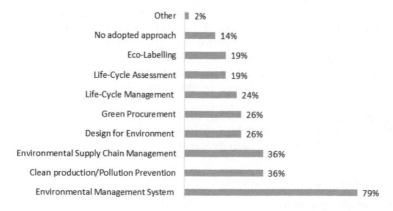

Figure 3.   Approaches used to support the process of improvement on environmental performance.

Respondents had to select from a list of eight approaches, the ones that were already in place to support the process of improvement of the environmental performance. The most reported approach was the environmental management system (79%), having a large difference to the remaining approaches, as shown in Figure 3.

3.4 Lean production implementation

The last question was intended to know if the companies implemented Lean Production. In order to clarify what Lean Production is, the following definition was presented: "*Lean production is a production model, with focus on the client, that promotes waste (activities that do not add value from the clients point of view) elimination, timely deliveries of quality products, at low cost and respecting people and the environment.*" (Maia et al. 2013).

Most companies reported to implement Lean Production (88%), with the remaining 12% reporting that they did not. Some reasons provided for justifying the lack of Lean implementation were: 1) "*The company has its production units outside the country*"; 2) "*We do not apply any production model*"; 3) "*Lean is rarely applied in this sector of activity*"; 4) "*We partially apply the Lean philosophy. We lack some consistent practices to systematically promote it*".

4 ANALISYS AND DISCUSSION

From the ten options given to respondents, regarding the production models, seven were Lean related. This was intended at knowing to what extent the respondents knew Lean. Most respondents, however, did not relate the seven options. Therefore, it seem that the respondents struggled to recognize or relate such concepts. For example, some of the options were related to tools used in the Lean context, such as Kaizen, Kanban, JIT.

Relating the companies concern, presented in section 3.2, the results point out that productivity and cost are still the leading aspects, followed by: people, wastes, environment and, finally, social responsibility. It is interesting to note that the wastes and the environment appear in 4th and 5th place, respectively. Attending to the benefits achieved by the companies, four of them selected all the sixteen benefits listed. These were large international companies from the Manufacturing sector, holding a certified management system.

A big agreement exists among respondents on their production model promoting sustainability (in section 3.3.). However, this seems rather incongruent with the results on the reported benefits, given that sustainability achieved a low score, while encompassing environmental wastes, as defined in the introduction, e.g. water consumption, toxic waste, among others. Additionally, since the sustainability indicators (in section 3.3.) are partly selected by the respondents (in average, only five indicators were selected; and just four companies selected all nine indicators, about 10%,) it is hard to understand how the companies measure such promotion on sustainability. All the companies considered that the production model adopted increased the concern about people, wastes and costs.

Considering the approaches for the improvement on the environmental performance (section 3.3), in average three approaches were adopted by the companies, and just three companies chose all the approaches. Five companies selected the five approaches holding the highest percentages. All these companies identified Lean Production and Kanban system as the adopted production model and considered that these promoted the company sustainability.

In spite of the above-mentioned, in section 3.4, respondents considered that implemented Lean, however it was not completely clear from the survey that they knew exactly what Lean was. Their answers seemed, in one point, contradictory. In addition, it is important to report that questions "*Do you consider that the company's production model promotes company's sustainability?*" and "*Do you consider that the company implements Lean Production?*" are independent, i.e. there is no association between them. This conclusion was based on Fisher test that given a $p = 0.119$ ($p>0.05$ means they are independent). Therefore, statistically, this result does not allow to infer that the Lean production model promotes sustainability.

A further study is necessary to arrive at a more conclusive result. Although the response rate was acceptable (11.8%), this was not appropriate because the answers have almost no variability. The authors highlight these as a limitation of the study.

5 CONCLUSION

This paper reports on a survey about Lean-Green synergy awareness in the North of Portugal. Forty-two (42) valid answers were considered from a sample of 357. The survey was intended to know if the companies recognize that Lean contributes to its sustainability. Most companies identified Lean as the adopted production model and, additionally, they recognize that this production model promotes sustainability. However, the statistical analysis of the results has shown a weak response variability combined with a low response rate. This lead to an inconclusive result. Furthermore, although the majority considered themselves Lean adopters, other responses results seem to indicate poor or misleading understandings on Lean. Additionally, the respective readings of the results of the study does not seem to clear out the impression that the respondents were not generally aware of the Lean-Green synergy, and probably this derives from lack of measurement and use, of sustainability indicators. Given the study limitations, future work will focus on selecting some of the respondents for interviews.

ACKNOWLEDGEMENTS

This work has been supported by COMPETE: POCI-01-0145-FEDER-007043 and FCT—Fundação para a Ciência e Tecnologia within the Project Scope: UID/CEC/00319/2013. The authors are also grateful to the surveyed companies.

REFERENCES

Alves, A.C., Dinis-Carvalho, J. and Sousa, R.M. 2012. 'Lean production as promoter of thinkers to achieve companies' agility'. *The Learning Organization* 19(3): 219–237.

Alves, A., Moreira, F., Abreu, F. and Colombo, C. 2016. *Sustainability, Lean and Eco-Efficiency Symbioses, Multiple Helix Ecosystems for Sustainable Competitiveness—Innovation, Technology, and Knowledge management.* Edited by M. Peris-Ortiz, J.J. Ferreira, L. Farinha, and N.O. Fernandes. Cham: Springer International Publishing (Innovation, Technology, and Knowledge Management).

EPA, U.S. 2003. *Lean manufacturing and the environment: Research on advanced manufacturing systems and the environment and recommendations for leveraging better environmental performance.* United States Environmental Protection Agency.

Found, P.A. 2009. Lean and Low Environmental Impact Manufacturing, *Production and Operations Management.* Proc. of 20th International Conference of Production and Operations Management, Orlando, Florida.

Just-in-Time Manufacturing and Pollution Prevention Generate Mutual Benefits in the Furniture Industry 2000. *Interfaces*: 95–106.

Larson, T. and Greenwood, R. 2004. Perfect complements: Synergies between lean production and eco-sustainability initiatives. *Environmental Quality Management* 13(4): 27–36. doi: 10.1002/tqem.20013.

Liker, J.K. 2004. *The Toyota Way: 14 Management Principles from the World's Greatest Manufacturer.* McGraw-Hill Education.

Maia, L.C., Alves, A.C. and Leão, C.P. 2013. Sustainable work environment with Lean Production in Textile and Clothing industry. *International Journal of Industrial Engineering and Management* 4: 183–190.

Maxwell, J., Briscoe, F., Schenk, B. and Rothenberg, S. 1998. Case study: Honda of America Manufacturing, Inc.: Can lean production practices increase environmental performance? Environ. Qual. Manage: 53–61. doi: 10.1002/tqem.3310080107.

Maxwell, J., Rothenberg, S. and Schenck, B. 1993. Does Lean Mean Green: The Implications of Lean Production for Environmental Management,. MIT, International Motor Vehicle Program, July.

Monden, Y. 1983. *Toyota Production System—An Integrated approach to Just-in-Time*. First Edit. Institute Industrial Engineers.

Moreira, F., Alves, A.C. and Sousa, R.M. 2010. *Towards eco-efficient lean production systems, IFIP Advances in Information and Communication Technology*. doi: 10.1007/978-3-642-14341-0_12.

Ohno, T. 1988. Toyota Production System: Beyond Large-Scale Production. *Productivity Press*: 152. doi: 10.1108/eb054703.

Pojasek, R.B. 2008. Framing Your Lean-to-Green Effort. *Environmental Quality Management*: 85–93. doi: 10.1002/tqem.

Rothenberg, S., Pil, F.K. and Maxwell, J. 2001. Lean, green, and the quest for superior environmental performance. *Production and Operations Management* 10(3): 228–243.

Saunders, M., Lewis, P. and Thornhill, A. 2008. *Research Methods for Business Students, Research methods for business students*. doi: 10.1007/s13398-014-0173-7.2.

UNEP 1996. *Cleaner Production: a training resource package, first edition. http://www.uneptie.org/shared/publications/pdf/WEBx0029xPA-CPtraining.pdf (accessed February 20, 2010)*.

US-EPA 2007. *The Lean and Environment Toolkit*. Available at: www.epa.gov/lean.

WBCSD/UNEP 1998. *Cleaner Production and Eco-efficiency: Complementary approaches to sustainable development*. WBSCD and UNEP edition.'

Womack, J.P., Jones, D.T., and Roos, D. 1990. *The Machine that changed the world*. New York: Rawson Associates.

Womack, J.P. and Jones, D.T. 1996. *Lean Thinking, The Library Quarterly*. doi: 10.1086/601582.

Efficiency of regeneration by solvent extraction for different types of waste oil

C.T. Pinheiro, M.J. Quina & L.M. Gando-Ferreira
CIEPQPF, Department of Chemical Engineering, University of Coimbra, Portugal

ABSTRACT: The main objective of this work was to evaluate the efficiency of the extraction-flocculation regeneration process for different types of Waste Lubricant Oil (WLO). The solvent used in the regeneration was 1-butanol. Two WLO samples with significant differences in physicochemical characteristics were selected. The effect of most critical parameters, namely the solvent/oil ratio, addition of KOH as flocculant agent and the Critical Clarifying Ratio (CCR) were investigated. The results showed that the equilibrium yield varied from 89 to 92% and the sludge removal from 6.5 to 8.5%, depending on the type of waste oil. The addition of 1 g/L of KOH in 1-butanol led to an increase of 13.1 and 32.9% in the sludge removal. The critical clarifying ratio varied from 1.4 and 2.1. This study demonstrated that different operation conditions could be used to optimize the efficiency of regeneration by adjusting the process to the type of waste oil.

1 INTRODUCTION

The European List of Waste (LoW), which is a key document for classification of waste, was established by the Commission Decision 2000/532/EC. More recently, this list was revised by the Decision 2014/955/EU, where waste lubricant oil (WLO) is classified as an absolute hazardous (AH) entry without any further assessment. The hazardous properties of WLO arise mainly from the presence of harmful components such as heavy metals, poly-chlorinated hydrocarbons (PCB) and poly-aromatic compounds (PAH) (Vazquez-Duhalt, 1989). Therefore, WLO requires proper management and treatment in order to avoid damage to the human health and environment. Estimates of the total worldwide demand for lubricants show that every year about 35 million tons of fresh lubricants are placed on the market, 50% of which generate WLO (Monier & Labouze, 2001; Statistic, 2015).

Regeneration represents the most favorable recovery option, supporting the circular economy, where resources are re-used or re-introduced in production processes, instead of being discarded after use (Ghisellini et al., 2016). It is based on the removal of contaminants and the recovery of undamaged hydrocarbons to yield base oils for the manufacture of lubricating products.

The earliest regeneration technology was the acid/clay process. Good quality base stocks were produced by this simple method. However, problems associated with secondary pollution as the disposal of spent clay and acid sludge besides the low yield (45–65%) discouraged this technology, which was banned since the early 1970s (Awaja & Pavel, 2006).

Other processes based on environmentally friendly technologies have been developed since then, namely solvent extraction has deserved special attention (Lukić et al., 2006). Propane is the most widely used solvent in industry to treat WLO (Rincón et al., 2003). A major drawback of this solvent is the flammability, increasing the risk of fire and explosion. In addition, the need to operate at high pressures makes the construction and safe operation of plants expensive (Awaja & Pavel, 2006). The extraction-flocculation process under mild operation conditions has been studied in the literature (Martins, 1997). Several organic solvents including alcohols, ketones and hydrocarbons are pointed as alternative solvents for the extraction process.

The technical performance of regeneration processes can be seriously affected by the chemical composition of the feedstock. Changes in engineering and lubricant technology has resulted in a significantly reduced volume of WLO produced. However, it has also led to the production of a great variety of synthetic oils which can affect the regeneration process. Pinheiro et al. (2017) showed that the characteristics of WLO can significantly vary for some key lubricant properties such as viscosity and saponification number. Although several studies based on the extraction-flocculation process have been described in the literature, none of them has addressed the effect of the feedstock in the process. Thus, the main objective of this study is to assess the influence of the WLO type in the efficiency of regeneration by solvent extraction-flocculation. The results of this work may be useful for their segregation according to the type to increase the performance of the extraction.

2 MATERIALS AND METHODS

2.1 Waste oil samples

In this work, two WLO samples with different physicochemical compositions were studied. The criterion for selecting the samples was based on the coagulation test previously described by Pinheiro et al. (2016). This is an empirical method often used in the regeneration industry to evaluate potential adverse reaction of WLO during alkaline pretreatment phase. Potassium hydroxide (KOH) is added in excess to ensure the detection of these reactions and avoid blocking problems in the regeneration plant. Differences in composition of oil lead to different outcomes of the coagulation test. The oil is categorized as class A if the oil matrix remains unaffected, B if a precipitate is formed or C if a gel-type oil is produced. Thus, two oil samples belonging to class A and C were collected, pretreated to remove water and sediments, and stored at room temperature in the dark to maintain their integrity until analysis. Further studies will be performed using a class B oil.

2.2 Solvent selection

The solvent selected was 1-butanol and this choice was based on: i) the Burrel's classification, which indicates that alcohols have a high removal capacity compared with ketones, with moderate removal capacity, and with hydrocarbons that have low removal capacity due to their ability to form hydrogen bonds (Burrell, 1968). Molecular weight of alcohols and ketones strongly influences the extraction process. Solvents with low molecular weights (<3 carbons) may not dissolve base oils and solvents with longer chain length (>5 carbons) might hinder the flocculation of impurities (Kamal & Khan, 2009). ii) The literature background provides detailed information on the solvents selectivity to extract base oil from the waste oil. Regarding single solvents, several researchers reported that among a great variety of alcohols and ketones, 1-butanol promotes the highest removal of impurities (Mohammed et al., 2013). iii) The GlaxoSmithKline (GSK) solvent selection guide ranked 1-butanol as one of the most 'greener' solvents with less environmental, health and safety issues when compared with others. Indeed, 1-butanol has a boiling point of 118°C that allows its recovery by distillation (Henderson et al., 2011).

2.3 Solvent extraction procedure

The extraction procedure was adapted from the centrifugal tube technique described by Reis & Jeronimo (1988). Previously weighed centrifugal tubes were filled with a mixture of 20 g of waste oil sample (W_{oil}) and solvent at a specified ratio. The samples were stirred by magnetic stirring at 500 rpm and during a specified time to ensure adequate mixing. The tubes were centrifuged at 4000 rpm for 20 min. After centrifugation, the wet sludge phase (additive, impurities and carbonaceous particles) was separated from the mixture solvent/oil. The solvent was recovered by vacuum distillation in a rotary evaporator. Distillation flasks were previously weighed, and the mass of the recovered oil and solvent was calculated by difference. The extraction yield was determined by Eq. (1) as the mass of recovered base oil, $W_{base\ oil}$, expressed in grams, separated from the initial waste oil, $W_{waste\ oil}$.

$$Yield\ (\%) = \frac{W_{base\ oil}}{W_{waste\ oil}} \times 100 \qquad (1)$$

The tubes were then placed in the oven for 1 h at 80°C to evaporate any remaining extraction solvent. The wet sludge was redispersed by addition of n-hexane followed by isopropanol (20/80 v/v%) added to separate the sludge again, using a solvent to sludge + oil ratio of 31/1. The addition of isopropanol immediately produces large flakes.

Finally, the washing liquid was discarded, and the tubes were introduced into the oven and dried for 1 h at 80°C ($W_{dry\ sludge}$). After cooling in a desiccator, the tubes were weighed and the percentage of sludge removal, *PSR*, was calculated according to Eq. (2).

$$PSR\ (\%) = \frac{W_{dry\ sludge}}{W_{waste\ oil}} \times 100 \qquad (2)$$

2.4 *Critical Clarifying Ratio (CCR)*

The procedure used for determining the critical clarifying ratio (CCR) was adapted from Reis & Jeronimo (1988). Centrifugal glass tubes with round bottom of 3.2 cm internal diameter and 10 cm height were filled with the specified proportion of waste oil and solvent and then centrifuged. To determine if the solution is clear, an arbitrary visual definition was accepted. The tube is checked at 20 cm distance from a 25 W cold light source. If a blue horizontal line is observed through the thicker part of the tube, the solution is considered clarified.

2.5 *Elemental content*

The elemental content of P, Ca, Cl, Zn, Fe and S was determined by Energy Dispersive X-Ray Fluorescence (EDXRF) using a Nex CG Rigaku spectrometer. About 4 g of lubricant was weighed to a sample cell Chemplex 1330 with a Polypropylene TF-240 film. Since no sample preparation was required, the cell was placed on the equipment and the analysis was immediately performed.

3 RESULTS AND DISCUSSION

3.1 *Waste lubricant oil characteristics*

Table 1 shows the physicochemical characteristics of the two WLO samples studied. While densities are similar, differences in viscosity are considerable, which might be related to the

Table 1. Waste oil properties.

Parameter	Test method/Analytical technique	Waste oil samples	
		Class A	Class C
Density at 15°C (kg/m^3)	ASTM D1298	870.0	876.2
Viscosity at 40°C (cSt)	ASTM D445	85.0	66.6
Total Acid number (mg KOH/g)	ASTM D664	0.48	1.87
Saponification number (mg KOH/g)	ASTM D94	7.4	19.6
Elemental content (ppm)	EDXRF		
P		761	717
Ca		1990	1620
Cl		123	202
Zn		1020	980
Fe		142	212
S (%)		1.02	1.13

type of base oil in each case. Indeed, the saponification number (a measure of the amount of saponifiable matter) indicates that the class C sample must have esters in their composition, added during the formulation of the lubricant to increase its performance.

3.2 Effect of contact time

Preliminary experiments were performed to determine the contact time required for the different waste oil systems to achieve equilibrium conditions. The contact time should allow the complete dissolution of hydrocarbons in the solvent and the segregation of non-polar macromolecules (degraded additives, carbonaceous particles and polymeric compounds) from the mixture by exerting an anti-solvent effect. The particles aggregate to sizes large enough that precipitate from the liquid phase by sedimentation.

Figure 1 shows that the extraction yield increases as the contact time increase, but becomes almost constant very fast at about 5–10 min, for both types of oils. To ensure equilibrium conditions, the subsequent experiments were performed with a contact time of 20 min. The optimum contact time is in good agreement with those found in the literature for 1-butanol (Mohammed et al., 2013).

3.3 Effect of solvent/oil ratio

High yields in the extraction process do not necessarily mean better efficiency of the solvent. On the contrary, high sludge removal results in an improved quality of the regenerated oil, which will reduce downstream stages such as vacuum distillation and finishing phases by activated clay adsorption or hydrofinishing.

Figure 2(a)-(b) illustrates the evolution of yield and sludge removal with the increase of the solvent/oil weight ratio. The trend found for the tested waste oils is similar, increasing both the yield and the sludge removal with the rise of the solvent/oil ratio up to a point where it stabilizes. This is because at the smaller ratios the saturation of the solvent occurs, which

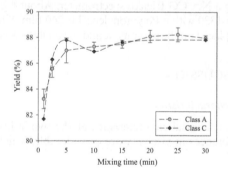

Figure 1. Effect of the contact time on the extraction yield for two types of waste oils. (Solvent/oil weight ratio 3/1, temperature 25°C, mixing speed 500 rpm).

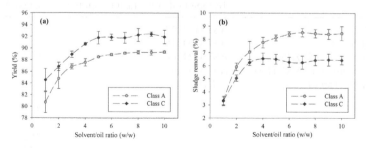

Figure 2. Effect of solvent to oil ratio on the extraction a) yield and b) sludge removal, for two types of waste oils. (Temperature 25°C, mixing speed 500 rpm and contact time 20 min).

136

is not capable to dissolve all base oil present. Increasing the solvent/oil ratio, the amount of base oil dissolved grows up to a ratio at which the solvent dissolution power is exhausted. Consequently, the increase of solvent/oil ratio will not continuously improve the extraction yield (Rincón et al., 2005). The efficiency of the extraction depends on the type of WLO. At equilibrium, an extraction yield of 89 and 92% was obtained for class A and C, respectively. Similarly for the sludge removal, 8.5 and 6.5%, was achieved for samples A and C, respectively. These variations are related to the characteristics of the WLO components. The class C base oil must be less soluble in the solvent, decreasing the antisolvent effect exerted on the impurities, or the contaminants are more soluble in the solvent, not precipitating.

3.4 Effect of KOH dosage

WLO is considered a colloidal system comprising a dispersion medium (the base oil) and a dispersive phase (the additives), which may accumulate contaminants such as oxidation products, metals, mechanical impurities and water. The addition of KOH dissolved in the solvent results in the neutralization of the electrostatic repulsion and a destabilization of the particles, that enhances the flocculation and increases the removal of impurities from the waste oil, as shown in Figure 3. The impact of the KOH addition on the extraction efficiency is more significant on the class A oil, increasing by 32.9% the sludge removal. The class C oil experienced a sludge removal increase of 13.1%. This can be explained by the fact that class C oil might be composed by esters, which interact with KOH through saponification reactions, decreasing the flocculation effect of the hydroxide. The increase on the KOH dosage from 1 to 6 g/L did not reveal a significant improvement on the sludge removal for both oils.

3.5 Critical clarifying ratio

The CCR is defined as the minimum solvent to oil weight ratio necessary to destabilize the dispersion contained in a column, producing aggregates that settle under centrifugal conditions equivalent to 24 h of gravity action and leaving a clear supernatant.

The results presented in Table 2 show that the CCR using the same solvent can be very different for different types of WLO. Class A sample exhibit a CCR of 1.4, comparable with the

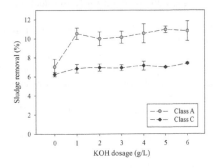

Figure 3. Effect of the addition of KOH on the sludge removal. (Temperature 25°C, mixing speed 500 rpm, contact time 20 min and solvent/oil weight ratio 3/1).

Table 2. Critical clarifying ratio for different types of waste oils.

Solvent	Critical clarifying ratio		
	Class A	Class C	Literature*
1-Butanol	1.4	2.1	1.2

*Reis & Jerónimo (1988).

137

1.2 determined by Reis & Jerónimo (1988) using 1-butanol, whereas class C has a higher CCR of 2.1. This result could be related to the polarity of the base oil that can be less miscible in the solvent, or the impurities present in this waste oil might have higher affinity with the solvent.

4 CONCLUSIONS

The extraction-flocculation regeneration process can be significantly affected by the waste oil type. The response variables such as the yield and sludge removal differ depending on the type of waste oil when critical parameters are evaluated. The operation conditions of the treatment can be adjusted according to the physicochemical characteristics of the oils, saving energy and materials. The addition of KOH to the alcohol increases the ability of the solvent to remove impurities from waste oils. However, the increase on the removal capacity is strongly dependent on the characteristics of the waste oil.

ACKNOWLEDGEMENTS

The authors gratefully acknowledge the financial support of SOGILUB—Sociedade de Gestão Integrada de Óleos Lubrificantes Usados, Lda.

REFERENCES

Awaja, F. & Pavel, D. 2006. Design Aspects of Used Lubricating Oil-Refining. Amsterdam: Elsevier.
Reis, M.A. & Jerónimo, M.S. 1988. Waste lubricating oil rerefining by extraction-flocculation. 1. A scientific basis to design efficient solvents. *Ind. Eng. Chem. Res.* 27: 1222–1228.
Burrell, H. 1968. The challenge of Solubility Parameter Concept. *J. Paint Technol.* 40: 197–208.
Ghisellini, P., Cialani, C. & Ulgiati, S. 2016. A review on circular economy: the expected transition to a balanced interplay of environmental and economic systems. *J. Clean. Prod.* 114: 11–32.
Henderson, R.K., Jiménez-González, C., Constable, D.J.C., Alston, S.R., Inglis, G.G.A., Fisher, G., Sherwood, J., Binks, S.P., Curzons, A.D., Perry, D.A. & Stefaniak, M. 2011. Expanding GSK's solvent selection guide – embedding sustainability into solvent selection starting at medicinal chemistry. *Green Chem.* 13: 854–862.
Kamal, A. & Khan, F. 2009. Effect of Extraction and Adsorption on Re-refining of Used Lubricating Oil. *Oil Gas Sci. Technol. - Rev. l'IFP* 64: 191–197.
Lukić, J., Orlović, A., Spiteller, M., Jovanović, J. & Skala, D. 2006. Re-refining of waste mineral insulating oil by extraction with N-methyl-2-pyrrolidone. *Sep. Purif. Technol.* 51: 150–156.
Martins, J.P. 1997. The Extraction-Flocculation Re-refining Lubricating Oil Process Using Ternary Organic Solvents. *Ind. Eng. Chem. Res.* 36: 3854–3858.
Mohammed, R.R., Ibrahim, I. a R., Taha, A.H. & McKay, G. 2013. Waste lubricating oil treatment by extraction and adsorption. *Chem. Eng. J.* 220: 343–351.
Monier, V. & Labouze, E. 2001. *Critical Review of Existing Studies and Life Cycle Analysis on the Regeneration and Incineration of Waste Oils.*
Pinheiro, C.T, Ascensão, V.R, Cardoso, C.M, Quina, M.J. & Gando-Ferreira, L.M. 2017. An overview of waste lubricant oil management system: Physicochemical characterization contribution for its improvement. *J. Clean. Prod.* 150: 301–308.
Pinheiro, C.T., Ascensão, V.R., Cardoso, M.M., Cardoso, C.M., Quina, M.J. & Gando-Ferreira, L.M. 2016. Physicochemical Characterization of Waste Lubricant Oils For Maximizing their Regeneration Potential. In *6th International Conference on Engineering for Waste and Biomass Valorisation, Albi, 23–26 May 2016.* pp. 374–358.
Rincón, J., Cañizares, P. & García, M.T. 2005. Waste oil recycling using mixtures of polar solvents. *Ind. Eng. Chem. Res.* 44: 7854–7859.
Rincón, J., Canizares, P., García, M.T. & Gracia, I. 2003. Regeneration of Used Lubricant Oil by Propane Extraction. *Ind. Eng. Chem. Res.* 42: 4867–4873.
Statistic 2015. *Global demand for lubricants from 2000 to 2015.* https://www.statista.com/statistics/411616/lubricants-demand-worldwide/ (accessed 3.12.17).
Vazquez-Duhalt, R. 1989. Environmental impact of used motor oil. *Sci. Total Environ.* 79: 1–23.

Diethylketone and Cd pilot-scale biosorption by a biofilm supported on vermiculite

F. Costa & T. Tavares
Centre of Biological Engineering, University of Minho, Braga, Portugal

ABSTRACT: The pilot-scale assays conducted with a joint bacteria-clay system—*Streptococcus equisimilis* (*S. equisimilis*) biofilm supported on vermiculite—were conducted in a close loop reactor aiming the decontamination of large volumes of binary aqueous solutions containing diethylketone and Cd^{2+}. The joint system employed proved to be able to simultaneously biodegrade diethylketone and biosorb Cd^{2+}: the removal (biodegradation and/or biosorption) percentage and the uptake increase through time, even with the replacement of the initial solution by new solutions.

1 INTRODUCTION

Contamination of aquatic systems has become a serious problem to society due to the ever-increasing use of modern agricultural practices, the rapid and growing industrialization, the illegal discharges of wastewater and the ruthless exploitation of natural resources (Bhuvaneshwari and Sivasubramanian 2014). These practices introduce dangerous and persistent compounds like organic solvents and metallic compounds in the environment, which tend to accumulate in and deteriorate the different environmental matrices (Murthy et al. 2012). On a small scale, both organic and inorganic pollutants can safely be separated from any water body and thus not affecting the aquatic communities. However, on a larger scale, the aquatic systems are unable, in due time, to successfully decontaminate the contaminated water bodies (Bhuvaneshwari and Sivasubramanian 2014). In this context, the understanding of the contaminants migration between the different phases and of the parameters affecting the decontamination processes (biodegradation, biosorption and sorption for example) and the development of techniques able to efficiently and effectively decontaminate wastewater are of utmost importance. In the past few years, several studies have proved that biological processes such as biodegradation or biosorption present various advantages over the traditional techniques and methods. These biological processes present an eco-friendly character since they do not produce nitrogen oxides and solid wastes, present high efficiency and reduced maintenance and operational costs (Costa et al. 2012).

Volatile organic compounds and metals are expected to be among the high variety of contaminants present in industrial and/or domestic effluents (Azeez et al. 2013). Several studies concerning the decontamination of such effluents and its optimization have been conducted and became of major importance and relevance not only for environment rehabilitation, but also aiming its economics sustainability.

In this work, a joint system that combines the properties of bacteria and clays was used to improve the decontaminations of binary aqueous solutions, composed by diethylketone and cadmium (Cd^{2+}). The scarcity of studies related to the decontamination of multi-component solutions highlights the importance of the work herein presented.

2 MATERIALS AND METHODS

2.1 *Support, microorganisms and chemicals*

Vermiculite was purchased from Sigma-Aldrich and it was used as a support for the bacterial biofilm establishment and development

The bacterium used was *Streptococcus equisimilis* and it was acquired from the Spanish Type Culture Collection, from the University of Valencia (reference CECT 926).

Diethylketone was acquired from Acros Organics (98% pure) and diluted in sterilized distilled water. Individual stock solutions of 1 g/L of cadmium ($CdSO_4 \cdot 8/3H_2O$, Riedel-de-Haën) were prepared by dissolving a correctly weighed amount of metal in sterilized distilled and deionised water. Previous assays performed at lab-scale in batch mode to assess the effect of the concentration of diethylketone, and Cd^{2+}, on singular and binary solutions, on the growth of *S. equisimilis* and on the sorption capacity of *S. equisimilis*, vermiculite, and a *S. equisimilis* biofilm supported on vermiculite are reported by our team in Costa et al. (2014) and Quintelas et al. (2011, 2012).

2.2 *Bioremoval of diethylketone and Cd^{2+} by a S. equisimilis biofilm supported in vermiculite—pilot scale experiments*

A compact polycarbonate acrylic column of 22.7 L with an internal diameter of 17 cm and a total height of 100 cm was used as a bioreactor in the pilot scale experiments. A maximum packing fraction of 1/3 of the bioreactor was filled with vermiculite (700 g). A 2 L Erlenmeyer flask containing 1 L of Brain Heart Infusion culture medium, previously sterilized at 121°C for 20 minutes was inoculated with *S. equisimilis* and incubated for 24 hours at 37°C and 150 rpm in an orbital shaker. The Erlenmeyer flask was capped with a cotton stopper in order to permit passive aeration. The inoculum culture was posteriorly transferred to the bioreactor setup and was pumped upwards at a flow rate of 250 mL/minute, during five days with total recirculation, in order to allow the biomass to attach to the vermiculite and form a biofilm. After the biofilm development, the bed was washed out and a 40 L solution containing 7.5 g/L of diethylketone and 100 mg/L of Cd^{2+} (S_1) was continuously pumped upwards in close loop through the bioreactor with a constant flow rate of 25 mL/minute. The first solution (S_1) was posteriorly replaced by new solutions (S_2 and S_3) with identical composition.

At pre-established time intervals, samples of the effluent (8 mL) were taken (Collection point, Figure 1), centrifuged at 1300 rpm for 10 minutes and the aqueous phase was analysed by Inductively Coupled Plasma Optical Emission Spectrometry (ICP-OES) and by Gas Chromatography Mass Spectrometry (GC-MS) and in order to determine respectively the concentration of Cd^{2+} and

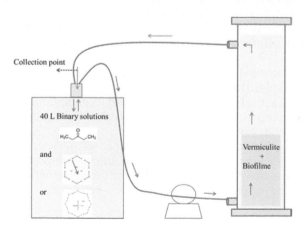

Figure 1.   Schematic set-up for the bioreactor column.

diethylketone through time. At the end, the bioreactor was washed out and samples of the effluent and of the support were inoculated in Petri plates with Brain Heart Infusion culture medium, in order to access about the metabolic activity of the bacteria. The pH was also monitored.

2.3 *Analytical procedures: ICP, GC, FTIR, XRD AND SEM*

An ICP-OES (Optima 8000, PerkinElmer) was used to determine the concentration of Cd^{2+}. All calibration solutions were prepared from a Cd^{2+} stock solution with a concentration of 1 g/L. All the samples were acidified with concentrated nitric acid (HNO_3 69%) and filtered before being analysed. A GC-MS Varian 4000, equipped with a flame ionization detector (FID), mass spectrometry (MS) and a ZB-WAXplus column (30 m × 0.53 mm × 1.0 μm) was used to determine the concentration of diethylketone in samples.

The functional surface groups implicated in the bioremoval (biodegradation or biosorption) of diethylketone and Cd^{2+} were determined using a Fourier Transform Infrared Spectrometer (FTIR BOMEM MB 104). The X-ray powder diffraction (XRD) analyses were performed using a Philips PW1710 diffractometer. Scans were taken at room temperature in a 2θ range between 5° and 60°, using CuK α radiation. Scanning Electron Microscopy (SEM) observations of vermiculite with and without previous contact with diethylketone and Cd^{2+}, with and without the biofilm, were performed on Leica Cambridge S360, in order to observe any morphological change on the sorbents properties.

2.4 *Breakthrough curves modelling*

In the present study, the Adams–Bohart, the Yoon and Nelson models and the Wolborska models were used to predict the breakthrough curves (Costa et al., 2017).

3 RESULTS AND DISCUSSION

In Figure 2 is possible to observe that the removal (biodegradation and/or biosorption) of diethylketone and Cd^{2+} is continuous and tends to increase through time and after the replacement of the first solution (S_1) by new solutions (S_2 and S_3).

This behaviour may be caused by the microbial growth, that in addition to increasing the substrate consumption it also increases the number of active sites of the biofilm, used for Cd^{2+} sorption, hampering the saturation of active sites present either on the biomass and vermiculite surface. The maximum removal percentage and uptake values obtained for diethylketone and Cd^{2+} were respectively, 98.2% and 421.0 mg/g and 87.6% and 5.0 mg/g. In Figure 2 and Table 1 is possible to observe that the removal percentage achieved for diethylketone is higher than the removal percentage obtained for Cd^{2+}, which corresponds to removal of 36.23 mol/g

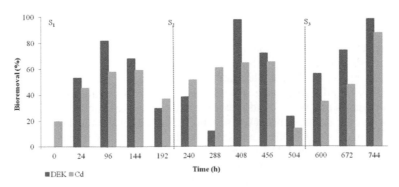

Figure 2. Bioremoval performance (%) obtained by a *S. equisimilis* biofilm supported on vermiculite for diethylketone (7.5 g/L) and Cd^{2+} (100 mg/L), at pilot plant scale.

Table 1. Biosorption performance of the bioreactor column for diethylketone (7.5 g/L) and for Cd^{2+} (100 mg/L).

Maximum bioremoval (%)		
Time (h)	DEK	Cd^{2+}
S_1 (0 h–192 h)	81.83	59.47
S_2 (192 h–504 h)	97.85	65.27
S_3 (504 h–744 h)	98.24	87.64

of diethylketone and 0.56 mol/g of Cd^{2+}. The results obtained may be explained since diethylketone can either be entrapped by the functional groups present on the biofilm and on the vermiculite surface or it can also suffer biodegradation by the microbial culture. The pH was found to freely range between 6 and 7 since the beginning of the experiment till its end, which according to the literature and the results obtained previously, is within the optimal range for Cd^{2+} sorption by vermiculite and is close to the optimum pH for the growth of *S. equisimilis* (7.4). Previous experiments conducted with different initial values of pH (pH 3, 5.5 and 8) and vermiculite revealed that the pH of the system has the capacity to return and maintain to pH values close to pH 7.This behaviour represents an additional significative and important advantage since no adjustment of pH is required.

The viability tests conducted at the end of the pilot scale assays revealed that *S. equisimilis* biofilm presents biological activity, which is an important advantage in the treatment of waste-water using microorganisms, since microbial culture resistance allows a continuous treatment of the contaminated solutions trough biological active and/or inactive mechanisms.

The solid residues generated by the term of the pilot-scale experiments may be sent to incineration, which ashes can be subsequently used or they can be reused in new decontamination systems. The latter hypothesis is achievable considering the results obtained in the cell viability tests. It is important to highlight that the purpose of this study and of this type of treatment, as well as its mode of use, consists in the decontaminations of aqueous solutions, such as real effluents, is continuous and open system, thus minimizing or even avoiding the production of solid waste.

3.1 *Breakthrough curves modelling*

The fitting parameters for the breakthrough curves as well as the predicted and experimental breakthrough curves are presented in Table 2 and in Figure 3.

The results obtained for diethylketone are appropriately described by any of the models employed ($R^2 < 0.75$), whereas for Cd^{2+} the experimental results are best fitted by the Adams-Bohart model ($R^2 > 0.80$).

3.2 *FTIR, XRD and SEM analysis*

The FTIR spectra obtained for the different samples analysed in the range of 500 cm^{-1} to 4000 cm^{-1} are shown in Figure 4. Succinctly, the clay used, vermiculite, presented several sorption peaks, reflecting the complex nature of the mineral clays.

Some of these band signals were also acknowledged by other authors (Eboigboduin and Biggs 2008) on clays and were found to correspond to surface functional groups responsible for the sorption of toxic substances (Volesky 2007). Samples of vermiculite exposed to diethyl-ketone and samples of the biofilm exposed to diethylketone and Cd^{2+} reveal the existence of several modifications on the FTIR spectra: either on the intensity, on the shape of the peaks and on the disappearance and/or formation of new peaks. After exposure to diethylketone, bands at 2400 cm^{-1} and 1400 cm^{-1} undergo important changes. The intensity of the bands detected at 3500 cm^{-1}, 1650 cm^{-1}, 1000 cm^{-1} and 675 cm^{-1}, was found to decrease considerably. All these modifications may be due to the interaction and participation of the functional

Table 2. Breakthrough parameters obtained for the pilot-scale experiments.

| | Adams-Bohart | | | Wolborska | | | Yoon and Nelson | |
	k_{AB} (L/mg h)	N_0 (mg/L)	R^2	β_a (h^{-1})	R^2	k_{YN} (h^{-1})	τ (h)	R^2
Cd^{2+}	9.953e-5	1.906	0.81	−1.446e-3	0.78	−3.189e-3	160	0.74

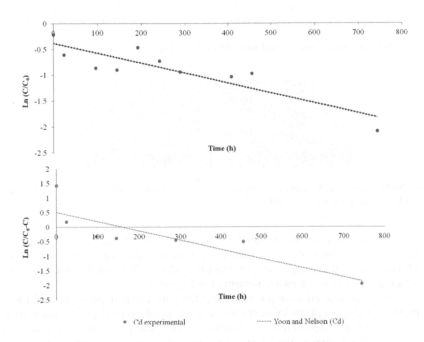

Figure 3. Predicted and experimental breakthrough curves for Cd^{2+} at pilot-scale.

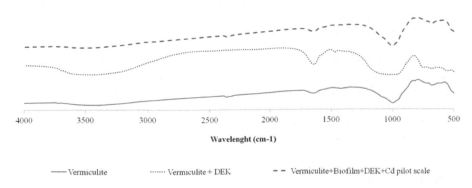

Figure 4. FTIR spectra of different samples: vermiculite unloaded, vermiculite exposed to diethylketone, *S. equisimilis* biofilm supported on vermiculite, loaded with diethylketone and Cd^{2+}.

groups present both in the vermiculite and in the biofilm surface with the functional groups (C = O) of diethylketone. Bands at 3500 cm^{-1} and at 2300 cm^{-1} were found to disappear on the *S. equisimilis* biofilm supported into vermiculite and exposed to diethylketone and to Cd^{2+} samples. According to Volesky (2007) the major functional groups accountable for biosorption processes are phosphodiester, carboxyl, amide, hydroxyl, phosphonate and carbonyl groups.

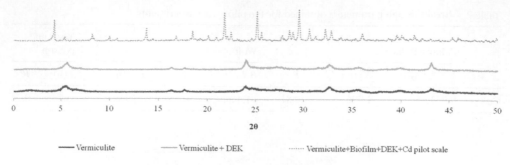

—— Vermiculite   —— Vermiculite + DEK   ······· Vermiculite+Biofilm+DEK+Cd pilot scale

Figure 5. XRD patterns of original and recovered vermiculite.

Figure 6. SEM images of *S. equisimilis* supported on vermiculite and exposed to diethylketone and Cd^2 for 456 hours (amplification of 3000x).

Some of these functional are present on the *Streptococcus* sp surface and are responsible for the biodegradation and biosorption of toxic compounds like diethylketone (Costa et al. 2014).

The XRD patterns of different samples were recorded at 2θ range between 5°C and 60°C and some representative patterns are presented in Figure 5.

The unloaded and loaded samples of vermiculite with diethylketone showed the characteristic pattern of clays, with no evident change in the position of the diffraction peaks after exposure to the ketone used. This resemblance between the two diffractograms suggests a similar crystallinity of vermiculite before and after exposure to the contaminants, revealing that no significant structural modification in the clay occurred. Nonetheless, the samples containing the biofilm supported into vermiculite, exposed simultaneously to diethylketone and Cd^{2+} revealed numerous changes, not only in their intensity but also on the position of the diffraction peaks. These changes are an evidence of the extensive actions on the clay structure.

SEM analyses confirm (a) that vermiculite exposure (without biofilm) to diethylketone and Cd^{2+} makes vermiculite surface more glazed and polished, with broken and worn leaves, (b) the presence of a well-developed bacterial biofilm supported on vermiculite (Figure 6) and (c) the presence of numerous white incrustations on the surface of the sorbent, which correspond to the binding of the metal ions to the surface of biomass and are a evidence of the biosorption of the metal by the biofilm (Sharma et al., 2008).

4 CONCLUSIONS

The joint system herein presented has proven to be capable to decontaminate efficiently and continuously, aqueous solutions containing high concentrations of diethylketone and Cd^{2+} without adjusting the pH. Both the bioremoval percentage and the uptake values increase through time, even after replacing the initial solution by new solutions. The results achieved for Cd^{2+} were found to be best described by the Adams-Bohart model. The FTIR and SEM analyses proved the existence of a well-established biofilm, with functional groups recognized as responsible for the sorption of toxic compounds. The XRD analyses demonstrated that the exposure of the support, with or without the biofilm, to both contaminants alters significantly the structure of vermiculite. It is therefore possible to infer that the developed joint system presents very major advantages concerning the simultaneous treatment of binary solutions.

ACKNOWLEDGEMENTS

This study was supported by the Portuguese Foundation for Science and Technology (FCT) under the scope of the research project PTDC/AAG-TEC/5269/2014, the strategic funding of UID/BIO/04469/2013 unit and COMPETE 2020 (POCI-01-0145-FEDER-006684) and Bio-TecNorte operation (NORTE-01-0145-FEDER-000004) funded by the European Regional Development Fund under the scope of Norte2020—Programa Operacional Regional do Norte. Filomena Costa thanks FCT for a PhD Grant (SFRH/BD/77666/2011).

REFERENCES

Azeez, L. Adeoye, M.D. Lawa, A.T. Idris, Z.A. Majolagbe, T.A. Agbaogun, B.K.O. Olaogur, M.A. 2013. Assessment of volatile organic compounds and heavy metals concentrations in some Nigerian-made cosmetics. *Analytical Chemistry: An Indian Journal* 12: 443–483.

Bhuvaneshwari, S. Sivasubramanian, V. 2014. Equilibrium, Kinetics, and Breakthrough Studies for Adsorption of Cr(VI) on Chitosan. *Chemical Engineering Communications* 201: 834–854.

Costa, F. Quintelas, C. Tavares, T. 2012. Kinetics of biodegradation of diethylketone by *Arthrobacter viscosus*. *Biodegradation* 23: 81–92.

Costa, F. Quintelas, C. Tavares, T. 2014. An approach to the metabolic degradation of diethylketone (DEK) by *Streptococcus equisimilis*: Effect of DEK on the growth, biodegradation kinetics and efficiency. *Ecological Engineering* 70: 183–188.

Costa, F. Tavares, T. 2017. Sorption studies of diethylketone in the presence of Al^{3+}, Cd^{2+}, Ni^{2+} and Mn^{2+}, from lab-scale to pilot scale. *Environmental Technology*. DOI: 10.1080/09593330.2016.1278462.

Eboigboduin, K.E. Biggs, C.A. 2008. Characterization of the extracellular polymeric substances produced by *Escheria coli* using infrared spectroscopic, proteomic and aggregation studies. *Biomacromolecules* 9: 686–695.

Murthy, S. Bali, G. Sarangi, S.K. 2012. Biosorption of Lead by *Bacillus cereus* Isolated from Industrial Effluents. *British Biotechnology Journal* 2: 73–84.

Quintelas, C. Costa, F. Tavares, T. 2012. Bioremoval of diethylketone by the synergistic combination of microorganisms and clays: Uptake, removal and kinetic studies. *Environmental Science Pollution Research* 20: 1374–1383.

Quintelas, C. Figueiredo, H. Tavares 2011. The effect of clay treatment on remediation of diethylketone contaminated wastewater: Uptake, equilibrium and kinetic studies. *Journal of Hazardous Materials* 186. 1241–1248.

Sharma, M., Kaushik, A., Somvir, B.K., Kamra, A. 2008. Sequestration of chromium by exopolysaccharides. *Journal of Hazardous Materials* 157: 315–318.

Volesky, B. 2007. Biosorption and me. *Water Research* 41: 4017–4029.

Removal of wastewater treatment sludge by incineration

S. Dursun, Z.C. Ayturan & G. Dinc
Environmental Engineering Department, Selcuk University, Konya, Turkey

ABSTRACT: Incineration is one of the mostly used techniques for the disposal of solid wastes applied in more than 40 European countries. There are several methods used for this purpose such as fluidized bed and rotary kiln incinerators. Several types of wastes can be reduced with incineration techniques. Incineration is a significant option disposing wastewater treatment sludge. However, specific features of sludge should be considered. Most of the treatment sludge originating from municipal wastewater treatment plants includes 95–99% of water, pathogenic microorganisms, food and large amounts of organic matter. According to the content of sludge, residues and air emissions show some differences. In this study, mainly wastewater treatment sludge incineration options were investigated and compared with respect to their efficiencies and sustainability. Moreover, the end products and emissions after application of incineration options were explored and precautions or modifications that can be applied to the systems was offered.

1 INTRODUCTION

Sludge from wastewater treatment plants is produced by clarifier units after operation of physical, chemical and biological treatment. In direct proportion to the population growth, the amount of wastewater sludge produced by the treatment plants increases. Globally, almost 330 km3 wastewater is produced every year (Mateo-Sagasta et al. 2015). Moreover, according to the assumptions of Global Water Intelligence, 83 million tons of sewage sludge would be generated in 2017 (Global Water Intelligence 2012). Disposal of treatment sludge in huge amounts may be a concern for the countries which does not have enough space to store their wastes so they prefer using incineration technologies. In European union countries, almost 60 million tons of solid wastes are disposed with incineration processes (Saltabaş et al. 2009).

Wastewater sewage sludges includes several materials and sources such as organic materials, water, nitrogen and phosphorous and energy. Especially nutrient materials found in wastewater sludge can be recovered with different methods (Mateo-Sagasta et al. 2015). There are many methods used for sludge handling distinctively incineration processes. Because of the low solids content, pathogenic and toxic substances in sewage sludge, it is very difficult to dispose. Main purpose of the sludge handling is reduction of pathogenic materials, organic materials, bad smell and water content (Werther & Ogada 1999). Thickening, stabilization, dewatering, conditioning, drying and other further thermal disposal processes are the mostly known sludge handling mechanisms. Mostly used classic methods for sludge handling are stabilization, dewatering and conditioning (Werther & Ogada 1999). Especially for the municipal wastewater sludges stabilization is preferable technique but for the industrial sludges it is not suggested. Stabilization technique based on organic material conversion to gas phase and production of stabilized residue by the activity of bacteria in aerobic and anaerobic conditions. Furthermore, with the application of lime into the stabilization process sludge condition to destroy microorganisms and lower water content (Werther & Ogada 1999). In order to prepare sludge for further treatment, dewatering is one of the necessary process. It is achieved with thickening and mechanical dewatering units (Werther & Ogada 1999). However, for most of the countries, the amount of sludge which can be landfilled is restricted with laws and regulations. For example, the directive of EU 99/13/EC implied that the amount of sewage sludges produced in 2010 should be reduced to 75% of the amount produced in 1995. Usage of the wastewater treatment sludge in agriculture could be solution for reducing landfill amount but it is not enough. Moreover, the land requirement for the landfilling is another obstacle for some countries (Stasta et al. 2006). Thus, thermal treatment methods are in inevitable for wastewater sludge handling.

Thermal treatment methods used for sludge disposal may be investigated in three main topics such as combustion, pyrolysis and gasification. Within the scope of this paper only combustion processes and end products resulting from combustion will be reviewed. While combustion process is based on burning of the wastes in an environment with high amount of oxygen, pyrolysis is thermal disposal in totally oxygen free environment. Gasification method is also occurred in oxygen free environment but small amount of air is given to the system (Saltabaş et al. 2009). Main advantages of incineration are very large amounts of volume reduction, destruction of organic toxics, recovery of energy, reduction of odor. However, after incineration process 30% of the solids stays as an ash so it is totally not a best option for sludge disposal (Fytili 2008). In this paper, main combustion processes such as multiple hearth incinerators, fluidized bed incinerators and rotary kiln incinerators was investigated and end products of these techniques was researched. Also, some suggestions were given for reduction of unwanted products from these techniques.

2 MAIN FEATURES OF SEWAGE SLUDGE

Sewage sludge produced by treatment plants mainly arisen by primary, secondary and tertiary treatment. Primary treatment includes physical or chemical operation, secondary treatment includes biological operation and tertiary treatment includes biological or chemical operation additional to secondary treatment (Fytili & Zabaniotou 2008). Type of the treatment method represents the main ingredients of sewage sludge. Pathogens, high amounts of water, high biochemical oxygen demand and disturbing odor are the main features of untreated sewage sludge. Moreover, some essential nutrients such as nitrogen and phosphorous that can be used as a plant fertilizer are found in sewage sludge. Also, organic carbon content of the sludge could be used for conditioner to improve soil structure and it could be returned to the energy with incineration (Mateo-Sagasta et al. 2015). Typical sludge characteristics of untreated and digested primary sludge are given in Table 1.

In addition to these main parameters contained in sewage sludge, it also may include some inorganic and organic materials such as heavy metals and polyaromatic hydrocarbons (PAHs), polychlorinated biphenyl (PCBs), hormones, pharmaceuticals, nanoparticles, adsorbable organohalogens (AOX), pesticides, surfactants, and many others (Siebielska 2014, Kacprzak et al. 2017). Furthermore, viruses, bacteria and other parasitic microorganisms found in sewage sludge can give severe damage to the human, plant and animal life (Kacprzak et al. 2017).

Table 1. Typical sludge characteristics of untreated and digested primary sludge (Tchobanoglous et al. 2003, Mateo-Sagasta et al. 2015).

Parameter (% dry weight)	Untreated primary sludge		Digested primary sludge	
	Range	Typical	Range	Typical
Total Dry Solids	2–8	5	6–12	10
Volatile Solids	60–80	65	30–60	40
N	1.5–4	2.5	1.6–6	3
P_2O_5	0.8–2.8	1.6	1.5–4	2.5
K_2O	0–1	0.4	0–3	1
pH	5–8	6	6.5–7.5	6

3 INCINERATION METHODS FOR WASTEWATER TREATMENT SLUDGE

There are three options which is possible to use as a combustion process namely multiple hearth incinerators, fluidized bed incinerators and rotary kiln incinerators. All of them simply are based on the burning of wastes in high temperatures with the presence of high amounts of oxygen. Especially in Europe, it is considered as very attractive waste disposal technique (Fytili 2008).

3.1 Multiple Hearth Incinerators (MHI)

Multiple hearth incinerator is one of the mostly used incineration device with the presence of seven to nine refractory hearths. It includes vertical steel furnace shell in the form of cylinder with diameter of 6–8 m long. In the center of the furnace vertical rotating shaft which is horizontally framed with rake linkage, made up from heat-resistant cast iron is placed. In each hearth, there are openings for transfer of waste (Turovskiy & Mathai 2006). Sludge is transported to the feeding point with conveyors and then moves into the upperpart of the first hearth of the furnace. With the help of rakes, sludge passes through one hearth to another from the openings and reached to the last heart. During this travel, combustion occurs and all sludge burned with the help of diesel burner or fuel gas and%50 percent of excess air supply. The vertical shaft and rakes are cooled by the blower system. The first hearth of the furnace works as a drier unit for sewage sludge. Sludge is burned in temperatures between 700°C and 900°C in middle part of the furnace and in the lower part of it ashes produced in middle part are cooled to 300°C. The reverse movement of gases and sludge in the furnace helps the combustion gases to cool. The gases produced by combustion processes was collected by another pipe and send to the control devices (Turovskiy & Mathai 2006).

The multiple hearth incinerator has one main advantage which is capability of usage of internal energy so the hot flue gas may directly contact with the sludge (Werther & Ogada 1999). Besides, several types and featured sludge can be disposed owing to the ease of operation. Large area requirement and high capital cost of the incinerator are the major disadvantages (Turovskiy & Mathai 2006).

3.2 Fluidized Bed Incinerators (FBI)

Fluidized bed incinerators are used in several types of industries for drying and combustion purposes. This incinerator includes vertical steel cylinder with refractory material, top of it has dome shape and low section has conical shape that supplies air distribution to the system. On the screen formation heat resistant diameter of 0.6 to 2.5 mm sand is placed to a depth of 0.8 to 1 m. In this system, fluidized bed should be turbulent which is achieved by the blowing air from the distribution screen and sand particles look like boiling. Air is heated by the recycling unit with the other gases leaving to the system to 600–700°C (Turovskiy & Mathai 2006). Sludge is given to the system from inlet structure and mixed with the fluidized sand. Then, it is taken to the furnace chamber with temperatures of 900–1000°C. As a result, agglomeration of sludge breaks, moisture of it evaporates, volatile organics separates, carbonized residue is combusted and mineral fraction of the sludge is burned. This process takes very short time like 2 minutes because of the huge heat and mass transfer. Gas temperature in the outlet of the incinerator is very high due to the volatile farction of the sludge. Temperature of the exhaust gas is decreased with the help of cold air to lower degrees before given to the atmosphere (Turovskiy & Mathai 2006).

There are many advantages of fluidized bed incinerators. First one is the combustion at low temperatures and low excess air owing to mixing in the bed which causes turbulence and increases the surface area suitable for heat transfer. Residence time of the sludge is long enough to burn of all sludge and free board provides complete reduction of organic materials. Second advantage is that the large heat reservoir provides stabilization for the short-time changes in the water content of the sludge so hot inert bed material may prevent instant temperature changes. Third advantage is the finish of combustion in very short time because of the low fuel and carbon amount in the bed. The last one is the long-lasting usage periods of refractory because

Table 2. Comparison of incineration techniques for wastewater treatment sludge (Werther & Ogada 1999, Van Caneghem et al. 2012, Mininni et al. 2003).

Criteria	MHI	Examples	FBI	Examples	RKI	Examples
Ash and Emission Production	High	200–215 mg/m^3 NO$_x$ emissions	Low	90–150 mg/m^3 NO$_x$ emissions	Low	60–150 mg/m^3 NO$_x$ emissions
Capital Cost	High	Additional fuel requirement	Low	No additional control device	High	Secondary firing requirement
Operation Simplicity	Simple	Efficient internal energy usage	Simple	Handling capability of sudden changes	Complex	Consisting of varying temperature zones
Operation Duration	Long	Need to maintain drying	Short	2 seconds residence time	Long	6–8 hours residence time
Area Requirement	High	8 m in diameter	Low	Internal diameter of 0.59 m	High	6 m of length 1.2 m of internal diameter

fluidized bed materials may prevent thermal shock and damage to the refractory. Therefore, the maintenance cost of fluidized bed incinerators is low (Werther & Ogada 1999).

3.3 *Rotary Kiln Incinerators (RKI)*

Like other incineration systems, rotary kiln incinerator also is used in several types of wastes produced from many industries. There is a drum in the system with an inclined angle of 2–4° and the external furnace is placed in the lower end. The shape of the furnace is cylindrical lined with refractory material. Sludge entering to the system is first dried at the dry zone and then burned in the incineration zone. Ashes produced in the furnace are collected from the opening in external furnace and sent to the handling operation (Turovskiy & Mathai 2006). For burning of the organic matter and deodorizing of the gases is achieved in another combustion chamber called afterburner. In the drying zone the water content of the sludges is decreased to almost 30%. Temperatures in combustion zone are reached to 900–1000°C. In drying zone both the temperatures of ashes and exhaust gas is decreased to acceptable levels (Turovskiy & Mathai 2006).

Rotary kiln incinerators have several advantages such as producing low emission of exhaust gases, disposal of sludges with high ash and moisture content and installation possibility of rotating section in the open air. However, it has some disadvantages like high capital cost, huge weight, operation complexity (Turovskiy & Mathai 2006).

4 COMPARISION OF INCINERATION METHODS

In Table 2 comparison of three incineration method used for wastewater treatment sludge disposal is given with respect to ash production, emission production, cost, operation simplicity, operation duration and area requirement. According to the comparison table of incineration methods for sewage sludge, fluidized bed incinerators are the most efficient way to dispose sludge within three combustion techniques.

5 INCINERATION END PRODUCTS

There are two types of end products produced during combustion process such as ash residues in the form of flying and settling, and gaseous emissions. The main gas emissions originating from combustion are sulphur oxides, nitrogen oxides, and mineral acids. Serious health issues can be arisen due to these gases with respect to dosage and exposure time. SO$_2$ gas is one of the most problematic one because it may lead to tissue damage, respiratory system failure (Ozturk & Dursun 2016). Furthermore, if the combusted sludge contains huge amount of nitrogen, NO_x and N_2O emissions increases (Murakami et al. 2009). Moreover, the uncomplete burning of the volatile compounds may cause the formation of carbon monoxide and organic materials such as polychlorinated dibenzo dioxin and dibenzo furans (PCDD and PCDF). Their formation mainly occurs at the condition of poor incineration operation (Niessen 2002). However, for the sludge incineration process the formation of PCDD and PCDF depends mainly on the reactor type and the other compounds found in the sludge like Cl_2, Cu, phenols which may react with volatile compounds (Fullana et al. 2004). With respect to report of Environmental Protection Agency (EPA), dioxin and furan are considered as toxic chemical materials which is responsible for severe public health issues and create cancer, damage to immune system and affect reproduction capability. The chemical characteristics of dioxin and furan are very similar with each other (McKay 2002). During combustion process, other important emission is the particulate matter (PM) which is referred as fly ash. The operation and design of the combustion facility influence the fly ash amount (Niessen 2002). The other types of ash produced from combustion are bottom ash, economizer ash, boiler ash (Hjelmar 1996). Bottom ash mainly includes coarse materials and unburned organics, boiler and economizer ash includes coarse particles carried by flue gas from combustion chamber while fly ash contains fine particles in the flue gas stream from heat recovery unit (Sabbas 2003).

Especially for the gaseous emissions and particulate matter produced during combustion, there are several technologies developed. Control devices for air emissions are Electrostatic Precipitator, Dry and Wet Scrubber, Fabric Filter, Circulating Fluidised Bed, Moving Bed Adsorber, Entrained Flow Adsorber, Selective Catalytic Reactor (McKay 2002). Working mechanism of electrostatic precipitator depends on electrostatic force to separate PM by charging particles. Fabric filter benefits from gravity force for removing particles. Dry and wet scrubber uses gravity and friction forces for catching particulates (Müezzinoğlu 1987). These three methods are not suitable for reduction of gaseous emissions because their temperature requirement is very low. Circulating fluidized bed makes use of hearth-oven coke (HOC) for adsorption of flue gas. Moving bed adsorber uses moving bed with continuous change of adsorbent for flue gas adsorption. Entrained flow adsorber injects powdered materials to the gas stream for reduction. Selective catalytic reactor uses catalyst for the reduction of the gaseous products. Especially, for the reduction of PCDD and PCDF circulating fluidised bed and selective catalytic reactors are preferred. Moreover, circulating fluidised bed is very effective for the reduction of HCl, HF and SO_2 (McKay 2002).

Moreover, during incineration process some other products can be formed such as grate siftings and scrubbing materials which cause problematic leachate formation in the system. Therefore, both ashes originating combustion and other residues should be controlled. The main methods used for reduction of these products are Physical and Chemical Separation, Solidification and Stabilisation and Thermal Treatment (Sabbas 2003). The quality of materials is not affected a lot from the physical and chemical separation methods. These methods are based on separation of the residues by using a bulk containing similar physical and chemical with them. The size of the particles is only important for physical separation. On the other hand, solidification and stabilization method is mainly used for the reduction reuse of the materials. In solidification process, more beneficial materials may be produced with better physical, chemical and mechanical properties. The most preferable solidification process for residues is hydraulic binder. The last option for removal of ash and residues is the thermal treatment which is conducted in very high temperatures. With this process, leachate formation result from residues is treated very efficiently (Sabbas 2003).

6 CONCLUSIONS

Wastewater treatment plant sludge production is increasing with the population growth and increasing of the consumption. It is becoming great concern because the disposal alternatives of the sludge are limited. Fundamentally, sludge disposal techniques are based on the reduction of pathogens, dewatering and drying. However, there is only one technique which make it possible all requirements, i.e. thermal treatment. In this review incineration method, one of the mostly used thermal treatment options, is investigated. Three incineration techniques are present for the wastewater treatment sludge disposal. These are multiple heart incinerators, fluidized bed incinerators and rotary kiln incinerators. When they are compared with each other, most suitable method is fluidized bed combustors with respect to operation simplicity, capital cost and operation duration. Furthermore, incineration end products such as gaseous emissions, particulate matter, residues (ash, grate siftings, scrubbing materials) were explored. For the removal of gaseous emissions circulating fluidised bed and selective catalytic reactors are best options. For the removal of particulate matter or fly ash electrostatic precipitator, dry and wet scrubber are the mostly used techniques. Finally, for the removal of other residues in the form of settlement physical and chemical separation, solidification and stabilisation and thermal treatment are the main options. Thermal treatment is the best one within these three options because it may reduce the leachate formation originating from residues of incineration.

REFERENCES

Fullana, A., Conesa, J., Font, R. & Sidhu, S. 2004. Formation and Destruction of Chlorinated Pollutants during Sewage Sludge Incineration. *Environ. Sci. Technol.* 38: 2953–2958.
Fytili, D. & Zabaniotou, A. 2008. Utilization of sewage sludge in EU application of old and new methods-A review. *Renewable and Sustainable Energy Reviews* 12 (2008):116–140.
Global Water Intelligence. 2012. *Sludge Management: Opportunities in growing volumes, disposal restrictions and energy recovery.* Global Water Intelligence. ISBN: 978-1-907467-21-9.
Hjelmar, O. 1996. Disposal strategies for municipal solid waste incineration residues. *Journal of Hazardous Materials* 47: 345–368.
Kacprzak, M., Neczaj, E., Fijałkowski, K., Grobelak, A., Grosser, A., Worwag, M., Rorat, A., Brattebo, H., Almås, Å. & Singh, B.B. 2017. Sewage sludge disposal strategies for sustainable development. *Environmental Research* 156 (2017): 39–46.
Mateo-Sagasta, J., Raschid-Sally, L. & Thebo, A. 2015. Global Wastewater and Sludge Production, Treatment and Use. *Wastewater*. Springer Science + Business Media Dordrecht. DOI 10.1007/978-94-017-9545-6_2: 15-24.
McKay, G. 2002. Dioxin characterisation, formation and minimisation during municipal solid waste (MSW) incineration: review. *Chemical Engineering Journal* 86: 343–368.
Mininni, G., Sbrilli, A., Guerriero, E. & Rotatori, M. 2003. Polycyclic Aromatic Hydrocarbons Formation in Sludge Incineration by Fluidised Bed and Rotary Kiln Furnace. *Water, Air, and Soil Pollution* 154: 3–18, 2004.
Müezzinoğlu, A. 1987. *Hava Kirliliği ve Kontrolü Esasları.* Dokuz Eylül Yayınları. İzmir.
Murakami, T., Suzuki, Y., Nagasawa H., Yamamoto, T., Koseki, T., Hirose, H. & Okamoto, S. 2009. Combustion characteristics of sewage sludge in an incineration plant for energy recovery. *Fuel Processing Technology* 90 (2009): 778–783.
Niessen, R.W. 2002. *Combustion and Incineration Processes.* Marcel Dekker, Inc. 3rd Ed, 354–357.
Ozturk, Z.C. & Dursun, S. 2016. Modelling of Atmospheric SO_2 Pollution in Seydişehir Town by Artificial Neural Networks. *J. Int. Environ. Appl. & Sci.* 11(1): 1–7.
Sabbas, T., Polettini, A., Pomi, R., Astrup, T., Astrup, T., Hjelmar, O., Mostbauer, P., Cappai, G., Magel, G., Salhofer, S., Speiser, C., Heuss-Assbichler, S., Klein, R. & Lechner, P. 2003. Management of municipal solid waste incineration residues. *Waste Management* 23: 61–88.
Saltabaş, F., Soysal, Y., Yıldız, Ş. & Balahorli, V. 2009. Municipal Solid Waste Thermal Disposal Methods and Its Applicability in Istanbul. *Türkiye'de Katı Atık Yönetimi Sempozyumu*, YTÜ, 15–17 Haziran 2009, İstanbul.
Siebielska, I., 2014. Comparison of changes in selected polycyclic aromatic hydrocarbons concentrations during the composting and anaerobic digestion processes of municipal waste and sewage sludge mixtures. *Water Sci. Tech.* 70: 1617–1624.
Stasta, P., Boran, J., Bebar, L., Stehlik, P. & Oral, J. 2006. Thermal processing of sewage sludge. *Applied Thermal Engineering* 26 (2006): 1420–1426.
Tchobanoglous, G., Burton, F. & Stensel, D. 2003. *Wastewater Engineering: Treatment and Reuse* 4th ed. McGraw Hill.
Turovskiy, I.S. & Mathai, P.K. 2006. *Wastewater Sludge Processing.* John Wiley & Sons Inc. Hokoben, New Jersey.
Van Caneghem, J., Brems, A., Lievens, P., Block, C., Billen, P., Vermeulen, I., Dewil, R., Baeyens, J. & Vandecasteele, C. 2012. Fluidized bed waste incinerators: Design, operational and environmental issues. *Progress in Energy and Combustion Science* 38 (2012) 551–582.
Werther, J. & Ogada, T. 1999. Sewage sludge combustion. *Progress in Energy and Combustion Science* 25: 55–116.

Double benefit biodiesel produced from waste frying oils and animal fats

M. Catarino & A.P. Soares Dias
LAETA, IDMEC, Instituto Superior Técnico, Universidade de Lisboa, Lisboa, Portugal

M. Ramos
LAETA, IDMEC, Instituto Superior Técnico, Universidade de Lisboa, Lisboa, Portugal
ISEL, Instituto Politécnico de Lisboa, Lisboa, Portugal

ABSTRACT: Biodiesel produced from oil crops raises up sustainability issues, being the use of arable lands the major drawback of such renewable fuel. Thus, alternative raw materials, such as non-edible animal fats and waste frying oils must be used. Biodiesel from waste frying oils, tallow and lard was produced over Ca based heterogeneous catalyst prepared using Ca rich scallop shells. In the tested conditions (methanol reflux temperature), the acidity and water content of low grade fats promoted a decrease of catalytic performances comparing with those obtained using soybean oil. Soap formation was observed in parallel with severe crystallographic modifications of catalyst: Ca diglyceroxide formation and amorphization. The use of isopropanol as co-solvent was able to improve the FAME yield reducing the soap formation. Additionally, isopropanol avoided modifications of catalyst during reaction, contributing to the catalyst stability. Glycerin phase quality was also improved in the presence of the co-solvent.

1 INTRODUCTION

Global warming and socio-economic issues related to the instability of oil and derivates market prompt the search for renewable and eco-friendly substitutes for fossil diesel. Furthermore, modern society is a huge waste producer. An approach that allows simultaneous resolution for both problems would have enormous environmental benefits. Moreover, according to EEC Renewable Energy Directive (2010/C 160/02) biofuels produced from wastes double count (double benefit) for demonstrating the compliance with the 10% renewable energy target by 2020.

Biodiesel, a fatty acid methyl esters mixture, is pointed out as a feasible substitute of fossil diesel (Ajala *et al.* 2015). By 2011, more than 95% of the biodiesel was produced from edible vegetable oils, thus creating competition between biofuel market and the food sector (Balat 2011). With nearly 11% of world population undernourished, the need for grains and other basic crops continues to be critical (Hunger Notes 2016). The use of waste frying oils (WFO) and non-edible animal fats as alternative feedstock would avoid this "food vs fuel" issue, making biodiesel production more sustainable. Moreover, the use of these low grade and low price raw materials would allow the decrease of biodiesel production costs, since the greatest barrier to its use is its high price compared with fossil diesel price (Balat 2011, Piker *et al.* 2016).

WFO and animal fats present high free fatty acids contents, moisture and other contaminants which reduce the efficiency of traditional transesterification methods (Canakci 2007). Pre-treatment of these low grade fats before basic catalyzed transesterification would increase the production costs (Rathore *et al.* 2016). A two-step process, starting with acid catalyzed esterification followed by the traditional basic catalyzed transesterification is usually reported as an effective way to minimize the undesirable effect of acid feedstocks (Dias *et al.* 2013, Cai *et al.* 2015,). However, a larger number of reaction steps (two-steps instead of a single step) means an increase in production costs (Dias *et al.* 2013, Piker *et al.* 2016), being advantageous to process low grade fats using the traditional single step reaction system, preferably with heterogeneous basic catalysts (Aransiola *et al.* 2014).

Among heterogeneous basic catalysts for the methanolysis of vegetable oils, lime catalyst has benefited from scientific as well as industrial attention because of its high basicity, low solubility, and its easy derivatization from natural resources (Avhad and Marchetti 2015). Food wastes such as eggshells, mollusks, and crustacean shells are characterized by a variable chemical composition and may contain a mixture of carbohydrates, lipids, proteins and inorganics, such as calcium (Girotto *et al.* 2015). These calcium rich food wastes can be used as cheap raw materials to produce CaO by a calcination process (Viriya-Empikul *et al.* 2010, Kumar *et al.* 2015, Sirisomboonchai *et al.* 2015). In fact, food wastes are a significant fraction of municipal solid wastes (Pham *et al.* 2015) and their use to produce biodiesel not only increases the process sustainability but also eliminates the need for dispose of these wastes.

Glycerin, the byproduct of fats transesterification, accounts for about 10% (W/W) of the final reaction mixture (Nanda *et al.* 2014) and the economic feasibility of biodiesel production process requires the valorization of this phase. Most commercial glycerin applications need a purity of, at least, 99.5%. Through a homogeneous transesterification process, the maximum purity achieved is approximately 80% but when a heterogeneous process is used the contamination of the final phases is minor, the purification steps are easy and the purity can reach a value as high as 95%. The final glycerin purity, in a biodiesel production facility, depends on the feedstock used (Gerpen 2007).

The addition of a co-solvent to the reactional system promotes a boost in reaction components miscibility (Roosta and Sabzpooshan 2016). The main concerns related with this approach are the increased complexity of phase purification systems and the co-solvent recovery. In order to overcome these drawbacks, the chosen co-solvent should have a boiling point close to the boiling point of the alcohol used in transesterification (Gerpen 2005). For methanolysis reactions THF is usually chosen (Roosta and Sabzpooshan 2016) raising additional concerns regarding the hazard level of the co-solvent. However, recent studies indicate that THF does not pose any risk to humans or to the environment (Fowles *et al.* 2013).

Few researchers reported the use of isopropanol (IPA) as co-solvent for oil methanolysis reactions. Lee *et al.* studied the use of branched-chain alcohols in the transesterification of oils and fats (Lee *et al.* 1995). Recently, Roschat *et al.* studied the effect of several co-solvents on FAME yield concluding that the use of IPA allows an increase in this (Roschat *et al.* 2016). IPA is an alcohol and thus can participate in the alcoholysis reactions producing propyl esters, however, the molecule shows great steric hindrance and therefore its contribution can be neglected (Maeda *et al.* 2011).

To improve the knowledge on the Ca based catalyst systems for biodiesel production from low grade fats, data on the methanolysis of WFO, lard and tallow over Ca catalysts using IPA as co-solvent are presented. Particular emphasis was given to the post reaction catalysts characterization in order to infer about the co-solvent effect on the catalyst stability.

2 MATERIALS AND METHODS

2.1 *Materials and catalyst preparation*

The methanolysis tests were carried out using soybean oil and lard (alimentary grades), from local producers (Sovena and Auchan, respectively), WFO collected from a restaurant, and tallow obtained by rendering of waste bovine tissues. The WFO samples were filtrated to remove solids. Fats acidity was evaluated by titration with ethanolic KOH solution (0.1 M) using phenolphthalein indicator (0.5% (W/W) ethanolic solution) (Puna *et al.* 2013).

Scallop shells were used as calcium rich material to prepare Ca based catalysts. The shells were washed, dried and reduced to powder using a ceramic mortar. The calcium catalyst was obtained by calcination of the powder in a muffle at 800°C for 3 h using a heating rate of 5°C/min. The calcination temperature was chosen from the thermal degradation profile obtained by thermogravimetry under air flow (30°C/min).

2.2 Catalyst characterization

The catalyst, fresh and post reaction, was characterized by X-ray diffraction (XRD) in order to identify the Ca crystalline phases and the degree of crystallinity of each phase. The diffractograms were recorded with a Rigaku Geigerflex diffractometer with Cu K_α radiation at 40 kV and 40 mA (2 °/min).

2.3 Reaction procedure

The methanolysis reactions, at methanol reflux temperature, were carried out using 5% (W/W) of catalyst (fat basis) and a methanol/fat molar ratio of 12. For each reaction batch, the catalyst was previously contacted with methanol for 1 h (65°C) and then pre-heated fat (100 g, 100°C) was added to the reaction flask. The co-solvent effect was evaluated using a methanol/IPA = 3 (V/V) ratio. IPA was added to the reaction mixture immediately before fat addition.

After reaction period (2.5 h), the slurry, containing the catalyst, the unreacted fat, the methyl esters, the glycerin and, when used, the co-solvent, was cooled down and the catalyst was removed by filtration. The collected liquid was transferred into a decantation funnel for gravity separation of the glycerin from the oily phase containing the biodiesel and the unreacted fat. Both phases were characterized by ATR-FTIR. The formed FAME was quantified based on the reflectance band centered at 1436 cm^{-1} (O'Donnell et al. 2013) and a calibration curve, obtained for mixtures of pure biodiesel and soybean oil. The post reaction catalyst was dried overnight at 105°C before characterization. More details on the reaction procedure are given elsewhere (Dias et al. 2016).

3 RESULTS AND DISCUSSION

3.1 Catalyst characterization

The chalky white fine powder, obtained by calcination of scallop shells, showed a XRD pattern (Figure 1) mainly ascribable to calcite slightly contained with lime. Higher calcination temperature is needed to totally convert calcite, the major component of scallop shells, into lime.

The post reaction catalysts, obtained in absence of IPA, presented XRD patterns (Figure 1) which depend on the processed fat. Soybean oil allowed to calcium diglyceroxide formation whereas calcite was almost vanished. When WFO was processed, no XRD lines of Ca-diglyceroxide were observed eventually due to solubilization of such phase in the reaction medium. The same was observed when lard and tallow were processed. Analogous effect was reported by Stojković et al. (2016), which underlined an increased Ca leaching for low grade fats.

All post reaction catalysts obtained when IPA was used as co-solvent revealed the absence of the Ca –diglyceroxide XRD lines (Figure 2). Such result seems to indicate that IPA avoid Ca-diglyceroxide formation eventually related to the fact that IPA helps to remove glycerin from the catalyst surface and thus reduces the occurrence of Ca-diglyceroxide formation. Thus, the use of IPA minimizes the undesirable homogeneous catalysis contribution caused by calcium diglyceroxide leaching.

The catalyst after lard methanolysis showed a conversion into low crystallinity portlandite ($Ca(OH)_2$) probably due to a particular conjugation of moist and acidity effects.

3.2 Methanolysis test

Since fat acidity is a relevant limitation to the basic catalyzed methanolysis reaction (Dias et al. 2013, Puna et al. 2013), such parameter was evaluated by colorimetric titration (Table 1). Lard acidity was lower than that reported in literature (Karmakar et al. 2010, Atadashi et al. 2012), however, the used lard was a commercial one and therefore had already suffered purification treatments.

The oily phase, separated from the glycerin phase by gravity settling and without further purification, was characterized by ATR-FTIR to evaluate FAME yield. Data in Table 1, as expected, show that methanolysis using low grade fats results in lower FAME yields due to the high free

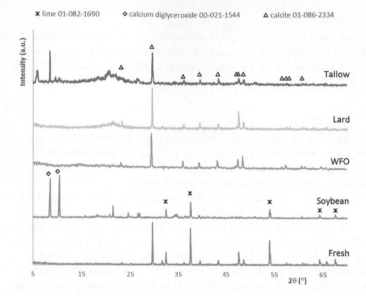

Figure 1. XRD diffractograms of fresh and post reaction catalysts, without IPA (phase identification using standard JCPDS files).

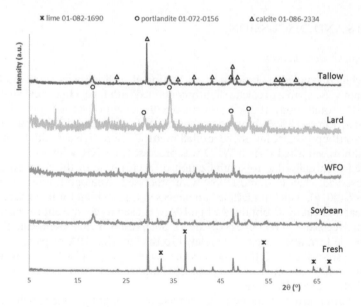

Figure 2. XRD diffractograms of post reaction catalysts, with IPA (phase identification using standard JCPDS files).

Table 1. Acid value for used feedstocks and FAME yield assessed by ATR-FTIR.

Feedstock	Acid value (mg_{KOH}/g_{fat})	FAME Yield (%)	
		Without IPA	With IPA
Soybean oil	0.6	88.4	77.4
Lard	1.4	81.6	66.7
WFO	2.2	76.3	91.7
Tallow	3.9	51.8	61.7

fatty acid content that promotes soap formation and, consequently, catalyst consumption (Dias *et al.* 2013). The use of IPA as co-solvent had a positive effect on the FAME yield for WFO and tallow, which have the highest acidity. It seems that IPA was able to remove free fatty acids from the catalyst surface thus minimizing the neutralization of the active sites. Additionally, the higher water content of these feedstocks, that reacts with the catalyst to produce calcium hydroxide, boosts the FAME yield since this calcium phase also shows high catalytic activity.

For the other raw fats, IPA reduced the FAME yield possibly due to the competitive adsorption of IPA to the active sites of the catalyst with the transesterification of this alcohol being slower than methanolysis (Maeda *et al.* 2011). The aforementioned $Ca(OH)_2$ production, for WFO and tallow methanolysis, can disguise the competitive adsorption of IPA, thus helping explaining the higher yields achieved.

Notwithstanding, as concluded from catalysts characterization, the use of IPA, enhancing the miscibility of the reactional mixture components, allows to minimize the contact time between the produced glycerin and the catalyst surface reducing calcium diglyceroxide production and therefore decreasing the undesirable homogeneous catalysis component. This result was also observed from the obtained glycerin color, since homogeneous catalysis produces a darker glycerin than the heterogeneous one.

3.3 *Glycerin phase characterization*

Glycerin is co-produced with biodiesel being 10% (W/W) of the raw oil. A high purity glycerin can contribute to the economic viability of the biodiesel process. As referred before (Kiss *et al.* 2010) heterogeneous catalysts, when compared with homogeneous ones, allow an improved glycerin quality. In order to evaluate the effect of low grade fats on the glycerin quality, the obtained phases without purification were characterized by FTIR. The spectra in Figure 3 show features similar to those of pure glycerin (used as standard) with the most intense band centered at 1030 cm⁻¹, however, slightly displaced toward lower wave

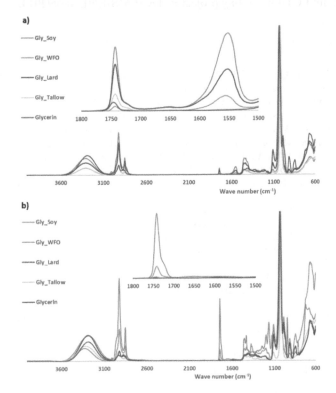

Figure 3.   Glycerins spectra obtained with (a) and without (b) IPA (pure glycerin as standard).

numbers due to methanol contamination. The glycerins obtained without IPA show a IR feature around 1580 cm⁻¹ due to carboxylate soap species (Nanda *et al.* 2014). Surprising, the soybean-derived glyrecin seems to have more soap than those obtained from animal fats. Such soap band is absent for glycerins obtained using IPA co-solvent but they are slightly contaminated with IPA.

4   CONCLUSIONS

In order to produce a more sustainable biodiesel, low grade fats, instead of oil crops, were used. The methanolysis was performed using lard, beef tallow and waste frying oils. The reactions were catalyzed through a Ca based catalyst obtained by calcination of scallop shells. The Ca rich powders presented XRD patterns belonging to lime and calcite. Usually, only lime is referred as active catalyst for fats alcoholysis but in the tested conditions (standard from the literature) the lime/calcite material was quite active.

As expected, the acidity and water content of the used fats promoted soap formation and lower FAME yield than soybean oil methanolysis.

To reduce the soap formation and improve the FAME yield, the methanolysis was carried out using isopropanol as co-solvent. The IPA co-solvent was effective reducing soap formation however, had a slight negative effect on the FAME yield. It seems that competitive adsorption of IPA on the active sites of the catalyst reduces the methanolysis rate. Also, IPA had a beneficial effect on the catalyst stability since decreases calcium dyglyceroxide formation.

ACKNOWLEDGEMENT

The authors thank FCT for funding project PTDC/EMS-ENE/4865/2014.

REFERENCES

Ajala, O.E., Aberuagba, F., Odetoye, T.E., Ajala, A.M. 2015. Sustainable energy replacement to pretroleum-based diesel fuel – A review. *ChemBioEng Reviews* 2: 145–156.

Aransiola, E., Ojumu, T., Oyekola, O., Madzimbamuto, T., Ikhu-Omoregbe, D. 2014. A review of current technology for biodiesel production: state of the art. *Biomass and Bioenergy* 61: 276–297.

Atadashi, I., Aroua, M., Aziz, A.A., Sulaiman, N. 2012. Production of biodiesel using high free fatty acid feedstocks. *Renewable and Sustainable Energy Reviews* 16(5): 3275–3285.

Avhad, M., Marchetti, J. 2015. A review on recent advancement in catalytic materials for biodiesel production. *Renewable and Sustainable Energy Reviews* 50: 696–718.

Balat, M. 2011. Potential alternatives to edible oils for biodiesel production – A review. *Energy conversion and management* 52: 1479–1492.

Cai, Z.-Z., Wang, Y., Teng, Y.-L., Chong, K.-M., Wang, J.-W., Zhang, J.-W., Yang, D.-P. 2015. A two-step biodiesel production process from waste cooking oil via recycling crude glycerol esterification catalyzed by alkali catalyst. *Fuel Processing Technology* 137: 186–193.

Canakci, M. 2007. The potential of restaurant waste lipids as biodiesel feedstocks. *Bioresource Technology* 98(1): 183–190.

Dias, A.P., Puna, J., Correia, M.J., Nogueira, I., Gomes, J., Bordado, J. 2013. Effect of oil acidity on the methanolysis performances of lime catalyst biodiesel from waste frying oils (WFO). *Fuel Processing Technology* 116: 94–100.

Dias, A.P.S., Puna, J., Gomes, J., Correia, M.J.N., Bordado, J. 2016. Biodiesel production over lime. Catalytic contributions of bulk phases and surface Ca species formed during reaction. *Renewable Energy* 99: 622–630.

Fowles, J., Boatman, R., Lewis, C., Morgott, D., Rushton, E., Rooij, J.V., Banton, M. 2013. A review of the toxicological and environmental hazards and risks of tetrahydrofuran. *Critical reviews in toxicology* 43(10): 811–828.

Gerpen, J.V. 2005. Biodiesel Processing and Production. *Fuel Processing Technology* 86(10): 1097–1107.

Gerpen, J.V. 2007. Biodiesel Production. In Ranalli, P. (ed), *Improvement of Crops Plants for Industrial Uses*: 281–289. The Netherlands: Springer.

Girotto, F., Alibardi, L., Cossu, R. 2015. Food waste generation and industrial uses: a review. *Waste Management* 45: 32–41.

Karmakar, A., Karmakar, S., Mukherjee, S. 2010. Properties of various plants and animals feedstocks for biodiesel production. *Bioresource Technology* 101(19): 7201–7210.

Kiss, F.E., Jovanović, M., Bošković, G.C. 2010. Economic and ecological aspects of biodiesel production over homogeneous and heterogeneous catalysts. *Fuel Processing Technology* 91(10): 1316–1320.

Kumar, P., Sarma, A.K., Jha, M.K., Bansal, A., Srivasatava, B. 2015. Utilization of renewable and waste materials for biodiesel production as catalyst. *Bulletin of Chemical Reaction Engineering and Catalysis* 10: 221–229.

Lee, I., Johnson, L.A., Hammond, E.G. 1995. Use of branched-chain esters to reduce the crystallization temperature of biodiesel. *Journal of the American Oil Chemists Society* 72(10): 1155–1160.

Maeda, Y., Imamura, K., Izutani, K., Okitsu, K., Boi, L.V., Lan, P.N., Tuan, N.C., Yoo, Y.E., Takenaka, N. 2011. New technology for the production of biodiesel fuel. *Green Chemistry* 13(5): 1124–1128.

Nanda, M.R., Yuan, Z., Qin, W., Poirier, M.A., Chunbao, X. 2014. Purification of Crude Glycerol using Acidification: Effects of Acid Types and Product Characterization. *Austin Chemical Engineering* 1(1): 1–7.

O'Donnell, S., Demshemino, I., Yahaya, M., Nwandike, I., Okoro, L. 2013. A review on the spectroscopic analyses of biodiesel. *European International Journal of Science and Technology* 2(7): 137–146.

Pham, T.P.T., Kaushik, R., Parshetti, G.K., Mahmood, R., Balasubramanian, R. 2015. Food waste-to-energy conversion technologies: current status and future directions. *Waste Management* 38: 399–408.

Piker, A., Tabah, B., Perkas, N., Gedanken, A. 2016. A green and low-cost room temperature biodiesel production method from waste oil using egg shells as catalyst. *Fuel* 182: 34–41.

Puna, J., Correia, M.J., Dias, A.P., Gomes, J., Bordado, J. 2013. Biodiesel production from waste frying oils over lime catalysts. *Reaction Kinetics, Mechanisms and Catalysis* 109: 405–415.

Rathore, V., Newalker, B.L., Badoni, R.P. 2016. Processing of vegetable oil for biofuel production through conventional and non-conventional routes. *Energy for Sustainable Development* 31: 24–49.

Roosta, A., Sabzpooshan, I. 2016. Modeling the Effects of Cosolvents on Biodiesel Production. *Fuel* 186: 779–786.

Roschat, W., Siritanon, T., Kaewpuang, T., Yoosuk, B., Promarak, V. 2016. Economical and green biodiesel production process using river snail shells-derived heterogeneous catalyst and co-solvent method. *Bioresource and Technology* 209: 343–350.

Sirisomboonchai, S., Abuduwayiti, M., Guan, G., Samart, C., Abliz, S., Hao, X., Kusakabe, K., Abuluda, A. 2015. Biodiesel production from waste cooking oil using calcined scallop shell as catalyst. *Energy Conversion and Management* 95: 242–247.

Stojković, I.J., Miladinović, M.R., Stamenković, O.S., Banković-Ilić, I.B., Povrenović, D.S., Veljković, V.B. 2016. Biodiesel production by methanolysis of waste lard from piglet roasting over quicklime. *Fuel* 182: 454–466.

Hunger Notes 2016, *World Hunger Educational Service*, accessed 22 March 2017, <http://www.world-hunger.org/2015-world-hunger-and-poverty-facts-and-statistics/>.

Viriya-Empikul, N., Krasae, P., Puttasawat, B., Yoosuk, B., Chollacoop, N., Faungnawakij, K. 2010. Waste shells of mollusk and egg as biodiesel production catalysts. *Bioresource Technology* 101: 3765–3767.

Alkali-activated cement using slags and fly ash

S. Rios, A. Viana da Fonseca & C. Pinheiro
CONSTRUCT-GEO, Department of Civil Engineering, Faculty of Engineering,
University of Porto, Portugal

S. Nunes
CONSTRUCT-LABEST, Department of Civil Engineering, Faculty of Engineering,
University of Porto, Portugal

N. Cristelo
CQVR, Department of Engineering, University of Trás-os-Montes e Alto Douro, Vila Real, Portugal

ABSTRACT: Alkali Activated Cements (AAC) are a very convenient alternative to common binders as waste materials like slag and fly ash are included in their production. In this paper, a response surface method is used to optimize an AAC made with fly ash, steel slag, sodium silicate and sodium hydroxide. For this purpose, an experimental plan contemplating 26 mixtures was developed, which included compression and flexural strength tests. The experimental data was then analyzed using regression analysis and ANOVA. The results indicate that the sodium hydroxide/sodium silicate solution ratio is the most relevant variable, followed by the ratio between the two solid components (slag and fly ash).

1 INTRODUCTION

Most binders used worldwide in the construction industry are based on Portland cement. However, society is more and more concerned about environmental issues, and the production of clinker involves a high amount of carbon dioxide (CO_2) released to the atmosphere, estimated as 5% to 8% of the overall CO_2 released to the atmosphere (Scrivener & Kirkpatrick, 2008). Therefore, there is an ongoing research effort targeting the development of more sustainable binders (Juenger et al., 2011). In particular, the use of waste materials is highly encouraged, since it allows an increase in resource efficiency, while contributing also to enhancing the circular economy.

The alkaline activation technique is particularly adequate to create binders based on residues, such as fly ash or ground granulated blast furnace slag, which constitute very effective options due to their amorphous aluminosilicate microstructure. It consists on a reaction between aluminosilicate materials and alkali or alkali-based earth substances, such as sodium (Na) or potassium (K), or an alkaline earth ion, such as calcium (Ca). The reactions can be summarized in the following sequence. First, there is the destruction, by the high hydroxyl (OH-) concentration in the alkaline medium, of the Si-O-Si, Al-O-Al and Al-O-Si covalent bonds present in the vitreous phase of the original semi-amorphous aluminosilicate (i.e. the precursor). The Si and Al ions are released into the solution as they become available and; at the same time, the alkaline cations—usually Na+ or K+, depending on the activator—compensate the excess negative charges associated with the modification of the aluminium coordination during the dissolution phase. The resulting products accumulate for a period of time, forming an ion-rich solution, which finally precipitates and reorganizes into more stable and ordered Si-O-Al and Si-O-Si structures. If calcium is predominant, relatively to the sodium or potassium, the dissolved Al-Si ions will diffuse from any solid surface, which favours the production of a C-S-H gel phase. Otherwise, the Si and Al ions will be able to accumulate around the nuclei points, sharing all the oxygen ions and forming a Si-O-Al and Si-O-Si three-dimensional structure (the formation of Al-O-Al is not favoured). The resulting product is an amorphous alumina-silicate gel, which evolves, with curing time and crystallization, from an Al-rich phase to a Si-rich phase. The crystallization, starting almost immediately after the precipitation, is responsible for the hardening of the gel, which eventually matures into alkaline cement, with pre-zeolite as secondary products (Fernández-Jiménez et al., 2005).

The precursor should always be submitted to a previous thermal treatment, capable of inducing the loss of constituent water and the subsequent re-coordination of the aluminium and oxygen ions, transforming an originally crystalline structure into an amorphous one, more susceptible to further chemical reactions. The relative presence of calcium in the precursor and/or in the activator is very important, since the speed of the reactions is highly dependent on the type of aluminosilicate gel being formed, either N-A-S-H or C-S-H. The former needs longer periods in order to mature into a stable and reliable matrix, while the latter has curing/developing periods similar to those obtained with cement-based binders (Dombrowski et al, 2007; Garcia-Lodeiro et al., 2013).

In some cases, constrains related with short deadlines often require tight construction periods, which hinders the use of alkali activated low calcium (class F) fly ash, based on the mentioned slower reaction kinetics of the N-A-S-H gel, when compared with the C-S-H gel. However, in the presence of enough calcium, the two systems are very compatible, and several studies have focused on the characterisation of their interaction and coexistence (Garcia-Lodeiro et al., 2009, 2011; Puligilla & Mondal, 2013). For this reason, in this work a mixture of low calcium fly ash and slag (very rich in calcium) was used to produce an alkaline activated cement (AAC) joining N-A-S-H and C-S-H gel. A central composite design was carried out to mathematically model the influence of mixture parameters and their coupled effects on compression strength and flexural strengths.

2 MATERIALS

2.1 Slag

The slag was collected at the Megasa Steel Industry of Maia, Portugal. This facility produces different types of slags some of them certified to be used in concrete as stone replacer. In this case, a white slag was used, which does not have, at the moment, any known application. Since this waste material is rich in calcium (Table 1) it was considered suitable for the production of alkaline activated cement.

2.2 Fly ash

The fly ash was collected at the PEGOP thermoelectric power plant at Pego, Portugal. It is classified as type F according to ASTM standard C 618 (ASTM, 2012) due to its low calcium content, as can be observed in the chemical composition presented in Table 2. The loss on ignition value was not specifically determined for the present fly ash, but it should be around 2.59, according to Cristelo et al. (2012) that has worked with a fly ash from the same thermoelectric power plant.

2.3 Alkaline activator

The alkaline activator solution was prepared by sodium hydroxide and sodium silicate. The sodium hydroxide was obtained in flake form, with a specific gravity of 2.13, at 20°C, and with

Table 1. Composition of the slag (wt%).

Element	SiO_2	Al_2O_3	CaO	MgO	MnO	Fe total	Cr_2O_3	Others
Slag	23.5	6.6	54.9	8.5	0.4	1.1	0	5

Table 2. Composition of the fly ash (wt%).

Element	SiO_2	Al_2O_3	Fe_2O_3	CaO	K_2O	TiO_2	MgO	Na_2O	SO_3	Others
Fly ash	54.84	19.46	10.73	4.68	4.26	1.40	1.79	1.65	0.7	0.5

95–99% purity. It was then dissolved in water up to the desired concentration. The sodium silicate was already in solution form with a specific gravity of 1.5 and SiO_2/Na_2O ratio of 2 by mass.

3 EXPERIMENTAL PLAN

3.1 *Design of experiments*

Response surface methods offer statistical design of experiment tools that lead to peak process performance. A precise map based on mathematical models is produced which can put all the responses together via sophisticated optimization approaches leading to the discovery of the best option (for example, meeting the required specifications at minimal cost).

These methods are very interesting for mixtures composed by several ingredients, as it is the case of the AAC, to determine the best performing combination. This is often achieved by changing the quantity of each ingredient, at a time, following the traditional one-factor-at-a-time approach. However, this methodology does not account for the interactions between variables which might be estimated using full factorial designs, symbolized mathematically as N^K, where N is the factorial design level and K is the number of variables. A three-level factorial design provides a good prediction. However, as K increases, the number of runs becomes excessive (e.g. with 5 factors, 243 runs are needed). For that reason, it is more convenient to use a two-level factorial design and try to improve it. Nevertheless, when approaching the optimum level of response, a two-level factorial design no longer provides sufficient information to adequately model the true response surface. Such difficulty can be overcome if center and axial points are added to the two-level factorial design, and thus a composite design is obtained, adequately suited to response surface methods (Anderson and Whitcomb, 2005). In this work, a face-centered composite design was used, which means that the axial points are at the center of each face of the factorial space.

3.2 *Key mixture parameters and experimental region*

Input variables for the present work could be the amount of each ingredient of the AAC, namely fly ash, slag, sodium silicate and sodium hydroxide. However, there are abundant works in the literature (e.g., Hardjito et al., 2004; Kupwade-Patil and Allouche, 2013; Cristelo et al., 2013; Rios et al., 2016) reporting correlations between certain parameters and mechanical properties of the AAC such as unconfined compression strength or elastic stiffness. These parameters generally involve: a parameter that relates the quantity of fluid in the mixture; a parameter relating the two solutions of the activator (the sodium silicate and the sodium hydroxide) and the sodium hydroxide concentration. In this work, another parameter was necessary, relating the amount of slag and fly ash. The selected variables can then be summarized as follows:

A. SS/(SS+SH) where SS means sodium silicate and SH means sodium hydroxide;
B. E/(E+C) where E means slag and C means fly ash
C. S/L where S is the solids weight (slag + fly ash) and L is the fluid weight (the activator made by SS and SH)
D. SH conc, meaning the sodium hydroxide concentration in molal

Please note that to apply the described response surface method it was necessary that the variables were continuous and without math indeterminations when some of the ingredients were null. The first variable (SS/(SS+SH)) range was defined between 0 and 1, where 0 corresponds to an activator only composed by sodium hydroxide and 1 to an activator only composed by sodium silicate. The second variable (E/(E+C)) range was defined between 0.1 and 1 where 1 corresponds to a mixture without fly ash. Since the mixtures without slag were slow to harden they were removed from the plan. The third variable (S/L) range was defined between 1.5 and 2.5, based on preliminary tests. Finally, the sodium hydroxide concentration varied between 5 and 12 molal, as currently found in the literature (Xu and van Deventer, 2003; Cristelo et al., 2013; Phummiphan et al., 2016). To facilitate the paper reading, these variables were named as A, B C and D. This lead to an experimental plan of 26 mixtures.

3.3 *Specimen's preparation and test methods*

To prepare the specimens, the solids (fly ash and slag) were first dry mixed and the solutions for the activator was also prepared. The sodium hydroxide pellets were dissolved in water to the desired molal concentration and let to cool down up to room temperature. Then, the sodium hydroxide was mixed with the sodium silicate solution according to the desired ratio (SS/(SS+SH)). Finally, the solids and the activator were mixed together in an automatic mixer for 10 minutes and molded in beams of $40 \times 40 \times 160$ mm^3 according to the ASTM D 1632 (2007). After two days, the specimens were consistent enough to be demolded, and so they were stored in a temperature controlled room at 20°C, to cure up to 14 days. After the curing period, the specimens were then tested for flexural strength, using three-point loading tests, according to ASTM D 1635 (2012). After leveling the surfaces, the two pieces resulting from the flexural test were also tested for unconfined compression, following the ASTM standard D 1633 (2007).

4 RESULTS

The results of unconfined compression strength (fc) and flexural strength (ffl), at 14 days, were then analyzed, namely, to fit models using regression analysis and ANOVA and to validate the models by examining the residuals for trends. Table 3 presents the results of the estimated models (both in terms of actual and coded values of mixture parameters), including the residual error term (ε), along with the correlation coefficients. In both cases, a variable transformation was used, as indicated in Table 3, in order to stabilize the response variance and improve the fit of the model to the data.

The estimates of the model coefficients, in terms of coded values, give an indication of the relative significance of the mixture parameters on each response. Higher values indicate greater influence of the mixture parameter in the response and on the other hand, a negative value reflects a response decrease to an increase in this parameter.

In this case, it is clear that A is the most important input variable for both response variables. This means that the relationship between the sodium silicate and the sodium hydroxide has a major effect on the mechanical behavior of the AAC. In fact, the quantity of sodium silicate will determine the amount of soluble silica incorporated in aluminosilicate gel, with

Table 3. Fitted numerical models in terms of actual and coded values of mixture parameters.

Response variable	Actual values		Coded variables	
	Log (fc)	√ ffl	Log (fc)	√ ffl
model terms	estimate	estimate	estimate	estimate
independent	−3.63262	+0.20951	+0.72	+1.44
× A	+2.10471	+0.87407	**+0.74**	**+0.66**
× B	−1.67121	+0.53578	−0.16	−0.10
× C	+3.59636	+0.30146	+0.21	+0.15
× A × B	+1.23311	+0.79508	+0.28	+0.18
× A × C	−0.27647	NS	−0.069	NS
× B × C	+0.35146	NS	+0.079	NS
× A^2	−0.74928	NS	−0.19	NS
× B^2	NS	−1.05356	NS	−0.21
× C^2	−0.80865	NS	−0.20	NS
residual error, ε *				
mean	0	0	0	0
standard deviation	0.13	0.17	0.13	0.17
$R^2/R^2_{adjusted}$	0.98/0.97	0.94/0.92	0.98/0.97	0.94/0.92

(NS) non-significant terms; (*) error term is a random and normally distributed variable.

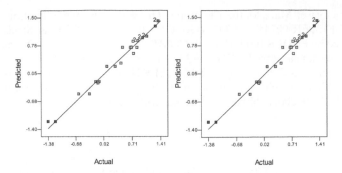

Figure 1. Relationships between predicted and experimental values for the two transformed response variables: a) Log(fc); b) √ ffl.

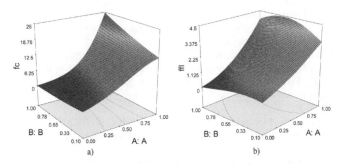

Figure 2. Response surfaces for each response variable: a) Log(fc); b) √ fl.

implications on the structure of the resulting gel (Silva et al., 2007). The second most important variable is the slag/ash ratio, which defines the amount of calcium present in the mixture and thus controls the relative amount of C-S-H and N-A-S-H gels formed, as explained above.

The numerical models presented in Table 3 are well adjusted to the experimental data, as expressed by the R^2, but also by the relationships between predicted and experimental values presented in Figure 1.

The 3D response surfaces associated to the proposed models are presented in Figure 2 for the two most important input variables (A and B), while keeping C = 2 and D = 8.5, which are the center values for these variables. From these surfaces, it is clear that higher values of A and B lead to higher compression and flexural strength values.

5 CONCLUSIONS

The paper presents the use of a response surface method to optimize an alkaline activated cement made by fly ash, slag, sodium silicate and sodium hydroxide. Design of experiments approach was very useful to define the experimental plan and also to analyze the results. Testing a total of 26 mixtures it was possible to obtain a numerical model of the two response variables which were very well adjusted to the experimental data. These models indicated that the best mixture composition is obtained by increasing the two most important variables: the ratio between the two solutions of the activator, and the ratio between the two solids (fly ash and slag).

ACKNOWLEDGMENTS

The authors would like to acknowledge the company Pegop—Energia Eléctrica SA, which runs the thermoelectric power plant in Pego, for the supply of fly ash, and the company Megasa, which runs the steel factory of Maia, for the slag supply. This work was financially supported by Project POCI-01-0145-FEDER-007457 – CONSTRUCT—Institute of R&D in Structures and Construction, funded by FEDER funds through COMPETE2020 – Programa Operacional Competitividade e Internacionalização (POCI)—and by national funds from the FCT—Fundação para a Ciência e a Tecnologia; Scholarship Reference: SFRH/BPD/85863/2012. It was also funded by CNPQ (the Brazilian council for scientific and technological development) for its financial support in 201465/2015-9 scholarship of the "Science without borders" program.

REFERENCES

ASTM D 1632. 2007. *Standard Practice for Making and Curing Soil-Cement Compression and Flexure Test Specimens in the Laboratory.* ASTM, United States.

ASTM D 1635. 2012. *Standard Test Method for Flexural Strength of Soil-Cement Using Simple Beam with Third-Point Loading. Annual Book of Standards.* Vol. 04.08. ASTM, United States.

ASTM D 1633. 2007. *Standard Test Methods for Compressive Strength of Molded Soil-Cement Cylinders.* ASTM, United States.

ASTM C 618. 2012. *Standard Specification for Coal Fly Ash and Raw or Calcined Natural Pozzolan for Use in Concrete.* ASTM, United States.

Anderson, M.J. & Whitcomb, P.J. 2005. RSM simplified. Optimizing processes using response surface methods for design of experiments. London, New York: CRS Press Boca Raton.

Cristelo, N., Soares, E., Rosa, I., Miranda, T., Oliveira, D., Silva, R.A. & Chaves, A. 2013. Rheological properties of alkaline activated fly ash used in jet grouting applications. *Construction and building materials* 48: 925–933.

Cristelo, N., Glendinning, S., Miranda, T., Oliveira, D., & Silva, R. 2012. Soil stabilisation using alkaline activation of fly ash for self-compacting rammed earth construction. *Construction and Building Materials* 36: 727–735.

Dombrowski, K., Buchwald, A. & Weil, M. 2006. The influence of calcium content on the structure and thermal performance of fly ash based geopolymers. *Journal of Materials Science* 42(9): 3033–3043. doi: 10.1007/s10853-006-0532-7.

Fernández-Jiménez, A., A. Palomo and M. Criado 2005. Microstructure development of alkali-activated fly ash cement: a descriptive model. *Cement and Concrete Research* 35(6): 1204–1209.

Garcia Lodeiro, I., Macphee, D.E., Palomo, A. & Fernández-Jiménez, A. 2009. Effect of alkalis on fresh C–S–H gels. FTIR analysis. *Cement and Concrete Research* 39(3): 147–153. doi: 10.1016/j.cemconres.2009.01.003.

Garcia-Lodeiro, I., Palomo, A., Fernández-Jiménez, A. & Macphee, D.E. 2011. Compatibility studies between N-A-S-H and C-A-S-H gels. Study in the ternary diagram Na2O–CaO–Al2O3–SiO2–H2O. *Cement and Concrete Research* 41(9): 923–931. doi:10.1016/j.cemconres.2011.05.006.

Garcia-Lodeiro I, Fernandez-Jimenez A and Palomo A. 2013. Variation in hybrid cements over time. Alkaline activation of fly ash-Portland cement blends. *Concrete Research* 52: 112–122. http://dx.doi.org/10.1016/j.cemconres.2013.03.022.

Hardjito, D., Wallah, S., Sumajouw D. & Rangan, B. 2004. Factors influencing the compressive strength of fly ash-based geopolymer concrete. *Civil Engineering Dimension* 6(2): 88–93.

Juenger, M.C.G., Winnefeld, F., Provis, J.L. & Ideker J.H. 2011. Advances in alternative cementitious binders. *Cement and Concrete Research* 41(12): 1232–1243. doi:10.1016/j.cemconres.2010.11.012.

Kupwade-Patil, K. & Allouche, E.N. 2013. Impact of Alkali Silica Reaction on Fly Ash-Based Geopolymer Concrete. *Journal of Materials in Civil Engineering* 25: 131–139.

Phummiphan, I., Horpibulsuk, S., Sukmak, P., Chinkulkijniwat, A., Arulrajah, A. & Shen, S.-L. 2016. Stabilisation of marginal lateritic soil using high calcium fly ash-based geopolymer. *Road Materials and Pavement Design* 17(4): 877–891, *DOI: 10.1080/14680629.2015.1132632.*

Puligilla, S. & Mondal, P. 2013. Role of slag in microstructural development and hardening of fly ash-slag geopolymer. *Cement and Concrete Research* 43: 70–80.

Rios, S., Cristelo, C., Viana da Fonseca, A. & Ferreira, C. 2016. Structural Performance of Alkali Activated Soil-Ash versus Soil-Cement. *Journal of Materials in Civil Engineering* 28(2). doi: 10.1061/(ASCE)MT.1943-5533.0001398.

Scrivener, K.L. & Kirkpatrick, R.J. 2008. Innovation in use and research on cementitious material. *Cement and Concrete Research* 38: 128–136. doi:10.1016/j.cemconres.2007.09.025.

Silva, P., Sagoe-Crenstil, K. & Sirivivatnanon, V. 2007. Kinetics of geopolymerization: Role of Al2O3 and SiO2. *Cement and Concrete Research* 37: 512–518.

Xu, H. & van Deventer, J.S.J. 2003. The effect of alkali metals on the formation of geopolymeric gels from alkali-feldspars. *Colloids and Surfaces A: Physicochemical Engineering Aspects* 216: 27–44.

WASTES – Solutions, Treatments and Opportunities II – Vilarinho, Castro & Lopes (Eds)
© 2018 Taylor & Francis Group, London, ISBN 978-1-138-19669-8

Waste of biodiesel production: Conversion of glycerol into biofuel additives

S. Carlota & J.E. Castanheiro
Centro de Química de Évora, Departamento de Química, Universidade de Évora, Évora, Portugal

A.P. Pinto
Departamento de Química, Universidade de Évora, Évora, Portugal
Instituto de Ciências Agrárias e Ambientais Mediterrânicas (ICAAM), Universidade de Évora,
CLAV, Évora, Portugal

ABSTRACT: The condensation of glycerol with acetone was performed using silicotungstic acid (SiW) immobilized in SBA-15, at 70°C. The main product of glycerol acetalization was solketal. Different techniques were used to characterize the catalysts. A series of catalysts with different heteropolyacid loading were prepared. It was observed that the catalytic activity increases with the amount of SiW immobilized in SBA-15, being the SiW2@SBA-15 (with 7.8 wt.%) the most active sample. All catalyst exhibited good values of selectivity to solketal (about 99% near complete conversion). To optimize the reaction conditions, the effect of different reaction parameters (catalyst loading, molar ratio of glycerol to acetone and temperature), on the glycerol acetalization over the SiW2@SBA-15 catalyst, was studied. Catalytic stability of the SiW2@SBA-15 was evaluated by performing consecutive batch runs with the same catalyst sample. It was observed that, after the third use, the catalytic activity stabilized.

1 INTRODUCTION

In the biodiesel production, the glycerol is a by-product. For every 900 kg of biodiesel produced, about 100 kg of a crude glycerol is formed. In the last years, the increase of use and production of biodiesel has resulted in an increase of glycerol production and a price decline. Therefore, it is imperative to develop new uses for glycerol to prevent environmental problems and to add value to the biodiesel production chain. The glycerol is a very promising low-cost feedstock or platform molecule for producing a wide variety of value-added special and fine chemicals (Avhad et al, 2016).

The reaction of glycerol with acetone gives a branched oxygen-containing compound and could be used as an additive in the biodiesel formulation, improving the cold properties and lowering the viscosity (Trifoi et al., 2016). The products of glycerol condensation with acetone are (2,2-Dimethyl-[1,3]dioxan-4-yl)-methanol (solketal) and 2,2-Dimethyl-[1,3]dioxan-5-ol. Solketal is an excellent compound for the formulation of gasoline, diesel and biodiesel fuels. This reaction has been studied over heterogeneous catalysts, including zeolites and Amberlyst (Deutsch et al., 2007), sulfonic mesostructured silica (Vicente et al., 2010) and PVA-SO$_3$H (Lopes et al., 2015).

Heteropolyacids (HPAs) are widely used as acid catalysts, due to their very strong Brönsted acidity and their structural properties. HPAs have been supported on different materials (Narkhede et al, 2016).

In this work, it is studied the condensation of glycerol with acetone over SiW immobilized in SBA–15. It is also studied the influence of different reaction parameters such as temperature, molar ratio of glycerol to acetone and catalyst loading on the activity of the most efficient catalyst.

2 EXPERIMENT

2.1 Catalysts preparation

The catalysts were prepared similarly to Luo et al., 2008. Firstly, template P123 (4.0 g) was dissolved in water (105 g) at room temperature, during 3 h. After this period, desired amount of SiW was added to mixture. Tetraethyl orthosilicate (TEOS) (8.55 g) and 0.75 cm^{-3} HCl (12 mol.L^{-1} were added to mixture. Subsequently, the mixture was stirred at 38°C during 20 h. The mixture was subjected to treatment at 96°C for 6 h. The dried gel was calcined at 100, °C for 2 h, in vacuum. The product was washed with dilute HCl (0.5 mol.L^{-1}) at 60°C for three times. The dried powder was refluxed in boiling absolute ethanol (30 mL) containing 1 mL of HCl (12 mol.L^{-1}) for 3 h.

2.2 Catalysts characterization

The textural characterization of the catalysts was based on the adsorption of N$_2$ at 77 K, with a Micromeritics ASAP 2010 apparatus.

The amount of SiW immobilized in SBA-15 was measured by dissolving the catalyst in H$_2$SO$_4$/HF 1:1 (v/v) and analyzing the obtained solution using inductively coupled plasma analysis (ICP), which was carried out in a Jobin-Yvon ULTIMA instrument.

FTIR spectra were recorded on a Bio-Rad FTS 155 FTIR spectrometer at room temperature in KBr pellets over range of 400–4000 cm^{-1} under atmospheric conditions.

The X-ray diffraction (XRD) patterns of the heteropolyacid, SBA-15 and catalysts were obtained by using a Rigaku Miniflex powder diffractometer with built-in recorder, using Cu Kα radiation, nickel filter, 30 mA and 40 kV in the high voltage source, and scanning angle between 0.5 and 10° of 2θ at a scanning rate of 1°/min.

Catalyst acidity was measured by means of potentiometric titration, according to Pizzio et al. 2003. A small quantity of n-butylamine solution (0.05 N) in acetonitrile was added to a known mass (0.05 g) of solid suspended in acetonitrile (90 cm^3), and shaken for 3 h. The suspension was then potentiometrically titrated with the same solution of *n*-butylamine in acetonitrile. The electrode potential variation was measured with a Crison micropH 2001 instrument.

2.3 Experiment

The catalytic experiments were carried out in a stirred batch reactor at 70°C. In a typical experiment, the reactor was loaded with 18.82 g of acetone and 4 g of glycerol (99%). Reactions were started by adding 0.2 g of catalyst.

Stability tests of the SiW2@SBA-15 were carried out by running four consecutive experiments, in the same reaction conditions. Between the catalytic experiments, the catalyst was separated from the reaction mixture by centrifugation, washed with acetone and dried at 120°C overnight.

Samples were taken periodically and analyzed by GC, using a KONIC HRGC-3000C instrument equipped with a 30 m × 0.25 mm DB-1 column.

3 RESULTS AND DISCUSSION

3.1 Characterization of catalysts

Table 1 shows the textural characterization of catalysts. The specific surface area (S$_{BET}$) of the catalysts was determined using the BET method. The total pore volume (V$_T$) estimated from the value corresponding to a ratio p/p° = 0.98. After immobilization of SiW in SBA-15, decrease of surface area and total pore volume were observed. Similar results were also observed by Gagea et al., 2009. The acidity of the SBA-15 and SiW@SBA-15 catalysts was measurement by potentiometric titration with n-butylamine. According to Pizzio, et al., 2003, the initial electrode potential (Ei) indicates the maximum acid strength of the surface sites. Materials with a Ei > 100 mV may be classified with very strong sites. The Ei of catalysts increased with the amount of SiW immobilized in SBA-15 (Table 1), which could be explained due to the increases of protons amount in silica.

FTIR spectra of the SBA-15 and SiW2@SBA-15 materials are shown in Figure 1. The bands at 982, 917 and 784 cm^{-1}, which are the same as those reported for the acid $H_4SiW_{12}O_{40}$, corresponding to the vibrations W = O (terminal oxygen), Si-O and W-O-W (edge-sharing oxygen), respectively (Bielanski et al., 2003).

Figure 2 shows the X-ray diffraction patterns of catalysts. A well resolved pattern with a prominent peak (100) and two weak peaks (110) and (200) are detected in the 2θ region of

Table 1.   Physicochemical characterization of SiW immobilized in SBA-15.

Sample	HPA load[a] (wt%)	Surface area[b] (m²/g)	V_T[c] (cm³/g)	E_i[d] (mV)
SBA-15	–	955	0.98	+105
SiW1@SBA-15	1.5	845	0.83	+254
SiW2@SBA-15	7.8	775	0.77	+415
SiW3@SBA-15	12.6	715	0.68	+532

[a]HPA load determined by ICP analysis; [b]BET; [c](p/p°) = 0.98; [d]Initial electrode potential.

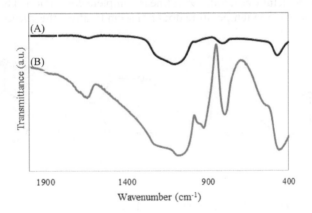

Figure 1.   FTIR spectra of catalyst: (A) SBA-15 and (B) SiW2@SBA-15.

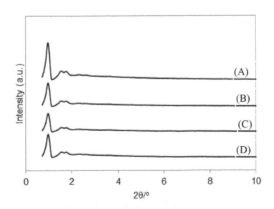

Figure 2.   X-ray diffractograms of catalysts: (A) SBA-15, (B) SiW1@SBA-15, (C) SiW2@SBA-15, (D) SiW3@SBA-15.

0.5° to 2°. These results suggest that the mesostructures are preserved after incorporation of heteropolyacid in SBA-15 (Chai et al., 2009).

3.2 *Catalytic experiments*

The condensation of glycerol with acetone was carried out over SiW incorporated in SBA-15. The main product of glycerol acetalization was the solketal (A) being also formed 2,2-Dimethyl-[1,3]dioxan-5-ol (B) (Figure 3).

Figure 4 shows the initial activity of SiW immobilized in SBA-15 on glycerol acetalization. It was observed that the catalytic activity increases with the amount of heteropolyacid immobilized in SBA-15 until a maximum, obtained with SiW2@SBA-15. This behavior can be due to the increase the amount of SiW immobilized in SBA-15 (Table 1). However, when the SiW amount increased from 7.8% (SiW2@SBA-15) to 12.6% (SiW3@SBA-15), a decrease of the catalytic activity was observed. A possible explanation of these results could be due to a decrease of surface area (S_{BET}) and total porous volume (Table 1) with the amount of heteropolyacid immobilized in SBA-15.

To study the effect of reaction temperature on acetalization of glycerol over SiW2@SBA-15, different experiments were carried out. The reaction temperature was changed from 40°C to 70°C, while the molar ratio glycerol:acetone and the amount of catalyst were kept constant. Figure 5 shows the effect of temperature on acetalization of glycerol. It was observed that the higher temperature yields the greater conversion of glycerol. Jermy and Pandurangan, 2006, observed similar results in the cyclohexanone acetalization with methanol over Al-MCM-41. The selectivity to solketal was about 99% at near complete conversion. This result can be an indication that the reaction temperature does not seem to affect the selectivity to solketal.

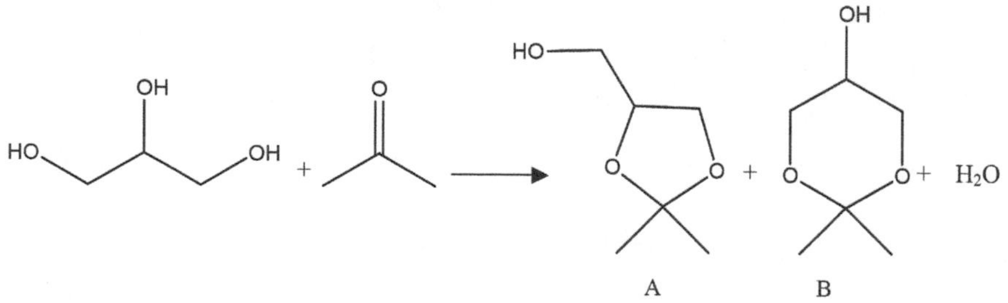

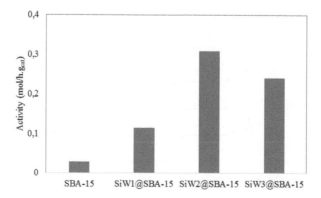

Figure 3.   Acid-catalyzed acetalization of glycerol with acetone.

Figure 4.   Acetalisation of glycerol over PW immobilized in SBA-15. Initial activities taken as the maximum observed reaction rate.

The effect of the amount of catalyst on the glycerol conversion is shown at Figure 6. The reaction has been performed by varying the amount of SiW2@SBA-15 catalyst between 0.1 g and 0.2 g keeping glycerol:acetone ratio fixed at 1:6, at 70°C. The increase in catalyst amount from 0.1 g to 0.2 g leads to an increase of the glycerol conversion. This behavior can be attributed to an increase in the availability and number of catalytically active sites to the reactant molecules. A similar behavior of HPW immobilized on MCM-41 in condensation of benzaldehyde with pentaerythritol was observed by Jermy and Pandurangan, 2005. It was also observed that the selectivity of SiW2@SBA-15 catalyst to solketal was not changed with the increase of catalyst loading (about 99% at near complete conversion).

The influence of molar ratio of glycerol to acetone on the glycerol conversion was also studied. The catalytic experiments were carried out at reaction temperature of 70°C, for sample SiW2@SBA-15, while the glycerol/acetone ratio was varied using the proportions 1:3, 1:6 and 1:12. Figure 7 shows the effect of the molar ratio glycerol to acetone on the glycerol conversion. An increase in the glycerol conversion with the molar ratio of glycerol to acetone was observed. This behavior can be explained due to the used excess of one reactant (acetone), reducing the backward reaction. However, when the molar ratio of glycerol to acetone increases (from 1:6 to 1:12) an increase of conversion is not observed. This behavior can be attributed with the occupancy of acetone over the active sites. Similar results were also observed by Jermy et al., 2006. Very good selectivity to the solketal was also observed (about 99% at near complete conversion), for all different molar ratio studied.

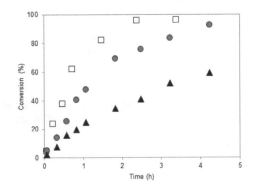

Figure 5. Acetalisation of glycerol over SiW2@SBA-15. Effect of the reaction temperature. Conversion (%) *versus* time (h): (▲) T = 40°C; (●) T = 55°C; (□) T = 70°C.

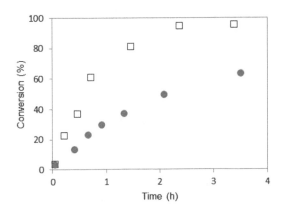

Figure 6. Acetalisation of glycerol over SiW2@SBA-15. Effect of the catalyst loading. Conversion (%) *versus* time (h): (●) m = 0.1 g; (□) m = 0.2 g.

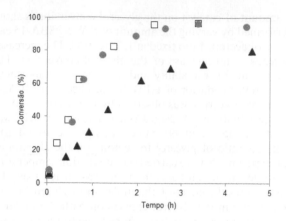

Figure 7. Acetalisation of glycerol over SiW2@SBA-15. Effect of the molar ratio of glycerol to acetone. Conversion (%) *versus* time (h): (▲) 1:3; (□) 1:6; (●) 1:12.

Table 2. Catalytic activity of SiW2@SBA-15. Study of stability.

	Catalytic activity (mol/h.g_{cat})
First use	0.303
Second use	0.299
Third use	0.287
Forth use	0.291

To study the catalytic stability of SiW2@SBA-15, consecutive batch runs with the same catalyst sample and in the same reaction conditions were performed. It was observed a small decrease of the catalytic activity from the first to the second use. However, after the third use, it was observed a stabilization of the catalytic activity (Table 2).

4 CONCLUSIONS

The condensation of glycerol with acetone was successfully performed over a series of SiW immobilized in SBA-15. Catalysts with different SiW loading in SBA-15, which was varied from 1.8 to 12.8 wt%, were prepared. SiW2@SBA-15 (with 7.6 wt.%) showed the highest catalytic activity. All catalysts exhibited high selectivity values to solketal. In optimized reaction conditions, it was found that at 70°C, with 0.2 g of SiW2@SBA-15 and with a molar ratio glycerol to acetone 1:6, a glycerol conversion of about 96% can be obtained, after 4 hours. SiW2@SBA-15 catalyst showed high catalytic stability.

REFERENCES

Avhad, M.R. and Marchetti, J.M. 2016. Innovation in solid heterogeneous catalysis for the generation of economically viable and ecofriendly biodiesel: A review. *Catal. Rev.* 58: 157–208.

Bielanski A., Lubanska, A., Pozniczek, J., Micek-Ilnicka, A. 2003. The formation of MTBE on supported and unsupported $H_4SiW_{12}O_{40}$. *Appl. Catal. A:Gen.* 256: 153–171.

Chai, S.-H., Wang, H.-P., Liang, Y., Xu, B.-Q. 2009. Sustainable production of acrolein: Preparation and characterization of zirconia-supported 12-tungstophosphoric acid catalyst for gas-phase dehydration of glycerol. *Appl. Catal. A: Gen.* 353: 213–222.

Deutsch, J., Martin, A., Lieske, H. 2007. Investigations on heterogeneously catalysed condensations of glycerol to cyclic acetals. *J. Catal.* 245: 428–435.

Gagea, B.C., Lorgouilloux, Y., Altintas, Y., Jacobs, P.A., Martens, J.A. 2009. Bifunctional conversion of n-decane over HPW heteropoly acid incorporated into SBA-15 during synthesis. *J. Catal.* 265: 99–108.

Jermy, B.R., Pandurangan, A. 2005. $H_3PW_{12}O_{40}$ supported on MCM-41 molecular sieves: An effective catalyst for acetal formation. *Appl. Catal. A: Gen.* 295: 185–192.

Jermy, B.R., Pandurangan, A. 2006. Al-MCM-41 as an efficient heterogeneous catalyst in the acetalization of cyclohexanone with methanol, ethylene glycol and pentaerythritol. *J. Mol. Catal. A: Chem.* 256: 184–192.

Kong, P.S., Aroua, M.K., Daud, W.M.A.W. 2016. Conversion of crude and pure glycerol into derivatives: A feasibility evaluation. *Renew. Sustain. Energy Rev.* 63: 533–555.

Lopes, N.F., Caiado, M., Canhao, P., Castanheiro, J.E. 2015. Synthesis of bio-fuel additives from glycerol over poly(vinyl alcohol) with sulfonic acid groups. *Energ. Source Part A: Recovery, Utilization and Environmental Effects* 37: 1928–1936.

Luo, Y., Hou, Z., Li, R., Zheng, X. 2008. Rapid synthesis of ordered mesoporous silica with the aid of heteropoly acids. *Micropor. Mesopor. Mater.* 109: 585–590.

Narkhede, N., Singh, S. and Patel, A. 2015. Recent progress on supported polyoxometalates for biodiesel synthesis via esterification and transesterification. *Green Chem.* 17: 89–107.

Pizzio, L.R., Vásquez, P.G., Cáceres, C.V., Blanco, M.N. 2003. Supported Keggin type heteropolycompounds for ecofriendly reactions. *Appl. Catal. A:Gen.* 256: 125–139.

Trifoi, A.R., Agachi, P.Ş., Pap, T. 2016. Glycerol acetals and ketals as possible diesel additives. A review of their synthesis protocols. *Renew. Sustain. Energy Rev.* 62: 804–814.

Vicente, G., Melero, J.A. Morales, G., Paniagua, M., Martín, E. 2010. Acetalisation of bio-glycerol with acetone to produce solketal over sulfonic mesostructured silicas. *Green Chem.* 12: 899–907.

WASTES – Solutions, Treatments and Opportunities II – Vilarinho, Castro & Lopes (Eds)
© *2018 Taylor & Francis Group, London, ISBN 978-1-138-19669-8*

Geotechnical characterization of recycled C&D wastes for use as trenches backfilling

C.S. Vieira & M.L. Lopes
CONSTRUCT, Faculty of Engineering, University of Porto, Porto, Portugal

N. Cristelo
CQVR, University of Trás-os-Montes e Alto Douro, Vila Real, Portugal

ABSTRACT: Construction and demolition wastes have been increasingly used as recycled aggregates in concrete production and roadway applications. However, the fine grain portion of these recycled aggregates is usually not considered as a suitable material for such applications, and is thus landfilled. This paper presents the geotechnical and geoenvironmental characterization of a fine grain non-selected C&DW, to assess its feasibility as backfilling material of trenches.

1 INTRODUCTION

Construction and Demolition Waste (C&DW) is one of the heaviest and most voluminous waste streams, representing approximately 25% to 30% of all waste generated in the European Union (Vieira and Pereira, 2015).

C&DW are usually defined as the residues from the operations of construction, reconstruction, maintenance and demolition of buildings and other civil infrastructures. These wastes are very heterogeneous and if a selective demolition is not carried out it is very hard to obtain recycled materials of good quality.

C&DW have been identified by the European Commission as a priority stream due to the large amounts generated and, additionally, their high potential for re-use and recycling. The re-use of C&DW reduces the need for natural resources (non-renewable) and avoids the landfill of inert materials. Despite these main advantages associated with the recycling of C&DW, some member states of the European Union still have low recycling rates, far below the minimum of 70% stipulated by the Waste Framework Directive of the European Parliament, to be achieved in 2020 (UE, 2008).

Recently, the Portuguese Environment Agency (APA) has promoted, together with the National Laboratory of Civil Engineering (LNEC), the publication of new technical specifications for the use of materials resulting from C&DW. One of these technical specifications is related to the use of materials resulting from C&DW in backfill of trenches (LNEC E485, 2016). This paper presents the geotechnical and geoenvironmental characterization of a fine grain non-selected C&DW, to assess whether the technical requirements laid by the LNEC E485 specification are fulfilled, which would allow its use as backfill of trenches.

2 LABORATORY CHARACTERIZATION

2.1 *Physical and geotechnical properties*

The recycled C&DW used in this study was collected from a single batch in a Portuguese recycling plant located in the central region of Portugal. It is a fine grain mixed recycled aggregate obtained from non-selected C&DW coming mainly from the demolition of small buildings and cleaning of lands with illegal deposition of C&DW. The recycling process involves the removal of unwanted materials (wood, plastics, cardboard,…), the fragmentation of the C&DW and the grain-size separation. Due to its granulometric fraction (fine grains), the material used in the study has a very heterogeneous composition.

The constituents of the recycled C&DW were evaluated following the test procedure laid out in the standard EN 933-11 (2009), but including some minor changes suggested by LNEC E485 (2016), particularly the separation of the soil portion from the constituent designated as "other materials". The constituents of the material under analysis are reproduced in Table 1. It is important to point out that this standard refers to the constituents of coarse recycled aggregates which means that only the mass comprised between the sieves 63 mm and 4 mm is analysed. The classification of the particles with dimensions lower than 4 mm is considered to be humanly impracticable.

The particle size distribution of the C&DW was determined by sieving and sedimentation (Figure 1). The gradation of the recycled material was firstly determined following the European Standard EN 933-1 (2012) for aggregates. However, since this material has a significant fines content (about 15.5%), the particle size distribution was also evaluated according to ISO/TS 17892-4 (2004), which was designed for soils. Based on the gradation curve, some grain size distribution parameters were determined and summarised in Table 2.

According to the Unified Soil Classification System (USCS) this recycled material can be classified as silty sand (SM).

Table 1. Classification of recycled C&DW constituents (Pereira et al., 2015).

Constituents	C&DW
Concrete, concrete products, mortar, concrete masonry units, R_c (%)	40.0
Unbound aggregate, natural stone, hydraulically bound aggregate, R_u (%)	36.5
Clay masonry units, calcium silicate masonry units, aerated non-floating concrete, R_b (%)	10.8
Bituminous materials, R_a (%)	0.5
Glass, R_g (%)	1.2
Soils, R_s (%)	10.8
Other materials, X (%)	0.1
Floating particles, FL (cm³/kg)	10.0

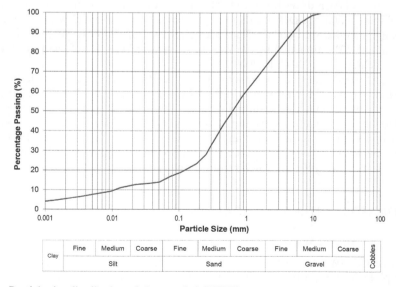

Figure 1. Particle size distribution of the recycled C&DW.

Modified compaction tests were carried out, according to EN 13286-2 (2010), to determine the moisture content—dry density curve of the recycled C&DW, allowing the subsequent determination of the maximum dry density, MDD (2.1 Mg/m³) and optimum moisture content, OMC (9.4%).

California bearing ratio (CBR) tests were performed in accordance with BS 1377-4 (1990). A 152 mm diameter by 178 mm high stainless steel cylinder was used to mould recycled C&DW specimens. The compaction was made in 5 layers, targeting a MDD of 95% of the Modified Proctor test result, and the corresponding OMC. CBR moulds were submerged in water for 96 hours.

After correction of concavity of the initial part of the stress-penetration curve, following the procedure laid out in BS 1377-4, the forces corresponding to 2.5 mm and 5 mm penetration were read and the CBR values were quantified. The higher CBR value was 112%.

Figure 2 presents the failure envelopes obtained by large-scale direct shear tests (Figure 2a) and triaxial compression tests (Figure 2b) carried out to evaluate the shear strength of the recycled C&DW. The direct shear tests were performed under normal stress of 25, 50, 100 and 150 kPa and triaxial compression tests were carried out for confining pressure of 25, 50, 100 and 200 kPa. C&DW samples were compacted, in all the tests, at 90% of maximum Modified Proctor dry density and at the optimum water content.

Direct shear tests led to a friction angle of 37.1° and a cohesion of 21.7 kPa, while the triaxial compression tests led to values of 36.8° and 23.2 kPa, respectively. Comparing the results obtained from both tests, it becomes apparent that the shear strength parameters are quite similar.

Table 2 summarises the values of some geotechnical parameters evaluated in this study.

Table 2. Summary of geotechnical properties of the recycled C&DW.

Parameter	
D_{max} (mm)	12.7
D_{50} (mm)	0.6
Coefficient of Uniformity, C_u	95.0
Coefficient of Curvature, C_c	7.7
Particles density, G_s	2.58
Minimum void ratio, e_{min}	0.434
Maximum void ratio, e_{max}	0.877
Modified Proctor OMC (%)	9.4
Modified Proctor MDD (Mg/m³)	2.1
CBR (%)	112
Direct shear tests, f' (°)	37.1
¢ (kPa)	21.7
Triaxial tests, f' (°)	36.8
¢ (kPa)	23.2

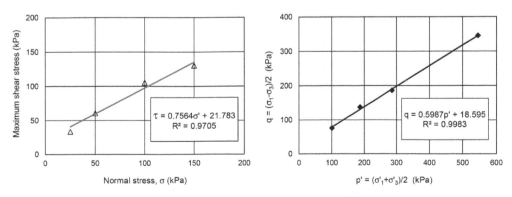

Figure 2. Shear strength envelopes for: (a) large-scale direct shear tests; (b) triaxial compression tests.

2.2 Chemical properties

Recycled materials used in construction projects must demonstrate an adequate leaching behaviour, complying with the limits imposed by the Directive 2003/33/EC (Council Decision 2003/33/EC). Laboratory leaching tests following the procedure described in the standard EN 12457-4 (2002) were carried out to assess the potential risk of ground contamination.

The limit values stipulated by the Council Decision 2003/33/EC for the acceptance of waste at inert landfills and the results of the laboratory leaching are presented in Table 3. The pH value of the eluate, measured at the time of the sampling, is also included in Table 3.

The recycled C&DW complies with the provisions of the European Council Decision 2003/33/EC for inert materials, with the exception of the sulphates (SO_4) and the total dissolved solids (TDS). Nevertheless, it is important to highlight the fact that the Directive 2003/33/EC allows the leaching of sulphates to reach a value as high as 6000 mg/kg (when performing a batch test with a liquid to solid ratio of 10 l/kg, which was the case in the present study), while, at the same time, the TDS evaluation is not compulsory.

Table 3. Results of laboratory leaching test (liquid to solid ratio of 10 l/kg).

Parameter	Value (mg/kg)	Acceptance criteria for leached concentrations—Inert landfill
Arsenic, As	0.021	0.5
Lead, Pb	<0.01	0.5
Cadmium, Cd	<0.003	0.04
Chromium, Cr	0.012	0.5
Copper, Cu	0.10	2
Nickel, Ni	0.011	0.4
Mercury, Hg	<0.002	0.01
Zinc, Zn	<0.1	4
Barium, Ba	0.11	20
Molybdenum, Mo	0.018	0.5
Antimony, Sb	<0.01	0.06
Selenium, Se	<0.02	0.1
Chloride, Cl	300	800
Fluoride, F	6.1	10
Sulphate, SO_4	3200	1000
Phenol index	<0.05	1
Dissolved Organic Carbon, DOC	220	500
Total Dissolved Solids, TDS	6580	4000
pH	8.2	–

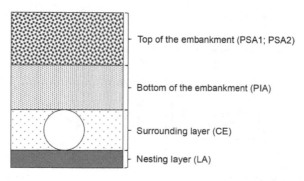

Figure 3. Representation of material categories in the trench (adapted from LNEC E485).

The evaluation of the organic content (humus) of the recycled C&DW was carried out following the disposed in Section 15 of the standard EN 1744–1 (2009). The method is based on the principle that humus develops a dark colour when it reacts with sodium hydroxide (NaOH). The intensity of the colour depends on the humus content. If the solution is clear or only slightly coloured, the aggregate does not contain a significant amount of humus (EN 1744-1, 2009). The colour of the solution containing the recycled C&DW was lighter than the standard colour, therefore the amount of humus (organic material) of this recycled material is negligible.

The content of water soluble sulphates was determined by spectrophotometry, following the Section 10 of the standard EN 1744–1 (2009) for recycled aggregates. The aggregate specimen was sieved through a 4 mm sieve and the oversized particles were crushed to pass the 4 mm sieve. The aggregate specimens were then mixed with hot water ($60° \pm 5°$), to extract water-soluble sulphate ions. Following the procedures mentioned in the above-mentioned standard, barium chloride was added so that sulphate ions precipitate as barium sulphate. The mean value of the spectrophotometer records was 0.14%.

The quality of fines of the recycled C&DW, expressed by the value of the methylene blue $MB_{0/D}$, was evaluated based on the standard EN 933-9 (2009). The value of the methylene blue multiplied by the fraction under 2 mm was 1.2 g/kg.

3 FEASIBILITY ASSESSMENT FOR USE IN TRENCHES

3.1 *Requirements from Portuguese specification LNEC E485*

Based on the relative proportions of the constituents, the specification LNEC E485 (2016) defines three classes (MR) of recycled materials resulting from C&DW (Table 4). The meaning of the parameters R_x, F_L and X in Table 4 is included in Table 1.

Different material categories are also defined by LNEC E485 (2016) depending on the position of the material in the trench (Figure 3). The properties and minimum requirements that each C&DW class needs to satisfy in order to be accepted in each of the trench categories are presented in Table 5.

3.2 *Assessment and discussion*

Comparing the proportion of constituents listed in Table 1 with the limits reported in Table 4, this recycled C&DW could be included in any of the three established classes (MR1, MR2 or MR3), if it wasn't for the high value of the floating particles, which is higher than the minimum value of 5 cm³/kg established for all the classes.

Furthermore, the fines content of 15.5% does not allow its use in nesting and surrounding layers (CE/LA) nor in the 'top of embankment' layer, when category PSA1 is required. Taking into account that the water soluble sulphates (0.14%) is lower than 0.7%, and also that the leaching meets the Directive 2003/33/EC requirements, this recycled material can be used as a PIA or PSA2 material (Table 5 and Figure 4), which means that it can be used to build the bottom layer of the embankment (under pavements of

Table 4. Classification of recycled materials coming from C&DW (adapted from LNEC E485).

| Material class | Proportion of constituents | | | | | |
	$R_c + R_u + R_g$ (%)	R_g (%)	R_a (%)	$R_b + R_s$ (%)	F_L (cm³/kg)	X (%)
MR1	≥ 70	≤ 25	≤ 30	≤ 30	≤ 5	≤ 1
MR2	No limit	≤ 25	No limit	No limit	≤ 5	≤ 1
MR3	No limit	≤ 5	≤ 10	No limit	≤ 5	≤ 1

Table 5. Properties and minimum requirements of C&DW classes for backfilling of trenches (adapted from LNEC E485).

Compliance requirements			Material category		
			CE/LA	PSA2/PIA	PSA1
Parameters	Property	Test standard	MR3	MR1, MR2, MR3	MR1
Geometry and nature	D_{max}	EN 933-1	≤ 20 mm, if $NPD^{(*)} \leq 200$ ≤ 40 mm, if $200 \leq NPD \leq 600$ ≤ 63 mm, if $NPD > 600$	≤ 180 mm	≤ 80 mm
	Fines content (≤ 0.063 mm)	EN 933-1	$\leq 12\%$	–	$\leq 12\%$
	Assessment of fines ($MB_{0/D}$)	EN 933-9	–	–	< 2.0
Mechanical behaviour	Resistance to wear (micro-Deval)	EN 1097-1	–	–	≤ 45
	Resistance to fragmentation (LA)	EN 1097-2	–	–	≤ 45
Chemical analysis	Water-soluble sulphates	EN 1744-1	$\leq 0.7\%$	$\leq 0.7\%$	$\leq 0.7\%$
	Release of dangerous substances	EN 12457-4	Classification as inert waste for landfill.		

$^{(*)}$ NPD – nominal pipe diameter (mm).

transport infrastructures, walkways and green spaces) and the top layer, except if the trench is located under pavements of transport infrastructures, in which case a PSA1 category is required.

4 CONCLUSIONS

The geotechnical and geoenvironmental characterization of a fine grain non-selected C&DW was presented and the feasibility of their use as backfilling material of trenches was assessed.

The high value of floating particles, resulting from impurities such as wood, plastics and foams, does not allow the integration of this C&DW in any of the classes defined in the Specification LNEC E485. This fact highlights the importance of selective demolitions.

If the limit of floating particles was neglected, this C&DW could be applied at the bottom or top layers of the trenches.

A proper mechanical behaviour, assessed through direct shear, triaxial and CBR tests, was also reached.

ACKNOWLEDGEMENTS

This work was financially supported by: Project POCI-01-0145-FEDER-007457 – CONSTRUCT—Institute of R&D In Structures and Construction funded by FEDER funds through COMPETE 2020 – Programa Operacional Competitividade e Internacionalização (POCI).

REFERENCES

BS 1377-4: 1990. Methods of test for soils for civil engineering purposes. Compaction-related tests. *British Standards Institution.*

Council Decision 2003/33/EC. Council Decision establishing criteria and procedures for the acceptance of waste at landfills pursuant to Article 16 of and Annex II to Directive 1999/31/EC. *Official Journal of European Union*: L11/27.

EN 933-1: 2012. *Tests for geometrical properties of aggregates—Part 1: Determination of particle size distribution—Sieving method.* CEN.

EN 933-9: 2009. *Tests for geometrical properties of aggregates. Assessment of fines. Methylene blue test.* CEN.

EN 933-11: 2009. *Tests for geometrical properties of aggregates* - Part 11: Classification test for the constituents of coarse recycled aggregate. CEN.

EN 1744-1: 2009. *Tests for chemical properties of aggregates.* Chemical analysis, CEN.

EN 12457-4: 2002. *Characterisation of waste—Leaching—Compliance test for leaching of granular waste materials and sludges—Part 4.* CEN.

EN 13286-2: 2010. *Unbound and hydraulically bound mixtures. Test methods for laboratory reference density and water content.* Proctor compaction. CEN.

ISO/TS 17892-4: 2004. *Geotechnical investigation and testing—Laboratory testing of soil* - Part 4: Determination of particle size distribution. ISO.

LNEC E485: 2016. Guide for the use of materials resulting from Construction and Demolition Waste in bakfill of trenches. *LNEC (Portuguese Laboratory of Civil Engineering).* 8p (in Portuguese).

Pereira, P.M., Vieira, C.S. & Lopes, M.L. 2015. Characterization of Construction and Demolition Wastes (C&DW)/geogrid interfaces. *3rd Edition of the International Conference on Wastes: Solutions, Treatments and Opportunities*: 215–220.

UE 2008. Directive 2008/98/EC of the European Parliament and of the Council of 19 November on waste and repealing certain Directives. *Official Journal of the European Union L312/3 of 22 November 2008.*

Vieira, C.S. & Pereira, P.M. 2015. Use of recycled construction and demolition materials in geotechnical applications: A review. *Resources, Conservation and Recycling* 103: 192–204.

Recycling of MSWI fly ash in clay bricks—effect of washing and electrodialytic treatment

W. Chen, E. Klupsch, G.M. Kirkelund, P.E. Jensen & L.M. Ottosen
Department of Civil Engineering, Technical University of Denmark, Lyngby, Denmark

C. Dias-Ferreira
Materials and Ceramic Engineering Department, CICECO, University of Aveiro, Aveiro, Portugal
CERNAS, College of Agriculture, Polytechnic Institute of Coimbra, Coimbra, Portugal

ABSTRACT: The feasibility of incorporating MSWI fly ash into sintered clay bricks was studied. Before being used in bricks, the ash was pre-treated by a combination of water washing and electrodialytic remediation to remove soluble salts and extract heavy metals. MSWI fly ash-clay bricks with 5%, 10% and 20% ash levels were handmade and fired at 1000°C for 6 h. The fired ash-clay bricks had higher porosity and lower compressive strength compared to the 100% clay brick, even though the washing-electrodialytic treatment improved the compressive strength and reduced the water absorption of the fired ash-clay bricks. This study indicates that fired ash-clay bricks with 5% treated ash may be feasible as construction materials, and the heavy metal leaching from these bricks in granular form (e.g. at the end of their service life) was low, indicating these demolished bricks could be reused again in construction work.

1 INTRODUCTION

Fly ash generated from municipal solid waste incineration (MSWI) is a hazardous waste due to presence and leachability of heavy metals and organic pollutants (e.g. dioxins and polycyclic aromatic hydrocarbons). In 2000, approximately 25 Mt/year of fly ash was generated in USA, Japan and EU (Reijnders 2005). Electrodialytic remediation (EDR) is one technique for MSWI fly ash treatment (Ferreira et al. 2005), where an electric DC field is applied to an ash-water suspension to extract and separate heavy metal by migration towards anode or cathode through ion exchange membranes. Ferreira et al. (2008) observed that in MSWI ash treated by water washing and EDR, metals were mainly in the strongly bonded and residual phases, indicating a reduction in the ash's environmental risk. Belmonte et al. (2016) made Greenlandic bricks (~2 g discs) containing 20% and 40% of EDR treated MSWI fly ash, and found that bricks had a low durability and high leaching of As and Cr. In the present study, fired fly ash-clay bricks with a larger size and with lower EDR-treated ash (water-washed before EDR) contents (5%, 10% and 20%) were made and characterized. These bricks were compared with 100% clay bricks and with bricks made from original MSWI fly ash at 20% substitution rate. The feasibility of incorporation of MSWI fly ash treated by combined washing and EDR in production of sintered clay bricks was investigated.

2 MATERIALS AND METHODS

2.1 Materials

The materials used were: C – clay (Wienerberger, Helsinge, Denmark), dried at 105°C for 24 h and dry-ring milled for 60 s; R – MSWI fly ash (from I/S Vestforbrænding, Glostrup, Denmark), collected after the electrostatic precipitator and before acid neutralization in the wet scrubber; T – treated MSWI fly ash obtained by a 3-step distilled water (DI) washing (circa 20 s vigorous end-over-end manual shaking at L/S 2.5 L/kg for each step) of the raw fly ash and then submitted to EDR. The electrodialytic cell used in EDR was cuboid-shaped and had two compartments, i.e. anode ($l \times w \times h = 37.5$ cm $\times$ 5.5 cm/6 cm to membrane $\times$ 27.5 cm) and cathode ($l \times w \times h = 37.5$ cm $\times$ 2.5 cm/3 cm to membrane $\times$ 27.5 cm), separated by a cation exchange membrane (effective area 500 cm^2). 1 kg of the dried water-washed ash suspended in 15 L DI was fed continuously into the anode compartment at 5 L/min and recirculated in the system (Fig. 1). An electrolyte solution (5 L 0.01 M NaNO$_3$, pH 2) was recirculated in the cathode compartment. Treatment of the suspension (1 kg ash in 15 L water) lasted 7 d at room temperature at a constant direct current (250 mA).

2.2 Characterization of MSWI fly ash

Heavy metals were determined by HNO$_3$ digestion (Dansk Standard 2003). 1 g (3–5 replicates) of dry ash was digested with 20 mL 1:1 HNO$_3$ at 200 kPa at 120°C for 30 min in an autoclave. This digestion procedure allows partially dissolution of the mineral phases, while silicate-bound metals remain undissolved. Digested samples were filtered (0.45 µm) and diluted to 100 mL. Metal concentrations were measured by inductively coupled plasma optical emission spectrometry (ICP-OES) (Varian 720-ES, Software vs: 1.1.0). Particle size distribution was measured using a Malvern Mastersizer 2000. Morphology was revealed by scanning electron microscope imaging, using a large field detector (20 kV). The crystalline phases were studied by X-ray diffraction using PANalytical X'pert Pro diffractometer (2θ mode, Cu Kα radiation with 40 mA and 45 kV).

2.3 Brick making and characterization of fired bricks

The bricks were handmade: (a) Mixing raw materials (Table 1) with tap water (water content 23–26 wt.%) in a blender; (b) Hand moulding into $5 \times 5 \times 5$ cm^3 blocks; (c) Drying at 40°C for 24 h and then at 105°C for 24 h; (d) Firing: The heating rate was 10°C/min. Temperature was kept at 200°C for 999 min to remove water and then increased to the sintering temperature 1000°C and kept for 6 h; (e) Cooling to room temperature.

Vacuum water absorption, apparent porosity and bulk density were determined (Dansk Standard 1997), with modifications of vacuum residence time and water-soaking time in order to sufficiently fill the open pores with water (Chen et al. 2014). Compressive strength

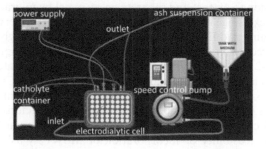

Figure 1. Flowsheet of the EDR treatment.

Table 1. The ratios of the raw materials by weight used for brick-making.

Brick name	C	R20	T5	T10	T20
Clay (wt.%)	100	80	95	90	80
Raw ash (wt.%)	–	20	–	–	–
Treated ash (wt.%)	–	–	5	10	20

without mortar joints was analysed (Shimadzu AG-A/AG-25TA). Soluble salts content (Dansk Standard 2016) 1 h DI extraction at L/S 10 L/kg on the crushed fired brick (<125 µm) and Na, K, Ca and Mg were determined in the eluent after filtration (0.45 µm).

Tank leaching test were conducted in which fired bricks were immersed in acidified DI at L/S 5 L/kg (pH 4, with HNO_3) during 64 d (NEN 7345 (Dutch Standard 1995)). Leachate samples were collected after 0.25, 1, 2.25, 4, 9, 16, 36 and 64 d, and after each sample collection the leachant was replaced by a fresh one. For batch leaching test, fired bricks were crushed to < 4 mm. The leachates were collected after 24-h agitated extraction using distilled water at L/S 2 L/kg (Dansk Standard, 2002). Metals in the leachates were analyzed by ICP-OES after filtering (0.45 µm).

3 RESULTS AND DISCUSSION

3.1 *Characteristics of raw materials*

The clay had a high $CaCO_3$ content (22 ± 0.2%), with a liquid and a plastic limit of 27% and 15%, respectively (Chen et al. 2014). Quartz, illite, dolomite, calcite and feldspar minerals (microcline and albite) were detected in the clay. Table 2 compares the HNO_3-digestible (not total) heavy metal concentrations of the ash before and after the washing-EDR treatment, and a slight reduction in As, Cd, Pb and Zn can be seen. The fluffy phase of small crystals, seen on the raw ash, was removed during treatment (Fig. 2). Mineral phases of anhydrite, calcite and quartz were detected in the ashes both before and after treatment. Hallite (NaCl) and sylvite (KCl) were detected in the raw fly ash, whereas gypsum ($CaSO_4 \cdot H_2O$) in the treated ash, which indicated a removal of alkaline metal chlorides during treatment. The ash particles were irregular in shape (Fig. 2), probably related to the composition of the ash particles. Carbon particles are bar-shaped, whereas siliceous particles are spherical (Keppert et al. 2015); both shapes were easily identified in the treated ash (Fig. 2, middle). In general, the ash particles were bigger than the clay's according to the particle size distribution given in Figure 2 (right).

3.2 *Properties of fired bricks*

The fired bricks exhibited a yellow color, which became increasingly lighter with increasing ash content. The lightest yellow was the brick with untreated ash (R20). The surface of bricks R20 was fluffy and particles peeled off, which is most likely ascribed to the high content of

Table 2. Trace elements in the MSWI fly ash. EC = electric conductivity.

	pH	EC mS/cm	As mg/kg	Cd mg/kg	Cr mg/kg	Cu mg/kg	Ni mg/kg	Pb g/kg	Zn g/kg
Raw ash	11.7 ± 0.1	61 ± 3	195 ± 15	144 ± 6	113 ± 6	680 ± 29	44 ± 2	5.3 ± 1.2	26 ± 8
Treated ash	9.7 ± 0	3.0 ± 0.1	172 ± 7	106 ± 3	119 ± 3	713 ± 26	45 ± 1	4.4 ± 0.1	19 ± 0.4

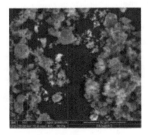  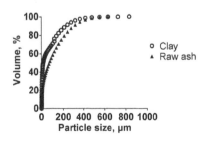

Figure 2. Microstructure of the raw MSWI fly ash (left) and treated fly ash (middle); Particle size distribution of clay and raw MSWI fly ash (right).

soluble salts. Clay bricks are defined by three categories (Dansk Standard 2015) based on the soluble salts content: S0: no requirement for soluble salts content (for use in situations whit a complete protection against water penetration); S1: 0.17% for the sum of Na^+ and K^+, and 0.08% for Mg^{2+}; and S2: 0.06% for the sum of Na^+ and K^+, and 0.03 g/l for Mg^{2+}. The soluble Na, K and Ca (Table 3) increased with increasing ash content in the bricks, except in the case of the bricks with 5% treated ash; the highest value was in 20% untreated ash, which indicated the highest risk of efflorescence. Bricks with up to 10% of treated fly ash obtained the best classification in this parameter (S2).

Mineralogy. Quartz, anorthite, augite, microcline, mullite and wollastonite were detected in all bricks (Fig. 3). A new phase, cordierite ($Mg_2AI_4Si_5O_{18}Ar_{0.625}$), was detected in the bricks with ash content > 5%, and its characteristic peaks became more obvious with increasing ash content. Cordierite was also observed by Zhang et al. (2011) in fired bricks made from MSWI fly ash (20%), red ceramic clay (60%), feldspar (10%) and gang sand (10%). Same mineral phases detected in the bricks T20 and R20 indicated that the ash treatment did not significantly affect the mineralogy of the fired bricks.

Weight loss, bulk density, vacuum water absorption and apparent porosity. The weight loss of all the bricks (Fig. 4a) during firing was < 15%, which is acceptable (Lin 2006). Substituting 20% clay with the treated ash slightly reduced the weight loss, whereas substituting 20% with the raw ash slightly increased weight loss. Burn-off of organic matter, dehydration of clay minerals or metal oxides, loss of volatile salts and/or loss of inorganic carbon in minerals were associated with the weight loss during firing (Heiri et al. 2001). The bulk density (Fig. 4b) was 1.44–1.65 g/cm^3, which is lower than that of normal bricks (1.8–2.0 g/cm^3) (Lin 2006). With increasing ash content, the bulk density decreased. However, the lowest density of the bricks was still acceptable, in comparison to London Stock bricks (hand-made bricks commonly made from yellow clay having bulk density of 1.39 g/cm^3 (Domone & Illston 2010)). Water absorption affects the durability of bricks and ranged 24.5% and 33.7% (Fig. 4d), within the range of London stock bricks: 22% to 37% (Domone & Illston 2010). Water absorption is closely related to the porosity of the bricks. The apparent porosity (Fig. 4c) was 40%-50% for all the bricks. Such high apparent porosity is also seen in London Stock bricks i.e. 36–50% (Domone & Illston 2010). Trapped air and low compaction during

Table 3. Soluble salts content (Dansk Standard 2016) in eluates from the fired bricks, in mg/l.

	Na	K	Ca	Mg	Category
C	6.5 ± 0.4	7.8 ± 0.6	22 ± 0.5	0.5 ± 0.0	S2
T5	5.6 ± 1.4	7.2 ± 2.0	83 ± 8	0.5 ± 0.1	S2
T10	14 ± 5	12 ± 3	187 ± 9	1.0 ± 0.1	S2
T20	69 ± 10	90 ± 20	259 ± 23	1.1 ± 0.1	S1
R20	272 ± 40	209 ± 12	292 ± 9	1.0 ± 0.1	S0

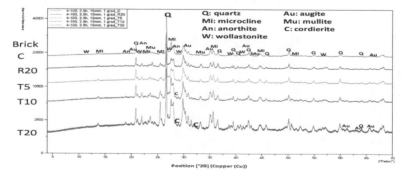

Figure 3. Mineralogy of the fired bricks.

hand-making of bricks, gaseous phase generation and release during firing, and the interactions between adjacent particles (clay-clay, clay-ash, and ash-ash) could affect the level of sintering and the development of voids in the bricks.

Compressive strength. Only the 100% clay bricks (Fig. 4e) met the minimum requirement for clay masonry (Dansk Standard 2015), i.e. 9 N/mm^2 for Group 2 bricks which have formed voids >25% and < 50% (the range that the studied bricks fell into). C, T5 and T10 had compressive strength of 5–15 N/mm^2, which fell into the range of compressive strength for London Stock bricks i.e. 5–20 N/mm^2 (Domone & Illston 2010). Making the bricks by hand was believed to be responsible for the significant deviation in the values. Increasing ash content decreased compressive strength (Fig. 4e). The bricks with 20% raw fly ash had lower compressive strength than those with 20% treated fly ash—thus, the washing-EDR treatment did improve the compressive strength.

Leaching properties. The Dutch building material decree (BMD 1999) regulates the leaching of construction materials (including secondary materials), and tank leaching, as a way to assess the release of inorganic substances from monolithic materials, was therefore conducted on the fired bricks in the present study. The acid leachant used represents contact with rain water during service life. Rain water has a natural pH of 4,7, but the values can be lower in case of acid rain, so the Dutch leaching tests recommends pH = 4 for the leachant. The results (Table 4) show that although compared to pure clay bricks, the ash-clay bricks had higher release of heavy metals, limits in the Dutch building material decree were still fulfilled. When increasing the addition of the treated fly ash, the release of almost all the heavy metals, except Cd, increased. With 20% ash addition, the release of As, Ba and Mo was higher from the bricks with treated ash than from those with untreated ash, whereas Cd, Cr, Cu, Ni, Pb and Zn decreased by treatment. By doing batch leaching at L/S of 2 L/kg the disposal of the fired ash-clay bricks was discussed: either being landfilled according to Danish Decree BEK nr. 252 on landfills, or recycled according to Danish Decree (BEK nr 1414 2015) for waste used in construction work. The results (Table 5) show that leaching of heavy metals from all the ash-clay bricks meets the limits for landfill of mineral waste in Class MA1. Brick T5 met the requirements of C3, suggesting it might be recycled again in construction work such as paths and foundations. It was mainly the batch leaching of As from T10, T20 and

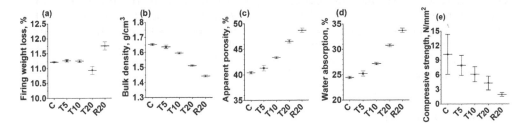

Figure 4. Firing weight loss, bulk density, apparent porosity, water absorption and compressive strength.

Table 4. Total leachability, in mg/m^2, of the heavy metals from the fired bricks determined by tank leaching test (Dutch standard 1995). (Assumption, the surface area was constant, 0.015 m^2). BMD = Building Material Decree (BMD 1999).

	As	Ba	Cd	Cr	Cu	Ni	Mo	Pb	Zn
C	2.1	0.06	0.15	2.1	0.6	0.8	3.5	0.8	0.6
T5	2.8 ± 0.7	3.2 ± 0.01	0.14 ± 0.07	2.5 ± 0.03	0.9 ± 0.1	1.4 ± 0.1	5.1 ± 0.2	1.2 ± 0.5	2.3 ± 0.5
T10	3.7 ± 0.8	8.7 ± 0.9	0.20 ± 0.15	4.3 ± 0.2	1.2 ± 0.4	1.5 ± 0.01	26 ± 21	2.4 ± 1.0	6.8 ± 1.5
T20	9.0 ± 2.9	11.3 ± 1.6	0.15 ± 0.15	16 ± 3	1.5 ± 0.2	2.0 ± 0.3	42 ± 2	1.9 ± 2.7	8.1 ± 6.0
R20	5.5	10.7	0.25	32	1.8	2.3	25.4	2.8	18
BMD	435	6300	12	1500	540	525	150	1275	2100

Table 5. Leaching from the fired bricks (mg/kg TS) in batch leaching test (DS/EN 12457–1). MA1-limits of leaching from coastal landfills for mineral waste in Class MA1 (BEK nr. 252). C3-converted limiting values in Category 3 (BEK nr 1414 2015) for waste used in construction, from liquid based µg/l to ash based mg/kg TS.

	pH	As	Ba	Cd	Cr	Cu	Mo	Ni	Pb	Zn
C	11.1	0.10	0.01	<d.l.	0.43	0.017	0.40	0.05	<d.l.	0.002
T5	10.5	0.06	0.11	<d.l.	0.27	0.022	0.55	0.02	0.007	0.001
T10	10.5	0.11	0.33	<d.l.	0.35	0.01	0.91	0.02	<d.l.	0.002
T20	10.3	0.19	0.23	<d.l.	0.95	0.01	3.55	0.04	<d.l.	0.003
R20	10.5	0.20	0.23	<d.l.	1.95	0.02	2.61	0.03	<d.l.	0.003
MA1		0.40	30	0.60	4	25	5	5	5	25
C3		0.10	8	0.08	1	4	–	0.14	0.2	3

R20 and Cr from R20, which increased with increasing ash content, that made those bricks fail to meet the recycling requirements (C3). The treatment of the raw ash by washing-EDR reduced leaching, especially of Cr, from the fired ash-clay bricks (T20 vs. R20).

4 CONCLUSIONS

The incorporation of the MSWI fly ash investigated in clay bricks reduced compressive strength and increased soluble salts content. Treatment of the MSWI fly ash by washing and EDR reduced soluble salts content in the fired fly ash-clay bricks, and reduced by half Cr leaching, compared to bricks with untreated ash. Tank leaching test showed that heavy metals leaching from monolithic ash-clay bricks did not exceed the limiting values regulated for Dutch building materials. Batch leaching test revealed that fired ash-clay bricks met the limits for waste landfill as mineral waste in granular form e.g. after being demolished at the end of their service life. On the contrary, 'demolished' bricks with 5% treated ash might be reused again in construction work considering low heavy metal leaching. Although the whole process, i.e. water washing (if necessary), electrodialytic remediation and brick production, would consume energy and resources/materials, this process maximizes the recovery from MSWI fly ash: recovering metals (e.g. generation of metal-bearing solutions from electrodialytic treatment), and recycling ash residue in building materials (bricks) — thus reducing the waste to be landfilled and then environmental impact from landfilling of fly ash. The environmental impact of the process, however, needs to be further studied.

ACKNOWLEDGEMENTS

The present work was financially supported by the Department of Civil Engineering at the Technical University of Denmark. C. Dias-Ferreira gratefully acknowledges FCT—Fundação para a Ciência e para a Tecnologia (SFRH/BPD/100717/2014).

REFERENCES

BEK nr 1414 2015. *Bekendtgørelse om anvendelse af restprodukter og jord til bygge—og anlægsarbejder og om anvendelse af sorteret, uforurenet bygge—og anlægsaffald.*
BEK nr 252 2009. *Bekendtgørelse om deponeringsanlæg.*
Belmonte, L.J., Ottosen, L.M., Kirkelund, G.M., Jensen, P.E., & Vestbø, A.P. 2016. Screening of heavy metal containing waste types for use as raw material in Arctic clay-based bricks. *Environmental Science and Pollution Research*: 1–13.
BMD Building Materials Decree 1999. *VROM Dutch Ministry of Housing, Spatial Planning and the Environment.* Sdu Uitgevers, The Hague.

Chen, W., Ottosen, L.M., Jensen, P.E., Kirkelund, G.M. & Schmidt, J.W. 2014. A comparative study on electrodialytically treated bio-ash and MSWI APC-residue for use in bricks. *Proceed. 5th Internat. Conf. on Engineer. for Waste and Biomass Valorization (wasteeng2014)*: 648–662.

Dansk Standard 1997. *DS/EN ISO 10545-3: Keramiske fliser—Del 3 : Bestemmelse af vandabsorption, åben porøsitet, synlig relativ densitet og volumendensitet.*

Dansk Standard 2002. *DS/EN 12457-1: Characterisation of waste—Leaching—Compliance test for leaching of granular waste materials and sludges—Part 1: One stage batch test at a liquid to solid ratio of 2 l/kg for materials with high solid content and with particle size below 4 mm.*

Dansk Standard 2003. *DS259: Vandundersøgelse—Bestemmelse af metaller i vand, jord, slam og sedimenter—Almene principper og retningslinjer for bestemmelse ved atomabsorptionsspektrofotometri i flamme.*

Dansk Standard 2015. *DS/EN 771-1:2011+A1:2015: Specification for masonry units—Part 1: Clay masonry units.*

Dansk Standard 2016. *DS/EN 772-5: Methods of test for masonry units—Part 5: Determination of the active soluble salts content of clay masonry units.*

Domone, P. & Illston, J. 2010. *Construction materials: their nature and behaviour.* London and New York: Spon Press.

Dutch Standard 1995. *NEN 7345. Leaching characteristics of building and solid waste materials—Leaching and determination of behavior of inorganic components from buildilng materials, monolithic waste and stabilized materials.*

Ferreira, C., Ribeiro, A. & Ottosen, L. 2005. Effect of Major Constituents of MSW Fly Ash During Electrodialytic Remediation of Heavy Metals. *Separation Science and Technology* 40(10): 2007–2019.

Ferreira, C.D., Jensen, P., Ottosen, L. & Ribeiro, A. 2008. Preliminary treatment of MSW fly ash as a way of improving electrodialytic remediation. *Journal of environmental science and health—Part A Toxic/hazardous substances & environmental engineering* 43(8): 837–843.

Heiri, O., Lotter, A.F. & Lemcke, G. 2001. Loss on ignition as a method for estimating organic and carbonate content in sediments: reproducibility and comparability of results. *Journal of paleolimnology* 25(1): 101–110.

Keppert, M., Siddique, J.A., Pavlík, Z. & Cerny, R. 2015. Wet-Treated MSWI Fly Ash Used as Supplementary Cementitious Material. *Advances in Materials Science and Engineering.*

Lin, K.L. 2006. Feasibility study of using brick made from municipal solid waste incinerator fly ash slag. *Journal of Hazardous Materials* 137(3): 1810–1816.

Reijnders, L. 2005. Disposal, uses and treatments of combustion ashes: A review. *Resources, Conservation and Recycling* 43(3): 313–336.

Zhang, H., Zhao, Y. & Qi, J. 2011. Utilization of municipal solid waste incineration (MSWI) fly ash in ceramic brick: Product characterization and environmental toxicity. *Waste Management* 31(2): 331–341.

Processing of metallurgical wastes with obtaining iron oxides nanopowders

I.Yu. Motovilov, V.A. Luganov, T.A. Chepushtanova & G.D. Guseynova
Non-Profit JSC, Kazakh National Research Technical University Named After
K.I. Satpaev, Kazakhstan

Sh.S. Itkulova
Institute of Fuel, Catalysis and Electrochemistry Named After D.V. Sokolsky, Kazakhstan

ABSTRACT: Hydrochloric acid waste etching solutions are formed as a result of metals pickling. The article suggests perspective directions for processing these wastes into iron oxide nanopowders. The proposed waste recycling technology has the potential to solve certain environmental problems and be economically more efficient than the current ones. High-temperature hydrolysishas been utilized (in the proposed technology). Results of microprobe, microscopic and X-ray analysis of the hydrolysis products show that the material represented by soot particles of less than 100 nm in size is composed of magnetite and hematite; magnetite—hematite ratio varies depending on the conditions of high temperature hydrolysis. The iron oxide nanopowders obtained by sol-gel method have been studied as well.

1 INTRODUCTION

1.1 Relevance

Development of metallurgical engineering is associated with acceleration of scientific and technological progress and improvement in product quality. New modern materials with unique properties are obtained by using economically feasible raw materials and relevant chemical and metallurgical methods (Petrova O.S., Chekanova A.E., Gudilin E.A., Zajtsev D.D.i dr. *Al'ternativnaja jenergetika i ekologija: Mezhdunarodnyj nauchnyj zhurnal 2007*). This principle applies to a number of powders of metal oxides, widely used in modern industry. Major distinction of the method of production of iron oxide nanopowders by reduction of iron chlorides is using solutions of ferric chloride as a starting material, subsequent reduction of ferric chloride by hydrogen, and precipitation of iron hydroxide from the solutions with aid of calcium oxide, magnesium and sodium, and calcination of the resulting iron hydroxides (Motovilov I.Yu., Chepushtanova T.A., Luganov V.A. Vestnik KazNTU 2013), (Motovilov I.Yu., Chepushtanova T.A., Luganov V.A. Gornui gurnal Kazakhstana 2015). One of the most perspective methods for obtaining nanopowders is the sol-gel method. Researchers used the combined Pechini method in order to obtain nanopowders of iron oxides from chloride solutions for comparing properties of powders obtained by different methods.

2 METHOD AND EXPERIMENTAL DATA

2.1 Method of reduction of iron chlorides

Thermodynamic analysis of hydrothermal decomposition of iron chlorides has been executed by using HSC program – 5.0 Company Outokumpu Ou in non-oxidizing environment and in the presence of an oxidant—oxygen. Hydrolysis process takes place in one step according to reaction (1) with the formation of FeO. Thermodynamically this reaction may flow at temperatures above 673 K.

$$2FeCl_2 + FeCl_3 + 4H_2O = Fe_3O_4 + +7HCl + 0,5H_2 \qquad (1)$$

In the presence of oxygen in the reaction the following products are possible—FeO, Fe_3O_4, Fe_2O_3, $FeCl_3$, HCl, and even—H_2, Cl_2 and thermodynamically more likely.

The processes of hydrolysis and reduction of iron chlorides occur at temperatures above 400°C. One of the main conditions of the process in the right direction is the control of the composition of the gas phase. These conditions are met when using tubular electric furnaces. The installation for hydrolysis and reduction of iron chlorides was mounted from a tubular electric furnace, and a quartz tube was used as the reactor. From both sides the quartz tube was hermetically sealed (by rubber stoppers with holes for gas injection and the withdrawal of gaseous and liquid reaction products). Depending on the conditions of the hydrolysis process, oxygen (oxidizing atmosphere), nitrogen (neutral atmosphere) was fed into the quartz reactor through the rheometer. Hydrogen needed for recovery was obtained from Kipp's apparatus. To control the process temperature, a thermocouple was inserted into the quartz reactor. The condensation of the reaction products was carried out in a refrigeration unit consisting of four reverse glass coolers; water was used as the cooling medium. The condensed products were fed into the acid collector, which was a conical flask with a hermetically sealed rubber stopper with holes for introducing the condensed phase and discharging the gaseous products. The acid collector was filled with NaOH solution to control the amount of the hydrochloric acid isolated by titration. To produce powders of iron oxides and regenerate hydrochloric acid, solutions of ferric chloride, obtained after purification from non-ferrous metals underwent hydrothermal decomposition. Figure 1 shows a diagram of a laboratory installation for the hydrothermal decomposition of ferric chloride.

A quartz tube with a diameter of 45 mm was used as the reaction tube. At both ends, the tube was closed with rubber stoppers with holes for introducing nitrogen and for removing gases from the reaction space. A thermocouple was inserted into one of the holes in the plug in a quartz case. In the pipe under the hot junction of the thermocouple in a porcelain boat the crystals of $FeCl_2 \cdot 4H_2O$ were formed. The tube was placed in an oven and the temperature was raised to a predetermined level. During the experiment, the temperature in the reaction tube was maintained with an accuracy of $\pm 5°$. Nitrogen was fed into the reaction space. At a given temperature, the time of the experiment was measured. The gases formed in the process were taken to the refrigeration unit. The refrigeration unit consisted of four reverse glass refrigerators connected in series. The water has been a cooling agent. Samples of hydrothermal decomposition of ferric chloride at 500 and 600°C have been obtained.

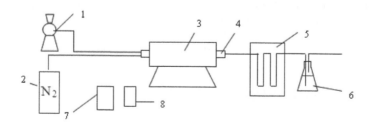

1 - Kipp apparatus; 2 - a cylinder with nitrogen; 3-stove; 4 - quartz tube;
5 - refrigerator; 6 - acid collector; 7 - the transformer; 8 - millivoltmeter

Figure 1. Scheme of installation for hydrothermal decomposition of ferric chloride.

2.2 Research results

The results of the study of the hydrolysis of $FeCl_3 \cdot 6H_2O$ showed that with an increase in temperature above 553 K doesn't have noticeable effect on the degree of decomposition. At 553 and 603 K within 50 minutes almost complete decomposition of $FeCl_3 \cdot 6H_2O$ takes place. Lowering the temperature of hydrolysis increases the decomposition time to 120 minutes at 503 K. Conducting the hydrolysis process at lower temperatures reduces the process speed significantly: at 453 K for 3 hours the degree of decomposition was 86.3%.

Increasing the hydrolysis temperature above 803 K doesn't have the noticeable effect on the degree of decomposition of $FeCl_2 \cdot 4H_2O$. Almost complete decomposition of $FeCl_2 \cdot 4H_2O$ occurs in 40 minutes at T = 803 to 903. Lowering hydrolysis temperature increases the decomposition time up to 180 minutes at 703 K. At lower temperatures 603 K the decomposition time is 180 minutes, and decomposition rate is 44.46%.

According to the Sakovich formula: $K = nk^n$ determines the reaction constants of the hydrolysis of chlorides (Fig. 2).

The dependence of the reaction constant on temperature obeys the Arrhenius equation— in the coordinates lgK – 1 / T results fit satisfactorily on a straight line (see Figure 2.). The graph has a fracture, indicating a change in the mechanism of the decomposition process of ferric chloride when the temperature changes. The temperature coefficients of the reaction constants $\gamma = K_{t+10}/K_t$, calculated according to the equation $lg\gamma = 2E/T^2$, equal to:

- Iron chloride (3+): $\gamma_{453-503} = 4{,}96$; $\gamma_{503-553} = 1{,}72$; $\gamma_{553-603} = 1{,}10$.
- Iron chloride (2+): $\gamma_{603-703} = 2{,}13$; $\gamma_{703-803} = 2{,}74$; $\gamma_{803-903} = 1{,}35$.

The values of apparent activation energy and temperature reaction constants show that at all studied temperatures the hydrothermal decomposition process of bivalent iron chloride and trivalent iron chloride takes place in the diffusion region.

Slowing of decomposition process for trivalent iron chloride at temperatures below 503 K is caused, presumably, by the rate of water dehydration (Fig. 2). The iron powders obtained as a result of bivalent iron chloride hydrolysis were subjected to an integrated validation of their properties.

2.3 The results of physical and chemical analysis methods

Electron microprobe analysis. The purpose of research—to determine the oxidized iron powder size; study the elemental composition of samples; study the morphological features of the powder: the geometric shape, the presence of pores and cracks.

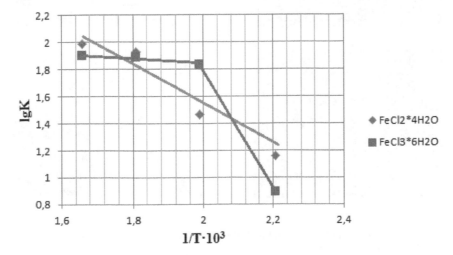

Figure 2. Dependence of 1 g K hydrothermal decomposition reaction of bivalent iron chloride and trivalent iron chloride and the reciprocal temperature.

The determination of particles size and morphological features of the oxidized iron powder has been carried out by a transmission electron microscope JEOL JEM-2100F.

The elemental composition of the samples with a resolution of depth was determined by electron X-ray microanalysis. Specter range has been recorded by energy dispersive spectrometry Inca X-act.

The analysis showed that at the temperature of the hydrolysis process of 500°C, hematite 53%, 30% iron oxides with different ratios of iron and oxygen, 8% hydroxide and 8% magnetite predominate in the powders obtained. With an increase in the temperature of the process to 600°C, the share of magnetite increases to 57%, 29% of iron oxides with different ratios of iron and oxygen and 14% of hematite, iron hydroxides are not detected. With a further increase in the hydrolysis temperature, the proportion of magnetite in the resulting powders will increase.

The materials obtained after hydrolysis are ultra-disperse powders of iron oxides of a size less than 50 microns. The results of X-ray diffraction studies. The aim of research was to determine the structure and composition of the samples. We used X-ray diffractometer XPertMPDPRO (PANalytical) and DRON-3 diffractometer (Cu_k – radiation, β-filter). The results (Table 1) show that optimal hydrolysis temperatures are 603–703 K.

At a temperature of 803 K ferric chloride is not found. The content pf magnetite increases from 28.5% to 66.5% with the increasing of temperature. At temperatures above 803 K hematite powder can be determined and at T = 903 K its content reaches 37.7%.

The bulk samples of the iron oxide powder are a fine-crystalline hematite, α- $α-Fe_2O_3$, with a share of finely crystalline magnetite content. Study of chemical composition showed that the peripheral layer of particles is represented by magnetite, and the basic layer is oxide $α-Fe_2O_3$.

The temperature of the experiments on the hydrolysis of divalent ferric chloride: 1) at 603 K; 2) at 703 K; 3) at 803 K; 4) at 903 K

Figure 3. Photographs of powders obtained by hydrolysis of divalent ferric chloride.

Table 1.   Results of semi quantitative XRD of hydrolysis products of bivalent iron chloride.

The composition of the products obtained	Chemical formula	Amount, %
Hydrolysis temperature 603 K		
Magnetite	Fe_3O_4	28,5
Iron Chloride Hydrate	$FeCl_2 \cdot 4H_2O$	71,5
Hydrolysis temperature 703 K		
Magnetite	Fe_3O_4	62,0
Iron Chloride Hydrate	$FeCl_2 \cdot 4H_2O$	38,0
Hydrolysis temperature 803 K		
Magnetite	Fe_3O_4	66,5
Hematite	Fe_2O_3	33,5
Hydrolysis temperature 903 K		
Magnetite	Fe_3O_4	62,3
Hematite	Fe_2O_3	37,7

2.4   Sol-gel method

The Pechini method is based on the ability of α-hydroxo-carboxylic acids: a) to form stable chelate complexes with many cations, b) to react with polycondensation (esterification) with polyhydric alcohols (Akbar S., Hasanain S.K., Azmat N. And Nadeem M. Journal of Applied Sciences Research 2008). This method uses citric acid because of its ability to form fairly stable complexes with metal cations, and as polyhydric alcohol—ethylene glycol, which acts as a solvent in the initial stages of synthesis and reagent for polyesterification in subsequent stages (Maksimov A. Technomedia, publishing Elmor 2007).

To synthesize nanodispersed powders, 4 samples were synthesized. The process of obtaining materials by the sol-gel method consists of several stages, the main ones of which are: synthesis of the sol; formation, aging (maturing), drying and heat treatment of the gel.

For the synthesis of samples 1 and 2, an aqueous solution of 0.1M ferric chloride $FeCl_3 \cdot 6H_2O$ (hp) was prepared as a precursor, and 0.2 M monohydrate of citric acid and water as a solvent were used as the chelating agent.

The stage of formation of the sol passed with constant stirring on a magnetic stirrer. To 800 ml of 0.2 M citric acid was added dropwise from the burette 200 ml 0.1 M ferric chloride solution at a rate of ~ 0.8 ml/min. In order to form a gel, the resulting solution was heated to 70°C for 2 hours, with constant stirring.

With the aging of the gel, a spontaneous decrease in the volume of the gel (accompanied by the liberation of the liquid) is inherent. This step involves deformation of the gel network and removal of fluid from the pores. Maturation of the gel was carried out until a sufficiently strong structure was formed. In other words, 8 hours of continuous mixing was required to separate water and produce the gel. The color of the solution gradually darkened to a reddish-brown gel.

The drying stage of the sample was carried out in a water bath at a temperature of 96°C with periodic manual mixing of 5–6 hours until the gel solidified. Using a vacuum oven at 100°C, the moisture residue was removed for 10 hours.

In the heat treatment stages, to prevent uncontrolled combustion of the samples, a gradual temperature rise was carried out. Samples were calcined in a muffle furnace.

For sample 1, the temperature was raised from room temperature to 300°C for 1 hour, at this temperature, held for 1 hour. Then the temperature was raised to 400°C for 1 hour and held at this temperature for 3 hours. Total calcination of sample 1 lasted 6 hours. Calcination of sample-2 was carried out similarly, but at a temperature of 500°C. For the synthesis of samples 3 and 4, aqueous solutions of more concentrated solutions of iron chloride $FeCl_3*6H_2O$ (hp) – 0.2 M and citric acid monohydrate 0.4 M were prepared. The stage of formation of the sol: 400 ml of 0.4 M citric acid was added dropwise with a syringe pump (unlike samples 1 and 2, where burettes were used) at a rate of ~ 0.8 ml/min. 100 ml 0.2 M ferric chloride. The gelling step is similar to the gelling step for samples 1 and 2, only the heating time is 1 h, and the total continuous mixing time is 4 h. Was established at high magnification, smaller particles with a size of less than 100 nm (~ 50–60 nm) are found, combining into larger structures.

2.5 *Magnetic investigation*

The results of the diffractometric analysis showed that the Fe_2O_3 nanopowders obtained by the sol-gel method and powders obtained by reduction method have the α-Fe_2O_3 structure. Following magnetic transitions have been determined by thermal analysis (by STA 409PC/ PG NETZSCH): α-Fe_2O_3 antiferromagnetic point at 340 K (Morin point), 611–640 K-Curie point the paramagnetic state of the samples and 862 and 951 K are transitions from the antiferromagnetic state to the ferromagnetic state. The temperatures of these transitions were confirmed by high temperature diffractometric analysis.

The magnetic investigation was studied on the SatisGeo KM-7 kappa-meter, the magnetic susceptibility of the samples obtained by the sol-gel method was 940–1159 × 10^{-6} CI/g, the magnetic susceptibility of the samples obtained by the reduction of iron chlorides 1250–1358 × 10^{-6} CI/g. The magnetic properties of α-Fe_2O_3 ultra-thin nanoparticles were investigated as a function of temperature and applied magnetic field (H). Magnetic measurements at low temperature were performed on a DC magnetometer SQUID-VSM; at a temperature of 300 K using the Microsense EV 7 VSM. There is an extremely small hysteresis loop with coercive force (~ 0.04 kOe) was established also weak residual magnetization (~ 0,28 electromagnetic units per gram, emu/g) are typical of superparamagnetic behavior was established.

3 CONCLUSION

Thus, it is established that the methods make it possible to achieve nanosized powders and a sufficiently developed specific surface. According to the results of transmission electron microscopy the powders obtained by the sol-gel method have a more developed specific surface, the structure is 90% spherical in its morphology. However, the magnetic properties of powders are inferior to powders obtained by the reduction of iron chlorides. All nanopowders possessing by magnetic properties from antiferromagnetic point to paramagnetic state, also was established small hysteresis loop that very positively describe the magnetic characteristics of such powders.

REFERENCES

Akbar S., Hasanain S.K., Azmat N. And Nadeem M. 2008. Synthesis of Fe_2O_3 nanoparticles by new Sol-Gel method and their structural and magnetic characterizations. *Journal of Applied Sciences Research.* 3(3): 417–433.

Maksimov A.I. 2007. Osnovu sol-gel technologii nanokompozitov/– SPb.: OOO «Technomedia»/Izd-vo «Elmor», 2007. P. 255.

Motovilov I.Yu., Chepushtanova T.A., Luganov V.A. 2013. Thermodynamicheski analys poluchenia poroshkov pyromethalurgicheskim methodom. *Vestnik KazNTU* 5: 220–225.

Motovilov I.Yu., Chepushtanova T.A., Luganov V.A. 2015. Complexnaya pererabotka ghelezosoderghachego suria s polucheniem poroshkov metallicheskogo I okislennogo gheleza. *Gornui ghyrnal Kazakhstana* 7: 36–38.

Petrova O.S., Chekanova A.E., Gudilin E.A., Zajtsev D.D.i dr. Sintez i harakterizatsija mezoporistyh nanochastic γ-Fe2O3. Al'ternativnaja jenergetika i jekologija: Mezhdunarodnyj nauchnyj zhurnal. 2007. № 1 (45). S. 70–73.

WASTES – Solutions, Treatments and Opportunities II – Vilarinho, Castro & Lopes (Eds)
© 2018 Taylor & Francis Group, London, ISBN 978-1-138-19669-8

A step forward on cleaner production: Remanufacturing and interchangeability

F. Moreira
Department of Production and Systems Engineering, ALGORITMI R&D Center, University of Minho, Guimarães, Portugal

ABSTRACT: This paper explores and develops further the link between the design of inter-changeable parts and the remanufacturing approach, with the aim to lay down the underlying rationale, and make evidence of the potential that such association might represent. The study is intended to groundwork the needs for further research work, prospectively targeting the development of standards for cross-sector product design and remanufacture. Such approach encompasses, not only clear environmental and economic benefits, but also social ones, which gear from a more labour intensive activity, which promotes local employment and income. The direct environment gains accrue from a wider reuse of end-of-life products, manufactur-ing processes energy related savings, raw materials savings, lower GHG emissions on several stages of the product' life-cycle, and an overall decline on the quantity of waste generated.

1   INTRODUCTION

A key contribution to a more sustainable future is that of providing valuable goods, which people truly desire and benefit, while providing the best environmental option from a cradle-to-grave perspective. Remanufacturing is one such approach, which is well known for being the ultimate form of recycling. Although acknowledged for being an excelled option for significant savings on energy, emissions and raw materials, while delivering fully functional goods at unmatched prices, the remanufacturing approach is still a marginal industrial activ-ity, which faces some problems relating to parts availability and public acceptance, among others. A key concept, that could unleash its hidden potential, is that of designing products bearing in mind its remanufacturability, which would likely translate into a broader availabil-ity of used interchangeable parts in the marketplace in the long run. Interchangeable parts simple but effective concept was a cornerstone of the industrialization era, which fundamen-tally shaped the making and remaking of things, and improved the access to commodities that smooth our daily living by enabling the large-scale manufacture of reliable and cheap products. The interchangeability concept was exploited on some product designs, but to a lesser extent on parts from distinct products and different industries.

The remanufacturing strategy brings about a number of economic and environmental benefits by allowing the materialisation of fully functional, like-new products which can provide value for consumers at a fraction of the original cost and unmatched environmental penalties [1]. On the top of that, remanufacturing is also known to be a labour intensive activity which promotes local employment, stimulates local economies and promotes the development of technical expertise and know how on products and on the technical details of the processes [2]. Therefore, remanufactur-ing can realistically provide a sustainable option for the making and delivery of valuable goods.

When products fail to deliver the original functionality, fail to do that at a given perform-ance level, or simply lose their appeal, bearing in mind that the consumer requirements evolve and that the competition initiatives might make a fully functional product obsolete, they can still be recovered in a number of ways. At a lower level, the materials that make-up the product can be recovered. This focus on delivering raw materials for new products and it is known as material recycling, or simply recycling. At intermediate level, product parts and modules can also be recovered from such products, cleaned and inspected and restored to like-new condition for reuse in remanufactured products. Finally, at the highest level, full worn-out or broken products can be remanufactured to original specifications or upgraded to new specifications, by making a set of operations that might require incorporation of remanufactured or new components/modules. For the purpose of this paper the term remanufacturing will be used indiscriminately for components/modules and full products recovery.

Although recycling activities bring about a number of benefits, many raw materials loose quality in the process and the recycling rate is rather small, e.g. 2/3 of the recycled metals have recycling rates below 50% [3]. This activity requires the full loss of the products' value, which was incorporated into the parts over a long chain of manufacturing activities which started with raw materials. The only intention of recycling is to partly recover those raw materials, and therefore save primary materials, and to avoid more waste generation. But secondary gains are also attained with recycling, such as those of energy savings, GHG emissions reduction and jobs creation relating the chain of activities required to actually put recycling into practice, from collection to transformation of materials into a basic form and shape that will subsequently supply the manufacturing industry. Overall the recycling activity is focused on material recovery, while remanufacturing is focused on value-added recovery, the later being fundamentally more environmental friendly since it not only promotes material recovery and less waste, but also shortens the chain of events and effort required to deliver valuable products into the marketplace. Remanufactured products normally cost about 60% to 70% less than new products [4] and according to Gray and Charter [5] offer a business model for sustainable prosperity.

Wider use of remanufacturing practices is notably compromised by a number of situations that narrow the possibility of reuse of EOL product components, and therefore do not grant fully development of such a business strategy. Since the design stage determines about two thirds of the remanufacturability of products [6], as well as the environmental impact during all stages of the products' life cycle [7], it is imperative that designers fully exploit all the design options that ensure remanufacturing easiness, require less skill to repair and provide maintenance, facilitate disassembly, as well as promoting a higher share of components that make it to the remanufacturing stage and are effectively used in remanufactured goods. Additionally, the design stage should promote the development of cross-sector product remanufacture by making interchangeable parts a common thread.

This paper exploits the potential offered by a broader use of interchangeable parts within a remanufacturing strategy. The study aims at making a contribution to the general framework of Cleaner Production, which essentially seeks to bring to the marketplace goods and services with minimum environmental impact, using a preventive approach to industrial activity.

2 LITERATURE REVIEW

2.1 *Remanufacturing*

Remanufacturing can be defined as [8, 9]:

"*Remanufacturing is an industrial process in which worn-out products are restored to like-new condition. Through a series of industrial processes in a factory environment, a discarded product is completely disassembled. Usable parts are cleaned, refurbished, and put into inventory. Then the new product is reassembled from both old and, where necessary, new parts to produce a unit fully equivalent-and sometimes superior-in performance and expected lifetime to the original new product. In contrast, a repaired or rebuilt product normally retains its identity, and only those parts that have failed or are badly worn are replaced or serviced.*"

According to the US International Trade Commission [10] The USA is the world's largest producer of remanufactured goods. The remanufacturing activity represented a hefty amount of USD $43 billion on the US economy in 2011 and was responsible for about 180,000 jobs. A significant part of this activity is granted by a big number of small and medium-sized enterprises. Overall, however, remanufactured goods represented only about

2% of total sales (of all products, new and remanufactured), which reveals that remanufacturing activity has a great margin for growth. The remanufacturing of products is performed on multiple industry sectors, but is more focused on sectors which manufacture products with long life cycles and capital intensive ones, such as in the aerospace and automotive industries, machinery, medical devices, among others.

A study from Boustead and Hancock [11] presented some findings on energy requirements for steel and cast iron metal products, on a cradle-to-gate basis. A later study by Adler et al. [12] estimated that energy requirements for remanufacturing such metal products would require a smaller fraction of the process chain requirements to make new products, starting on ore extraction right down to manufacturing. Sutherland et al. [13] examined the benefits of remanufacturing diesel engines taking into account that existing design analysis methods are generally focused on a single-use lifecycle for new products. Figure 1 illustrates the full life-cycle stages that products undergo, and the alternative options after an initial utilization stage (used products), namely reuse, recycle, repair and remanufacturing.

Haynsworth and Lyons [8] reported back in 1987some challenges that the remanufacturing activity faced: a) difficulties in establishing efficient channels for the collection and distribution of cores (main parts to be recovered from EOL products); b) scarcity of cores that limited the viability of the business; c) lack of effort by OEM designers on facilitating remanufacturing; d) new parts non-interchangeable with old ones. They also pointed out the societal benefits of extensive remanufacturing, i.e. a) savings in energy use and raw materials consumption; b) waste disposal and landfills; c) reduced costs for durable goods that enabled a greater consumer choice and higher standard of living; d) employment opportunities since remanufacturing activity is labour intensive. These authors consider as well that the design stage should improve a number of characteristics that would likely improve the remanufacture stage. Among those, the authors pinpoint that distinct models of a given product should contain the maximum number of interchangeable parts that is practicable, and suggest that remanufacturing could be included in the product life cycle concept. Figure 2 depicts the main stages required in a remanufacturing strategy. Here components considered unrecovered might undergo a material recovery operation (recycling) which provide raw materials for new parts, or otherwise adequately disposed. On the product remanufacturing process, new parts might need to be incorporated into fully functional products, which partly retain their original value.

Matsumoto and Umeda [15] have studied a number of remanufacturing practices in Japan and identified requirements for remanufacturing and a number of efforts made towards its effectiveness. They also found out that some OEMs engaged in remanufacturing incorporate remanufactured components into new products and that recycling laws in Japan enforce material recovery, but a much smaller stimulus is put on the remanufacturing activity.

According to Steinhilper [6] (pp. 86) intelligent designs can substantially improve the effectiveness of some operations normally required by remanufacturing, namely disassembly, cleaning and inspection. Therefore, as with design for manufacturability and assembly

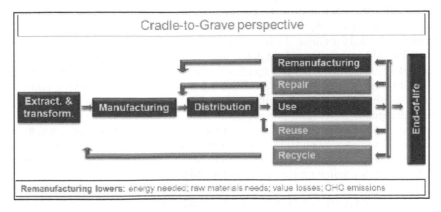

Figure 1.   Cradle-to-grave perspective on products and remanufacturing contribution.

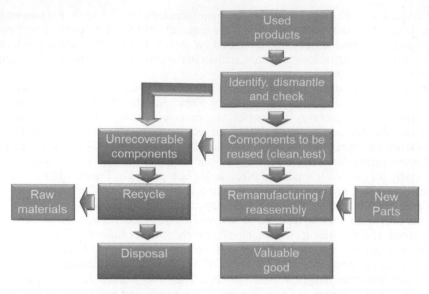

Figure 2.   Product remanufacturing stages.

(DFMA), the design for recycling and remanufacturing could also be considered at design stage. This could incorporate the selection of materials, modular product structures and joining for easy disassembly. A number of tools are suggested to support such endeavour, such as standards, checklists and guidelines. Although these aspects are essential to justify such a strategy and to facilitate and reduce the cost associated with the remanufacturing operations, the more fundamental aspect of cores availability needs to be endorsed. The remanufacturing alternative relies heavily on a consistent flow of cores from a closed-loop supply chain of used/discarded items. These cores are the prime items to be remanufactured. Core scarcity might represent an issue on the overall remanufacturing strategy, as presented above. Such a challenge could potentially be tackled by using a concept that is undoubtedly linked with the development of the industrial era—interchangeable parts.

2.2   *Interchangeable parts*

Interchangeable parts simple concept was a fundamental development for manufacturing [16]. Interchangeability required parts to be uniform, which in turn required stable standard work processes. This led the way to the economies of scale, enabled by labor specialization. The concept was envisioned for producing artillery pieces, which could be repaired on the battlefield. After the US war of 1812 a large quantity of damaged guns were sent back to the armouries for repair [16]. These would pile up in large quantities. This vision triggered the fundamental need for remanufacturing valuable items out of a huge number of cores that were available. The same vision could potentially be used today with a broader view over the multitude of EOL products that pile up all over the world at a faster pace than ever. Bearing in mind that the technological issues for delivering uniform parts and provide standard work is not the issue that it was in the beginning of the 19th century, the main challenge to achieve truly interchangeable parts (and provide some standardized ones) across industrial sectors relies on the product design activity and the proprietary issues. These parts, which could be interchanged amongst a much wider range of product types and makes, would predictively retain more value over extended periods of time, and could be fast, easily and inexpensively dismantled from its original products and snapped-in into remanufactured products.

Interchangeable parts can be designed within the framework of individual products, product families or a multitude of rather independent and different product categories. Although remanufacturing activity can effectively benefit from individual product designs that consider the possibility of exchanging parts among different product generations, its expansion to entire product families inherently advances further and magnifies that possibility, but requires some form of clustering methods that are capable of capturing a given degree of commonality. In the case of cross industry products, although part interchangeability seems possible, especially if product design is a common and shared activity, it is still not the current practice within industry. Some examples of such a strategy can be observed, e.g. on the automotive industry, mainly with partnerships or within different brands that pertain to the same group, that use common platforms and key cores, used on competing products, but differentiate from one another with other key aspect, e.g. design, quality, branding. Part interchangeability across industries is fundamentally limited, given that the design activity itself, which dictates its future prospects, does not take into consideration such possibility.

3 DISCUSSION

It seems conceivable that the societal benefits of expansion of the remanufacturing activity along Lean [17] could bring a number of economic, social and environmental gains. This activity is partly conducted by a large number of SMEs that focus their activity in remanufacturing cores and full products that were not designed for more than a single lifetime. Companies already established on such a business, do that profitably, at a very competitive price and facing important challenges dictated by poor designs, which normally do not consider disassembly and reuse a major thread of a product life cycle. Since production and inspection technology does not represent an issue anymore for achieving truly interchangeable parts, the problem seems to rely on both the design and proprietary issues. By one side the companies could design interchangeable parts among different product models and versions of a given model, so that at EOL the issue of unavailability of cores for remanufacturing would not be raised. Interchangeable parts use, amongst individual OEMs competing products, should be stimulated, while distinctive features of the product should not constrain the ability/easiness to remanufacture them out of a number of parts from alternative manufacturers. Ultimately, designers could exploit the opportunity to incorporate components/modules into their designs that pertain to different industries. This holds the potential to truly trigger interchangeable parts use across products and industries, and would likely considerably raise the possibility of use of cores in a second or more lives, thereby making a strong contribution to conservation of energy and materials.

4 CONCLUSIONS

The foregoing discussion presented the rationale beyond a possible strategy to boost the remanufacturing activity. This industrial activity holds a great potential to provide goods at a fraction of the environmental penalties only, when compared to that of manufacturing, while promoting local economies and employment. Interchangeable parts concept, and the design activity in particular, seem to hold the key for the industrialization of the remanufacturing activity. Interchangeable parts among succeeding models and across families of products could potentially boom the remanufacturing of used products from individual original equipment manufacturers. The possibility of using components originated from different industries seems to represent a challenge of a higher level, with a number of issues to be solved first, namely proprietary issues, but potentially, the reuse of end-of-life products could rise accordingly. Remanufacturing pose remarkable challenges to companies, but, in contrast, also stands out as a noteworthy opportunity to scale up an environmentally sound activity for the making of things.

NOMENCLATURE

Core   Used good that is the primary component input for remanufactured goods
EOL   End Of Life

DFMA Design for Manufacturability and Assembly
GHG Green House Gas
LCA Life Cycle Assessment/Analysis
OEM Original Equipment Manufacturer
SME Small and Medium-sized Enterprise
USA United States of America

ACKNOWLEDGMENTS

This work has been supported by COMPETE: POCI-01-0145-FEDER-007043 and the Portuguese Foundation for Science and Technology within the Project Scope: UID/CEC/00319/2013.

REFERENCES

[1] Souza, G.C., Ketzenberg, M.E., Guide, V.D. 2002. Capacitated remanufacturing with service level constraints. *Production and Operations Management* 11(2): 232–248.

[2] Lund, R. and Hauser, W. 2010. Remanufacturing—An American Perspective. In *Proceedings of the 5th International Conference on Responsive Manufacturing—Green Manufacturing (ICRM 2010)*, January 2010 p.1–6.

[3] Graedel, T.E., Allwood, J., Birat, J.P., Buchert, M., Hagelüken, C., Reck, B.K., Sibley, S.F., Sonnemann, G.2013. What Do We Know About Metal Recycling Rates?. USGS Staff—Published Research. Paper 596, 2011. Available: http://digitalcommons.unl.edu/usgsstaffpub/596 [Accessed: 22-May-2013].

[4] Ilgin M., and Gupta S. 2012. *Remanufacturing Modeling and Analysis*. CRC Press. P.3.

[5] Gray, C. and Charter, M. 2007. *Remanufacturing and Product Design: Designing for the 7th Generation, The Centre for Sustainable Design.*

[6] Steinhilper, R. 1998. *Remanufacturing: the ultimate form of recycling*. Fraunhofer IRB Verlag.

[7] WBCSD, Eco-efficiency learning module. World Business Council for Sustainable Development/Five Winds International, 2006. ISBN: 2-940240-84-1.

[8] Haynsworth, H.C., and Lyons, T. 1987. Remanufacturing by Design, the missing link. *Production and Inventory Management Journal*, second quarter: 24–28.

[9] Lund, R. 1984. Remanufacturing. *Technology Review* February/March: 19–29.

[10] USITC 2012. *Remanufactured Goods: An Overview of the U.S. and Global Industries, Markets, and Trade*. United States International Trade Commission, Investigation No. 332–525, USITC Publication 4356, October 2012.

[11] Boustead, I., Hancock, G. 1979. Handbook of Industrial Energy Analysis. NY: Halsted Press.

[12] Adler, D.P., Kumar V., Ludewig, P.A., Sutherland, J.W. 2007. Comparing Energy and Other Measures of Environmental Performance in the Original Manufacturing and Remanufacturing of Engine Components. *Proceedings of the 2007 ASME International Manufacturing Science and Engineering Conference, Atlanta, GA.*

[13] Sutherland, J.W., Adler, D.P., Haapala, K.R., Kumar, V. 2008. A comparison of manufacturing and remanufacturing energy intensities with application to diesel engine production. *CIRP Annals—Manufacturing Technology* 57 (2008): 5–8.

[14] Hammond, R., Amezquita, T., Bras, B.A. 1998. Issues in the Automotive Parts Remanufacturing Industry: Discussion of Results from Surveys Performed among Remanufacturers. *International Journal of Engineering Design and Automation*. Special Issue on Environmentally Conscious Design and Manufacturing 4(1): 27–46.

[15] Matsumoto, M., and Umeda, Y. 2011. An analysis of remanufacturing practices in Japan. *Journal of Remanufacturing* 1: 2. doi:10.1186/2210-4690-1–2.

[16] Ford, R.C. 2005. The Springfield Armory's role in developing interchangeable parts. *Management Decision* 43(2): 265–77. ISSN: 0025–1747.

[17] Moreira, F., Alves, A.C. and Sousa, R.M. 2010. *Towards eco-efficient lean production systems, IFIP Advances in Information and Communication Technology*. doi: 10.1007/978-3-642-14341-0_12.

WASTES – Solutions, Treatments and Opportunities II – Vilarinho, Castro & Lopes (Eds)
© *2018 Taylor & Francis Group, London, ISBN 978-1-138-19669-8*

Review of potential ways for resource recovery from human urine

J. Santos, E. Cifrian, T. Llano, C. Rico & A. Andrés
University of Cantabria, Santander, Cantabria, Spain

C. Alegría
Soluciones Sanitarias Modulares, Vitoria Gasteiz, Spain

ABSTRACT: More than a half of the nutrient load of the domestic wastewater streams is caused by urine, that representing only 1% of the total volume of the stream, contains 40% of phosphorus, 69% nitrogen and 60% of potassium of it. The objective of this study is to perfomance a review of process of urine valorization for the recovery of these macronutrients. The potential ways for the management of the human urine can be divided into two large groups: the use of urine as fertilizer and the recovery of nutrients cultivating microalgae. The urine to the plants for its direct recovery can be carried out by using different technologies. The growth of different microalgae (in real and synthetic urine) has been studied aiming for different products with interesting results. At the highest level of technology, it is possible to use the urine as fuel to feed Microbial Fuel Cells (MFC's).

1 INTRODUCTION

Human urine is a natural resource produced worldwide. Nowadays, there are over 7.3 billion people in the world and each human produces approximately 1.2 L of urine per day. Urine contains significant quantities of the main macronutrients required by plant, Nitrogen (N), and Phosphorus (P). Such macronutrients are the main cause of eutrophication of aquatic systems and their recovery from urine would alleviate this problem. Furthermore, safe pure water and food availability will be two of the main challenges of this century and for this reason nutrient recovery from wastewater has gained increasing attention in wastewater treatment plants due to it may represent a source for the constantly increasing demand on these elements (Jaatinen et al, 2015; Jönsson et al., 2004).

2 METHODOLOGY

To make a review of the main recovery solutions and valorization options of human urine, more than 104 documents were reviewed -61 scientific papers, 26 press articles and 16 patents— due to its relevance in the research field. In a second step, a comprehensive analysis to determine the feasibility of the urine recovery has been done. From the bibliography revised, only a small part was selected and included in the reference section.

3 RESULTS

3.1 *Contextualization*

As was stated before, more than 100 publications were analyzed. An overview of the different technologies studied, permit authors to divide in four the ways of recovery of human urine due to its volume of researches: (i) the direct use of human urine as fertilizer, (ii) the struvite precipitation, (iii) the microalgae cultivation and (iv) the use as fuel for Microbial Fuel Cells (MFC's). In Figure 1a, is possible to distinguish that the use of the urine as fertilizer it seems to be the most common due to its low technological requirement. Furthermore, the interest in microalgae cultivation has increased, especially in recent years with the development of MFC's.

The map of Figure 1b shows the location of the different technologies studies. The most common location of studies about the use of human urine as fertilizer (Pronk and Koné, 2009) have been those countries with food security problems (mainly Africa and India). The simplicity of the technology used for this purpose and the immediate approach for the self-management are the main reason. Microalgae cultivation is a popular management for different reasons. In Europe, it has been developed with the main purpose of fuel generation (Rawat et al., 2011) and in other regions, as China, the nutrient production has been the main motor for this technology. Due to the complex technology involved, the researches in the MFC's are located in the northern regions.

From the patent analysis it should be highlighted the Microbial Fuel Cells predominance (Patent CN104466212). As it is the newest technology, the researches in this area increase faster. If the average number of patents of the period were analyzed, the results would show that the use of urine as fertilizer is the most patented. This is probably due to the fact that this use is the oldest (and so the most extended) and the simplest management for urine valorization.

3.2 Composition of human urine

Human urine is a complex water solution, containing a variety of water-soluble compounds, which are eliminated from the human bloodstream. Fresh urine contains a large quantity of urea, which will eventually completely transform to ammonia; and phosphorus is available as phosphate. After a storage the ions presented in urine are Na^+, K^+, NH_4^+, Ca^{2+}, Cl^-, PO_4^{3-}, SO_4^{2-} and HCO_3^-. In Figure 2, the average composition of urine is shown and its standard deviation obtained from the literature (Udert et al., 2006; Rose, et al., 2015; Simha and Ganesapillai, 2017). The high standard deviation in some compounds is due to different factors as the separate collection, and geographical, social and economic aspects.

Urine also contains heavy metals in a ppb-range and pharmaceuticals, as approximately 60–70% of human pharmaceuticals and hormones (the so-called emerging contaminants) are excreted in urine.

3.3 Human urine pre-treatments

Different pre-treatment techniques can be applied to the stored urine. A relation of this technologies and their objective are shown in the Table 1.

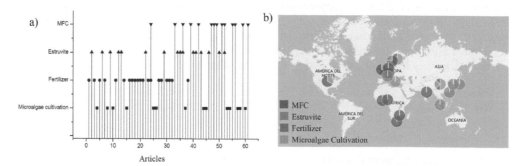

Figure 1. Human urine management contextualization: a) Types of management b) Distribution of kind of management depending of the location.

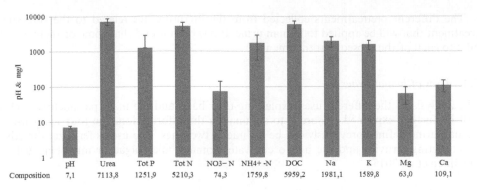

Composition	pH	Urea	Tot P	Tot N	NO3–N	NH4+ -N	DOC	Na	K	Mg	Ca
	7,1	7113,8	1251,9	5210,3	74,3	1759,8	5959,2	1981,1	1589,8	63,0	109,1

Figure 2. Average composition of human urine.

Table 1. Overview of the pretreatment process.

Pre treatment process	Aim of treatment	Application scale	Reference
Acidification	Stabilization (prevention of precipitation, degradation and volatilization)	lab	Maurer et al., 2006
Adsorption	Ammonia and phosphate removal from urine with e.g. zeolite	lab	Tettenborn et al., 2007
Crystallization	Crystal fertilizer product formation	lab	Tettenborn et al., 2007
Electrochemical oxidation on graphite	Ammonia removal	lab	Udert et al., 2006 Zöllig et al., 2015
Electrodialysis	Salt concentration, pharmaceutical reduction	lab and pilot plant	Dodd et al., 2008 Pronk et al., 2007
Evaporation	Volume reduction and laboratory concentration of nutrients, pharmaceutical concentration and/or reduction	pilot plant	Tettenborn et al., 2007
Freeze-thaw	Volume reduction	lab	Maurer et al., 2006
Ion exchange	Nitrogen recovery	lab	Maurer et al., 2006
Microbial electrolysis	Hydrogen production and ammonium recovery	lab	Kuntke et al., 2012
Microbial fuel cells	Energy production, N recovery, NH_4^+ recovery, N & P removal, pharmaceutical removal	lab	Kuntke et al., 2012 Ledezma et al., 2015
Micro- and nanofiltration	Turbidity removal, separation of nutrients from pharmaceuticals	lab	Pronk et al., 2007
Nitrification/ Distillation	Nutrient recovery, partial degradation of pharmaceuticals	lab	Bischel et al., 2015 Udert et al., 2006
Ozonation	Pharmaceutical concentration and/or reduction	lab	Dodd et al., 2008 Tettenborn et al., 2007 Pronk et al., 2007
Precipitation	Solid fertilizer, mainly struvite	lab and pilot plant	Maurer et al., 2006
Reverse Osmosis	Volume reduction	lab and pilot plant	Maurer et al., 2006
Steam Stripping	Nutrients recovery	lab	Tettenborn et al., 2007
Storage	Hygienization	lab and pilot plant	Maurer et al., 2006
UV-radiation	Pharmaceutical concentration and/or reduction	lab and pilot plant	Tettenborn et al., 2007

The different pretreatments collected from the literature are related to the subsequent treatment that will be applied to human urine. It has been found that some of them are specific to each of the subsequent treatments.

3.4 Uses of human urine

The analysis of the different uses, explaining their basis and the main parameters to check will be further explained. In order to summarize the information, the four main uses of human urine defined previously can be include in two ones: The use as fertilizer directly or precipitating struvite, and the use to cultivate biomass, microalgae or microbial fuel cells (Jaatinen et al, 2015).

3.4.1 Use of human urine as fertilizer

Urine is a well-balanced nitrogen-rich quick acting liquid fertilizer. Urea/ammonium in urine and urea/ammonium in artificial fertilizers are similar; 90–100% of urine N is in the form of either urea or ammonium, as has been verified in fertilizing experiment. P and K contents in urine are almost totally (95–100%) in an inorganic form, that is plant-available. Applying urine to plants for its direct recovery is a very adaptable process that can be done with complex or simple technologies (Germer et al., 2011).

The easiest way to urine recovery is to apply it after storage. This kind of fertilizers are an easy help for the economic and fossil fuel depending regions. This simple method has been studied and tested all around the planet and in all kind of plants, and the results usually show nutrient recovery. The yields of these experiments are similar to the mineral fertilizer ones. To solve the inadequate macronutrient relation, different additives can be added to the urine, but it requires a higher technological level. The most popular components to add are wood ashes, TSP or KCl. In these cases, the yields obtained are similar and even better than the mineral commercial fertilizer. When a source of magnesium is added to the stored human urine a salt called struvite ($MgNH_4PO_4 \cdot 6H_2O$) precipitates (Patent WO9951522; Patent US2016185633). Through a basic precipitation reaction, the majority of phosphorus in urine can be crystallized into a white, odorless powder. The struvite can be used as a slow release fertilizer, and different studies show better results than commercial fertilizers, and reducing the volume of the flow to the 2% (Jaatinen et al, 2015; Germer et al., 2011; Liu et al., 2013; Kataki et al., 2016).

The main advantage for the use of the human urine as fertilizer is the adaptability to the technologic level. The big range of pretreatment that exits permits to adapt its use depending on the environment. Nevertheless, the use of urine as fertilizer has some problems that should be taken into account. The huge volume produced that must be transported makes it expensive and less economic competitive, and depending on several factors such as the time of storage or the diet of the individuals, some pollutants could persist.

3.4.2 Use of human urine for biomass cultivation

The cultivation of microalgae in human urine shows high efficiency for removing phosphorus, nitrogen and trace metals from wastewater. The growth of different microalgae in real and synthetic human urine has been studied using cyanobacterium *Spirulina Platensis* and microalgae *Chlorella Sorokiniana* and *Scenedesmus Acuminatus*. Results showed yields of 90%–99% in nutrient recovery. After harvesting, microalgae biomass can be utilized in anaerobic digestion to produce methane, in biodiesel production by extracting the algae, or as another type of products, such as nutritional supplements, pharmaceutical products, or aquaculture feed. There are several factors, such as light intensity, possible water source, cultivation strategies, climate conditions, existing infrastructure and logistic considerations or additional supply of carbon dioxide that may affect the final product and it is important to take in consideration (Jaatinen et al., 2015; Coppens et al., 2016).

As the highest level of technology, it is possible to use the urine as fuel to feed Microbial Fuel Cells (MFC's). The MFC is a technology in which microorganisms employ an electrode

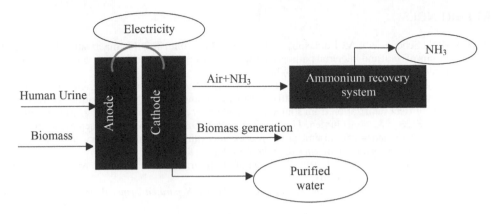

Figure 3. Scheme of a combined treatment technology.

(the anode) as the end-terminal electron acceptor in their electroactive anaerobic respiration. This results in the direct transformation of chemical energy (reduced organic matter) into electrical energy. The latest researches in the field include reactor designs, component optimization and stack configuration.

Microbial fuel cells are an extremely attractive technology for the generation of clean electricity from a range of waste streams. The most promising researches aim to develop small-scale devices and arrange multiple units in stacks. With this technology self-sustainable system directly powered by wearable MFCs have been performed.

In order to improve these methods of urine valorization some of these concepts can be combined. As it is shown in the Figure 3, is possible to use the human urine feeding a MFC while the ammonium is recovered in the upstream (Zang et al., 2011; Walter et al., 2016).

4 CONCLUSIONS

Human urine is a rich stream in macronutrients, mainly N and P, and for this reason can cause eutrophication in aquatic system if is not treated adequately. Phosphorus is a high value element due to its scarcity and growing demand. The present review concludes that there are two main management options for human urine recovery. On one hand, the use as fertilizer presents the advantage of the high technological adaptability and it represents a better yield for growing plants. It is recommended to add different components to achieve mineral fertilizer yields. It is possible to precipitate a solid fertilizer, mainly struvite, to improve such yields. There are problems such as transportation or storage (due to its volume). On the other hand, is possible to use human urine for biomass cultivation. The use of human urine as a source of nutrients to cultivate microalgae has huge interest because the wide range of different products that can be obtained, since biofuels (lipids) to nutrition supplements (proteins). Microbial Fuel Cells are a relatively new technology in this research field. Studies have demonstrated that it is capable to generate electricity. Even the most recent works study the possibility of combining energy production, biomass and nitrogen recovery.

As a general conclusion, the possibility of valorisation of the components of human urine is being realized in some countries, and causes a great number of advantages. Depending on the area or the technological development, a management strategy or another can be chosen, obtaining a benefit that varies from the water purification, to the fertilization of crops, nutritional or pharmaceutical products and even with the technological implementation, clean energy can be obtained. The great challenge is to establish a system of separate collection of this stream.

REFERENCES

Bischel, H.N., Schertenleib, A., Fumasoli, A., Udertb, K.M., Kohn, T. 2015. Inactivation kinetics and mechanisms of viral and bacterial pathogen surrogates during urine nitrification. *Environmental Science Water Research & Technology* 1: 65–76.

Coppens, J., Lindeboom, R., Muys, M., Coessens, W., Alloul, A., Meerbergen, K., Lievens, B., Clauwaert, P., Boon, N., Vlaeminck, S.E. 2016. Nitrification and microalgae cultivation for two-stage biological nutrient valorization from source separated urine. *Bioresource Technology* 211: 41–50.

Dodd, M.C., Zuleeg, S., von Gunten, U., Pronk, W. 2008. Ozonation of source-separated urine for resource recovery and waste minimization: process modeling, reaction chemistry, and operational considerations. *Environmental Science & Technology* 42(24): 9329–37.

Germer, J., Addai, S., Sauerborn, J. 2011. Response of grain sorghum to fertilisation with human urine. *Field Crops Research* 122: 234–241.

Jaatinen, S. 2015. *Characterization and Potential Use of Source-Separated Urine*. Tampere University of Technology. Publication; Vol. 1391.

Jönsson, H., Stinzing, A.R., Vinnerås, B., Salomon, E. 2004. Guidelines on the Use of Urine and Faeces in Crop Production. *EcoSanRes Publication Series Report 2004–2*. Stockholm Environment Institute, Sweden.

Kataki, S., West, H., Clarke, M., Baruah, D.C. 2016. Phosphorus recovery as struvite: Recent concerns for use of seed, alternative Mg source, nitrogen conservation and fertilizer potential. *Resources, Conservation and Recycling* 107: 142–156.

Kuntke, P., Smiech, K.M., Bruning, H., Zeeman, G., Saakes, M., Sleutels, T.H.J.A., Hamelers, H.V.M., Buisman, C.J.N. 2012. Ammonium recovery and energy production from urine by a microbial fuel cell. *Water research* 46: 2627–2636.

Ledezma, P., Kunte, P., Buisman, C.J.N, Keller, J., Freguia, S. 2015. Source-separated urine opens Golden opportunities for microbial electrochemical technologies. *Trends in Biotechnology* 33(4): 214–220.

Liu, B., Giannis, A., Zhang, J., Chang, V., Wang J.Y. 2013. Characterization of induced struvite formation from source-separated urine using seawater and brine as magnesium sources. *Chemosphere* 93: 2738–2747.

Maurer, M., Pronk, W., Larsen, T.A. 2006. Treatment processes for source-separated urine. *Water research* 40: 3151–3166.

Patent CN104466212. *Microbial fuel cell based ammonia recovery unit.*

Patent US2016185633. *Recovery of nutrients from water and wastewater by precipitation as struvite.*

Patent WO9951522. *Nutrient recovery from human urine.*

Pronk, W., Koné, D. 2009. Options for urine treatment in developing countries. *Desalination* 248: 360–368.

Pronk, W., Zuleeg, S., Lienert, J., Escher, B., Koller, M., Berner, A., Koch, G., Boller, M. 2007. Pilot experiments with electrodialysis and ozonation for the production of a fertiliser from urine. *Water Science & Technology* 56(5): 219–27.

Rawat, I., Kumar, R., Mutanda, T., Bux, F. 2011. Dual role of microalgae: Phycoremediation of domestic wastewater and biomass production for sustainable biofuels production. *Applied Energy* 88: 3411–3424.

Rose, C., Parker, A. Jefferson, B., Cartmell, E. 2015. The characterization of feces and urine: A review of the literature to inform advanced treatment technology. *Critical Reviews in Environmental Science and Technology* 45(17): 1827–1879.

Simha, P., Ganesapillai, M. 2017. Ecological Sanitation and nutrient recovery from human urine: How far have we come? A review. *Sustainable Environment Research* 27: 107–116.

Tettenborn, F., Behrendt, J., Otterpohl, J. 2007. *Resource recovery and removal of pharmaceutical residues Treatment of separate collected urine.* Doctoral Thesis. University of Technology of Hamburg.

Udert, K.M., Larsen, T.A., Gujer, W. 2006. Fate of major compounds in source-separated urine. *Water Science & Technology* 54(11–12): 413–420.

Walter, X.A., Gajda, I., Forbes, S., Winfield, J., Greenman, J., Ieropoulos, I. 2016. Scaling-up of a novel, simplified MFC stack based on a self-stratifying urine column. *Biotechnology Biofuels* 9(93).

Zang, G.L., Sheng, G.P., Li, W.W., Tong, Z.H., Zeng, R.J., Shi, C., and Yu, H.Q. 2011. Nutrient removal and energy production in a urine treatment process using magnesium ammonium phosphate precipitation and a microbial fuel cell technique. *Physical Chemistry Chemical Physics* 14: 1978–1984.

Zöllig, H. et al. 2015. Direct Electrochemical Oxidation of Ammonia on Graphite as a Treatment Option for Stored Source-Separated Urine. *Water Research* 69: 284–294.

Portugal lacks refuse derived fuel production from municipal solid waste

P.C. Berardi
LEPABE, Department of Metallurgical and Materials Engineering, Faculty of Engineering,
University of Porto, Porto, Portugal
Faculty of Business Administration, Superior School of Advertising and Marketing ESPM,
São Paulo, Brazil

M.F. Almeida & J.M. Dias
LEPABE, Department of Metallurgical and Materials Engineering, Faculty of Engineering,
University of Porto, Porto, Portugal

M.L. Lopes
CONSTRUCT-GEO, Department of Civil Engineering, Faculty of Engineering, University of Porto,
Porto, Portugal

ABSTRACT: The present study provides an overview on municipal waste management in Portugal, with focus on current status of Refuse Derived Fuel Recovery (RDF), forecasts set by the National strategy and 2020 targets. Considering the official reports on municipal waste management and related legislation, although the capacity (considering MT and MBT) would allow obtaining material for RDF production (high calorific fraction) within the range predicted in the National strategy (0.95–1.2 Mt for 2013), the effective amount sent for RDF production was residual (107,623 t), from which only 29,770 t of RDF were obtained in 2015. A lack of economic viable recovery solutions exists, due to political, economic, legal and technical constraints. The recovery of the high calorific fraction including RDF production is essential to fulfill National environmental targets, thus, it is urgent the revision of current strategies to provide viable alternatives for recovery at a National level.

1 INTRODUCTION

The increase in population and consumption practices in the last decades led to a huge impact with the consequent increase in municipal waste (MW) generation, which has become a critical environmental problem. MW management is therefore a central theme in the world's agenda being closely related to the Sustainable Development Goals (UN, 2015) and the Circular Economy premises. On average, an European citizen produced 460 kg/year of waste in the 1990s whereas in the following decade the amount increased to 520 kg/year and previous projections indicated that by 2020such indicator could reach around 680 kg/year (EEA, 2008). Although being of unquestionable relevance to develop effective waste management practices, in 2014 the generation of waste per capita was not as high as previously predicted being 478 kg/person/year) (EUROSTAT, 2017), mostly due to economic, environmental and social factors.

As a milestone in the current European waste policy, in 2008 the parliament approved the waste framework directive (2008/98/EC), in which the waste management hierarchy was defined to promote waste reduction and reuse as well as proper waste recovery avoiding as much as possible disposal solutions such as landfilling. The guidance goes beyond reducing environmental impacts, since it seeks to promote economic activities by stimulating local economies to reuse and recover such resources and it also introduces the possibility to apply end-of-waste criteria for certain materials (EU, 2008). In 2015, the European Commission presented the European Union's Plan of Action for the Circular Economy, which must be followed by all Member States and aims to minimize the production of waste and, above all, maximize the use of waste as a resource (EC, 2015).

After one year, the European Commission produced a report that brought notes to relevant issues, with energy production from waste being one of these highlights. This communication clearly emphasizes the need to ensure that the energy recovery of waste generated in the European Union is in line with the Circular Economy, the waste management hierarchy and the strategies of the respective Member States (EC, 2017).

Portugal strategy on MW is currently defined by PERSU 2020 (Ordinance No.187-A/2014) and aims to reduce environmental impacts and promote a better use of the socio-economic value of wastes, associated with the progressive diversion of recoverable materials from landfills. Many Member States have adopted Mechanical (MT) or Mechanical Biological Treatment (MBT) of MW to comply with European Union directive targets on landfill diversion (Garg et al, 2007). Portugal is not an exception, having also established, through Order No. 21295/2009, a national Strategy for Refuse Derived Fuel, that might be obtained from a significant amount of the outputs of such technologies, and which recovery is fundamental to achieve such goals.

2 MUNICIPAL WASTE MANAGEMENT IN PORTUGAL

Based on the Annual Municipal Waste Report released by the end of 2016, concerning 2015 data, an analysis was made to the 23 MW management operating systems of Continental Portugal.

In 2015, 4.523 Mt of MW were produced which is equivalent to 459 kg/person/year, below the European average of 478 kg/person/year (APA, 2016, EUROSTAT, 2017). The following weight of direct destinies is reported: 34% sent to landfill, 23% underwent mechanical biological treatment (MBT), 20% had energy recovery, 10% undergone mechanical treatment (MT), 10% were sent for material recovery and 2% for organic recovery (difference to 100% relates to rounding). Among the 66% of wastes that weren't directly sent to landfill, a significant fraction had further such destiny. In agreement, the material effectively sent to landfill in 2015 represented 51% of the waste produced (APA, 2016), corresponding to 2.306 Mt, and mostly associated with MT and MBT operations. Such fact shows that a relevant fraction of recoverable material is still being discarded.

Among the alternatives to reduce landfilling and promote energy recovery from waste that could not be recycled is the use of Refuse Derived Fuel (RDF). It has been classified by the European List of Waste (Decision 2014/955/EU) with the code 191210 being a non-hazardous combustible fraction resulting from the mechanical treatment of waste. At an European level, EN 15357:2011 on Solid Recovered Fuel (SRF) was released, defining it as "a solid prepared from non-hazardous waste, to be utilized for energy recovery in incineration or co-incineration plants". In Portugal, NP 4486:2008 defines refuse derived fuels as a solid waste prepared from non-hazardous waste to be used in incineration and co-incineration plants with energy recovery, and which fulfills the requirements set by the standard.

In Portugal, Order No. 21295/2009 established forecasts of RDF production potential for the five regions of the country. This document, in agreement with the MW management planning, envisaged the start-up of new MT and MBT plants in 2012 and 2013, which could result in a production between 0.95 and 1.2 Mt of material for RDF in 2013, according to Table 1. The strategy of highlighting the use of RDF is vital aiming to comply with the Portuguese and European targets related to waste diversion from landfilling. Note that the Order. 21295/2009 considered for the goals up to 2020 that from the municipal waste that enters a mechanical biological treatment plant, 40–55% can be for RDF.

In 2015, from the total MW produced in mainland Portugal, 1.436 Mt (32%) went through MT or MBT plants (Table 2).

Table 1. Forecast of material for Refuse Derived Fuel (RDF) production in Portugal by region, for 2013, according to Order No. 21295/2009.

	Alentejo	Algarve	Centre	Lisboa and Vale do Tejo	North
MW (kt)	322	397	809	1 879	1 613
Material for RDF (kt)	62–82	64–71	248–327	389–486	187–241

MW – Municipal Waste.
RDF – Refuse Derived Fuel.

Table 2. Refuse derived fuel production by region, in Portugal, in 2015 (APA, 2016).

Category	Alentejo	Algarve	Centre	Lisboa and Vale do Tejo	North	Total
MW (t)	311 143	347 981	701 662	1 678 013	1 483 814	4 522 613
% Total MW	6	8	16	37	33	100
Input MT+MBT (t)	161 171	46 433	556 341	406 676	265 773	1 436 394
% Input MT+MBT	52	13	79	24	18	32
RDF Production (t)	2 613	0	17 898	9 259	0	29 770
% RDF produced from MT+MBT	2	0	3	2	0	2

MW – Municipal Waste.
RDF – Refuse Derived Fuel.
MT – Mechanical Treatment.
MBT – Mechanical Biological Treatment.

The Annual MW report clearly indicates, for each system, the amount of material sent for RDF production from MRF and MT+MBT (directed to an RDF production plant) as well as the RDF produced. Table 3 summarizes information on the systems which sent material for RDF production. The material sent for RDF production represented 19.5% of MT + MBT inputs, which were 884 156 t, showing that only some plants sent material for recovery. Analyzing each region, it can be verified that the amount of RDF produced in 2015 is far below the potential, although the installed capacity on MT + MBT would allow results within the expected range for 2013, being estimated 983 170 t of potential material for RDF, based in the inputs of MT (482 883 t) and MBT (953 511 t) alone and the Order No. 21295/2009 premises (considering 95% MT and 55% MBT). In the Northern region, from the eight municipal waste management systems, only three operated MT or MBT plants and there was no material sent for RDF production. There were five systems for the region comprising Lisboa and Vale do Tejo, where three of them operated with MT or MBT, which sent 50,851 t of material for RDF. There were four systems in the central region, but only one sent 20,465 t for RDF production. Of the five existing systems in the Alentejo region, only one did not operate MT or MBT. Three of the four systems sent 34,999 t of material for RDF production. At Algarve region, no material was sent for RDF preparation, despite having a MT plant in operation. It is worth noting that the amount of RDF obtained by each of the systems varies according to the types of treatment applied in each process (between 1 and 38% for MT or MBT and between 1 and 7% for source segregated collection, Table 3). From the 23 MW management systems, where 15 have MT or MTB plants, only seven had material resulting from MBT or MT which was sent to RDF production, totalizing 106,315 t.

Only 2 systems (Centre and Alentejo) sent material from Materials Recovery Facilities (MRF) for RDF production, representing a further 1,308 t from the selective collection, of which 84% came from the central region system and 16% from the Alentejo system. According to this analysis, it can be seen that from the total waste treated through MT or MBT at those seven systems (884 156 t in 2015), the material sent for RDF production was equivalent

Table 3. Quantitative information on Municipal Waste Management Systems which sent material for Refuse Derived Fuel production in 2015, by source (APA, 2016).

Category [Region] System	Lisboa and Vale do Tejo					Centre			Alentejo							Total (t)
	Amarsul		Resitejo	Tratolixo	Total Region	ERSUC		Total Region	Ambilital		Gesamb		Valnor		Total Region	
	t	%RDF Prod	t	t		t	%RDF Prod		t	%RDF Prod	t	%RDF Prod	t	%RDF Prod		
MW	391 599		89 902	398 139	879 640	385 705		385 705	61 614		76 809		113 305		251 728	1 517 073
1. Source Segregated Collection																
1.1 MRF	23 011		7 471	12 166	42 648	15 726		15 726	4 633		4 757		19 042		28 432	86 806
1.1.1 Sent for RDF Production						1 100	7.0%	1 100					208	1.1%	208	1 308
1.2 Landfill	5265				5 265				201		118				319	5 584
1.3 Incineration																
1.4 Others																
2. Input MT + MBT	138 386		80 468	187 822	406 676	328 731		328 731	3 418		56 953		88 378		148 749	884 156
2.1 Sent for RDF Production	32 857		893	17 101	50 851	20 465		20 465	396		1 050		33 553		34 999	106 315
2.1.1 RDF produced	9 259	28.2%			9 259	17 898	87.5%	17 898	294	74.2%	580	55.2%	1 739	5.2%	2 613	29 770
2.1.2 Landfill	14 622	44.5%			14 622	3 600	17.6%	3 600	19						19	18 241
3. Landfill	66 954		51 165		118 119	170 650		170 650	1 537		26 345		1 459		29 341	318 110
4. Others				117 655	117 655								6 649		6 649	124 304
Total Sent for RDF Production																107 623

MW – Municipal Waste.
MRF – Materials Recovery Facilities.
RDF – Refuse Derived Fuel.
MT – Mechanical Treatment.
MBT – Mechanical Biological Treatment.
Prod – Production.

to 12% (107 623 t) and a significant amount is being send to landfills. Regarding all the material sent for RDF production from both the source segregated collection and MT/MBT plants, only 27.7% were effectively transformed into RDF.

From the total material sent for RDF production (107,623 t), 29,770 t of RDF was produced in 2015 throughout the country. Table 3 presents the representation of each region in total MW and the respective amount of RDF production in 2015. The Central region was the one that assigned the most waste to MT or MBT with 79% of the total produced and the Alentejo region allocated more than half of its MW to these treatments. Algarve was the region that treated less waste through MT or MBT representing 13%. The Northern region, which has the second largest MW production in the country, did not obtain any production of RDF and only one system from the Central region produced RDF, which represented the largest production for the country in 2015, of 17,898 t, with low amounts of residuals from preparation, if moisture losses are considered.

From the Lisboa and Vale do Tejo region, from which was expected the largest generation of material for RDF production, concerning the Order No. 21295/2009 (between 389 to 486 k t per year), insignificant production occurred. In several systems, the amount of RDF produced clearly shows absence of processing, not expected due to technical reasons but occurring due to economic reasons.

Feedbacks of the systems and end-users attribute the current status of lack of RDF production to absence of economic viable recovery solutions namely because: i) the only end users are the cement kilns, where viability strongly depends upon location and a strong competition exists with non-domestic product of higher quality and low cost, due to high discrepancies concerning landfill tax at other countries; ii) there is a lack of co-incineration and other alternatives that could be natural clients to the potential production of RDF, namely in the ceramic, ironmaking and steelmaking sectors; iii) no dedicated alternatives exist for energy recovery, except two mass-burn incinerators for MSW (Portugal Continental area); iv) there are difficulties to apply the end-of-waste criteria.

Due to the major relevance of RDF recovery to fulfill National environmental targets, it is urgent to revise current strategies, to provide the means to develop at a National level environmental and economic viable alternatives for obtaining and using such energetic product.

3 CONCLUSIONS

The need to increase waste recovery and avoid landfilling is imperative and needs to be addressed by all Member States of the European Union. Deadlines and quantitative limits were set as goals to be achieved by 2020, and they depend upon the effective use of the MT and MBT plants functioning in the country. The current status shows that still much remains to be done to achieve such goals.

In 2015, 29 770 t of RDF from Municipal Waste were produced, resulting from 107 623 t of material sent for RDF production and representing around 12% of the input of MT and MBT related systems. Overall, 32% of the total MW (around 4.5 Mt) went through MT or MBT and according to the Order No. 21295/2009, it was expected that 55% and 95% of the material which entered MBT and MT plants, respectively, could be sent for RDF production. Thus, 983 170 t of RDF could have been sent for RDF production in 2015, based in MT + MBT only.

The strategy on RDF assumed that the market would easily absorb the valuable energy resource resulting from MT and MBT plants; however, reality shows high economic constrains to process and use this material, which requires the environmental policy to be put into action to promote the effective use of the resources and of the investments made so far on waste management facilities. In agreement, the National Strategy on RDF needs to be revised urgently.

As long as waste is perceived as a liability rather than as resource, the reduction of environmental impacts is compromised and stays far from the assumptions of the circular economy.

ACKNOWLEDGEMENTS

This work was financially supported by: Project UID/EQU/00511/2013-LEPABE (Laboratory for Process Engineering, Environment, Biotechnology and Energy—EQU/00511) and Project POCI-01–0145-FEDER-007457-CONSTRUCT (R&D Institute for Structures and Construction) by FEDER funds through Programa Operacional Competitividade e Internacionalização—COMPETE2020 and by national funds through FCT—Fundação para a Ciência e a Tecnologia.

REFERENCES

APA 2016. *Resíduos Urbanos*. Relatório Anual 2015, Agência Portuguesa do Ambiente, I.P. Departamento de Resíduos.

APA 2016. *Resíduos Urbanos*. Relatório Anual 2015, Agência Portuguesa do Ambiente, I.P. Departamento de Resíduos. RARU2015_Fichas_SGRU.

EC 2015. *Proposal for a Directive of the European Parliament and of the Council amending Directive 2008/98/EC on waste*. COM/2015/0595 final - 2015/0275 (COD). European Commission, Directorate-General for Environment.

EC 2017. *Report from the Commission to the European Parliament, the Council, the European Economic and Social Committee and the Committee of the Regions on the implementation of the Circular Economy Action Plan*. COM (2017) 33 final. European Commission.

EEA 2008. *Better management of waste management will reduce greenhouse gas emissions, Supporting document to EEA 2008/01*. Copenhagen: European Environment Agency.

EU 2008. *Directive 2008/98/EC of the European Parliament and of the Council of 19 November 2008 on waste*. European Parliament/Council of the European Union.

EUROSTAT 2017. European Commission. *Waste generation. Municipal waste generation and treatment, by type of treatment method*. Code: tsdpc240.

Garg, A., Smith, R., Hill, D., Simms, N., Pollard S. 2007. Wastes as co-fuels: the policy framework for solid recovered fuel (SRF) in Europe, with UK implications. *Environmental science & technology* 41(14): 4868–4874.

Order No. 21295/2009. Diário da República - 2ª. Série, nº. 184 de 22.09.2009. *Estratégia para os Combustíveis Derivados de Resíduos*. Ministério do Ambiente, do Ordenamento do Território e do Desenvolvimento Regional e da Economia e da Inovação.

Ordinance No. 187-A/2014. Diário da República No. 179/2014, Série I de 2014–09–17. *PERSU 2020 - Plano Estratégico para os Resíduos Urbanos*. Ministério do Ambiente, Ordenamento do Território e Energia.

UN 2015. *Time for global action for people and planet*. United Nations Department of Economic and Social Affairs. New York: United Nations.

Removal of Cr(III) from aqueous solutions by modified lignocellulosic waste

A.L. Arim
UNIPAMPA—Federal University of Pampa, RS, Brazil

D.F.M. Cecílio, M.J. Quina & L.M. Gando-Ferreira
CIEQPF—Centre of Chemical Processes Engineering and Forest Products, Department of Chemical Engineering, University of Coimbra, Coimbra, Portugal

ABSTRACT: Pine bark (NPB) is a common agricultural waste in Portugal and was investigated as biosorbent in its natural state and after sodium hydroxide treatment (MPB). Surface modification was carried out using NaOH solutions with concentration 10, 20 and 30% (wt%). The best adsorption capacity of NPB was 8.35 mg g^{-1}, while with MPB the maximum was 17.06 mg g^{-1}. A Box-Behnken design was used to obtain the response surface and identify the optimal operating region as well as the variables with the most influence on Cr(III) removal efficiency. The results showed that the Cr(III) adsorption is favored in the case of the initial pH 4.8, initial Cr(III) concentration of 137.3 mg L^{-1}, liquid-to-solid ratio of 200. The temperature did not affect the process. As main conclusion, this study demonstrated that this low cost biosorbent has good potential to recover chromium from wastewaters, and at the same time protect environment.

1 INTRODUCTION

The electroplating industry is responsible for the generation of effluents containing heavy metals, due to the use of metal solutions and a considerable amount of washing water in electrodeposition processes (Fu & Wang, 2011). Chromium is a common type of pollutant in the effluents of these industries and can exist in various oxidation states (di-, tri-, penta-, and hexa). Even though Cr(VI) and Cr(III) are the most stable forms present in the aqueous environment (Fonseca-Correa et al., 2013). Cr(III) is an essential nutrient for humans and its lack can cause heart problems, metabolic disorders and diabetes, while long term exposure may cause allergic skin reactions and cancer (Vilar, et al., 2012). Adsorption is one of the most prevalent methods for wastewater treatment allowing the removal of metal ions from dilute solutions. Frequently, the regeneration of the adsorbent allows a reduction in the overall cost of the procedure (Abdolali et al., 2014; O'Connell et al., 2008).

Agro-industrial or lignocellulosic residues have been studied as low-cost adsorbents for uptaking heavy metal from wastewaters (Feng et al., 2011). They are eco-friendly materials with promising adsorption capacities, available in large quantity and with low cost (Oliveira, 1999). Lignocellulosic residues may have a natural capacity to remove some heavy metals, due to their functional surface groups. However, their uptake capacity is often very low and their physical stability is variable. To overcome those problems physical and/or chemical modifications can be made to the material. Alkali treatment also known as mercerization, involves strong alkali solutions, normally sodium hydroxide (NaOH), put in contact with the lignocellulosic residue (Ofomaja et al., 2009). This method changes the crystallinity of cellulose structures present on the material, increases the specific surface area and makes the hydroxyl groups more available for adsorption (Abdolali et al., 2014; Gurgel et al., 2008; O'Connell et al., 2008). One of the most abundant agro-industrial residues in Portugal is pine bark (*Pinus pinaster*), which has limited applications, namely as soil fertilizer (after biological decomposition) or as fuel. The main objective of this work was to optimize the chemical modification of pine bark and the operation conditions, to maximize its capacity for adsorption of Cr(III) from an aqueous solutions.

2 METHODOLOGIES

2.1 *Materials and reagents*

Pine bark from species *Pinus pinaster* was collected in North of Portugal. In the laboratory the pine bark was washed with distilled water several times to remove the dirt, lichens and resin. Then it was dried an oven at 60°C during 24 h to remove superficial water. The bark was grinded in a Reischt MUHLER, model 5657 HAAN and sieved using JULABO model SW-21C automatic siever to achieve adequate particle diameter (250–595 μm). Then, it was dried again at 80°C for 24–48 h, and stored in dry conditions until used, with the reference NPB (nature pine bark).

The solutions of Cr(III) were prepared with chromium nitrate nonahydrated ($CrN_3O_9.9H_2O$) from Alfa Aesar and ultrapure water. For chemical modifications of pine bark, solutions of NaOH were utilized. The pH was controlled with solutions of 0.01–0.1 mol L^{-1} of NaOH or 0.01–0.1 mol L^{-1} of HCl as required.

2.2 *Chemical modification of pine bark*

About 30 g of NPB were treated with 900 mL of NaOH solutions with concentrations of 10, 20 and 30% (wt%) at 25°C and magnetic stirred at 1200 rpm for 16 h. Then, each of the samples was filtered and washed with ultrapure water until pH constant. The material was then dried in an oven at 80°C for 48 h (Gurgel and Gil, 2009). These samples were labelled as MPB1_10%, MPB1_20% and MPB1_30%, respectively. Samples mercerized twice from the dry MPB were referred as MPB2_10%, MPB2_20% and MPB2_30%. The modified pine bark (MPB) was then filtered using ASHLESS filter paper 40 and washed with distilled water until constant and neutral pH. Finally, MPB was dried at 90°C for 48 h and stored in a dry conditions until utilization.

2.3 *Adsorption studies*

The adsorption equilibrium studies in batch conditions were conducted for testing the effect of different variables. The amount of metal adsorbed (q_e, mg g^{-1}) and removal efficiency (R,%) were calculated as follows:

$$q_e = \left(\frac{C_o - C_e}{m}\right) \times V \tag{1}$$

$$R(\%) = \left(\frac{C_0 - C_e}{C_o}\right) \tag{2}$$

where C_o is the initial concentration of the adsorbate (mg L^{-1}), C_e is the final concentration of adsorbate (mg L^{-1}), m is the mass of the adsorbent (g) and V the volume of the solution (L). Cr(III) concentration was determined using a flame atomic absorption spectrometry (FAAS), Analytickjena—Contra 300, with wavelength of 357 nm and flame of acetylene/air.

2.4 *Design of experiments*

Design of experiments (DOE) based on Box-Benken design (BBD) was implemented to obtain the response surface, using as independent variables (factors) the initial concentration,

Table 1. Factors and levels tested in the Box-Benken design.

Factor	−1	0	+1
Initial concentration, C_o (mg L^{-1})	137.3	364.7	466.3
pH	2.3	3.4	4.8
Temperature, T (°C)	25	40	55
Liquid-to-solid ratio, L/S	70	135	200

pH, temperature and liquid-to-solid ratio (L/S). The method involved an incomplete three-level factorial, which will allow the determination of the coefficients of a second-order model by fitting the experimental data. The levels tested for each factor are presented in Table 1.

Accordingly, 27 experiments were generated on Statistica 7 software, which the total number was estimated using Eq. (3):

$$N = 2 \times f \times (f - 1) + CP \qquad (3)$$

where N is the number of experiments, f is the number of factors and CP is the number of central points (3 replicas). The experimental tests were performed using 25 mL of Cr(III) solution agitated for 3 h in contact with the adsorbent.

3 RESULTS AND DISCUSSION

3.1 Effect of chemical treatment on the adsorbent performance

Figs. 1 (a) and (b) shows the effect of the chemical treatment with NaOH on the sorption capacity and on the removal efficiency of Cr(III), respectively. NPB is an limited adsorbent for the removal of Cr(III) ions in aqueous solutions achieving a removal efficiency of 22.80%. The alkali treatment improved the sorption capacity in all conditions tested, resulting efficiencies higher than 40% in all cases.

Figs. 1 (a) and (b) reveal that in case of pine bark was treated twice with NaOH solution, the capacity or the removal efficiency did not improve when compared with only one step. It should be emphasized that the mercerization process allows the transformation of cellulose I be converted into cellulose II. The latter contains hydroxyl groups more accessible, turning the material more reactive and thus the adsorption process becomes more efficient. The modification of cellulose I to cellulose II is irreversible, due to the fact that cellulose II is thermodynamically more stable than cellulose I (Roy et al., 2009). The results demonstrate that the first treatment with NaOH doubled the adsorption capacity of NPB, which may have occurred due to change cellulose structure from parallel chains (cellulose I) to antiparallel chains (cellulose II) (Gurgel et al., 2008). The second treatment with the same modifying agent seems not improve this (Hill, 2006; Oudiani et al., 2011; Rowell, 2005). Moreover, a little increase on adsorption capacity was observed when the concentration NaOH increased from 10% to 20%, as already observed by Oudiani (2011). However, 30% of alkali tends to diminish the adsorption capacity. This occurs because degradation of cellulose starts to happen (Oudiani et al., 2011). In mercerization treatment, the concentration of NaOH must be optimized, since degradation reactions may occur if the concentration is too high. Indeed, swelling and dissolution from amorphous cellulose II may take place in these conditions (Oudiani et al., 2011; Rowell, 2005). Based on experimental results, it was concluded that the best alkali treatment for further studies is a treatment with NaOH 10% (v/v) applied all in one single step.

3.2 Design of experiments results

The Box-Behnken design was used in order to extract information with a reduced the number of experiments, namely understand which operational variable would have a major impact on

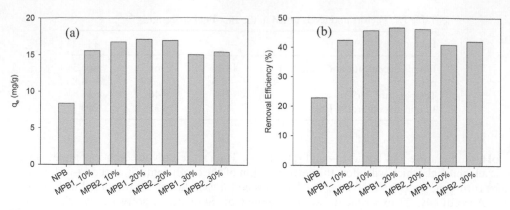

Figure 1. Effect of concentration of NaOH solution for pine bark modification on (a) adsorption capacity; (b) removal efficiency.

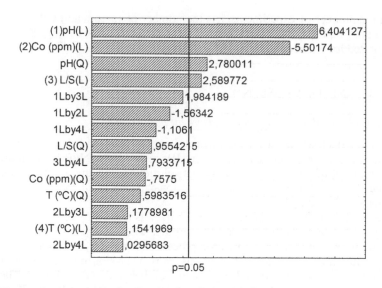

Figure 2. Pareto chart obtained after Box-Benken design results.

adsorption process, how they interact between each other and develop a mathematical model (Ferreira et al., 2007). As variable response was chosen the removal efficiency of Cr(III), while the factors were the initial solution pH, initial concentration of the adsorbate (C_o), liquid-to-solid ratio (L/R) and temperature. One advantage of this type of design is the fact that it does not contain combinations in which all factors are simultaneously tested under extreme conditions (Ferreira et al., 2007). Figure 2 represents the Pareto chart of standardized effects at p = 0.05.

Analysing the results plotted in Fig. 2 it can be concluded that the pH is the most significant effect on the removal efficiency of Cr(III), followed by initial concentration (C_o) and L/S ratio. Temperature does not affect the process as much as expected and this factor may be removed from the design of experiments. The dependence between the pH and the adsorbent performance is positive, which means that the removal efficiency increases with the increase of pH. Indeed, pH of the solution influences the charges of the active sites present in the adsorbent surface. At low pH, the binding sites will be most likely protonated, resulting in poor heavy metal binding. It should be highlighted that at high pH (mostly in alkaline conditions), there will be a higher concentration of OH⁻ on the solution that will compete with Cr(III) ions (Anirudhan and Radhakrishnan, 2007; Pehlivan et al., 2012). The predictive model obtained is represented by Eq. (4) and Fig. 3 shows the values observed as function of predicted values. As it can be seen, the values observed are close to the predicted ones (coefficient of determination, $R^2 = 0{,}906$).

$$R(\%) = -210.36 + 120.55pH - 11.57pH^2 - 0.120C_o - 0.036(L/S) - 0.002(L/S)^2 \quad (4)$$

From the surface responses, Figs. 4 (a) and (b), it can be observed that the highest removal efficiency of Cr(III) occurs for high pH values as suggested by Pareto chart results. Lower C_o-favors the removal efficiency because there is less Cr(III) ions to uptake with the same adsorbent mass, obtaining a higher removal efficiency. Higher removal efficiency values are obtained for high L/S as observed.

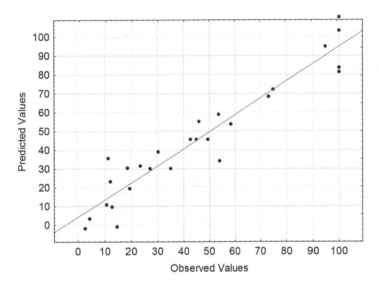

Figure 3.   Observed values as a function of predicted values.

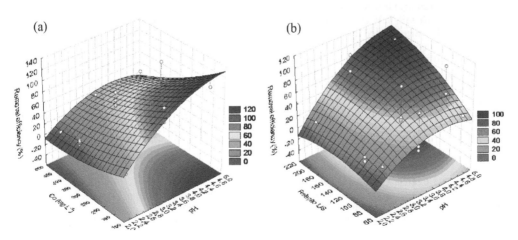

Figure 4.   Surface response of the Box-Benken design (a) influence of C_o and pH; (b) the influence of L/S and pH.

4 CONCLUSION

The alkali treatment of the adsorbent enabled to improve the uptake of Cr(III) and the most successful treatment was a NaOH concentration of 10% (v/v) used only in one step. A Box-Benken design was employed for optimization of the batch adsorption of Cr(III) using mercerized pine bark. According to the significance effect obtained in variance analysis, the initial pH was found to have the most significant effect on the adsorption of Cr(III), followed by initial concentration and L/S ratio. Temperature revealed to be a factor of weak influence on the removal efficiency. Experimental removal efficiencies of Cr(III) were well predicted by a quadratic model, where the independent variables are initial pH, initial concentration Cr(III) and L/S ratio. The results have demonstrated that mercerized pine bark is a promising adsorbent for removing Cr(III) ions from aqueous solutions.

ACKNOWLEDGEMENT

Aline L. Arim gratefully acknowledges Conselho Nacional de Desenvolvimento Científico e Tecnológico (Cnpq) by the financial support under (201264/2014-5).

REFERENCES

Abdolali, A., Guo, W.S., Ngo, H.H., Chen, S.S., Nguyen, N.C. & Tung, K.L. 2014. Typical lignocellulosic wastes and by-products for biosorption process in water and wastewater treatment: A critical review. *Bioresour. Technol.* 160: 57–66.

Anirudhan, T.S.& Radhakrishnan, P.G. 2007. Chromium(III) removal from water and wastewater using a carboxylate-functionalized cation exchanger prepared from a lignocellulosic residue. *J. Colloid Interface Sci.* 316: 268–276.

Feng, N., Guo, X., Liang, S., Zhu, Y. & Liu, J. 2011. Biosorption of heavy metals from aqueous solutions by chemically modified orange peel. *J. Hazard. Mater.* 185: 49–54.

Ferreira, S.L.C., Bruns, R.E., Ferreira, H.S., Matos, G.D., David, J.M., Brandão, G.C., da Silva, E.G.P., Portugal, L.A., dos Reis, P.S., Souza, A.S. & dos Santos, W.N.L. 2007. Box-Behnken design: An alternative for the optimization of analytical methods. *Anal. Chim. Acta* 597: 179–186.

Fonseca-Correa, R., Giraldo, L. & Moreno-Piraján, J.C. 2013. Trivalent chromium removal from aqueous solution with physically and chemically modified corncob waste. *J. Anal. Appl. Pyrolysis* 101: 132–141.

Fu, F. & Wang, Q. 2011. Removal of heavy metal ions from wastewaters: A review. *J. Environ. Manage* 92: 407–418.

Gurgel, L.V.A., Freitas, R.P. de & Gil, L.F. 2008. Adsorption of Cu(II), Cd(II), and Pb(II) from aqueous single metal solutions by sugarcane bagasse and mercerized sugarcane bagasse chemically modified with succinic anhydride. *Carbohydr. Polym.* 74: 922–929.

Gurgel, L.V.A. & Gil, L.F. 2009. Adsorption of Cu(II), Cd(II) and Pb(II) from aqueous single metal solutions by succinylated twice-mercerized sugarcane bagasse functionalized with triethylenetetramine. *Water Res.* 43: 4479–4488.

Hill, C.A.S. 2006. *Wood modification: Chemical, Thermal and Other Processes.* Chichester: John Wiley & Sons, Ltd.

O´Connell, D.W., Birkinshaw, C. & Dwyer, T.F. 2008. Heavy metal adsorbents prepared from the modification of cellulose: A review. *Bioresour. Technol.* 99: 6709–6724.

Ofomaja, A.E., Naidoo, E.B. & Modise, S.J. 2009. Removal of copper(II) from aqueous solution by pine and base modified pine cone powder as biosorbent. *J. Hazard. Mater.* 168: 909–917.

Oliveira, A., 1999. Boas Práticas Florestais para o Pinheiro Bravo. Centro Pinus—Associação para a Valorização da Floresta.

Oudiani, A. El, Chaabouni, Y., Msahli, S. & Sakli, F. 2011. Crystal transition from cellulose i to cellulose II in NaOH treated Agave americana L. fibre. *Carbohydr. Polym.*

Pehlivan, E., Altun, T. & Parlayici, S. 2012. Modified barley straw as a potential biosorbent for removal of copper ions from aqueous solution. *Food Chem.* 135: 2229–2234.

Rowell, R.M. 2005. *Handbook of Wood Chemistry and Wood Composites.* Madison: CRC Press Taylor & Francis Group.

Roy, D., Semsarilar, M., Guthrie, J.T., & Perrier, S. 2009. Cellulose modification by polymer grafting: a review. *Chemical Society Reviews* 2046–2064.

Valorization of residues from fig processing industry by anaerobic digestion

D.P. Rodrigues, C.I. Alves, R.C. Martins & M.J. Quina
CIEPQPF, Department of Chemical Engineering, University of Coimbra, Coimbra, Portugal

A. Klepacz-Smolka
Faculty of Process and Environmental Engineering, Lodz University of Technology, Lodz, Poland

M.N. Coelho Pinheiro
Department of Chemical and Biological Engineering, Coimbra Superior Institute of Engineering, Polytechnic of Coimbra, Coimbra, Portugal

L.M. Castro
CIEPQPF, Department of Chemical Engineering, University of Coimbra, Coimbra, Portugal
Department of Chemical and Biological Engineering, Coimbra Superior Institute of Engineering, Polytechnic of Coimbra, Coimbra, Portugal

ABSTRACT: There is a growing interest in the economic benefits from the management of seasonal fruit wastes by anaerobic digestion (AD) processes, instead of landfilling the organic matter. The aim of the present study is to investigate the valorization of a fig processing industry waste (FgW) by AD. To the best of our knowledge, this is the first time this residue is considered as a substrate. A batch reactor with an operating volume of 3.75 L, at $37 \pm 2°C$ (mesophilic operation) and a substrate to inoculum (S/I) ratio within the range 0.2 to 1.2 g VS/ gVS0 were used. The best specific methane production achieved at lab scale was $0.284 \ Nm^3 \ CH4/kg$. The results reveal that FgW can be valorized by AD and the biogas used as a source of energy in the small to medium enterprise, where the waste has been produced.

1 INTRODUCTION

The European Union (EU) created recently "The Circular Economy Package" which includes specific proposals to improve waste management practices and limit the use of landfilling. The proposals will provide a clear and stable long-term policy focused on prevention, reuse and recycling. In the key elements of the proposal, there is an important objective "binding landfill reduction target of 10% by 2030". Thus, it is strategic to find alternatives to landfill disposal of organic wastes, namely those produced in food and drink processing industry, which accounts for about 12.5% of manufacturing industry waste (Arhoun et al., 2013). Anaerobic digestion (AD) of waste may have an important contribution to this goal. European policies aimed at promoting that at least 25% of bioenergy is from biogas. In Italy, 3405 GWh of electricity were produced from biogas in 2011. Germany reveals high interest in this technology with almost 4000 anaerobic digestion units for biogas production in agro-industries (Mao et al., 2015). Indeed, Europe and Asia have the impressive number of more than 30 million AD units of both large and small-scale (Fagbohungbe et al., 2017). In Portugal, the National Renewable Energy Action Plan prepared on the basis of the Commission Decision of 30 June 2009, according to the Directive 2009/28/EC of the European Parliament and of the Council, of 23 April 2009, imposed the important goal of increasing the energy production from biogas to 413 GWh until 2020 (PNAER, 2013).

Within this context, and although AD is not a new technology, it is clear that research efforts should be performed in order to optimize this technology so that European goals can be reached. Thus, the interest on the application of this technology for solid wastes processing is expressively growing. Anaerobic digestion consists in the biological stabilization of organic matter with the production of biogas (methane and carbon dioxide) that can be used as renewable energy source. Moreover, the other major by-product is the digestate, which is rich in nutrients and often used in agriculture as a soil improver. Important challenges AD still faces are related to operational instability due to load and substrate selection, as well as the quality of the digestate produced (Fagbohungbe et al., 2017).

Residues from food industry have been used to produce biogas by AD and thus contribute to reducing the amount of waste from this sector of economic activity, replacing fossil fuels. This strategy also contributes to the climate EU targets, allowing the reduction of greenhouse gas emissions, as well as enlarging the share of renewable energy and cutting the use of primary energy down (Bayard, 2016).

In a Portuguese food processing SME (small and medium-size enterprise) the production of fig vinegar generates a great amount of waste that nowadays is mainly sent to landfill. This work aims to assess the possibility of valorizing fig processing industry waste by AD, avoiding the adverse environmental impacts related to the unstable organic matter. To the best of our knowledge, this is the first time this residue is considered as a substrate for AD. This strategy will contribute to the implementation of the circular economy in SME company. With the valorization of the fig waste, the SME will be able to reduce the amount of waste sent to landfill, reducing the associated economic and environmental costs, and producing a significant amount of renewable energy that can be used as heat and electricity source in the production process of fig vinegar. Consequently, the environmental footprint of the company can be reduced, making their products more sustainable.

2 MATERIALS AND METHODS

2.1 *Substrate and inoculum*

The fig waste (FgW) was collected from a Portuguese industry that produces fig vinegar and sauces with an annual production of about 200 Mg (wet basis). FgW was pre-treated mechanically in order to reduce the size of the particles (dp≤1.68 mm). The inoculum was collected from anaerobic digesters of a municipal wastewater treatment plant, where primary and secondary sewage sludge were used as the substrate to AD.

2.2 *Experimental set-up*

AD was carried out in a batch reactor, Figure 1, with a global volume of 5 L and an operating volume of 3.75 L. The reactor temperature was controlled at $37 \pm 2°C$ (mesophilic operation) by means of a jacket that involves the vessel, heated by water bath (with a set-point at 38.5°C). The reactor is agitated magnetically and the biogas is collected in a graduated vessel that moves freely in a container filled with salty water (with NaCl) to minimize CO_2 absorption.

At lab scale different starting conditions were tested, namely the reactor substrate to inoculum (S/I) ratio in the range 0.2 to 1.2 g VS/gVS_0. During the experiments, the supernatant was analyzed for COD, pH, total alkalinity (TA) and volatile fatty acids (VFA). The volume of biogas produced and its composition (%CH_4, in volume) were determined a daily basis.

2.3 *Analytical techniques*

Waste and inoculum were characterized regarding several parameters. Moisture, total solids (TS), volatile solids (VS), suspended solids (SS) and total chemical oxygen demand (tCOD) were determined according to the Standard Methods (APHA, 1992). pH was measured by CRISON micro pH 2002. Total alkalinity (TA) and volatile fatty acids (VFA) were determined by titration based on the method proposed by Purser et al. (2014). Lignin content was measured gravimetrically by the Klason method after extracting the sample with sulphuric acid 72% to dissolve out the other components.

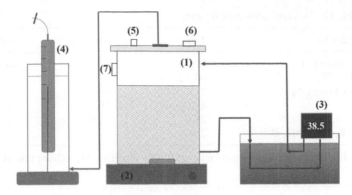

Figure 1. Scheme of AD laboratory plant: (1) reactor, (2) agitation plate, (3) thermal bath, (4) biogas meter, (5) sample manifold, (7) feeding valve.

Biochemical methane potential (BMP) was estimated theoretically (ThBMP) in mL CH$_4$/g VS at PTN conditions (273 K, 101.3 kPa) according to the Eq. (1) and Eq. (2) (Vazifehkhoran et al., 2016):

$$C_nH_aO_bN_c + \left(n - \frac{a}{4} - \frac{b}{2} + \frac{3c}{4}\right)H_2O \rightarrow \left(\frac{n}{2} + \frac{a}{8} - \frac{b}{4} - \frac{3c}{8}\right)CH_4 + \left(\frac{n}{2} - \frac{a}{8} + \frac{b}{4} + \frac{3c}{8}\right)CO_2 + cNH_3$$

(1)

$$ThBMP = 22400 \frac{\dfrac{n}{2} + \dfrac{a}{8} - \dfrac{b}{4} - \dfrac{3c}{8}}{12n + a + 16b + 14c}$$

(2)

Elemental composition of substrates (CHNS) in percentage (w/w) was measured by Elemental Analyzer NA 2500. The oxygen content was calculated from the mass balance. Moreover, NIRFlex N500 (BÜCHI, Switzerland) was used to obtain the NIR spectra of the FgW, and then a calibration model delivered by producer was utilized to interpret sample data and estimate the BMP (hereafter referred as BMP$_{NIR}$).

The methane concentration in the biogas was measured according to the procedure described by Abdel-Hadi (2008), assuming a binary mixture of CH$_4$ and CO$_2$. For the absorption of CO$_2$ a solution of NaOH 4M was used.

3 RESULTS AND DISCUSSION

Table 1 summarizes the main characteristics of FgW sample.

The percentage of volatile solids (VS) in FgW was about 90% (which is within the values reported by Gunaseelan, 2004, for several fruit and vegetable wastes) and tCOD was 1130 mg O$_2$ g^{-1}. The waste has acidic properties, which requires pH control in the context of AD. Regarding C/N ratio obtained from elemental analysis, FgW revealed a value of 22.4, which is within the optimal range (16–25:1) reported by Abbasi et al. (2012). The elemental composition indicated in Table 1 allows the representation of the organic matter of FgW as C$_{26}$H$_{42}$O$_{17}$ N. From Eq. (1) and (2) ThBMP was estimated as 238 mL CH$_4$/g SV. Furthermore, BMP$_{NIR}$ was of about 220 mLCH$_4$/gVS which may indicate that this material can be considered as an important substrate for methane production. This value is within the range usually reported in the literature (Gunaseelan, 2004) for fruit wastes (180–730 mLCH$_4$/gVS) and for vegetable wastes (190–400 mLCH$_4$/gVS). Nonetheless, the lignin content for FgW is about 25%TS, which has a strong influence on the biodegradable fraction of the waste (Triolo et al., 2011).

AD was carried out using several ratios of substrate to inoculum (S/I) (g VS/gVS$_0$): 0.2, 0.4, 0.6, 0.8 and 1.2 (represented in Figure 2 as F1 to F5, respectively). Figure 2a)–b) shows the

Table 1. Characterization of fig waste.

Parameter	Value	Parameter	Value
Moisture (%)	87.0 ± 0.1	C (%TS)	44.1
VS (%TS)	90.3 ± 0.4	N (%TS)	1.97
tCOD (mg O_2 g^{-1})	1130 ± 30	O (%TS)	38.3
pH	4.01 ± 0.02	H (%TS)	5.96
Lignin (%TS)	24.6	C/N	22.4
BMP_{NIR} (mLCH$_4$ gVS^{-1})	220		

tCOD – total chemical oxygen demand; BMP – biochemical methane potential.

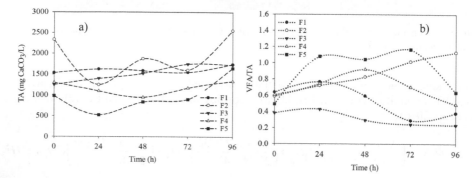

Figure 2. Evolution of a) total alkalinity; b) VFA/TA ratio over time.

evolution of TA and VFA/TA during 96 h of operation for each experiment. Indeed, the evaluation of TA is important to predict the buffering capacity of the digester, and anticipate any potential adverse effects of VFA accumulation. According to Owamah and Izinyon (2015) AD reactors stability is ensured when VFA/TA < 0.4. However, for ratios below 0.8, some AD systems can still maintain stability, while above 0.8 the instability may end up the biogas production. Figure 2b) reveals that only F1 and F3 led to stable conditions with VFA/TA values within the range of 0.29–0.77 and 0.23–0.43, respectively. In fact, only in F3, which corresponds to a 0.6 g VS/gVS$_0$, no intervention was required during all the experiment time, so that biogas production did not stop. For the other experiments, pH corrections were always needed.

Figure 3 illustrates the specific methane production (SMP) and methane mean composition as a function of S/I ratio. Results indicate that 0.6 seems to be the optimal S/I for the AD of FgW.SMP corresponds to the cumulative production of methane in 5 days.

These results show that for low S/I (experiments F1 and F2) biogas production was low, corresponding to an accumulated SMP of 74.4 mL/gVS and 101.6 mL/gVS, respectively. This may be related to the low amount of substrate available for biogas production. The increasing of S/I to 0.6 led to the maximum biogas production (458.8 mL/g VS), which is near the value obtained by Chen et al. (2016) for tomato waste (416 mL/g VS). The accumulation of VFA in experiments F4 and F5, due to the excess of substrate available, increased the reactor instability inhibiting this way biogas production (Chen et al., 2014).

In what regards %CH$_4$ in the biogas, Figure 3 reveals that this parameter ranges from 49% (for S/I 0.2) and 61% (for S/I 0.6). This corresponds to an increase of 12% on biogas methane richness when S/I ratio varies from 0.2 to 0.6. For even higher S/I ratios, methane concentration decreases, which can be attributed to the reactor acidification promoted by the excessive production of VFA. In fact, for acidic conditions methanogen bacteria present metabolic detrimental effects, with inhibitory consequences on methane production (Buyukkamaci and Filibeli 2004).

Biochemical methane potential (BMP) can be an interesting parameter to infer about the anaerobic biodegradability of a substrate (Labatut et al., 2011; Triolo et al., 2011). Figure 4 compares the SMP obtained in the lab AD reactor for FgW using several S/I ratios with the BMP$_{NIR}$ and ThBMP.

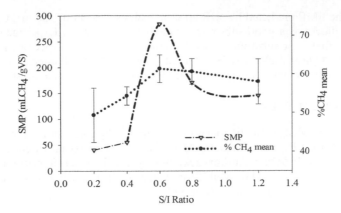

Figure 3. Specific methane production as a function of S/I ratio.

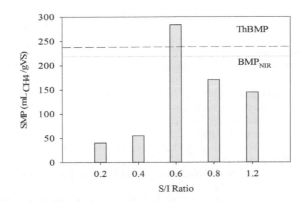

Figure 4. Specific methane production as a function of S/I ratio.

The ThBMP and BMP$_{NIR}$ estimated were 238 and 220 mLCH$_4$/gVS, respectively. Considering that Eq. (1) and Eq. (2) do not differentiate between biodegradable and non-biodegradable matter nor the fact that part of the biodegradable organic matter is used by the bacteria to grow, it can be concluded that both methods are well in agreement with each other. For the optimal S/I ratio (0.6), the specific methane production obtained in the AD reactor was 284 mLCH$_4$/gVS, which is 16% and 22% higher than ThBMP and BMP$_{NIR}$. This means that theoretical values and other expedited methods must be used with precaution to conclude about the production at higher scale.

Admitting the AD reactor at industrial scale may produce 284 mLCH$_4$/gVS FgW, it can be concluded that an annual waste production of 200 Mg (wet basis) would have a potential production of 6668 Nm^3/year of methane. Considering a lower calorific value of methane of 50 MJ/kg (Ragland and Bryden, 2011) and 0.716 for the methane density (at PTN conditions), the generated biogas would correspond to a production of 238 GJ. Assuming the efficiency for the production of electrical energy provided in Annex II of Directive 2006/32/EC of 40%, would give an annual production of 26 MW.h of electric energy that will give an interesting supplementary profit to be added to the stabilization of the organic residue.

4 CONCLUSIONS

In this work, the possibility of valorizing FgW by AD and to contribute to the implementation of the circular economy in an SME company was assessed. The characterization of FgW revealed that pH is acidic, but the C/N ratio is in the optimal range usually required for AD. The organic matter content (measured as VS) is high (about 90%), but about 25%TS is lignin.

Accordingly, the BMP predicted by NIR spectra analysis was about 220 mL CH$_4$ gVS^{-1} what allow for regarding it as a good substrate for methane production. Experimental tests at lab scale confirmed that if the substrate to inoculum ratio is optimized, a specific methane production of 284 mLCH$_4$/gVS FgW may be achieved, which could be of interest for the industrial plants integrating waste management with energy production after scale-up studies.

ACKNOWLEDGEMENTS

R.C. Martins acknowledges the financial support under the contract FCT investigator 2014 Program (IF/00215/2014) with financing from the European Social Fund and the Human Potential Operational Program. Authors acknowledge the financial support under the project POCI-01–0145-FEDER-016403—MultiBiorefinary, with financing from FEEI through COMPETE 2020 and National funds from FCT.

REFERENCES

Abbasi, T., Tauseef, S.M, Abbasi, S.A. 2012. Anaerobic digestion for global warming control and energy generation—An overview. *Renewable and Sustainable Energy Reviews* 16: 3228–3242.

Abdel-Hadi, M.A. 2008. A simple apparatus for biogas quality determination. Misr J. Ag. Eng., *Biological Engineering* 25(3): 1055–1066.

APHA 1992. *Standard Methods for the Examination of Water and Wastewater*. AWWA, WPCF, 18th ed., Washington DC.

Arhoun, B., Bakkali, A., El Mail, R., Rodriguez-Maroto, J.M., Garcia-Herruzo, F. 2013. Biogas production from pear residues using sludge from a wastewater treatment plant digester. Influence of the feed delivery procedure. *Bioresource Technology* 127: 242–247.

Bayard, R., Liu, X., Benbelkacem, H., Buffiere, P., Gourdon, R. 2016. Can biomethane potential (BMP) be predicted from other variables such as biochemical composition in lignocellulosic biomass and related organic residues?. *BioEnergy Research*, 9 (2): 610–623.

Buyukkamaci, N., and Filibeli, A. 2004. Volatile Fatty Acid Formation in an Anaerobic Hybrid Reactor. *Process Biochemistry* 39 (11): 1491–1494.

Chen, L., Jian, S., Bi, J., Li, Y., Chang, Z., He, J, Ye, X. 2016. Anaerobic Digestion in Mesophilic and Room Temperature Conditions: Digestion Performance and Soil-Borne Pathogen Survival. *Journal of Environmental Sciences* 43: 224–233.

Chen, X., Yan, W., Sheng, K., Samati, M. 2014. Comparison of high-solids to liquid anaerobic co-digestion of food waste and green waste. *Bioresource Technology* 152: 215–221.

Fagbohungbe, M.O., Herbert, B.M., Hurst, L., Ibeto, C.N., Li, H., Usmani, S.Q., Semple, K.T. 2017. The challenges of anaerobic digestion and the role of biochar in optimizing anaerobic digestion. *Waste Management* 61: 236–249.

Gunaseelan, V.N. 2004. Biochemical methane potencial of fruits and vegetable solid waste feedstocks. *Biomass and Bioenergy* 26 (4): 389–399.

Labatut, R.A., Angenent, L.T., Scott, N.R. 2011. Biochemical Methane Potential and Biodegradability of Complex Organic Substrates. *Bioresource Technology* 102(3): 2255–64.

Mao, C., Feng, Y., Wang, X., Ren, G. 2015. Review on research achievements of biogas from anaerobic digestion. *Renewable and Sustainable Energy Reviews* 45: 540–555.

Owamah, H.I., and O.C. Izinyon. 2015. The Effect of Organic Loading Rates (OLRs) on the Performances of Food Wastes and Maize Husks Anaerobic Co-Digestion in Continuous Mode. *Sustainable Energy Technologies and Assessments* 11: 71–76.

PNAER 2013. *Plano Nacional de Ação para as Energias Renováveis. Resolução do Conselho de Ministros* n.º 20/2013 de 10 de abril. Diário da República, 1.ª série, N.º 70.

Purser, B.J., Thai S.M., Fritz, T., Esteves, S.R., Dinsdale, R.M., Guwy, A.J. 2014. An Improved Titration Model Reducing over Estimation of Total Volatile Fatty Acids in Anaerobic Digestion of Energy Crop, Animal Slurry and Food Waste. *Water Research* 61: 162–70.

Ragland, K., and Bryden, K. 2011. Combustion engineering, Second edition. Boca Raton: Taylor and Francis.

Triolo, J.M.; Sommer, S.G.; Møller, H.B.; Weisbjerg, M.R.; Jiang, X.Y. 2011. A new algorithm to characterize biodegradability of biomass during anaerobic digestion: Influence of lignin concentration on methane production potential. *Bioresour. Technol.* 102: 9395–9402.

Vazifehkhoran, A.H., Triolo, J.M., Larsen, S.U., Stefanek, K., Sommer, S.G. 2016. Assessment of the Variability of Biogas Production from Sugar Beet Silage as Affected by Movement and Loss of the Produced Alcohols and Organic Acids. *Energies* 9: 368. doi:10.3390/en9050368.

WASTES – Solutions, Treatments and Opportunities II – Vilarinho, Castro & Lopes (Eds)
© 2018 Taylor & Francis Group, London, ISBN 978-1-138-19669-8

Hazards identification in waste collection systems: A case study

B. Rani-Borges & J.M.P. Vieira
University of Minho, Braga, Portugal

ABSTRACT: The advances and the use of new technologies for adequate management of solid wastes in large urban centers in the last few years has been remarkable. Such practices are motivated by the need for minimize and avoid possible negative impacts on public health and environment caused by waste management. This paper presents a research work on the hazards and control measures during the steps of collection, transport, storage and sorting of waste in two utilities in the north of Portugal. The purpose of this research is gathering information on impacts caused by the current operational system management. The methodology applied was based on face-to-face interviews with the top managers of waste companies and in analyzing the records of previously occurred hazardous events. Data collection was followed by the construction and analysis of two tables containing essential information for a future elaboration of a solid wastes safety plan.

1 INTRODUCTION

The increase of municipal solid waste (MSW) generation and the way how it is treated are topics of constant debate in various governmental bodies. This concern happens due to the negative impacts caused by the lack of treatment and/or inadequate final disposal of waste. These practices cause damage to public health and reduce the environmental quality, also resulting in financial losses to society (Cossu 2013, Wilson & Velis 2015).

Currently, the efficiency of MSW management system is considered as one of the parameters for assessing the country's level development (Silva et al. 2015). In Portugal, according to data recently published by ERSAR, about 34% of waste generated in the country is prepared for reuse and recycling, while 41% of biodegradable waste is disposed of in landfills (ERSAR 2016).

The provision of MSW management services is an essential element to promote public health policies and protect the environment. Therefore, it is imperative to provide a service of quality and efficiency for the entire population. (Gouveia 2012). MSW management systems include hazards and risks, which must be identified in order to better predict and minimize possible risk scenarios and then, consequently, promote an increasingly optimized system.

This paper is a small part of a larger research project whose objective is to prepare a methodology for the design of a solid waste safety plan. This methodology incorporates hazard identification and risk management leading to more efficient and less vulnerable waste systems. Also, it will serve as a guide to monitoring and assessing all the system steps in order to contribute to the establishment of control measures and necessary actions to ensure the safety of these infrastructures in relation to public health and environment. To elaborate a waste safety plan all components of the flux diagram of a MSW system should be studied and as much data as possible need to be analyzed, including the identification of hazards and the control measures applied in collection, transport, storage, sorting, and treatment (Rani-Borges & Vieira 2016).

This work presents the results obtained in identifying hazards and control measures applied in waste collection, transport, storage, and sorting for two large MSW systems in the north of Portugal.

2 METHODOLOGY

The methodology adopted for data collection consisted of two approaches: face-to-face interviews with top managers and high technical staff of two waste companies, and search for internal documentation on recorded hazardous events (Figure 1). Especially, this paper addresses the hazards and the control measures during the steps of collection, transport, storage, and sorting in two MSW systems in the north of Portugal. The criteria adopted in selection of the systems were: dimension (population served); technological infrastructure; and waste treatment scheme in place. Two large MSW systems were chosen: the metropolitan area of Porto, managed by the company Lipor; and the Baixo Cávado region, managed by the company Braval. (Figure 2)

This work focused only in four systems components (collection, transport, storage, and sorting). The reason for that is that the remaining components—treatment and final disposal—are responsibility of other different entities (companies and municipalities). These aspects will be addressed elsewhere in specific publications.

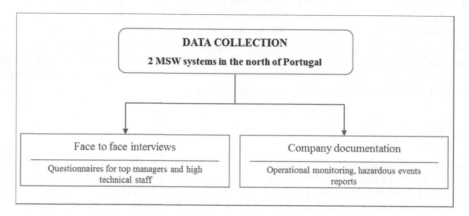

Figure 1.   Methodology applied in data collection.

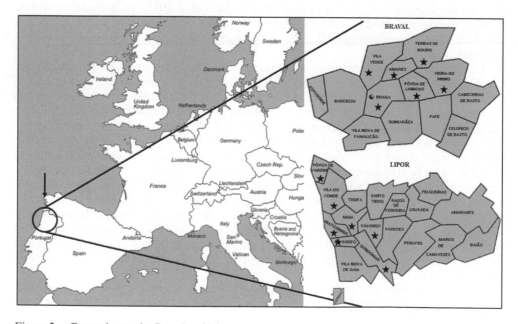

Figure 2.   Covered areas by Braval and Lipor.

The face-to-face interviews with top managers and high technical staff were based on a questionnaire containing two types of items: (i) systems components, quantities of waste collected, waste characteristics, strategic planning for improving the system; and (ii) identification of hazards, frequency of hazardous events; control measures adopted. Internal reports on operational monitoring, hazardous events, and strategies to achieve established goals by the Portuguese Environmental Agency were also analyzed.

After information collection, the data of the two companies were crossed with the aim to build a complete matrix with hazards and control measures applied in the different waste systems components. These data will support the development and implementation of a specific solid waste safety plan.

The most important difference between the two companies is that the Braval sorting process is manually carried out by professionals. After the separation process, the material is baled. In Lipor, the sorting is performed in the same way but, in addition to this procedure, there is also an automatic sorting, where an equipment separates the material according to its characteristics. Then the materials go straight to the baling process without going through a manual procedure. As the automatic sorting of Lipor does not present risks to public health and the environment the data referring to this process are not relevant to the present work.

3 CASE STUDY

3.1 The system of Baixo Cávado

The region of Baixo Cávado, Portugal, including the cities of Braga, Póvoa de Lanhoso, Vieira do Minho, Amares, Vila Verde and Terras de Bouro, has the waste collection and treatment system managed by the company Braval. The system has the components: collection, mechanical treatment, composting, leachate treatment, biogas burner system and sanitary landfill.

In 1996, the company was created with the purpose of closing the open dumping sites in the region with subsequent environmental recovery, landscape integration of degraded areas, building landfills and implementing the selective collection providing the ecopoint system. Currently, Braval attends a population of circa 290,000 inhabitants. (Table 1).

Table 1.   General aspects of waste system of Baixo Cávado region.

System components	Recyclables collection, transportation, sorting, storage, composting, incineration and landfill
Waste characteristics	Non-hazardous waste
Population served	290,407 inhabitants
MSW production	1.1 kg/inhabitant/day
MSW recyclable	15,000 ton/year
Landfill	98,978 ton/year
Total of MSW	113,978 ton/year

Table 2.   General aspects of the waste system of metropolitan area of Porto.

System components	Recyclables collection, transportation, sorting, storage, composting, incineration and landfill
Waste characteristics	Non-hazardous waste
Population served	1 million inhabitants
MSW production	1.38 kg/inhabitant/day
MSW recyclable	41,800,000 ton/year
Landfill	5,000 ton/year
Total of MSW	500,000 ton/year

Table 3. Hazards and control measures in collection and transport steps.

Hazards	Control measures										
	Use of personal protection equipment	Cargo handling training	Road code fulfillment	Drivers have a proper driver's license	Inspection and maintenance of vehicles	Organization of waste and bags	Safe access to the tray and cargo area	Medical office: vaccines, appointments, exams	Lumbar tapes	Anti-fatigue footwear	Mandatory workout program
Moving on sharp surfaces	X					X		X			
Exposure to odors	X										
Exposure to vibrations and noise	X				X						
Handling of cargoes		X							X	X	X
Exposure to adverse temperature	X	X									
Waste handling	X	X						X	X	X	X
Contact with sharp objects	X							X			
Exposure to biological agents	X							X			
Road traffic	X	X	X	X							
Injury and muscle pain	X	X						X	X	X	X
Charging the truck	X	X			X	X	X				
Oil or fuel spill on public roads			X		X						
Vehicle accident					X			X			

230

Table 4. Identified hazards and control measures applied by both companies in storage and sorting steps.

Hazards	Use of personal protection equipment	Ensure safe distance to the vehicle	Periodic checking and maintenance of machinery and equipment	Security signs	Cleaning of discharge zone	Maintenance of air treatment system	Formation of handling machines	Alcohol dispenser for disinfecting hands	Medical office: vaccines, appointments and exams	Vibration and noise analysis	Thermal comfort analysis	Mandatory workout program
Drop of discharge box	X	X	X									
Machines and vehicles				X								
Displacement at discharge dock	X		X									
Exposure to biological agents	X				X	X		X	X			
Use of equipment	X		X		X	X	X					
Waste contact	X				X		X	X	X			
Exposure to adverse temperature	X					X					X	
Exposure to dust and powder	X					X						
Exposure to odors	X					X						
Exposure to vibration and noise	X		X				X			X		
Exposure to radiation	X		X					X	X			
Container dump	X						X		X			
Exposure and inhalation of gases	X					X			X			
Contact with sharp objects	X						X					
Handling of bales	X		X		X		X					
Handling heavy objects	X				X	X						X
Contact with organic material	X							X				

Control measures

231

Lipor is an entity responsible for the management, collection and treatment of MSW produced in the metropolitan area of Porto, Portugal. The company began operating in 1982 and since then has been implementing integrated waste management with a system composed of recyclables collection, transportation, sorting, storage, composting, incineration and sanitary landfill.

The cities served by Lipor are: Espinho, Gondomar, Maia, Matosinhos, Porto, Póvoa de Varzim, Valongo and Vila do Conde. Lipor treats about 500 thousand tons per year of MSW produced by approximately 1 million inhabitants.

The company's system has the strategic central goal of minimize landfill disposal. In order to achieve it, Lipor implemented an integrated strategy for treatment of waste, based on three main components: recycling, composting and incineration. The landfill completes the process being used only to receive the rejected, which today amounts to 1% of all waste produced in the region. General aspects of the system are shown in Table 2.

4   RESULTS AND DISCUSSION

This research work gathered information related to hazards associated to the system components of collection and transport of recyclable waste, storage and sorting. All these steps are carried out in the visited infrastructures and, therefore, the hazards and hazardous events

Table 5.   Hazards registered with higher frequency in Braval and Lipor systems.

Step	Hazard	Frequency
Collection and Transport	Exposure to odors	Frequent / once a week
	Exposure to vibrations and noise	Frequent / once a week
	Exposure to adverse temperatures	Frequent / once a week
	Injury and muscle pain	Occasional / once every six months
	Vehicle accident	Occasional / once every six months
	Contact with sharp objects	Occasional / once every six months
	Handling of cargoes	Unlikely / once a year
	Waste handling	Unlikely / once a year
	Moving on sharp surfaces	Unlikely / once a year
	Road traffic	Unlikely / once a year
	Charging the truck	Unlikely / once a year
	Oil or fuel spill on public roads	Unlikely / once a year
	Exposure to biological agents	Very unlikely / once in five years or more
Storage and Sorting	Exposure to odors	Frequent / once a week
	Exposure to vibrations and noise	Frequent / once a week
	Contact with organic material	Frequent / once a week
	Exposure to dust and powder	Frequent / once a week
	Waste contact	Frequent / once a week
	Exposure to adverse temperatures	Probable / once a month
	Contact with sharp objects	Probable / once a month
	Use of equipment	Occasional / once every six months
	Machines and vehicles	Occasional / once every six months
	Handling of bales	Occasional / once every six months
	Drop of discharge box	Unlikely / once a year
	Exposure to biological agents	Unlikely / once a year
	Exposure to radiation	Unlikely / once a year
	Handling heavy objects	Unlikely / once a year
	Exposure and inhalation of gases	Unlikely / once a year
	Displacement at discharge dock	Very unlikely / once in five years or more
	Container dump	Very unlikely / once in five years or more

are responsibility of the two companies. Risk assessment and risk management policies have been implemented by both companies: since 2004 in Lipor, and 2005 in Braval.

According to data collected in Braval and Lipor, the control measures applied to minimize and/or avoid hazardous events are quite similar, since they obey very strict standards of quality based on health, security, and environmental concerns.

The main hazards considered in collection and transport and the related control measures implemented by Braval and Lipor can be found in Table 3.

Tables 4 and 5 present hazards and control measures during the steps of storage and sorting, and the most frequent hazards registered in both systems, respectively.

Although the companies have little information on which events stopped to occur or occur less frequently after security plan implementation, the working conditions have improved significantly in recent years. Hence, if preventive safety plans could be implemented it is expected that security levels can increase more and more.

5 CONCLUSION

Waste systems are sanitary infrastructures that present broad beneficial impacts at the social, economic and environmental dimensions, and are therefore an essential element of public health policies. MSW management is composed of several steps, which may involve hazards e hazardous events that constitute threats to public health, occupational hygiene and environment integrity. Hazards and hazardous events should be identified and risk assessment and risk management methodologies must be implemented in order to define control measures to avoid or minimize such threats to human health and environmental integrity.

This work focused on data collection of hazards and control measures in the steps of collection, transport, storage and sorting in two waste management systems in the north of Portugal. The results obtained will be inserted in a wide methodology for the design of a solid waste safety plan whose purpose is to establish a methodology for risk assessment and risk management in MSW systems providing more efficiency to the system. This tool can be also useful to understand and circumvent the systems vulnerabilities to potential hazards and hazardous events, contributing to increase the resilience and safety of these infrastructures.

ACKNOWLEDGEMENT

Financial support from CAPES scholarship and Science Without Borders program.

REFERENCES

Cossu R. 2013. Groundwater contamination from landfill leachate: when appearances are deceiving! *Waste Management* 33: 1793–1794.

ERSAR. 2016. *Relatório Anual dos Serviços de Águas e Resíduos em Portugal Volume 1 – Caraterização do setor de águas e resíduos.*

Gouveia N. 2012. Resíduos sólidos urbanos: impactos socioambientais e perspectiva de manejo sustentável com inclusão social. *Ciência & Saúde Coletiva* 17(6): 1503–1510.

Rani-Borges B. & Vieira J.M.P. 2016. Avaliação de perigos e eventos perigosos em sistemas de gestão de resíduos sólidos urbanos. APESB. *17.º Encontro Nacional de Engenharia Sanitária e Ambiental, Guimarães, 14-16 Setembro 2016*: 367–374.

Silva, M.E.F., Ferreira, B.S.T. & Brás, I.P.L. 2015. Indicadores de Qualidade na Prestação de Serviços de Gestão de Resíduos Urbanos—Caso de Estudo. *Millenium* 48: 91–109.

Wilson, D.C. & Velis, C.A. 2015. Waste management—still a global challenge in the 21st century: An evidence-based call for action. *Waste Management & Research* 33(12): 1049–1051.

Outlining strategies to improve eco-efficiency and efficiency performance

E.J. Lourenço, A.J. Baptista & J.P. Pereira
INEGI—Institute of Science and Innovation in Mechanical and Industrial Engineering, Porto, Portugal

Célia Dias-Ferreira
CERNAS—Research Center for Natural Resources, Environment and Society, Polytechnic Institute of Coimbra, College of Agriculture, Coimbra, Portugal
Materials and Ceramic Engineering Department, CICECO, University of Aveiro, Aveiro, Portugal

ABSTRACT: Nowadays achieving sustainable development is a global concern. Economic and environmental sustainability can be driven by assessing and improving industrial production system's performance. An evaluation that assesses if materials, energy and resources are used to their full potential is a powerful tool for improving economic and environmental performance, and consequently supports the identification of all types of waste and inefficiencies along the production system. The goal of this work is to assess overall production system's efficiency and eco-efficiency using Multi Layer Stream Mapping (MSM). The outputs of this approach is used to scrutinize "where" and "how much" can a unit process and/or a production system improve its financial, environmental and overall efficiency, thereby being of great importance for decision-making and correct implementation of improvement actions. This paper highlights the results from the application of the MSM methodology in a real industrial case regarding a painting unit.

1 INTRODUCTION

Sustainability is one of the major priorities nowadays, due to population growth, scarcity of resources and their rising prices. However, measuring sustainability and assessing its evolution is an ambiguous and difficult task, taking into account that this concept, other than economic and ecological elements also embraces the social needs and social well-being. Eco-efficiency is intrinsically related to the concept of sustainability. The main purpose is to maximize value creation and minimize environmental burdens (Verfaillie and Bidwell, 2000). The term eco-efficiency was proposed in 1990 by two Swiss researchers, Schaltegger and Sturm (Czaplicka-Kolarz et al., 2010). The concept was first adopted by the World Business Council for Sustainable Development (WBCSD). Eco-efficiency is intended to measure the relationship between economic growth and environmental pressure, and is generally expressed as represented in Equation 1 below:

$$\text{Eco-efficiency} = \text{Production Or Service Value} \Big/ \text{Environmental Influence} \qquad (1)$$

The WBCSD established a framework to assess eco-efficiency, leaving aside the social dimension, that is part of the sustainability structure (Lehni et al., 2000). Eco-efficiency analysis enables the assessment of products and processes so, it is progressively becoming more common and widely used in several sectors and with different purposes leading to win-win situation in terms of costs and environmental improvement (Czaplicka-Kolarz et al., 2010; Michelsen et al., 2006). The eco-efficiency concept initially focused on companies. Later, it was improved to assess policy strategies and their possible macroeconomic outcomes. According to the WBCSD the two most common goals of eco-efficiency assessments are (i) measuring progress and (ii) internal and external communication (Michelsen et al., 2006).

As shown in Equation 1 environmental influence is required when assessing eco-efficiency. One of the most used and useful tools for quantify the environmental influence is Life Cycle Assessment (LCA). The product or service value may be determined by monetary indicators, which are easy to understand (Michelsen et al., 2006). For instance, the costs, net sales or gross value added, or even by the functional value (e.g. durability, luminous flux) (Baptista et al., 2016). Eco-efficiency and environmental performance assessment of industrial processes are powerful tools for achieving sustainability, especially for decompiling economic performance from environmental burdens (Baptista et al., 2016), i.e., maintain economic growth and reduce environmental impacts (Verfaillie and Bidwell, 2000). On the other hand, by assessing effectiveness and "descanting" eco-efficiency, it does not mean that environmental and economic improvements will not be achieved. For example, Despeisse et al. (2012) states that efficient and effective material and energy use can reduce natural resource inputs and waste or pollutant outputs. Yet, one can effectively use materials, energy or resources and not take into account the related environmental impact, and consequently never look for alternatives and improvements that can enhance economic and environmental performance.

2 GOAL AND MAIN FOCUS

The goal of this work is to quantify the efficiency and eco-efficiency of a panting unit of the metalworking industry using MSM as base approach. This seeks to assess the production system's efficiency, so one can act in order to reduce wastes and inefficiencies. This work also focuses on the importance of presenting the environmental issues and the eco-efficiency performance in a simple manner which considers wastes and inefficiencies, in order to simplify the decision making process. Ultimately, the goal is also to make use of eco-efficiency and efficiency results in order to improved decision making in order to improve both eco-efficiency and resource efficiency

3 THE MSM APPROACH

This paper presents an approach on how unit processes and the global production system's efficiency and eco-efficiency can be assessed and interpreted in an expeditious manner. As presented in Lourenço et al. (2013), the MSM, approach comprises several layers, i.e. one layer for each input or output that's part of the production system (for each variable). The combined use of the layers emerges in order to "see beyond" the global performance. The MSM approach evaluates the resource efficiency taking into account the material, energy and time consumed in each unit process. This enables identifying and quantifying inefficiencies and wastes among the production system (Lourenço et al., 2013).

MSM is a lean based resource efficiency assessment methodology, that starts from the classic VA vs NVA segments graph of Value Stream Mapping (VSM) and takes into account the dichotomy between value and waste in a given production system. In order to identify and quantify, at each stage of the process system, all "value adding" (VA) and "non-value adding" (NVA) actions will be accounted for, as well as, all types of waste and inefficiencies. In MSM approach, the idea is to assess, simultaneously, the unit process efficiency and eco-efficiency performance, for all relevant process variables, which can be used to quantify in detail process efficiency.

The MSM methodology resembles a matrix (m × n), where "n" is the number of variables evaluated and "m" the number of steps of the production system (i.e. unit processes). In order to apply the MSM, the following steps should be carried out:

- Identification of the system boundaries;
- Identification of the unit processes;
- Identification of all relevant process variables and parameters;
- Definition of the associated KPI to each variable, always to be maximized and with values ranging between [0–100%];
- Analysis of the results and identification of the variables and unit process with lower efficiency results;

- Study and prioritization the improvement actions;
- Implementation of improvement actions and assessment of the efficiency gains evolution and cost reductions.

It is important to notice that, for MSM to work all variables should be addressed in order to maximize performance (increased efficiency). The dimensionless character of the efficiency ratios of variables allows them to be combined and aggregated in order to compute a global efficiency for a part or for all the system. Therefore, the efficiency of the variables that characterize the production system are calculated as the ratio between the portion of the "variable that adds value" to the product and the "total of the variable that enters the unit process" (see equation 2)

$$\text{Efficiency Ratio} = \frac{\text{Value added Fraction}}{\text{Value added fraction} + \text{Non-value added fraction}} \quad (2)$$

Assessing the eco-efficiency performance, in accordance with Equation 2, of a unit process or a production system is possible if one considers the environmental influence and costs related to the NA and NVA fraction of all variables. Nevertheless, just by analyzing the results of the MSM approach, i.e. evaluate the efficiency results over time, it is possible to analyze if efficiency performance is increasing. Which would mean that the eco-efficiency performance is also improving (considering that the economic variables are constant and the same materials and resources are being used), since the same value is being added to the product but more efficiently, and therefore fulfilling the main goal of eco-efficiency, which is, "doing more with less".

4 CASE STUDY

4.1 *Scope*

To understand the real benefits of applying the MSM tool to a production system, a particular case study was performed in an industrial environment. The company where the study occurred belongs to the metal working industry, being specialized in the conception and assembly of large metal structures. For this case study, the MSM approach was applied to a wind power tower painting unit. For this study, the functional unit was considered as application of the primer coat using airless spray coating on a 514 m² WPTS.

The painting unit limits the system boundaries of the production system. All processes, equipment, energy consumption, materials used, etc. that are within these boundaries will be considered for the assessment.

The production system under study comprises six unit processes, described, as follow:

- Cleaning – Consist in cleaning the WPTS before applying the primer coat;
- Coating bolt holes – This process consist of manually painting the bolt holes;
- Mixing paint – Mixing the paint with the solvents by using a pneumatic mixer. This tasks occurs occasionally during the coat application, i.e. according to the painter's necessity;
- Primer coat – The application of the primer coat; it takes about 1,5 hours and it's done by airless spray painting;
- Drying – This process is to dry and cure the paint with the heat that is generated by the boiler. The drying process takes place during and after the application of the top coat;
- Inspection – Here the painting crew inspects the WPTS for any imperfection. This is done after the WPTS is dry.

4.2 *MSM inventory*

Table 1 presents the inventory of the overall consumption of materials and energy, and also quantifies the VA and NVA fractions for each variable in each process step. All values presented are according to the functional unit. To evaluate operation time efficiency, the unit processes within the production system were analyzed in terms of the time spent on VA and NVA activities. For this particular case, the NVA time that was registered was mainly needed for setting up. Energy data (electricity and fuel) was obtained from consumption records provided by the company under assessment. The energy used during VA and NVA time is quantified and is respectively allocated to each unit process. Besides the inefficiency of the machines, all energy used during NVA time is considered as energy that adds no value (NVA). The diesel consumption, VA and NVA portion, was calculated in the same manner as the electrical energy consumption. The main and most important material used is the paint, and according to the Reference Document on Best Available Techniques on Surface Treatment using Organic Solvents (2007), the usage efficiency of airless spray coating is around 60%. This means that 40% of the paint is considered as NVA.

4.3 *Environmental and economic assessment via MSM approach*

Here the purpose of using a Life Cycle Impact Assessment method is to establish a valid comparison between the several unit processes under study and compare the environmental impacts. It should be noticed that the environmental impacts of the WPTS is not considered since the scope of this work is to assess the production system's efficiency and eco-efficiency performance, and not the product. Table 2 lists the single score impact points related to material or energy consumption in each unit process. The unit process with the highest single score impact points is the application of the primer coat. This is related to the fact that during the application of the primer coat a relatively high amount of VA and NVA paint, diesel and electrical energy are used.

The labor, energy and paint costs are shown in Table 3. The cost data was obtained by the factorizations of the cost for all VA and NVA variables identified in Table 1. The base cost data, for such assessment, was collected by analyzing the materials and energy invoices and the pay slips of the workers. The MSM cost assessment, on its own, allows identifying the major costs contribution, namely the NVA related costs which are waste, since these costs add no value.

Table 1. MSM inventory.

Unit process	Labour [h]		Electrical energy [kWh]		Diesel [kg]		Paint & Curing agent & Diluent [kg]	
	VA	NVA	VA	NVA	VA	NVA	VA	NVA
Cleaning WPTS	0.8	0.4	28.4	12.5	0.0	0.0	0.0	0.0
Coating bolt holes	0.5	0.0	1.40	0.7	0.0	0.0	1.7	0.2
Mixing paint	0.5	0.2	28.6	12.5	0.0	0.0	0.0	0.0
Applying primer coat	1.5	0.2	97.4	45.4	41.3	7.5	84.7	56.5
Drying	3.0	0.2	77.1	38.6	82.5	10.6	0.0	0.0
Inspection	0.5	0.1	1.40	0.70	0.0	0.0	0.0	0.0
Total	7.7		344.7		141.9		143.1	

Table 2. Single score impact (Pt) for each variable in each unit process.

Unit process	Electrical energy [Pt]		Diesel [Pt]		Paint & Curing agent & Diluent [Pt]		Total [Pt]	
	VA	NVA	VA	NVA	VA	NVA	VA	NVA
Cleaning WPTS	1.03	0.45	0.00	0.00	0.00	0.00	1.03	0.45
Coating bolt holes	0.05	0.03	0.00	0.00	0.38	0.04	0.43	0.07
Mixing paint	1.04	0.45	0.00	0.00	0.00	0.00	1.04	0.45
Applying primer coat	3.55	1.65	1.59	0.29	18.55	12.37	23.69	14.31
Drying	2.81	1.41	3.18	0.41	0.00	0.00	5.99	1.81
Inspection	0.05	0.03	0.00	0.00	0.00	0.00	0.05	0.03
Total	12.55		5.46		31.34		49.36	

Table 3. MSM Cost analysis for each variable in each unit process.

Unit process	Labour [€]		Electrical energy [€]		Diesel [€]		Paint & Curing agent & Diluent [€]		Total [€]	
	VA	NVA	VA	NVA	VA	NVA	VA	NVA	VA	NVA
Cleaning WPTS	9.5	4.8	2.5	1.1	0.0	0.0	0.0	0.0	12.0	5.9
Coating bolt holes	6.4	0.4	0.1	0.1	0.0	0.0	33.3	3.7	39.8	4.2
Mixing paint	6.4	2.5	2.5	1.1	0.0	0.0	0.0	0.0	8.9	3.6
Applying primer coat	19.1	1.9	8.5	4.0	51.9	9.4	1622.7	1081.8	1702.3	1097.2
Drying	38.2	1.9	6.8	3.4	103.8	13.3	0.0	0.0	148.8	18.6
Inspection	6.4	0.8	0.1	0.1	0.0	0.0	0.0	0.0	6.5	0.8
Total	98.2		30.2		178.5		2 741.6		3 048.5	

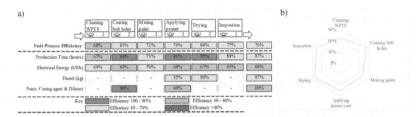

Figure 1. a) MSM dashboard for the panting unit efficiency. b) MSM efficiency fingerprint.

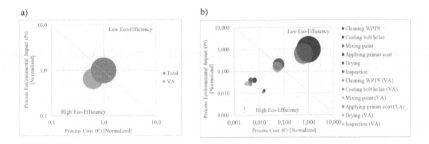

Figure 2. Eco-efficiency plots a) Total eco-efficiency performance of the painting unit vs. the VA fraction eco-efficiency. b) Total unit process eco-efficiency vs. the VA fraction eco-efficiency.

4.4 The MSM efficiency performance overview for the production system

The production system's unit process and overall efficiency is presented in Figure 1. This dashboard presents the outcomes of the MSM approach, and quantifies the global and unit process efficiency for each variable. The results shown in Figure 1, and the data in the previous tables, are of great importance, and usefulness for assessing efficiency as well as for quantifying and allocating inefficiencies and wastes. The unit process efficiency is determined by the average efficiency value of all variables. The variable efficiency is determined by the total VA of the variable over the total VA and NVA of the variable (e.g. paint efficiency 60%). The overall efficiency value is calculated by the average efficiency value of all unit process (67%). These results help to identify the major inefficacies and wastes along the process.

Figure 2 depicts the MSM eco-efficiency analysis, where it is possible to see the eco-efficiency performance if the painting unit had no NVA actions. Moreover, with such combined analysis, eco-efficiency and efficiency performance, it is possible to understand if the efficiency performance is high while eco-efficiency is low, or vice-versa. For example, when evaluating the primer coat application, it is noticeable that the coating process is more efficient than the cleaning and the mixing paint processes, but the coating has the lowest

239

eco-efficiency performance. Therefore, the strategy to enhance overall eco-efficiency could start by improving the coating's eco-efficiency without compromising the efficiency. The eco-efficiency performance results in Figure 2 a) help to define the path for an improved sate in terms of eco-efficiency by improving efficiency, through the reduction and/or elimination of NVA activities.

5 CONCLUSIONS

The focus of this paper brings a new and fresh perspective for quantifying the eco-efficiency and efficiency of a production system or its unit processes. This new approach can be very useful to underpin the efficient use of resources and materials within the productions systems, thereby leading towards sustainable production. One of the major benefit of using such method relates to the fact that it is possible to assess eco-efficiency and efficiency in such a manner that one can trace and allocate the major influence in terms of eco-efficiency, efficiency, costs or environmental impacts to each unit process or even to a specific material, or energy type. In the case study presented, it was possible to determine that the use of paint has the highest environmental impacts. The coat application is not the unit processes with the lowest efficiency performances, yet is the one with the lowest eco-efficiency value. These results indicate that the coating unit processes should be subject of further studies and optimization in order to improve its performance and reduce costs and environmental impacts (enhance eco-efficiency performance). It is worth mentioning that the cleaning unit process is the less efficient, even though it is a simple process, due to the fact that it relies on compressed air which is one of the less efficient forms of energy. Yet, it has a quite good eco-efficiency performance. As future work, the MSM approach should be extended to the whole production system for the WPTS and ultimately, used to evaluate strategies to improve overall eco-efficiency and efficiency.

ACKNOWLEDGMENTS

This work was supported by the European Union's Horizon 2020 research and innovation program through the MAESTRI project (grant no. 680570). The authors gratefully acknowledge the funding of Project NORTE-010145-FEDER-000022 - SciTech—Science and Technology for Competitive and Sustainable Industries, cofinanced by Programa Operacional Regional do Norte (NORTE2020), through Fundo Europeu de Desenvolvimento Regional (FEDER). C. Dias-Ferreira gratefully acknowledge FCT—Fundação para a Ciência e para a Tecnologia (SFRH/BPD/100717/2014).

REFERENCES

A.J. Baptista, E.J. Lourenço, J.P. Pereira, F. Cunha, E.J. Silva, P. Peças 2016. ecoPROSYS: An Eco-efficiency Framework Applied to a Medium Density Fiberboard Finishing Line. *Procedia CIRP* 48: 170–175. ISSN 2212–8271. http://dx.doi.org/10.1016/j.procir.2016.04.061.

Czaplicka-Kolarz, K., Burchart-Korol, D., Krawczyk, P. 2010. Eco-efficiency analysis methodology on the example of the chosen polyolefins production. *Journal of Achievements in Materials and Manufacturing Engineering* 43.

Despeisse, M., Ball, P.D., Evans, S., Levers, A. 2012. Industrial ecology at factory level—a conceptual model. *Journal of Cleaner Production* 31: 30–39. http://dx.doi.org/10.1016/j.jclepro.2012.02.027.

Lehni, M., Schmidheiny, S., Stigson, B. 2000. *Eco-efficiency: creating more value with less impact*. World Business Council for Sustainable Development, Geneva. 2-940240-17-5.

Lourenço, E.J., Baptista, A.J., Pereira, J.P., Dias-Ferreira, C. 2013. Multi-Layer Stream Mapping as a Combined Approach for Industrial Processes Eco-efficiency Assessment. *Re-engineering Manufacturing for Sustainability*: 427–433. http://dx.doi.org/10.1007/978-981-4451-48-2_70.

Michelsen, O., Fet, A.M., Dahlsrud, A. 2006. Eco-efficiency in extended supply chains: a case study of furniture production. *Journal of environmental management* 79: 290–297. http://dx.doi.org/10.1016/j.jenvman.2005.07.007.

Verfaillie, H.A., Bidwell, R. 2000. *Measuring Eco-Efficiency: A Guide to Reporting Company Performance*. World Business Council for Sustainable Development, Geneva. 2-940240-14-0.

WASTES – Solutions, Treatments and Opportunities II – Vilarinho, Castro & Lopes (Eds)
© 2018 Taylor & Francis Group, London, ISBN 978-1-138-19669-8

Enzymatic esterification of pre-treated and untreated acid oil soapstock

J. Borges, C. Alvim-Ferraz, M.F. Almeida & J.M. Dias
LEPABE, Faculty of Engineering, Oporto University, Porto, Portugal

S. Budžaki
Faculty of Food Technology, Josip Juraj Strossmayer University of Osijek, Osijek, Croatia

ABSTRACT: The soapstock oil presents high Free Fatty Acid (FFA) content, being potentially a sustainable alternative feedstock for biodiesel production. In the present work, enzymatic esterification using a suspension of *Thermomyces lanuginosus* lipase was conducted on pre-treated and untreated acid oil from soapstock derived from the refining of vegetable oil mixtures. The enzyme suspension was prepared by mixing it with a 1 M phosphate buffer (1:10 V/V, enzyme:buffer). Esterification was conducted in an orbital shaking incubator at 40.0°C using a molar ratio of acid to methanol of 1:3, during 24 h at 200 rpm; monitoring was performed at 0, 7 and 24 h. Around 80% FFA reduction was obtained using raw and pre-treated oil, showing that the use of suspended enzyme avoids the need for pretreatment. Fatty Acid Methyl Ester content was determined, indicating that both esterification and transesterification reactions occurred.

1 INTRODUCTION

The current exploitation of fossil fuels is a major contributor to the increase of the greenhouse gases (GHG) in atmosphere, which are directly associated with the accelerated global warming (Suganya et al., 2016). The adverse effects of GHG on the environment, combined with the decrease of petroleum reserves, highlights the importance of reaching sustainable and economically feasible fuels based on "green" sources of energy, named biofuels (Babaki et al., 2016), especially those derived from renewable sources (Scott et al., 2010; Silitonga et al., 2016). The "first generation" biofuels, derived from edible oils, can offer some emissions benefits, but the problems regarding the impact on feedstock source, mainly on biodiversity, land use and competition with food crops as well as high production costs (Hajamini et al., 2016; Suganya et al., 2016), leads to a necessary search for cheaper, competitive and sustainable alternative sources. The renewable energy directive (Directive 2009/28/EC), amended by Directive 2015/1513, encourage the production of biofuels from "advanced" raw materials such as wastes, through incentive mechanisms, such as double and quadruple counting, depending on the feedstock, to meet the incorporation targets. The use of feedstocks such as waste cooking oils, animal fats, soapstock oil, amongst others, might be included under such framework.

The soapstock is a waste material resulting from the neutralization of free fatty acids during the refining of raw vegetable oils. Such waste is usually chemically treated by acidification using a strong mineral acid, which then leads to the production of a high acidic oil, comprised mostly of free fatty acids and other lipid constituents such as mono, di and triglycerides, among others (Basheer and Watanabe, 2016; Park et al., 2008).

Enzymatic esterification appears as a relevant technology to convert FFA into mixture of Fatty Acid Methyl Esters (biodiesel), namely to replace conventional homogeneous pre-treatment routes (homogeneous acid catalysed esterification). The following advantages are highlighted, compared to the conventional processes: i) possibility of catalyst recovery, ii) environmental acceptability; ii) simultaneous conversion of glycerides and free fatty acids; and iii) lower reaction temperature (Bhuiya et al., 2016; Lopresto et al., 2015). The enzyme cost, high reaction times and enzyme inhibition (namely depending on the pH alcohol type and concentration) are reported as fundamental drawbacks of the process (Bhuiya et al., 2016). Previous studies showed that the pre-treatment of the mineral acidity present in acid oil from soapstock, by neutralization, was imperative to enable enzymatic esterification (Cruz et al., 2016). According to Budžaki et al. (2015), it is possible to reduce the inhibitory effects on enzyme (for transesterification) by diluting it on a phosphate-buffered saline (PBS) solution and using the suspension as a catalyst; however, no studies were found on the use of such an enzyme suspension for FFA esterification.

In agreement, in the present work, enzymatic esterification of pre-treated and untreated acid oil from soapstock obtained from vegetable oil refining (oil mixtures) was evaluated using a suspension of *Thermomyces lanuginosus* lipase, in order to estimate the effectiveness of the buffer and the application of such process towards FFA esterification.

2 EXPERIMENT

2.1 *Feedstock*

The acid oil from soapstock (mixture of vegetable oils) was provided by Nature Light, S. A. In order to avoid the problems associated with mineral acidity, a pre-treatment was performed by neutralization using NaOH according to Cruz *et al.* (2016), which consisted in one washing 1:1 V/V with NaOH (750 ppm) and two washings with distilled water, followed by the separation of the organic fraction. The lipase from *Thermomyces lanuginosus* (Lipolase 100 L), was purchased from Sigma-Aldrich Handels GmbH (Vienna, Austria). The phosphate buffer—PBS (pH 7.4, 1 M) was prepared from KH_2PO_4 (3.41 g ~25 mL H_2O) and K_2HPO_4 (17.6 g ~100 mL H_2O). The solution of KH_2PO_4 (25 mL) was gradually added to the 100 mL solution of K_2HPO_4 until pH was 7.4. The enzyme was mixed with the buffer at an 1:10 V/V enzyme:buffer solution ratio.

2.2 *Experimental set-up*

The enzymatic esterification of the treated and non-treated acid oil was performed in 250 mL flasks, using an orbital shaking incubator (Agitorb 200 IC), at 40.0°C.

The acid oil was added to each flask (100 g) and the following conditions were established: molar ratio of acid to methanol of 1:3, constant stirring of 200 rpm, and, enzyme suspension concentration of 10 wt%, relative to oil. The experiments were performed during 24 h and samples (4 g per sample) were taken at time 0, 7 and 24 h, to monitor the progression of the esterification reaction. After the reaction, the mixture was centrifuged to allow enzyme recovery. To evaluate methyl ester content, excess methanol was recovered in a rotary evaporator at 65°C, for 30 min and then the product was oven-dried until constant weight. The experiments were performed in duplicate.

2.3 *Analytical methods*

The acidity of the raw materials and of the reaction mixture was measured by volumetric titration, according to the standard NP EN ISO 660 (2009). Methyl ester content was determined by gas chromatography according to EN 14103 (2003), only on the final purified product (after 24 h reaction).

3 RESULTS AND DISCUSSION

3.1 *Monitoring of esterification reaction*

The acid oil from soapstock presented an acidity of 63.72 ± 0.76 wt% (relative to oleic acid); after pre-treatment with NaOH and water washing, the acidity was reduced to 49.55 ± 0.16 wt% (relative to oleic acid).

The results of the esterification reaction are presented in Figure 1. The acidity reductions at the different times are presented in Table 1.

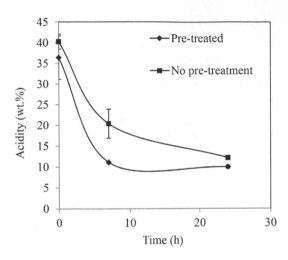

Figure 1. Evolution of acidity during enzymatic esterification of pre-treated and untreated soapstock oil ($T = 40°C$; 10 wt.% enzyme suspension, relative to oil; 200 rpm; 1:3 acid:methanol molar ratio).

Table 1. Free Fatty Acid (FFA) reduction (%) relative to feedstock, during enzymatic esterification of soapstock oil without pre-treatment (No pre-treatment) and with pre-treatment (Pre-treated).

| Time (h) | FFA reduction, % ($FFA_{xh} - FFA_{Feedstock}$)/ $FFA_{Feedstock} \times 100$ | |
	No pre-treatment	Pre-treated
0	37	27
7	68	78
24	81	80

The acidity reductions at zero time relate with the dilution effect of the reactants and the rate of the reaction progression. The monitoring of the reaction showed that after 7 h most of the conversion occurred. In fact, the reaction progressed faster with the pre-treated oil, which after 7 h reached close to the maximum acidity reduction. After 24 h, the results are similar using both the treated and untreated oil, with an acidity reduction around 80% being obtained for both raw materials. The final acidity of the products was of around 10 wt.% for the pre-treated oil and around 12 wt% for the untreated oil.

The results indicate that the use of the suspended enzyme allows achieving high acidity reductions without the need for a pre-treatment, when performing the reaction during 24 h, and therefore it might be an effective alternative to overcome the observed enzyme inhibition with such raw material, due to mineral acidity (Cruz et al., 2016).

Taking into account the pre-treatment costs and expected product losses, the use of the suspended enzyme appears as an effective alternative towards a more sustainable application of enzymatic esterification of highly acidic raw materials.

3.2 *Fatty acid methyl ester content of the products*

The Fatty Acid Methyl Esters contents of the products are presented in Table 2. Taking into account the initial FFA of the raw-materials and the reduction of acidity obtained, of around 80% (ca. 40 wt.% FFA converted for pre-treated oil and ca. 51 wt.% FFA for raw soapstock oil). The results show that in some extent the transesterification of mono–, di—and/or triglycerides present also occurred, with more expression in the case of the raw soapstock oil.

Table 2. Methyl ester content of the final products, after enzymatic esterification of the pre-treated and untreated soapstock oil.

Raw material	Free fatty acid content of raw-material (wt.%)	Fatty acid methyl ester content of the product* (wt.%)
Pre-treated soapstock oil	49.55 ± 0.16	43.07 ± 0.42
Untreated soapstock oil	63.72 ± 0.76	60.39 ± 0.59

*Obtained after purification of the esterified product.

4 CONCLUSIONS

The use of a *Thermomyces lanuginosus* lipase suspension, by enzyme dilution with a phosphate buffer solution was effective to avoid enzyme inhibition by mineral acidity of raw soapstock oil.

Under the studied conditions (T = 40°C; 10 wt.% enzyme suspension, relative to oil; 200 rpm; 1:3 acid:methanol molar ratio) it was possible to achieve the same acidity reduction, of 80 wt.% after enzymatic esterification using both the pre-treated and untreated oil. Using the pre-treated oil, the reaction progressed faster; however, taking into account pre-treatment costs and expected loss of product, the use of suspended enzyme as described appears to be promising towards avoiding enzyme inactivation and effective esterification of free fatty acids.

REFERENCES

Babaki, M., Yousefi, M., Habibi, Z., Mohammadi, M., Yousefi, P., Mohammadi, J., Brask, J., 2016. Enzymatic production of biodiesel using lipases immobilized on silica nanoparticles as highly reusable biocatalysts: effect ofwater, t-butanol and blue silica gel contents. *Renewable Energy* 91: 196–206.
Basheer, S., Watanabe, Y., 2016. Enzymatic conversion of acid oils to biodiesel. *Lipid Technology* 28: 16–18.
Bhuiya, M.M.K., Rasul, M.G., Khan, M.M.K., Ashwath, N., Azada, A.K., 2016. Prospects of 2nd generation biodiesel as a sustainable fuel—Part 1: selection of feedstocks, oil extraction techniques and conve sion technologies. *Renewable and Sustainable Energy Reviews* 55: 1109–1128.
Budžaki, S., Šalić, A., Zelić, B., Tišma, M., 2015. Enzyme-catalysed Biodiesel Production from Edible and Waste Cooking Oils. *Chem. Biochem. Eng. Q.* 29 (3): 329–333.
Cruz, M., Pinho, S.C., Mota, R., Almeida, M.F., Dias, J.M., 2016. Enzymatic Esterification of Acid Oil from Soapstocks Obtained from Vegetable Oil Refining: Effect of Enzyme Concentration. Chapter 439., Venice 2016—*Sixth International Symposium on Energy from Biomass and Waste.* CISA Publisher. ISBN code 9788862650090.
Hajamini, Z., Sobati, M.A., Shahhosseini, S., Ghobadian, B., 2016. Waste fish oil (WFO) esterification catalyzed by sulfonated activated carbon under ultrasound irradiation. *Applied Thermal Engineering* 94: 141–150.
Lopresto, C.G., Naccarato, S., Albo, L., Paola, M.G.D., Chakraborty, S., Curcio, S., Calabrò, V., 2015. Enzymatic transesterification of waste vegetable oil to produce biodiesel. *Ecotoxicology and Environmental Safety*: 229–235.
Park, J.-Y., Kim, D.-K., Wang, Z.-M., Lee, J.-P., Park, S.-C., Lee, J.-S., 2008. Production of biodiesel from soapstock using an ion-exchange resin catalyst. *Korean Journal of Chemical Engineering* 25: 1350–1354.
Scott, S.A., Davey, M.P., Dennis, J.S., Horst, I., Howe, C.J., Lea-Smith, D.J., Smith, A.G., 2010. Biodiesel from algae: challenges and prospects. *Current Opinion in Biotechnology* 21: 277–286.
Silitonga, A.S., Masjuki, H.H., Ong, H.C., Yusaf, T., Kusumo, F., Mahlia, T.M.I., 2016. Synthesis and optimization of *Hevea brasiliensis* and *Ricinus communis* as feedstock for biodiesel production: A comparative study. *Industrial Crops and Products* 85: 274–286.
Suganya, T., Varman, M., Masjuki, H.H., Renganathan, S., 2016. Macroalgae and microalgae as a potential source for commercial applications along with biofuels production: A biorefinery approach. *Renewable and Sustainable Energy Reviews* 55: 909–941.

Formulation of waste mixtures towards effective composting: A case study

M.J. Fernandes & F.C. Pires
Gintegral—Gestão Ambiental, S.A., Póvoa de Varzim, Portugal

J.M. Dias
LEPABE, DEMM, Faculdade de Engenharia, Universidade do Porto, Porto, Portugal

ABSTRACT: The main objective of this study was to evaluate the potential of organic waste streams for composting. For that purpose, different blends were used, with prevalence of waste-water treatment sludge, considering: 30 or 50% sludge and 50 or 70% of other wastes such as corn silage, expired animal feed and biomass ash. Composting in turned piles of around 1 m³ was performed. The process was monitored during 73 days. The obtained product was analyzed regarding physical, chemical, microbiological and phytotoxicity parameters. The best results were obtained using 50% sludge, 30% of wood chips, 5% of ash, 5% of animal feed and 10% of corn silage; however, high losses of heat occurred, hygienization temperatures were not reached and thus microbiological parameters weren't fulfilled. A class II type of material (Decree-Law No. 103/2015 on fertilizing materials) is expected to be obtained at a higher scale.

1 INTRODUCTION

The recovery of biodegradable waste is of undeniable relevance, taking into account the impacts of its inappropriate management on public health and the environment.

Due to the evolution, growth and development of the society, there has been an increasing production of wastewater, whose treatment prior to discharge includes the elimination or reduction of the substances in suspension and organic content to acceptable discharge limits in the receptor water resources. These processes of municipal and industrial wastewater treatment, which occur in wastewater treatment plants using physicochemical and/or biological processes, imply the separation and concentration of solids to obtain sewage sludge (Mendes 2014, Candeias 2008, Gutiérrez, M.C. et al 2017). This waste material presents high variations in quantity and quality, depending on the size of the plant and the type of treated effluent.

Amongst the parameters of concern, according to different authors and entities (Gonçalves, M.S. 2005, Rasquilha 2010, European Commission 2008) are pathogenic microorganisms, heavy metals, organic micropollutants and phytotoxic metabolites, which, together with the large quantities generated, led to a great mobilization of the responsible entities and environmental institutions. In agreement, the definition of treatment solutions that reduce the quantity and increase the final quality of the SS are fundamental to optimize SS management, consequently leading to cost reduction, also allowing the preservation of the public health and the environment (Rasquilha 2010). The agricultural interest of this type of organic waste should be emphasized taking into account the need for organic correction of national agricultural soils, similarly to the rest of the European Union, which turns SS a waste with high potential for compost production. Such material should be applied as an environmentally sustainable measure, providing that it does not present polluting potential (Gonçalves, M.S. 2005).

Composting is a biological process, mediated by microorganisms, which allows the biodegradation of the organic matter under controlled environmental conditions. Due to ventilation and exothermic reactions, such process is known to allow sewage sludge sanitation (elimination of pathogens), reduction of volatile matter and moisture loss, thus converting the SS into a stabilized organic rich material (Gutiérrez, M.C. et al 2017). In agreement, properly composted organic waste becomes aesthetically acceptable, free of pathogens, easy to handle and applicable to improve soil structure, water retention and nutrients uptake (Gonçalves, M.S. 2005).

The present study aimed to recover sewage sludge through composting; taking into account the required substrate properties, different mixtures were evaluated using corn silage, expired animal food and biomass ash and as structural material a mixture of eucalyptus and pine bark, branches and dried leaves. A small-scale composting process was adopted, in piles, in order to study and select the best solution for integration into an industrial-scale composting plant.

2 MATERIALS AND METHODS

2.1 Raw-materials

For the formulation of the mixture the following waste materials were used: sludge resulting from the treatment of urban wastewater (Fátima wastewater treatment), corn silage from *Pioneer Sementes*, expired animal food from *Avenal rações*, biomass ash and woodchips from *Capitão & Lima, Lda*. Sludge resulting from wastewater treatment had a high pH = 9, an organic matter of 62 wt.% (dry basis) and a moisture content of 78 wt.%. Regarding heavy metals, laboratorial analysis revealed a copper and zinc concentration of 142 mg kg^{-1} (dry basis) and 340 mg kg^{-1} (dry basis), respectively. Corn silage dry matter content was around 35 wt.% (dry basis). This residue additionally had an inoculant function, since it contained some developed microbial fauna. As regards the structuring material, a mixture of pine bark and eucalyptus, dry branches and leaves with a moisture content of about 30% was used. Expired animal food had particles size between 0.5 and 1 cm and a moisture content of about 10%. Concerning the biomass ash characteristics, it presented a low moisture content (12%) and a high pH = 11.9.

2.2 Experimental procedures

2.2.1 Construction of compost piles

The formulation of the mixtures was carried out to obtain a moisture content of 50 – 60%, ideal for the development of the composting process (Turovskiy & Mathai 2006). Other properties of the materials, such as particle size (between 1 e 10 cm) and pH (between 5.5 and 7.5), were also taken into account. The pH was a limiting factor in the ash, so it was restricted to 5%. The selected mixtures were 70% of sludge/30% other materials to the mixture 1(M1) and 50% of sludge/50% other residues in the case of mixture 2 (M2) and 3 (M3) (Table 1).

The moisture content and density of the mixtures performed were: M1 – 64.6% and 848.7 kg m^{-3}; M2 – 55.6% and 747.9 kg m^{-3}; M3 – 53.6% and 758.5 kg m^{-3}. Piles of 1.14 m^3 were built, taking into account the literature (Wilkinson 2011) and the amount of available material.

The trapezoidal shape was considered, taking into account the advantages reported on using a flattened pile top (Kreith & Tchobanoglous 2002). For each mixture, two compost piles were made (a and b), in order to allow the comparison of results and to observe possible anomalous events (Figure 1).

Table 1. Proportion of waste materials used in mixtures 1, 2 and 3 (M1, M2 and M3), and physical properties of each waste material (Density and Moisture content).

Portion	Material	Density (kg.m^{-3})	Moisture (wt.%)	M1 (wt.%)	M2 (wt.%)	M3 (wt.%)
Sludge	Sludge	1000	78	70	50	50
Other	Woodchips	500	30	18	30	47.5
Residues	Ash	840	12	3	5	2.5
	Expired animal food	650	10	3	5	0
	Corn silage	234	65	6	10	0

Figure 1.   Trapezoidal shaped piles built for composting of Mixture 1, 2 and 3, and replicas (a, b).

The process of building the piles was started by applying a pre-mix sludge base with 18% woodchips and about 15 centimeters height; the other components were added in layers and the process was further repeated until the pile reached 1 m height, with the appropriate amount of each component. Revolving was performed manually (shovel).

2.2.2   Process control

In order to control the development of the process, temperature and humidity were measured over time. The temperature measurement was performed using a set of thermocouple sensors, connected to a data acquisition system, with measurements every 15 minutes. The values obtained were used to define the aeration needs, which led to manual revolving. Thermographic imaging equipment was also used to enable visualization of temperature distribution at the pile surface. The analysis of the samples moisture content was performed according to standard EN 13040: 1999 by gravimetry (drying at $100 \pm 5°C$ in oven until constant weight). The aeration of the compost piles was made if: T > 55°C; unexpected temperature drop.

2.2.3   Characterization of the produced material

Physical, chemical, microbiological and phytotoxicity analyzes were conducted in the product. Physical and chemical analyzes included: moisture content (standard EN 13040: 1999 by gravimetry), particle size distribution (mechanical sieving), mercury (cold-vapor atomic absorption and reading at 253.7 nm), organic matter (Standard EN 12879: 2000), chemical composition (through an X-ray fluorescence analyzer (XRF)) including heavy metals, and, organic micropollutants (PCBs and PAHs) by Gas Chromatography Mass Spectrometry. Microbiological analyzes were based on the standard ISO 9308-1: 2000—"Water quality—Detection and enumeration of Escherichia coli and coliform bacteria. Different soil/compound/turf compositions were used (33%/33%/33%; 60%/30%/10%; 30%/60%/10%) to determine phytotoxicity.

3   RESULTS AND DISCUSSION

The maximum temperature reached was close to 50°C (Fig. 2). The reason why higher temperatures could not be reached was attributed to the excessive heat loss due to reduced pile size (United States Environmental Protection Agency 2010). In the first phase, prior to the first turning, it is possible to observe an initial rise in temperature in all piles, more or less accentuated according to the mixture of materials with a more pronounced heating in mixtures 2a and 2b, which may be being driven by the use of corn silage, which stimulated the initial development of the microbial flora (Fig. 1A). This phenomenon is supported by comparison with the M3 piles (3a and 3b), since both mixtures have a 50% ratio between sludge/remaining residues, but M2 presents 10% of corn silage in the final mixture and the M3 pile does not have silage.

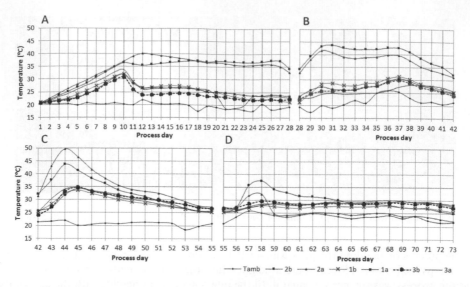

Figure 2.   Temperatures measured since the starting of the composting process until: A – the first stirring (day 28); B – the second stirring (day 42); C – the third stirring (day 55); D – the end of the experiment (day 73) in the center of each pile and room temperature (Tamb). Note: changes observed in the temperature measured in sample 1a between day 19 and 22 resulted from a wrong positioning of the probe.

It is also possible to observe that the addition of silage in the M2 pile had more influence on the starting thrust of the process than the structuring residues (47.5% of the M3 formation).

All replicates showed similar behavior.

After 8 days, pile 2a had already reached 35°C (day 9), and 40°C at 12 days of composting, the highest temperature reached between the various compost piles before the first settling, entering the thermophilic phase. Regarding M3 piles, since those have the most amount of structuring material, they suffer some influence from the environment that surrounds them.

After the first turning performed, 28 days after the start of the process, the drop in temperatures was not very significant in piles with M2, being of about 5°C, and in the other mixtures, the influence is hardly visible (Fig. 1B). In the M1 piles, from the day 28 to the 42th, the temperature variation was different, a phenomenon that can be explained by the lack of structure of this mixture, which causes the oxygen access to the microbial flora to be very dependent on the revolving. At temperatures of the cells constructed with M2, similar oscillations occurred, although with a temperature difference of 2.5 or 3°C between them.

After revolting on the 42th day of the process, all the piles showed a rise in temperature (Fig. 1C). The piles with M1 and M3 showed very similar values of temperature and oscillations of the same order of magnitude during the 13 days until the next stirring. All piles had the highest temperature in the range of 33 to 35°C. Among the piles made up of M2, the replica "a" was the one that reached higher temperatures, of around 50°C. Pile 2b reached the maximum temperature of 44°C. These two replicates of the same mixture had similar temperature oscillations, with the maximum difference being 7°C. After the 44th day, when the highest temperatures in the M2 cells were reached (50°C), the temperature drop to 27 and 25°C in piles 2a and 2b, respectively.

At the day 55, the last stirring was carried out until the interruption of the process, which took place on day 73. The temperature of the piles 1 and 3 increased slightly (about 4°C) with temperatures ranging from 27.5 to 30°C thereafter. The temperatures presented in all piles were further stabilized (Fig. 1D). The temperature monitoring indicates that the pile 2a came closest to the end of the composting process (T close to ambient). Taking into account the obtained results, the following reasons are attributed to the difficulties in reaching hygienization temperatures: i) reduced microbial population—highlighted by increased activity when corn silage residues were used, which might have functioned as inoculant; ii) Lack of appropriate structure—Using corn silage residues affects the structure of the mixture which might impair the circulation of air and consequent distribution of oxygen, making important to supplement with structure material (M2 compared to M1); iii) small size of piles, leading to heat loss and difficulties in maintaining thermophilic temperatures.

The analysis of the moisture content at each pile throughout the process was performed in three moments and the results are presented in Table 2.

The moisture content of the piles was similar to that predicted. The results show a higher homogeneous nature of the materials initially, reflected by low values of standard deviation. In the second measurement, the sample had already undergone two heating phases which corresponded to different temperatures, leading to different moisture contents and higher dispersion of the results. At the last measurement, the material was again more homogeneous, and the lower moisture contents indicate a more stabilized material. At pile 2, the results agree with the 40% limit value imposed by Decree-Law No. 103/2015, on fertilizing materials.

Taking into account the best results obtained (pile 2), the material was further analysed.

Table 3 presents the particle size distribution of the material, revealing slight heterogeneity. All the material presents a size less than 25 mm in agreement with Decree-Law No. 103/2015. The particle size less than 2 mm shows the highest expression, but the material is fairly distributed between 4 particle size ranges (> 2 mm, between 2 and 5 mm, 5 and 14 mm, 14 and 25).

The organic matter content was 25 wt.% for material from the 2a pile and 35 wt.% for material from the 2b pile. Values higher than 30% in agreement with Decree-Law No. 103/2015 are required. The observed deviation from the legal reference limit (30%) may be due to the fact that mineralization and humification was not completed, indicating uncomplete maturation phase. Concentrations of cadmium, lead, copper, chromium, mercury and nickel are below the maximum limit of Class I organic compounds. On the other hand, the zinc concentration in the material exceeds the maximum limit for Class I (general application in agriculture using a maximum of 50 t/ha/year), being around 200 ppm, becoming within the limits imposed for Class II (general application in agriculture using a maximum of 25 t/ha/year) according to Decree-Law No. 103/2015.

Regarding the analysis of organic micropollutants, PCB, which have a high tendency to be adsorbed by solids and accumulate in sludge from wastewater treatment plants (IC Consulants, Ltd. 2001), presented values in piles 2a and b below the established legal limit of 0.8 mg kg^{-1} (dry

Table 2. Moisture content (wt.%) in compost piles, at different periods the process.

Pile	Initial Moisture content (wt.%, wet basis)[a]	31th day Average moisture content (Mean ± SD, wt.%, wet basis)	41th day Average moisture content (Mean ± SD, wt.%, wet basis)	59th day Average moisture content (Mean ± SD, wt.%, wet basis)
1a	65	69.5 ± 0.8	60.0 ± 3.9	–
1b		70.0 ± 0.3	67.3 ± 1.2	–
2a	56	58.6 ± 4.9	50.9 ± 6.6	36.6 ± 0.3
2b		59.3 ± 1.7	46.7 ± 6.6	31.6 ± 0.5
3a	56	57.8 ± 2.2	54.0 ± 3.5	–
3b		60.2 ± 3.6	59.3 ± 1.7	

[a]Estimated; SD – Standard deviation.

Table 3. Percentage of sample retained in each sieve.

Pile	Mesh opening (mm), retained mass (wt.%)					
	31.5	25	14	5	2	< 2
2a	0	0	7.8	31.4	27.1	33.8
2b	0	0	6.2	27.2	32.3	34.2

Note: Sum might note equal 100% due to rounding.

matter). Also the PAH concentrations were below the established limit of 6 mg kg⁻¹ (dry matter) according to Decree-Law No. 103/2015. Other metals such as iron, calcium, potassium and sulfur were found and although there are no legal limitations, it is important to consider for further use. The pH value was slightly below 7 (6.9) in pile 2a and was of 7.2 in pile 2b possibly reflecting the differences in the maturation. In terms of phytotoxicity, the mixture with the highest amount of organic material (60% organic material, 30% soil and 10% peat) was the most favorable for lettuce growth. No phytotoxicity occurred by the use of any of the mixtures.

The values obtained after the microbiological analyzes carried out on the samples of the piles 2a and 2b were 1 125.2 CFU/g and 3 910.8 CFU/g, respectively, higher than the legal limit of 1000 CFU/g for *Escherichia Coli* according to Decree-Law No. 103/2015. The fact that high temperatures have not been reached, capable to sanitize the material produced, should be the main reason why this parameter was exceeded.

4 CONCLUSIONS

The study showed that the mixture with the best behavior when submitted to the composting process was M2 (50% of sewage sludge, 30% of wood chips, 5% of ash, 5% of expired animal food and 10% of corn silage). The moisture content of the final product was 34.1%, consistent with the legislation on fertilizing materials (<4 0%); particle size was also within the requirements (>99% passed the square mesh with 25 mm opening). The analysis of organic matter content led to different results for the two replicates of M2, with one not fulfilling the requirements (<30%), probably, due to incomplete stabilization. Taking into account zinc concentration, the product might be classified as Class II (application in agriculture with an annual maximum of 25 t/ha). The values obtained for HAPs and PCBs were below the limit as well as the pH, which is within the required range. No phytotoxicity occurred in the use of any of the mixtures. The microbiological analysis of *E. Coli* showed negative results (>1000 CFU/g), attributed to difficulties in reaching hygienization temperatures, due to heat losses related with the small scale of process studied. At a higher scale, a Class II material is expected to be obtained.

REFERENCES

Candeias, M. 2008. *Gestão de Lamas de Depuração – Urbanas e de composição similar.* Ministério do Ambiente e do Ordenamento do Território e do Desenvolvimento Regional—Inspecção-Geral do Ambiente e do Ordenamento do Território.

Decree-Law No. 103/2015. (15 June 2015). *Diário da República* 1st series (11): 3756–3788.

European Commission 2008. *Part I: Overview Report. Environmental, economic and social impacts of the use of sewage sludge on land.*

Gonçalves, M. 2005. *Gestão de resíduos orgânicos.* SPI—Sociedade Portuguesa de Inovação. S. João do Estoril: Principia, Publicações Universitárias e Científicas.

Gutiérrez, M.C., A. Serrano, J.A. Siles, A.F. Chica, M.A. Martín 2017. Centralized management of sewage sludge and agro-industrial waste through co-composting. *Journal of Environmental Management* 196 (1): 387–393.

IC Consulants, Ltd. 2001. *Pollutants in urban waste water and sewage sludge.* Luxembourg: Technical University Munich, IRSA Rome, and ECA Barcelona.

Kreith, F., & Tchobanoglous, G. 2002. *Handbook of solid waste management* (2ª ed.). McGraw-Hill.

Rasquilha, F. 2010. *Contribuição para o tratamento e gestão das lamas em excesso das estações de tratamento de águas residuais—caso de estudo de optimização para a 4 Etar do Conselho de Elvas.* Lisboa: Tese de Mestrado apresentada na Faculdade de Ciências e Tecnologia Lisboa—Universidade Nove de Lisboa.

Turovskiy, I., & Mathai, P. 2006. Wastewater sludge processing. *Wiley-Interscience.*

United States Environmental Protection Agency 2010. *Chapter 2 Composting, in National Engineering Handbook—Part 637 Environmental Engineering.* Washington, DC: US: United States Department of Agriculture (USDA) and Natural Resources Conservation Service (NRCS).

WASTES – *Solutions, Treatments and Opportunities II – Vilarinho, Castro & Lopes (Eds)*
© *2018 Taylor & Francis Group, London, ISBN 978-1-138-19669-8*

Selective extraction of lithium from spent lithium-ion batteries

N. Vieceli & F. Margarido
*Center for Innovation, Technology and Policy Research—IN+, Instituto Superior Técnico,
University of Lisbon, Lisbon, Portugal*

M.F.C. Pereira, F. Durão & C. Guimarães
CERENA, Instituto Superior Técnico, University of Lisbon, Lisbon, Portugal

C.A. Nogueira
Laboratório Nacional de Energia e Geologia, I.P. (LNEG), Lisbon, Portugal

ABSTRACT: Several processes for extraction of metals from LIBs have been developed. However, although it could represent a simple and efficient approach, selective leaching of lithium is being rarely investigated. Therefore, in this study, several additives were tested for the selective extraction of lithium from samples (<6.7 mm) of spent LIBs. The effect of $NaHCO_3$ and H_2SO_4 concentrations, combined with milling (disc mill, 2–10 min) was investigated with more details, and allow obtaining a selective extraction of lithium higher than 60%, with a temperature 120°C, for 3.5 h, a L/S ratio of 10 L/kg and a molar ratio of $NaHCO_3/H_2SO_4$ of 20. Further investigation at higher pressures and temperatures could result in an increase in the extraction. Therefore, the method studied could represent an alternative approach for the selective leaching of LIBs.

1 INTRODUCTION

Lithium has several uses but one of the most important is in high energy-density rechargeable lithium-ion batteries (LIBs), which already have a considerable market, powering laptop computers, cordless heavy-duty power tools, and portable electronic devices. Moreover, because of concerns about carbon dioxide footprint and increasing hydrocarbon fuel cost, lithium is becoming even more important in large batteries for powering electric and hybrid vehicles and also for load leveling in solar and wind-powered electricity generation systems (Goonan, 2012).

Lithium demand rose 26% in 2016, is predicted to rise another 39% in 2018 and by 2025 the demand is projected to increase by 73%, as electric vehicles become more viable and as more countries restrict gas and diesel powered vehicles (Freeman, 2017). Lithium batteries are an alternative to reduce the current dependence on fossil fuels, however, to allow labelling LIBs as a green energy source, the evaluation of the entire life cycle is necessary. Therefore, recycling of these batteries is mandatory, and thinking efficient recycling process is essential considering economic, strategic, environmental and safety aspects (Joulié et al. 2014).

Several processes for battery recycling already exist and they can be divided basically into hydro and pyrometallurgical. Moreover, some recycling processes combine both steps and frequently have integrated pre-treatment operations, such as mechanical processing, crushing, pyrolysis and material separation (Georgi-Maschler et al. 2012). Although traditional pyrometallurgical processes can burn off organic electrolyte and binder, lithium is not recovered. Beyond that, these processes involve emission of hazardous gases, dust and are high energy consumption. Hydrometallurgical processes represent an alternative to recover metals from batteries with low energy requirements and no gas emissions. In these processes, the dismantled electrodes have to be treated using aqueous solutions containing acids, bases, oxidants/reducers, or others, and, therefore, wastewater is produced requiring adequate treatment (Meshram et al. 2014, Joulié et al. 2014). Thus, hydrometallurgical processes are considered more suitable for recycling metals from LIBs, if compared with pyrometallurgical process, since they allow complete recovery of high purity metals (Li et al. 2015). In this context, acid leaching plays an important role in the hydrometallurgical processes promoting the solubilisation of metals to the leaching solution (Fan et al. 2016). Several studies focused on the leaching of batteries have been developed, using strong inorganic acids such as H_2SO_4 (Joulié et al. 2014, Grazt et al. 2014, Meshram et al. 2015), HNO_3 (Joulié et al 2014), HCl (Joulié et al. 2014, Takacova et al. 2016) and more recently, eco-friendly approaches have been developed using organic acids, such as DL-malic acid (Li et al. 2010), citric acid (Chen & Zou 2014, Nayaka et al. 2015, Fan et al. 2016), succinic acid (Li et al. 2015) and oxalic acid (Zeng et al. 2015). However, in many of these studies, large amounts of acids are required and normally, there is no selectivity in the recovery of lithium and cobalt, the extraction reached being usually above 90% for both metals. Therefore, the main objective of this study was to evaluate the use of different leaching agents in the treatment of LIBs from portable computers, trying to obtain a selective recovery of lithium along with a reduced consumption of acid, since a selective method can represent a simple and less expensive approach to recover metals from LIBs.

2 METHODOLOGY

The initial sample, constituted by 279 batteries from laptops, was characterised identifying their model, brand and average weight. Considering the representativeness of each model and brand, a sample of 40 batteries were selected and manually dismantled with the aid of pliers. The exterior plastic case, the microcontroller board and the connector were removed. Each battery contained six lithium ion cells, whose potential was measured using a multimeter, and then they were stored for subsequent tests. Contestabile et al. (2001) had already observed that cutting the cell cases to extract their active material produced a strong heating, due to the internal short circuit of the cell. For this reason, this author recommended on an industrial scale, a cryogenic treatment during this operation, in order to prevent accidents related to flames and explosions. Therefore, a cryogenic procedure was adopted in this study before crushing the cells. The cells were immersed in liquid nitrogen during 4–6 min and after that they were shredded using a grab shredder (Erdwich EWZ 2000) with a 6 mm bottom discharge grid.

After shredding, the material was sieved at 6.7 mm, and only the fraction with size <6.7 mm (which corresponds to 82.6% of the sample) was used in the leaching tests. This fraction was utilized because it concentrated the metals of interest (active electrode material) having also a lower content of iron. An additional fine crushing was applied, using a cutting mill (Retsch SM2000) with a discharge grid of 2 mm, in order to reduce the particle size and to allow a more efficient sampling, given the heterogeneity of the material. Finally, this material was milled in a high energy disk mill (N.V. Tema) under different time conditions, and sampled for leaching tests using a rotary sample splitter. The leaching tests were performed in closed glass flasks placed in a thermostatic orbital shaker (at 100 rpm). Leaching conditions, such as leaching temperature and time, liquid/solid ratio (L/S) and the amount of reagents were diverse in the experiments. The conditions of each test are detailed in the results section.

After the leaching procedure a filtration was performed (Whatman 50 filter papers) and the solid fraction obtained was dried (50°C for 24 h) and stored for eventual additional analysis to confirm the metal contents. The elemental analysis of the initial solids as well as the leachates was carried out by atomic absorption spectrometry (AAS, Thermo Elemental SOLAR 969 AA spectrometer). In the case of the solids, the samples were previously dissolved by appropriate chemical attack. The metals recovery was determined relating the concentration in the leaching solutions with the initial content of metals. Experimental errors were estimated to be 3–5%. All chemicals used were of analytical grade and demineralized water was used to prepare the leaching solutions and also for the analytical procedures.

3 RESULTS AND DISCUSSION

3.1 *Initial sample characterization*

The distribution of the initial sample according to the battery brands is shown in Figure 1a, as well as their average weight and standard deviation (Figure. 1b). It is verified that the majority are from the Asus brand. Batteries average weight (including the plastic case) varied from about 280 g (Toshiba) to 350 g (HP). Given the reduction in the weight and size of the current laptops, it is also expected a weight reduction in the batteries. A high standard deviation of the weight was observed for most of the brands and can be related to the large variety of models and computers on the market.

The elemental content of materials used in the leaching tests was: 2.5% Li, 8.1% Co, 7.4% Cu and 9% Fe. Other metals such as Ni and Mn also exist in the battery electrodes, but only cobalt was considered in the analysis since this metal is usually considered the most frequent and was chosen as representative of the transition metals for evaluation of the Li selectivity in the leaching operation. The most common cathodes of LIBs are constituted by the phase $LiMO_2$, M being Co or mixtures of Co, Ni and Mn.

3.2 *Evaluation of different leaching agents*

The approach pursued to attain Li selectivity in leaching consisted of using non-acidic or low acidic leachants. So, several salts were chosen, together with some weak acids and also alkaline agents (Table 1). In all tests, H_2O_2 (6%v/v) was used as a reducer, except when referred to the contrary. The milling time was varied in some tests.

When the leachants NaOH, $Ca(OH)_2$, Na_2CO_3, Na_2SO_4, K_2SO_4, and only H_2O_2 were used, besides the selectivity in relation to Co, the Li extractions were low (about 10%). When $NaHCO_3$ and H_2SO_4 were used the Li extraction increased and Co was not detected. Lithium extraction attained was 15% and 46%, respectively for 2 min and 10 min of milling. High Li extraction were obtained using H_3PO_4 and CH_3COOH, however, without selectivity.

3.3 *Leaching with the mixture $NaHCO_3/H_2SO_4$*

Taking into account the previous results, the effect of the use of $NaHCO_3$ and H_2SO_4 was then studied in more detail. The extraction of lithium and cobalt was followed, since the main objective was to get selectivity between them. The promising results obtained with the mixture of these two reagents were related to the control of the pH of the leachant. Bicarbonate leaching allows the formation of highly soluble $LiHCO_3$, providing that the pH is low enough to attack the oxide of the sample, but not so low that can dissolve cobalt. As the ratio

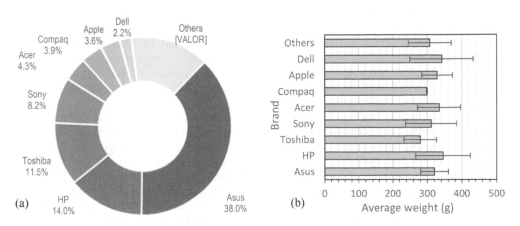

Figure 1. Distribution of the initial sample by brands (a) and their average weight (b).

Table 1. Results of the preliminary tests. Leaching time: 6.5 h, leaching temperature: 90°C.

Conditions		Extraction (%)		
Milling	Leaching agents	Li	Co	Cu
10 min (disc mill)	Na_2CO_3 (1 M)	5.7	nd	nd
	$NaHCO_3$ (1 M)	13.5	nd	2.4
	Na_2SO_4 (1 M)	8.1	nd	nd
	K_2SO_4 (1M)	7.7	nd	nd
	H_3PO_4 (1 M)	60.3	27.3	5.4
	CH_3COOH (1 M)	44.4	34.1	18.0
	H_2O_2	9.7	nd	nd
	Na_2SO_4 (2 M)	7.7	nd	nd
	$NaHCO_3$ (2 M)	23.7	nd	10.7
	CH_3COOH (2 M)	67.2	69.8	37.2
	$NaHCO_3$ (2 M) + H_2SO_4 (0.1 M)	46.0	nd	10.0
2 min (disc mill)	NaOH (1 M)	9.6	nd	nd
	NaOH (2 M)	10.3	nd	nd
	H_3PO_4 (1 M) ♦	61.8	34.0	8.5
	H_3PO_4 (1 M)	63.4	29.1	8.0
	H_3PO_4 (2 M)	87.2	70.3	32.6
	$NaHCO_3$ (2 M) + H_2SO_4 (0.1 M)	14.5	nd	9.3
	$Ca(OH)_2$ (2 M)	9.3	nd	nd

♦Without H_2O_2. nd – values lower than the analytical detection limit.

between bicarbonate and acid concentrations is the crucial factor, the following results will be expressed in terms of this ratio. Basically, a 2 M $NaHCO_3$ solution was thought and the acid hence adjusted. As milling seemed to influence the extraction of lithium using $NaHCO_3$ and H_2SO_4, a test varying the milling conditions was carried out and the $NaHCO_3/H_2SO_4$ molar concentration ratio was fixed at 20, the leaching time in 6.5 h, the leaching temperature was 90°C and the L/S 10 L/kg. Results are presented in Figure 2. When a disc mill was used, lithium extraction was higher when the milling time increased from 2 to 10 min, however, increasing the time to 15 min did not enhance the lithium extraction. An additional test was done using a mortar grinder of agate and milling the sample for 10 min (label 10* in Figure 2) but the result was worse than using the disc mill, indicating that the type of milling can affect the extraction. Co was not detected in the leaching solutions. According to these results, a constant milling time of 10 min was applied in the subsequent tests.

Some tests varying the $NaHCO_3/H_2SO_4$ molar ratios were carried out at 60°C (Figure 3a) and at 90°C (Fig. 3b). In both tests leaching time was fixed at 6.5 h and L/S in 10 L/kg. When the temperature was 60°C, at low $NaHCO_3/H_2SO_4$ ratios (1.3–1.5), higher extractions of Li and Co were obtained, indicating that the excess of acid could have worsened the selectivity. When higher ratios were used, the lithium extraction was reduced, however, selectivity in relation to cobalt was obtained. When the temperature was 90°C, and low ratios were applied, initially some cobalt was also leached, probably due to an excess of acid in solution. For ratios ranging from 4 to 20, an increase in the Li extraction was observed as well as a selectivity in relation to Co and after that, Li extraction decreased again.

A test varying the leaching time (Figure 4a) and another one varying the L/S ratio (Figure 4b), were carried out at a fixed temperature of 90°C and $NaHCO_3/H_2SO_4$ molar ratio of 20. In the first test the L/S was fixed in 10 L/kg. As noted in Figure 4a, Li extraction seems to increase until 1.5 h, after that presented a slight decrease and then increased again, becoming stable. Therefore, for these conditions, leaching times above 1.5 h seem to be unnecessary. Regarding the L/S ratio (Figure 4b), for the value 3.5 L/kg, some Co content was also detected in the solutions, which was not verified when the ratios were 10 or 20 L/kg. However, the ratio of 10 L/kg seems to be better than 20 L/kg since allowed to obtain higher Li extractions, good selectivity, and a more concentrated Li solution.

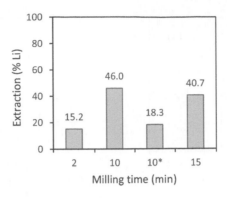

Figure 2.　Extraction of lithium according to the milling time. 10 min*: mortar grinder (agate).

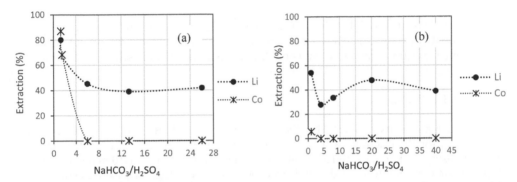

Figure 3.　Extraction of Li and Co according to the $NaHCO_3/H_2SO_4$ molar ratios at 60°C (a) and 90°C (b).

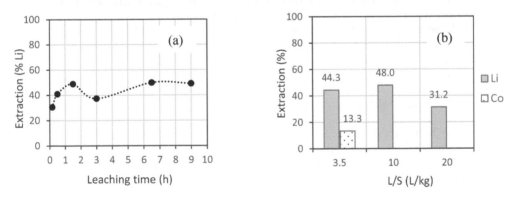

Figure 4.　Extractions according to the leaching time (a) and the L/S ratio (b).

　　Tests varying the leaching temperature were made to verify its influence on the extraction of Li and its selectivity in relation to Co. The molar ratio of $NaHCO_3/H_2SO_4$ employed was 20 L/kg, the leaching time was fixed at 6.5 h and the L/S in 10 L/kg. Selectivity of Li in relation to Co was achieved in all tests. In a first phase, tests were performed at 30, 60 and 90°C, for 6.5 h. Lithium extraction (Figure 5a) seems to increase with temperature and the maximum extraction was obtained at 90°C (49.9% Li). For this reason, in a second phase an additional test at 97°C for 3.5 h was performed in the thermostatic orbital shaker (Figure 5b). However, in this case Li extraction was similar to that obtained at 60 and 90°C. Additional tests at 120 and 150°C for 3.5 h were carried out in a furnace using a teflon pressure vessel,

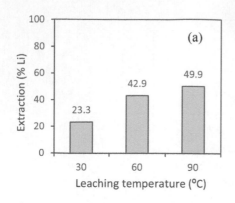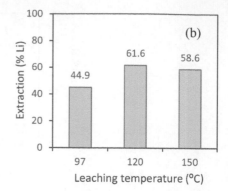

Figure 5. Effect of the leaching temperature in the Li extraction. First phase (a) and second phase (b).

obtaining Li extractions higher than 50%, and maintaining the selectivity in relation to Co. Thus, it could be expected that a process combining temperatures above 100°C and high pressures, using for example an autoclave, could lead to an increase in the selective Li extraction.

4 CONCLUSIONS

Several leaching agents were tested in the treatment of spent LIBs, trying to obtain a selective method for the leaching of lithium. Selective extraction of lithium of about 50% was obtained using a mixture of $NaHCO_3$ and H_2SO_4 and a milling time of 10 min in a disc mill. Type and time of milling seemed to influence the process. The highest selective extraction of lithium (above 60% Li) was obtained when the sample was leached in a pressure vessel during 3.5 h at 120°C using a $NaHCO_3/H_2SO_4$ molar ratio of 20 and an L/S ratio of 10 L/kg. Additional investigation in order to identify the factors that significantly affect the process and, to optimize it, should be performed. The proposed method represents a simple alternative for the selective extraction of lithium from spent LIBs, and additional tests combining temperatures higher than 100°C and high pressures reached for example in an autoclave, could increase the obtained results and is a potential research topic.

ACKNOWLEDGEMENTS

The author N. Vieceli acknowledges the doctorate grant ref. 9244/13-1 supplied by CAPES Foundation, Ministry of Education of Brazil. The research work here presented is related to the activities included in the project proposal MITEXPL/SUS/0020/2017 (Fundação para a Ciência e a Tecnologia, Portugal).

REFERENCES

Chen, X. & Zhou, T. 2014. Hydrometallurgical process for the recovery of metal values from spent lithium-ion batteries in citric acid media. *Waste Management & Research* 32(11) 1083–1093.
Contestabile, M., Panero, S. & Scrosati, B. 2001. A laboratory-scale lithium-ion battery recycling process. *Journal of Power Sources* 92: 65–69.
Fan, B., Chen, X., Zhou, T., Zhang, J. & Xu, B. 2016. A sustainable process for the recovery of valuable metals from spent lithium-ion batteries. *Waste Management & Research:* 1–8.
Freeman. S. Financial Post, Mining. 2017. *Lithium is the latest hot metal commodity, but investor fever could be cooling.* In < http://business.financialpost.com/news/mining/lithium-is-the-latest-hot-metal-commodity-but-investor-fever-could-be-cooling>. Acessed on 21 Mar. 2017.

Georgi-Maschler, T., Friedrich, B., Weyhe, R., Heegn, H. & Rutz, M. 2012. Development of a recycling process for Li-ion batteries. *Journal of Power Sources* 207: 173–182 .

Goonan, T.G. 2012. *Lithium use in batteries*. Circular 1371. U.S. Department of the Interior. U.S. Geological Survey. In: <https://pubs.usgs.gov/circ/1371/>. Acessed on 21 Mar. 2017.

Gratz, E., Sa, Q., Apelian, D. & Wang, Y. 2014. A closed loop process for recycling spent lithium ion batteries. *Journal of Power Sources* 262: 255–262.

Joulié, M., Laucournet, R. & Billy, E. 2014. Hydrometallurgical process for the recovery of high value metals from spent lithium nickel cobalt aluminum oxide based lithium-ion batteries. *Journal of Power Sources* 247:551–555.

Li, L., Ge, J., Chen, R., Wua, F., Chen, S. & Zhang, X. 2010. Environmental friendly leaching reagent for cobalt and lithium recovery from spent lithium-ion batteries. *Waste Management* 30: 2615–2621.

Li, L., Qu, W., Zhang, X., Lu, J., Chen, R., Wu, F. & Amine, K. 2015. Succinic acid-based leaching system: A sustainable process for recovery of valuable metals from spent Li-ion batteries. *Journal of Power Sources* 282: 544–551.

Meshram, P., Pandey, B.D. & Mankhand, T.R. 2014. Extraction of lithium from primary and secondary sources by pre-treatment, leaching and separation: A comprehensive review. *Hydrometallurgy* 150: 192–208.

Meshram, P., Pandey, B.D. & Mankhand, T.R. 2015. Recovery of valuable metals from cathodic active material of spent lithium ion batteries: Leaching and kinetic aspects. *Waste Management* 45:306–313.

Nayaka, G.P., Manjanna, J., Pai, K.V.,Vadavi, R., Keny, S.J. & Tripathi, V.S. 2015. Recovery of valuable metal ions from the spent lithium-ion battery using aqueous mixture of mild organic acids as alternative to mineral acids. *Hydrometallurgy 151*: 73–77.

Takacova, Z., Havlik, T., Kukurugya, F. & Orac, D. 2016. Cobalt and lithium recovery from active mass of spent Li-ion batteries: Theoretical and experimental approach. *Hydrometallurgy 163:* 9–17.

Zeng, X., Li, J. & Shen, B. 2015. Novel approach to recover cobalt and lithium from spent lithium-ion battery using oxalic acid. *Journal of Hazardous Materials* 295: 112–118.

The main environmental impacts of a university restaurant and the search for solutions

M.C. Rizk, D.B. Nascimento & B.A. Perão
*Departamento de Planejamento, Urbanismo e Ambiente, Faculdade de Ciência e Tecnologia,
Universidade Estadual Paulista "Júlio de Mesquita Filho", São Paulo, Brasil*

F.P. Camacho
Departamento de Engenharia Química, Universidade Estadual de Maringá, Maringá, Paraná, Brasil

ABSTRACT: The objective of this study was to evaluate the food and water waste in a university restaurant. During the evaluated period, 18,083.08 kg of food were produced of which 17.85% had become solid waste that was disposed in the environment, being that US $ 1,079.65 of what was invested for production was lost. The consumption of water for the production of meals was 29.71 L, above the recommended level, contributes to a reduction of this natural resource, besides generating an additional amount of wastewater, increasing the volume of effluents dumped in the rivers. Thus, it is evident the need to develop actions aimed at optimizing the management of natural resources, reducing the waste and, consequently, the negative impacts caused by the restaurant's activities. So, in order to minimize the impacts, environmental programs were designed to incorporate the principles of 4R (reduce, reuse, recycle and re-think).

1 INTRODUCTION

Davies & Konisky (2000) apud Wang et al. (2013) related that the direct impacts of restaurants include energy consumption, solid waste generation, atmospheric emissions, wastewater generation and food diseases.

According to Beretta et al. (2012) food loss over the entire food productive chain represents a significant loss of resources invested in food production, transport and storage.

The issue is linked to several quite significant aspects related to environmental and food issues, such as the control of natural resources, the loss of biodiversity, climate change, soil erosion, the reduction of scarce resources such as agricultural soil and water, and the loss of natural habitats (Lucifero, 2016).

The environmental diagnosis in restaurants is essential, through it is identified the solid wastes and wastewater generated in all stages of the productive process of meals, permitting an adoption of actions that minimize the solid waste generation and the rational use of water.

Therefore, the aim of this study was evaluate the food waste and water loss at the University Restaurant of UNESP-FCT, Brazil, and propose alternatives to minimize the main environmental impacts.

2 MATERIALS AND METHODS

The quantification of solid waste was made by weighing the wastes after its generation. It was registered the daily quantity of meals produced and the quantity of service waste and costumer leftovers. The data collection occurred in the months of January, March, May and June 2015.

The evaluation of food waste was based on service waste and costumer leftovers. Service waste can be divided into two categories: reusable and non-reusable. The reusable service waste are those foods produced that are not placed on the distribution counter, remaining in the kitchen—these foods can be reused or donated to charities, under strict hygienic-sanitary control measures. The non-reusable service waste are those that were placed in the distribution counter where customers can serve themselves, when the food is exposed, it can not be guaranteed its hygienic-sanitary quality, so these food should be discarded.

The costumer leftovers are related to the food distributed, but not consumed by costumers, being discarded.

The per capita consumption was calculated by equation 1 and the per capita waste was calculated by equation 2, as established by Varela et al. (2015).

$$\text{per capita consumption} = \left(\frac{\text{distributed meal weight}}{\text{numbers of meals served}} \right) \tag{1}$$

$$\text{waste per capita} = \left(\frac{\text{weight of the remains}}{\text{numbers of meals served}} \right) \tag{2}$$

Based on the unit price of meal production in the studied months, it was possible to calculate monetary losses due to food waste. To this, it was used the equations 3 and 4, according to Varela et al. (2015).

$$\text{per capta cost of food waste} = \left(\frac{\text{unit price of production of meals with food} \times \text{waste per capita}}{\text{per capita consumption}} \right) \tag{3}$$

$$\text{total cost of food waste} = \text{unit price of meal production} \times \text{number of wasted meals} \tag{4}$$

To determine the losses of water, parameters of water consumption from restaurants of the same size as the university restaurant were collected in the literature, being established an average volume standard. This value was subtracted from the measured value under monitoring in the university restaurant, according to equation (5).

$$\text{waste of water} = \\ \text{volume measured in the university restaurant - average standar volume for restaurants} \tag{5}$$

Then, it was considered the tariff defined by the municipal sanitation company, and, based on the volume of water consumed, the associated costs were established, both for consumption and for losses.

The estimated consumption of water was made in the kitchen (five taps) and in the utensils hygiene sector (three taps and a dishwasher), because these are the most significant sectors of water consumption.

The estimative was made with the average flow rate of the taps and other sources of consumption (Q), the average frequency of use of the sources (f) and the average time of each use (t), according to equation (6) (Fasola et al., 2011).

$$\text{water consumption} = f \times t \times Q \tag{6}$$

Finally, in order to determine the quantity of meals that could be produced with the amount currently spent in the productive process with wastes, the total was calculated with wastes and divided by the unit price of meals in relation to water and food, according to equation (7).

$$\text{quantity of meals} = \left(\frac{\text{amount spent in excess}}{\text{unit value of meals}} \right) \qquad (7)$$

3 RESULTS AND DISCUSSION

During the months of January, May, March and June 2015 it was produced 18,083.08 kg of meals, of which 3,228.95 kg of food were discarded, what means that 17.85% of all production of food was thrown away.

Analyzing the Table 1, it is observed that the per capita waste is above the values recommended in the literature. Teixeira et al. (2006) recommended values between 15 to 45 grams of costumer leftovers per person. In this study, it was observed an average of 66.23 grams of costumer leftovers per person.

The causes include inadequate portion sizes of main dish and trim, utensils such as stainless trays that induce consumer to serve larger quantities than they can consume, food preferences and the acceptability of the menu offered.

The calculation of the cost of waste was done considering only the food production; these value could be higher if it would be considered energy, pay and taxes.

According to the *Associação Brasileira de Restaurantes e Empresas de Entretenimentos – ABRASEL*, the establishment that has strict control of its kitchen discards around 15% of what is prepared to serve the clients, which corresponds to approximately 5% of gross revenue. However, more than 50% of the preparation can be lost, equivalent to 15% of the month's invoice (ABRASEL, 2002 apud SOARES et al., 2011).

The data presented in Table 2 show that more than 15% of food produced in the university restaurant is discarded. If it will be considered that 15% of the production can be discarded, in four months there would be a loss of US$ 1,079.65 of the costs of the production of meals that include purchase of food, consumer materials and cooking gas. Considering that per month the university restaurant loses about US$ 269.91, in one year the economic losses

Table 1. Average daily amount of waste discarded.

Variables	January	May	March	June
Average weight of costumer leftovers (kg)	19.65	16.89	18.04	19.74
Average weight of distributed meals (kg)	207.36	195.01	206.83	208.90
Average number of meals served	285	267	290	279
Consumption per capita (g)	726.84	729.12	713.35	747.73
Waste per capita (g)	68.88	63.14	62.22	70.67
Number of waste meals per day	27	23	25	26
Cost of waste per capita (US$)	0.18	0.15	0.14	0.16
Cost total of waste (US$)	51.08	38.57	39.65	44.73

Table 2. Monetary losses due the food waste.

Variables	January	May	March	June
Total cost of meal production (US$)	7,939.73	10,343.98	9,404.82	10,320.79
Total weight of meals produced (kg)	3410.18	5002.80	4715.90	4954.20
Total weight of discarded food (kg)	605.98	934.45	829.77	858.75
% of discarded food	17.77	18.68	17.60	17.33
Monetary losses (US$)*	219.90	380.51	244.07	240.84

* Considering that 15% of production can be discarded.

would be around US$ 3,238.95. These values could be invested in improvements to the productive process, such as purchases of equipment and increase of employees, these actions can contribute to increase the number of meals that is currently insufficient to meet all the demand of the university.

The service waste from salads were served in the next day, as the rice and beans, while the main dish and trim were frozen for later consumption. This practice brings benefits not only from the economic point of view, avoiding the resources and inputs used for food production to be wasted, but also prevents more solid waste from being generated.

The reusable service waste being reincorporated into the food production process were not considered for the calculation of the food waste impacts in the university restaurant.

The university restaurant is a public institution and it is not intended to make a profit from the sales of meals, but its limited economic resources and the growing demand for spending could encourage the restaurant to optimize its processes and improving the quality of its services.

The waste of food does not only cause economic impacts, but the environmental impacts resulting from the generation of solid waste are worrisome. The final disposal of waste solids is one of the main challenges today; when inadequately disposed can cause serious problems to the environment, such as the production of greenhouse gases, contamination of the soil and water by the decomposition of organic matter.

In addition to issues related to solid waste, Table 3 shows the results regarding the consumption of water daily through the monitoring of kitchen and utensils hygiene sector.

Beyond the taps, there is in the sanitations sector a washing machine dishes with 800 L/h capacity. Each cycle of machine have about one minute and washes ten trays. As it had been served 300 meals by day, it was used 300 trays, therefore the machine is used 30 minutes per day and consume around 400 L daily

Thus, the estimated daily consumption was 8,912.70 liters, representing a consumption per meal of 29.71 L.

According to Nunes (2006) restaurants have a daily average water consumption between 20–30 L per meal produced. However, the Company of Basic Sanitation of the São Paulo State (Sabesb), which is the company responsible for water supply of the city, esteem in its Technical Standard n° 181 (2012) that the average build consumption of restaurants is 25 L by meal.

So, it can be observed that the studied university restaurant showed a high consumption. Considering that the service is to students of a Public Higher Education Institutions and that in the Campus there is the Environmental Engineering course, the water consumption should be considered as reference for other restaurants, and not with bad values.

Table 3. Water consumption monitoring results.

Tap's kitchen	Flow rate (L/s)	Frequency (x/day)	Average time (s)	Volume consumed (L)
1	0.17	19	108.80	351.42
2	0.2	5	60	60.00
3	0.25	12	99.17	297.51
4	0.26	21	108.10	590.3
5	0.28	9	115.56	291.21
Total consumption in the kitchen = 1,590.39 L (Equation 6)				
Tap's sanitation	Flow rate (L/s)	Frequency (x/day)	Average time (s)	Volume consumed (L)
1	0.22	25	431.40	2,372.70
2	0.28	25	414.23	2,899.61
3	0.25	5	1,320	1,650.00
Total consumption in the sanitation sector = 6,922.31 L (Equation 6)				

The results obtained in equation (5) pointing that have been consumed 1,412.70 liters daily in excess, extremely significant value, because considering 20 working days in a month, the monthly waste is of 28,254 L.

The monthly consumption (considering 20 working days) was of 178,254 L and the applicable tariff according to Sabesp (2016) would be US\$ 3.18/m^3. The monthly cost of water supply would be about US\$ 567.38, and this value US\$ 89.93 would be spend unnecessarily, representing 15.85% of the total.

Even more aggravating than the financial waste is the significant environmental impact of the reducing the supply of natural resource and of the significant portion of sewage that is being generated and discharged into the environment.

The most contribution of losses is in the cleaning activity of food and utensils, since a lot of items that need to be rapidly sanitized, keeping the taps constantly open, even when they are lathering the utensils or even exchanging small information among themselves.

According to the estimates of this study, due to the losses during the production process are spent excessively US\$ 361.26 per month.

In a controlled setting for the production of a meal should be spent US\$ 0.08 with water and US\$ 1.71 with food ingredients, so with the amount of waste could be produced 201 meals (Equation 7), representing an increase of 67% relative to current production.

Thus, in order to minimize the impacts caused by the university restaurant, some environmental programs were designed to incorporate the principles of 3R (reduce, reuse and recycle). In addition to the three principles, one can incorporate a fourth R, rethink. It is important rethink the current consumption model and how the usual practices can interfere in the environment.

The Waste Reduction Program aims to reduce the food waste that occurs daily; the Composting Program will be designed to the treatment of organic waste generated; the Selective Collection Program will be essential to improve the segregation and quality of the recyclable waste generated. The Practicality Program is designed to replace taps with threaded closures by 90° levers (1-handle pull-down), in order to reduce waste when opening and closing the taps and to encourage employees always close them at the inter-vals of rinsing. The Pre-Dry Selection Program aims to suppress the washing of food that will be discarded later and to minimize the use of washing with running water. Finally, the Environmental Education Program will be the basis of all programs, which will promote, through various actions, the awareness of the users and employees of the restaurant.

4 CONCLUSIONS

The large demand for natural resources and the generation of solid and liquids waste are intrinsic to the productive process of meals, mainly on a large scale, but the results of the study show that unnecessary and significant losses are currently occurring in the university restaurant. These losses can be environmental, such as reducing the supply of food, and the generation of solid and liquids waste with polluting potential, or financial, both of which can be avoided by adopting sustainable practices aimed at waste control.

Food business managers should increasingly seek to understand their procedures in order to be able to identify the deficiencies in the infrastructure and vices in the way employees work, and to be proactive in minimizing their liabilities, as each unidentified fragility, or identified and uncorrected, represents a source of loss.

Sometimes the resistance in environmental adaptations of productive systems refers to the initial investments, but it was possible to verify the financial amount wasted with poorly controlled processes. The financial gains obtained with the adoption of sustainable practices can be reversed in improvements in the establishment, the cheapening of the features for the users or even in the increase of the number of meals served, as discussed in this work.

ACKNOWLEDGEMENTS

The authors would like to thank PIBIC (Programa de Iniciação Científica da UNESP) Processes n. 31440/2014, n. 34315/2015 and n. 39346/2016 for financial support.

REFERENCES

Beretta, C., Stoessel, F., Baier, U., Hellweg, S. 2012. Quantifying food losses and the potential for reduction in Switzerland. *Waste Management* 33: 764–773.

Companhia de Saneamento Básico do Estado de São Paulo—SABESP 2012. *Norma Técnica Sabesp NTS 181*. 3 v. São Paulo. Available at: < http://www2.sabesp.com.br/normas/nts/NTS181.pdf>. Accessed on: 02 Feb. 2017.

Fasola, G.B., Ghisi, E., Marinoski, A.K., Borinelli, J. 2011. B. Potencial de economia de água em duas escolas em Florianópolis, SC. *Ambiente Construído* 11: 65–78.

Kummu, M., Moel, H., Porkka, M., Siebert, S., Varis, O., Ward, P.J. 2012. Lost food, wasted resources: Global food supply chain losses and their impacts on freshwater, cropland, and fertiliser use. *Science of the Total Environment* 438: 477–489.

Lucifero, N. 2016. Food loss and waste in the EU law between sustainability of well-being and the implications on food system and on environment. *Agriculture and Agricultural Science Procedia* 8: 282–289.

Nunes, Riano T. Santiago 2006. *Conservação da água em edifícios comerciais: Potencial de uso racional e reúso em shopping center*. 157 f. Tese (Mestrado em Ciência em planejamento energético. Universidade Federal Do Rio De Janeiro. Rio de Janeiro).

Soares, I.C.C., Silva, E.R., Priore, S.E., et al. 2011. Quantificação e análise do custo da sobra limpa em unidades de alimentação e nutrição de uma empresa de grande porte. *Revista de Nutrição* 24 (4): 593–604.

Teixeira, S., Oliveira, Z.M.C., Rego, J.C. 2006. *Administração aplicada às unidades de alimentação e nutrição*. 1ª Edição. São Paulo: Atheneu.

Varela, M.C., Carvalho, D.R.C., Oliveira, R.M.A., Dantas, M.G.S. 2015. O custo dos desperdícios: um estudo de caso no restaurante universitário da Universidade Federal do Rio Grande do Norte. In: *XXII Congresso Brasileiro de Custos*. Foz do Iguaçu.

Wang, Y., Chen, S., Lee, Y., Tsai, C. 2013. Developing green management standards for restaurants: An application of green supply chain management. *International Journal of Hospitality Management* 34: 263–273.

Treatment of food waste from a university restaurant added to sugarcane bagasse

M.C. Rizk, I.P. Bonalumi & T.S. Almeida
Departamento de Planejamento, Urbanismo e Ambiente, Faculdade de Ciência e Tecnologia, Universidade Estadual Paulista "Júlio de Mesquita Filho", São Paulo, Brasil

F.P. Camacho
Departamento de Engenharia Química, Universidade Estadual de Maringá, Maringá, Paraná, Brasil

ABSTRACT: Restaurants produce large-scale meals, generating large amounts of solid waste, which if not properly managed cause environmental and sanitary problems. Composting is one of the most recommended methods for the treatment of organic solid waste, since the wastes that would be disposed off in the environment, causing environmental and public health problems, are transformed into organic fertilizers. Thus, the present study had the objective of monitoring some physical-chemical parameters of the composting process of food waste from a university restaurant mixed to the sugarcane bagasse. The final values indicated that the moisture content is above that determined in the legislation, while the pH and the C/N ratio presented values very close to the reference values. In this way, it's concluded that it is necessary more time of composting to adjust some parameters, so it will be possible to obtain an organic compost suitable for use in soil.

1 INTRODUCTION

According to Vavouraki et al. (2014), residential kitchens, restaurants, wholesale and retail food distributors and other similar food processing activities are considered the main contributors to the generation of food waste.

Treatment of food wastes is becoming a stern issue worldwide due to their generation in significant quantities (Chan et al., 2015).

Composting provides an effective alternative for food waste management by stabilizing organic matters in food waste and converting it to a soil conditioner or fertilizer (Chan et al., 2015).

Composting consists of two narrowly linked processes, degradation and humification: a part of compost organic matter is aerobically mineralized by microorganisms, whereas the partially decomposed or undegraded part leads to the formation of humic substances. The final product, compost, is a soil amendment, an efficient tool for bioremediation processes or soil restoration, and has capability for improving soil's physical, chemical and biological properties (Benlboukht et al., 2016).

In order to the composting process proceed satisfactorily, some physical-chemical parameters need to be respected, allowing favorable conditions to develop and transforming the organic matter into a fertilizer.

The acceptable ranges for the composting process are: temperature of 54–60°C, C/N of 25–30/1, aeration (oxygen percentage) >5%, moisture content of 50–60%, porosity of 30–36% and pH of 6.5–7.5 (Shafawati & Siddiquee, 2013).

Thus, the present study aimed to study the composting process in the treatment of food waste generated at a university restaurant incorporated to the residues of sugarcane bagasse, aiming the production of organic fertilizer and recognizing the importance of organic waste management systems to reduce environmental and health impacts.

2 MATERIALS AND METHODS

The food waste used in this study were collected at the University Restaurant of UNESP-FCT, Brazil and the sugarcane bagasse was collected in a sugarcane processing industry. The sugarcane bagasse was used as a structuring residue in the mixture with the food waste, in order to adjust the optimal parameters of the composting process.

The wastes were characterized in terms of pH, moisture content, organic matter, total ash, organic carbon, total nitrogen (organic and ammonia nitrogen) and C/N ratio. The pH was determined in calcium chloride solution ($CaCl_2$), using a bench pHmeter, brand HANNA—model HI-221 (Kiehl, 1985). The percentage of moisture, organic matter, mineral residue and organic carbon were determined by the calcination method (Kiehl, 1985). The Kjeldahl nitrogen was determined by the methodology described by IAL (1985).

The experiment was carried out, during 75 days, in a composting pile with a total weight of 100 kg (50 kg of food waste from the university restaurant and 50 kg of sugarcane bagasse).

The pile was developed above a plastic tarp, in which the wastes were intercalated. In cases of rain, the pile was covered.

During the composting process, the pile was manually mixed with shovels. At the beginning of the composting, the pile was every 3 days mixed during the first 15 days. After, the pile was manually mixed once a week.

The pile temperature was measured by a mercury bulb glass thermometer. The temperature measurements were done before the pile mixing.

The same parameters used in the characterization, above mentioned, were monitored every 15 days.

The obtained results were compared to the Normative Instruction (NI) 25/2009 from the Ministry of Agriculture, Livestock and Supply of Brazil (Brazil, 2009), which establishes the reference values to commercialize organic compost from waste treated.

3 RESULTS AND DISCUSSION

The results, obtained in the initial characterization, are shown in Table 1.

Analyzing Table 1, it can be noted that the pH of the food waste was very low when compared to the pH of sugarcane bagasse. The initial pH levels below 5.0 cause a significant reduction in microbiological activity, inhibiting the thermophilic phase (Sundberg et al., 2004). However, after mixing the residues, the pH value reached an acceptable value.

In accordance to Shafawati & Siddiquee (2013), the adequate moisture content is between 50–60%. Analyzing the values obtained, it is noticed that the moisture content of the food waste and the sugarcane bagasse not within the recommended range. However, after mixing the residues, the value obtained was close to the ideal.

The high concentrations of organic matter of 94.7% and 89.1% obtained for the food waste and sugarcane bagasse, respectively, indicate that biological processes can decompose the residues.

Table 1. Initial characterization of food waste and sugarcane bagasse.

Parameters	Food waste	Sugarcane bagasse	50% of food waste + 50% of sugarcane bagasse
pH	4.7	6.2	5.1
Moisture Content (%)	65.8	43.0	47.5
Organic matter (%)	94.7	89.1	90.8
Total ash (%)	5.3	11.0	9.2
Organic Carbon (%)	52.6	49.5	50.4
Total Nitrogen (%)	7.3	0.5	3.1
C/N ratio	7.2	110.4	16.2

Silva et al. (2002) classify organic matter as: optimal (greater than 60%); good (50–60%); low (less than 50%). In this way, the organic matter content obtained was considered optimal, since it was greater than 60%.

The total ash, 5.3% for the food waste and 11.0% for the sugarcane bagasse, is considered optimal according to Silva et al. (2002), who classified the values as: optimal (less than 20%); good (20–40%); undesirable (greater than 40%).

The C/N ratio was very low for the food waste and very high for the sugarcane bagasse, however the mixture of the wastes presented value ideal for the beginning of composting (Qian et al., 2014).

3.1 Composting monitoring

Figure 1 shows a variation of temperature during the composting process.

It can be observed that it was necessary just 1 day for the pile enter in the thermophilic phase, this is the phase that the material reaches maximum temperatures, occurring a rapid degradation of the material. It is possible to observe that the pile remained at this stage until the 13th day. It is worth mentioning that the maximum temperature reached was 58.4°C. Later on, there was a reduction from the temperature and the pile entered the mesophilic phase. This phase is characterized by higher temperatures than the ambient temperature. The pile remained during 27 days in this phase, with temperatures varying between 31.6°C to 40.5°C. After the mesophilic phase, the cooling phase occurs and the temperature is equals or very close to the ambient temperature. It can be observed that from the 52th day the temperature variation was small with values close to the ambient temperature. On the 75th day, the pile presented a temperature of 26.8°C, almost equal to the ambient temperature (Bernal et al., 1998).

The behavior of the pH during the composting process can be observed in Figure 2.

During the monitoring of the process, the pH had a slight increase in the first days of composting and, after, a slight reduction between the 15th day and the 30th day. Afterwards, the pH remained practically constant.

According to Shafawati & Siddiquee (2013) the acceptable range for the pH is 6.5–7.5. Thus, the pile had the pH considered optimum for the development of the microorganisms responsible for composting.

Figure 3 shows the variation of moisture content, organic matter and total ash during the composting.

At the beginning of the study, the moisture was at the optimal range (47.5%), however, between the 1st and 30th day there was a decrement in moisture (21.8%). In order to increase the moisture, it was added water to the residues. Thus, between the 30th and 45th days there

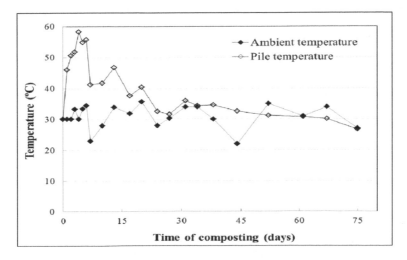

Figure 1. Temperature monitoring.

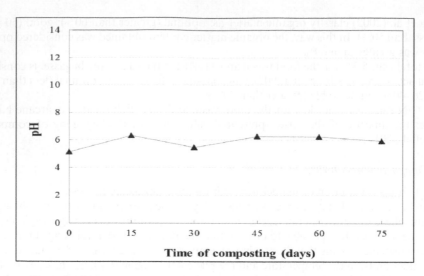

Figure 2. pH monitoring.

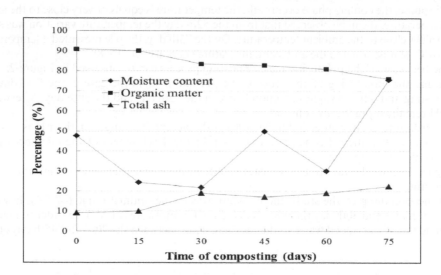

Figure 3. Moisture content, organic matter and total ash monitoring.

was a significant increase in humidity (49.8%). After 45 days, there was a new decrement in moisture, due to the high temperatures in the period, and it was necessary a new addition of water. However, between the 60th and 75th day there was an intense rainfall, and then the parameter had a significant increase.

Regarding the organic matter, it is observed that at the beginning of the process it was high, 90.8%. With the time, there was a reduction of the organic matter content and at the 75th day the value was approximately 76.1%.

In general, the total ash content increased until the 75th day of the composting process. Analyzing the behavior of the total ash, it is observed that up to the 60th day it was classified as optimal, since it was less than 20%. However, as there was a slight increase between the 60th and 75th day, the parameter was classified as good.

The behavior of the C/N ratio during the composting is showed in Figure 4.

According to D'Almeida and Vilhena (2000), the C/N ratio to apply the organic compost on soil in the agriculture not should be greater than 18/1 (organic compost semi-cured).

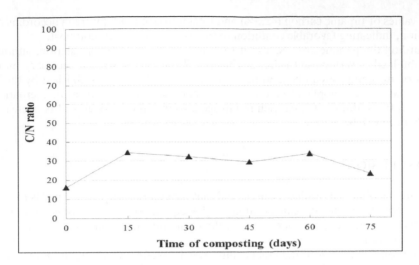

Figure 4.   C/N ratio monitoring.

Table 2.   Final characterization of the organic compost produced by the mixing of food waste and sugarcane bagasse.

Parameters	Pile	N. I. 25/2009 Class A
pH	5.9	Minimum of 6
Moisture Content (%)	75.5	Maximum of 50%
Organic matter (%)	76.1	—
Total ash (%)	22.5	—
Organic carbon (%)	42.3	Minimum 10%
Total nitrogen (%)	1.8	Minimum 0.5%
C/N ratio	22.9	Maximum of 20/1

When this ratio is less than 12/1, the organic compost is classified as cured and shows an optimal condition to be applied.

As it can be observed in Figure 4, at the beginning of the process there was a significant increase in the C/N ratio, followed by a decrease that remained until the 75th day, reaching the value of 22.9/1. Thus, it was observed that in the analyzed period of time the organic compost produced could be not applied on soil, it being necessary a longer composting time to obtain the desired product.

3.2   Characterization of compost after 75 days of composting

The Table 2 shows the parameters of the compost obtained after 75 days of composting and the values established by the Normative Instruction nº 25 de 23/07/2009 of the Ministry of Agriculture, Livestock and Supply (Brazil) to apply the organic compost.

According to Table 2, it is noted that the pH value should be at least 6 and the obtained value is very close to the established, being necessary a little bit adjustment, for example, with the addition of lime to the organic compost.

The moisture content is also not in compliance, because the value is higher than the one established by the legislation (50%). This parameter needs to be adjusted, for example, by the drying of the compost.

The values of organic matter and total ash are not parameters determined by the Normative Instruction. However, according to Silva et al. (2002), the organic matter is considered optimal and the total ash is considered good.

The values of organic carbon (42.3%) and total nitrogen (1.8%) were in accordance to the legislation, indicating favorable condition to the use of the organic compost.

Regarding the parameter C/N, also it isn't in accordance to the legislation, although the value obtained is very close to that established. Therefore, it will be necessary more time to the organic compost can be maturated and reaching the value established by the legislation, so that it can be applied to the soil. If this parameter won't reduce, an alternative to be adopted is the addition of urea, rich in nitrogen, so that the nitrogen will be increased and the C/N ratio will be reduced.

4 CONCLUSION

Analyzing the obtained results, it can be said that the composting process needs to be longer than 75 days, since some parameters aren't still adequate to the soil's application. However, it is noted that the values are very close to the ideal, indicating that the organic compost is close to maturation.

In addition, it has been verified that composting is an effective alternative that can be used in the treatment of food waste, since it transforms waste into organic compost, which has great applicability on soil and on different crop breeding.

ACKNOWLEDGEMENTS

The authors would like to thank PIBIC (Programa de Iniciação Científica da UNESP) Process n. 39788/2016 for financial support.

REFERENCES

Benlboukht, F., Lemee, L., Amir, S., Ambles, A., Hafidi, M. 2016. Biotransformation of organic matter during composting of solid wastes from traditional tanneries by thermochemolysis coupled with gas chromatography and mass spectrometry. *Ecological Engineering* 90: 87–95.

Bernal, M.P.; Sánchez-Monedero, M.A.; Paredes, C., et al. 1998. Carbon mineralization from organic wastes at different composting stages during their incubation with soil. *Agriculture Ecosystems & Environment* 69: 175–189.

Brazil. Ministério da Agricultura, Pecuária e Abastecimento. Secretaria de Defesa Agropecuária. Instrução Normativa n° 25, de 23 de julho de 2009. *Diário Oficial da República Federativa do Brasil.* Poder Executivo, Brasília, DF, 2009.

Chan, M.T., Selvam, A., Wong, J.W.C. 2015. Reducing nitrogen loss and salinity during 'struvite' food waste composting by zeolite amendment. *Bioresource Technology* 200: 838–844.

D' Almeida, M.L.O.; Vilhena, A. 2000. *Lixo Municipal: Manual de Gerenciamento de Resíduos.* 2ª ed. São Paulo: IPT/CEMPRE.

IAL 1985. Normas Analíticas do Instituto Adolfo Lutz. *Métodos químicos e físicos para análises de alimentos.* 3. ed. São Paulo: Editoração Débora D. Estrella Rebocho.

Kiehl, E.J. 1985. Fertilizantes orgânicos. *Piracicaba: Agronômica Ceres.* 492 p.

Qian, X., Shen, G., Wang, Z., Guo, C., Liu, Y., Lei, Z., Zhang, Z. 2014. Co-composting of livestock manure with rice straw: Characterization and establishment of maturity evaluation system. *Waste Management* 34: 530–535.

Shafawati, S.N., Siddiquee, S. 2013. Composting of oil palm fibres and Trichoderma spp. as the biological control agent: A review. *International Biodeterioration & Biodegradation* 85: 243–253.

Silva, F.C., Berton, R.S., Chitolina, J.C., et al. 2002. Recomendações Técnicas para o Uso Agrícola do Composto de Lixo Urbano no Estado de São Paulo. *Embrapa–Circular Técnica* 3: 1–17.

Sundberg, C., S. Smärs, Jonsson, H. 2004. Low pH as an inhibiting factor in the transition from mesophilic to thermophilic phase in composting. *Bioresource Technology* 95: 145–150.

Vavouraki, A.I., Volioti, V., Kornaros, M.E. 2014. Optimization of thermo-chemical pretreatment and enzymatic hydrolysis of kitchen wastes. *Waste Management* 34: 167–173.

Biosolids production and COD removal in activated sludge and moving bed biofilm reactors

Rui A. Dias & Rui C. Martins
CIEPQPF, Department of Chemical Engineering, University of Coimbra, Coimbra, Portugal

Luis M. Castro
Department of Chemical and Biological Engineering, Coimbra Superior Institute of Engineering, Polytechnic of Coimbra, Coimbra, Portugal

Rosa M. Quinta-Ferreira
CIEPQPF, Department of Chemical Engineering, University of Coimbra, Coimbra, Portugal

ABSTRACT: The present work aims to compare two types of biological treatment. The Activated Sludge (AS) process and the Moving Bed Biofilm Reactor (MBBR). The latter is an emerging technology that has been proving quite effective in the removal of high organic strength wastewaters as well as for having diminished operational technicalities. An experimental arrangement was set up: One MBBR and one AS reactor fed continuously with synthetical dairy wastewater. The organic load was regulated by adjusting the milk ratio in the dilution, resulting in a chemical oxygen demand ranging from an average of 582 mg/L to 2646 mg/L. The experimental results obtained showed very high removal capabilities in both treatments, although the MBBR had slightly better results. COD removal efficiencies were, 92.6%, and 95.2%, for the AS, and the MBBR, respectively. Concerning the quantification of excess sludge, the MBBR produced roughly 50% of the amount produced in the AS reactor.

1 INTRODUCTION

An increasing awareness about the environmental impact of discharges is leading many investors to build new wastewater treatment plants (WWTP) or upgrade the existing ones. However, increased urbanization reduces the area available for building these new plants. In addition, the space required for the conventional activated sludge treatment (AS) would be excessive in case of very high pollutant removal efficiency requirements. In order to improve the quality of treated wastewater comply with environmental regulations, implementation of advanced technologies for treatment is required (Di Trapani et al. 2010).

Moreover, with the rising costs of sludge disposal, minimizing sludge production has become increasingly important. According to Egemen et al. (2001), the cost of excess sludge treatment has been estimated to be 50–60% of the total cost of municipal wastewater treatment. Therefore, modifications to existing aerobic treatment processes capable of reducing biosolids production are promising and of great interest (Kulikowska et al. 2007, Ødegaard 2004).

Therefore, the Moving Bed Biofilm technology is an interesting approach that can overcome the shortcomings exhibited of both suspended-growth and attached-growth technologies, by having the biomass attached to carrier elements, forming a biofilm, that moves freely along with the water in the reactor. Due to its small footprint and ease of operation, it is already being considered as an upgrade option for an increasing number of wastewater treatment facilities (Forrest et al. 2016). However, in Portugal, where most of the wastewater treatment plants are based in activated sludge processes, the application of this technology is still incipient.

The aim of this work is to compare the conventional Activated Sludge process and the Moving Bed Biofilm Reactor as concerns organic matter removal and the total sludge produced. In order to evaluate the performance of the technologies under discussion, a system with two independent reactors was implemented. One activated sludge reactor (AS), and one Moving Bed Biofilm Reactor (MBBR) operated in a continuous flow mode.

2 MATERIALS AND METHODS

The laboratory scale reactors were made of plexiglass with a total liquid volume of 3.5 L. The sedimentation unit had a liquid volume estimated at 3 L. The reactors were continuously fed with synthetical wastewater, by diluting market low-fat milk in water. The organic load was regulated by adjusting the milk ratio in the dilution. Flow rate was set to 7 L/d, which is equivalent to a hydraulic retention time of 12 hours. Synthetic dairy wastewater was used because of its simplicity, since it can be approximated to a real milk processing industry wastewater and it provides stable loading rates within limited variations.

The experiment was divided in three parts. The first part, Period A, with a duration of 17 days, corresponded to a low-fat milk dilution of 1/200, followed by Period B, lasting 14 days, in which milk concentration was doubled in the dilution, i.e., 2/200. Period C, was characterized by the highest organic loading which resulted from yet another duplication of the milk concentration, i.e., 4/200, lasting 9 days. The total duration of this experience was, 40 days, starting from November 4th and ending on December 14th.

For this investigation, mixed liquors, and carrier samples were analysed for organic matter content, and suspended solids. The influent and effluent wastewater was measured in terms of the total chemical oxygen demand (totCOD). For these tests, samples were analysed in duplicate to reduce the effect of experimental errors. Effluent samples are composite from a time period of 36 to 48 hours. COD and suspended matter content were measured according to Standard Methods for Water and Wastewater Examination (APHA et al., 2005).

2.1 Biocarriers

The biocarriers contained in the Moving Bed Biofilm Reactor were the bioflow 9, whose dimensions are 9 mm × 7 mm made from High Density Polyethilene (HDPE) and with a bulk density of 145 kg/m^3. The total surface area available for biofilm growth specified by the manufacturer is 800 m^2/m^3. The carrier filling fraction was set to be roughly 50% of the reactor liquid volume, which is in accordance to literature values, setting up the specific surface area to 0.875 m^2.

2.2 Seeding and start-up

The Activated sludge reactor (AS) was inoculated with activated sludge from the wastewater treatment plant of Ribeira de Frades, in Coimbra, which operates as a traditional activated sludge WWTP. The MBBR was seeded with pre-inoculated carriers from the wastewater treatment plant of Arzila, Coimbra. The start-up period lasted roughly two weeks, allowing the microbian communities to adapt to their new environment and to acclimate to the milked influent.

3 RESULTS AND DISCUSSION

3.1 Wastewater

As described above, the organic loading rate was gradually increased, by increasing the milk ratio in the wastewater dilution, which was done by successively doubling the concentration, from 1/200 to 4/200. The COD of the wastewater was determined everytime the feed tank was filled. Table 1 provides the measured influent COD values for each dilution, which also characterizes each period.

3.2   Organic matter removal

The results from the comparison experiments are presented in Figure 1, which represents the totCOD values in the effluent per sampling day, and considers the amount of particulate organic matter (TSS) present in the effluent.

Starting from Period A, the Moving Bed Biofilm Reactor shows very high totCOD removal performances. AS lower efficiencies could be explained by the low biomass concentration, observable in Figure 2 (average mixed liquor suspended solids (MLTSS) of 980 mg/L in this period) in the reactor and poor settleability resulting in too much biomass being lost with the effluent and thus contributing to relatively high effluent totCOD values. New biomass was then added into the AS reactor, upping its MLTSS concentration to an average value of 3 490 mg/L for the subsequent periods (Fig. 2).

In Period B a slight increase in performance is observed as concerns both the AS and the MBBR. The totCOD of the wastewater entering the system was averaging roughly 1 397 mg O2/L (Table 1) with a very high removal efficiency period from day November 23rd to December 2nd. Despite the increase in the organic load, the reactors were able to adequately degrade the pollutants, resulting in effluent totCOD values which were not much higher than in the previous period.

Another increase in the organic load led to Period C of the experiment. The MBBR performance remained fairly constant, while the AS did not. It is also important to note that throughout the entire experience, the AS reactor operated with a much higher average suspended solid concentration (Fig. 2) when compared to the MBBR reactor (Fig. 3).

Table 1.   Influent wastewater total COD.

Dilution	Period	Date	Average (ST.D*)	Min	Max
1/200	A	4/11 to 21/11	582 (± 65)	462	647
2/200	B	21/11 to 5/12	1397 (± 131)	1277	1572
4/200	C	5/12 to 14/12	2646 (± 276)	2303	2874

*Standard Deviation.

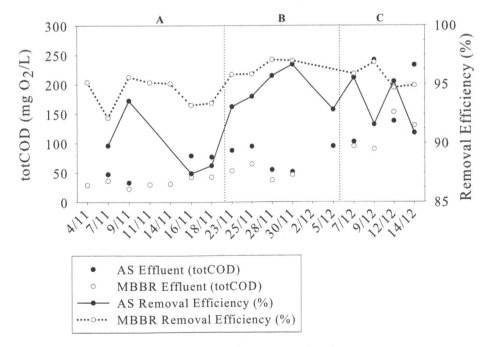

Figure 1.   Comparison of COD values in the effluent per sampling day.

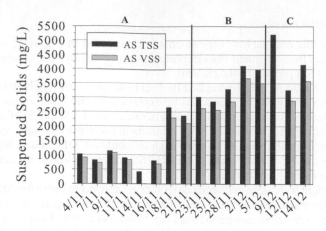

Figure 2. Total suspended and volatile suspended solids in the AS reactor.

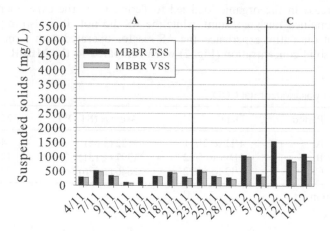

Figure 3. Total suspended and volatile suspended solids in the MBBR.

Table 2. Effluent wastewater characterization.

Efficient	Parameter	Units	Average (ST.D*)	Min	Max	Efficiency
AS	totCOD	mg/L	103 (± 66)	33	241	92.6%
	TSS	mg/L	60 (± 20)	7	88	–
MBBR	totCOD	mg/L	60 (± 39)	23	152	95.2(%)
	TSS	mg/L	10 (± 6)	3	27	–

*Standard Deviation.

The **MBBR** total biomass is composed of the suspended biomass and the biomass within the carrier, the latter's values being equivalent to an average of 851 mg/L.

Ultimately, this means that even though the **MBBR** operated with lower total biomass, it did not compromise the effectiveness of the degradation process, which confers high activity for the biofilm present in the carriers. In fact, having continuous **MBBR**s operating with lower suspended biomass concentrations, gives the possibility of having smaller clarifiers, without compromising the treatment quality. Moreover, the simplicity of the process, could offer the opportunity in reducing investment and operational costs.

Table 2 gives an overview of the data obtained for each parameter in respect to the final effluent of each reactor as well as for the totality of the experiment.

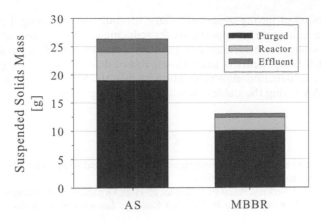

Figure 4. Comparison of the produced solids.

3.3 *Sludge production*

Total solids production was calculated in a period in which stable or pseudo-stable conditions were found, which was from November 18th to December 2nd. Figure 4 represents the solids production in each reactor, broken down per purged solids, which are the solids removed through sampling and wasting, broken down per difference in the reactor solids between the first day and the last day of the abovementioned period, and per solids contained in the effluent (the total mass of suspended solids discharged inthe treated effluent).

The total produced solids for AS and MBBR was 26.26 and 13.03 g TSS, which is equivalent to 1.88 and 0.93 g TSS/d, respectively. If we relate the solid production per amount of substrate removed then we have 0.25 and 0.12, g TSS/(g removed COD), respectively. The daily amount of COD entering the system was calculated by the average of the totCOD values in the respective period multiplied by the flow rate (7 L/d), which was 8.26 g totCOD/d. The removed totCOD was calculated by the average efficiencies for each system in the abovementioned period.

Aygun et al. (2008) studied the production of solids in a MBBR with an average of 2.85 g COD/d and found that the MBBR produced 0.35 g TSS/d which is equivalent to 0.12 g TSS/(g removed COD), the same value found for the MBBR in this experiment. This validates the values found in the present work. The value found for the MBBR is certainly lower than the common values for the activated sludge process.

Decreasing the amount of excess sludge produced and, therefore, the amount of excess sludge disposed of could bring considerable economic advantages, lowering the costs of wastewater treatment plants and reducing the environmental burden of biological treatment plants.

4 CONCLUSIONS

Our findings lead us to conclude that both systems offered very good treatment capabilities for a dairy industry effluent. Indeed, in terms of the organic removal capabilities, both technologies showed similar performances, with the MBBR being slightly better.

With a HRT of 12 hours, the two reactors were able to provide the necessary conditions for a direct discharge onto the aquatic environment, in accordance with the current Portuguese legislation (Decree-law No. 236/98, 1 of August), with an influent totCOD concentration of up to 2 647 mgO_2/L (average). However, during the period of higher organic load (period C) the effluent totCOD concentration approached the maximum discharge values (150 mg O_2/L).

The MBBR offered more simplicity in the treatment process, by having the sludge recirculation line removed. Moreover, operating with lower suspended biomass concentration, grants the possibility of having smaller clarifiers, without compromising the treatment quality. These reasons constitute an opportunity for reducing investment and operational costs.

Considering treatment performance in terms of the quantity of biomass present in the reactor, the MBBR brings benefits, since it is capable of dealing with very high loads, without necessarily meaning higher biomass production. In fact, the MBBR produced 50% less excess sludge than the AS during the stable conditions of the experiment.

REFERENCES

APHA. 2005. Standard methods for the examination of water and wastewater 21st edn. Washington, DC: American Public Health Association.

Aygun, A., Nas, B., and Berktay, A. 2008. Influence of High Organic Loading Rates on COD Removal and Sludge Production in Moving Bed Bioflm Reactor. *Environmental Engineering Science* 25(9): 1311–1316.

Di Trapani, D., Mannina, G., Torregrossa, M., & Viviani, G. 2010. Quantification of kinetic parameters for heterotrophic bacteria via respirometry in a hybrid reactor. *Water Science and Technology* 61(7): 1757–1766.

Egemen, E., Corpening, J., & Nirmalakhandan, N. 2001. Evaluation of an ozonation system for reduced waste sludge generation. *Water Science and Technology* 44(2–3): 445–452.

Forrest, D., Delatolla, R., & Kennedy, K. 2016. Carrier effects on tertiary nitrifying moving bed biofilm reactor: An examination of performance, biofilm and biologically produced solids. *Environmental Technology* 37(6): 662–671.

Kulikowska, D., Klimiuk, E., and Drzewicki, A. 2007. BOD5 and COD removal and sludge production in SBR working with or without anoxic phase. *Bioresource Technology* 98(7): 1426–1432.

Ødegaard, H. 2004. Sludge minimization technologies—An overview. *Water Science and Technology* 49(10): 31–40.

Anaerobic digestion impact on the adaptation to climate change in São Tomé and Príncipe

J.F. Pesqueira, M.F. Almeida & J.M. Dias
Faculty of Engineering, University of Porto, Porto, Portugal

D. Carneiro, A. Justo & M.J. Martins
Ecovisão—Tecnologias do Meio Ambiente, Porto, Portugal

ABSTRACT: The present study evaluated a pilot scale project's impact—"Bioenergy in São Tomé and Príncipe (STP)", to install small anaerobic digesters—on greenhouse gases emissions. At STP, wood is a subsistence fuel source and it is imperative to develop more sustainable alternatives to it and for waste management. For this study, 3 communities were chosen: *Santa Jenny*, *Novo Destino*, and *Mendes da Silva*. Emissions were estimated before and after the implementation using the Intergovernmental Panel on Climate Change's methodology for national inventories, considering the categories: Energy; Agriculture, Forest, and Other Land Use; Waste. Under the established conditions and for said population and sectors, a reduction of 20% of yearly emissions is expected. The use of human waste could increase the reduction to 58%. Expansion to the remaining families of these communities could imply a 97% diminishment and to other similar communities, a 25% reduction of national emissions was predicted.

1 INTRODUCTION

Similarly to what happens in other isolated developing countries, São Tomé and Príncipe's overall environmental situation can be characterized as good, when compared to strongly urbanized and industrialized areas. Nevertheless, anthropogenic impact is a reality, and most of it is caused by habitat destruction due to forest exploitation and consequent deforestation, along with the absence of systems for solid waste and wastewater management, which not only threatens biodiversity, but also public health (RDSTP, 2012).

These environmental impacts are mainly the result of an attempt to find a means of subsistence in a developing country where the economical level, especially in rural areas, is low. At São Tomé and Príncipe, out of the total population of 178 739 inhabitants (in 2012, according to their National Statistics Institute), only approximately 49% of the family aggregates have access to electricity services. Such services present a coverage of 62% on urban areas and 34% on rural areas. Merely 30% of the families have access to suitable sanitation systems. Consequently, 78% of the families use coal and fuelwood to prepare their meals, and only 7% adequately dispose their waste on containers from the public collection system. The inappropriate sanitation levels are associated with an increase of diseases, such as malaria and diarrheic pathologies, which to this day, are still one of the main causes of mortality (RDSTP, 2012).

This country is particularly vulnerable to climate change as it is an insulate state of reduced size, with fragile ecosystems and a strong anthropogenic pressure on their limited land resources (RDSTP, 2015). A temperature increase of about 1.15°C has been registered between 1951 and 2010, along with a decrease in rainfall of 1.7 mm year^{-1}, with a small increase during the dry season. Extreme events such as floods have been registered as well, along with coastal erosion, increase of the savannah area and sea level rise. These events affect important sectors such as the agricultural sector, which is of major worth on developing countries, since population's subsistence depends on it (RDSTP, 2012). Mitigation and adaptation techniques must be established to allow country growth and to fight poverty, without generating negative contributions for climate change. STP's current contribution, in terms of equivalent CO_2 is, up to this moment, negative. That is, the country's uptake of equivalent CO_2 is higher than its emissions. Nonetheless, from all the evaluated sectors, the energy sector is the responsible for the biggest fraction of national greenhouse gases (GHG) emissions, mainly due to the residential subsector's use of fossil fuels. The second most representative sector is the Forest sector, due to the use of fuelwood (RDSTP, 2012). Therefore, it is important to diversify the countries energetic mix, simultaneously guaranteeing that the established alternatives are more sustainable at an economic, environmental and social level. In parallel, actions must be taken to improve the population's health; which, no matter what the strategy is, must include an adequate waste management plan, which presents high impact on health and on the environment.

Organic waste is generated in both urban and rural areas, in developed and developing countries, since it's a result of human and animal life (Seadi, 2008). On developing countries, it's the most representative type of waste (mass-wise). The recovery of biodegradable organic waste might be achieved through different ways. If the intention is to take advantage of the organic and nutrient content of the waste, aerobic biodegradation, or composting, is the best alternative. If the intent is to generate a fuel, anaerobic digestion is the best option (Goodwin, 1997).

On STP, it's important not only to manage and treat the waste but to generate energy for the most isolated communities. Using bioenergy, generated by materials other than wood, appears, in addition, as an adequate mitigation measure towards GHG emissions. Waste management can be integrated with its production by the collection and management of the organic waste via anaerobic digestion.

2 THE PROJECT: BIOENERGY IN SÃO TOMÉ AND PRÍNCIPE

Operating since 2014, the project "Bioenergy in São Tomé and Príncipe" aims to test the viability of applying anaerobic digestion for the treatment of organic waste of communities in rural areas. At the moment, it's in a pilot phase, being tested on the island of São Tomé. Ecovisão, an Environmental Technology company, is responsible for the execution and implementation of the project, in partnership with São Tomé and Príncipe's Directorate General of Environment. It's supported by Camões—Institute for Cooperation and Language, and the Portuguese Environment Agency, with funding from the Portuguese Carbon Fund (DGA, 2013).

Its main aim is to improve life quality of small rural communities of the island by providing a more sustainable energy alternative, and to preserve the forest area of STP, namely, the Obô Natural Park. This park has several communities living on its fringes and threatening it due to their dependency on fuelwood. Forest exploitation is not controlled because authorities are incapable of inspecting the whole area, due to a shortage of people and means.

On the pilot phase, the intention was to install small scale unities on communities, with their active participation. Two manuals were also elaborated: a Management, Operation and Maintenance Manual, and one about constructing anaerobic digesters using local resources. The population also received training to raise their awareness on environmental issues and to promote anaerobic digestion. The communities were chosen based on visits to the communities and the following criteria: small, isolated, poor rural communities; no current (or predicted at a short/medium term) access to the public electricity grid and solid waste management systems; significant agriculture and animal production (so the waste generated by these activities can be used in the anaerobic digestion process); number and size of the family aggregates (to evaluate the amount of produced waste); access to the community; organization of the houses (to facilitate the digesters and grid implementation); proximity to the Obô Natural Park and its buffer zone.

A total of 33 communities were visited by the Ecovisão and DGA teams and 4 communities were chosen: *Santa Jenny* (*Lembá* district), *Dona Augusta* (*Caué*), *Novo Destino* (*Mé-zóchi*), and *Mendes da Silva* (*Cantagalo*). *Dona Augusta* was later excluded, due to low motivation of the community and waste production. Afterwards, the families of the 3 remaining communities were inquired and characterized. The project and process were explained to the families and the community leader. Each family then received a bag to collect the organic waste that was produced and

278

that they were willing to send to the digester (domestic, agriculture, animals), during one week on July 2015 (dry season). The production of waste per capita was quantified, and this information, along with a description of the waste sources and family size and activities can be found in Table 1.

None of the communities accepted to use domestic human waste (i.e., urine and feces) for biogas production. It's necessary to highlight the fact that part of the waste from meals and agricultural practices is used to feed the animals or as manure. Thus, the waste per capita reported represents the amount of available waste (discarded) for the process, and not total waste.

Interviews have shown that oil is the main fuel used in term of general energy production. When it comes to cooking, the main fuel is fuelwood. Fuelwood is used exclusively in most cases (68% of families in *Mendes da Silva*, 62% in *Novo Destino* and 81% in *Santa Jenny*) or combined with oil (approximately 1 L/meal, as indicated by some families), mainly during the cold rainfall-rich period (243 days/year). Combustion is made on open fireplaces with no emission control.

Table 1. Data obtained for each community.

Parameter\Community		*Mendes da Silva*	*Santa Jenny*	*Novo Destino*
Population		147	62	160
Human waste	Pit Latrine	32	0	30
	Forest	2	16	4
Organic Domestic Waste	Waste management	Animal feeding (62%), Forest (35%)	Animal feeding (50%), Forest (44%)	Animal feeding (23%), Forest (65%)
Agricultural activity	Main crops	Banana, Cocoyam, Cocoa	Banana, Cocoyam, Cocoa, Breadfruit	Banana, Cocoyam
	Waste management	Forest (67%), Manure (21%), Animal feeding (5%), Burnt (3%), Nothing (3%)	Forest (39%), Manure (36%), Owned land (21%), Animal feeding (4%)	Owned land (16%), Manure (40%), Forest (40%), Burnt (4%)
Animal production	Number	Chicken (218) | Pigs (70) | Goats (30) Ducks (30)	Chicken (105) | Pigs (59)	Chicken (112) | Pigs (79) Goats (14) | Ducks (60)
	Roaming area	Closed; Most of the day	Closed; Most or part of the day	Closed (70%), Open (30%)
	Waste management	Forest (100%)	Forest (58%), Manure (25%)	By the pen (43%), Forest (22%)
Organic waste collection	Participants	97	41	107
	kg waste/ week	890	208	597
	kg waste/ person/day	1.88	1.05	0.80
	Type of waste	50% banana peels, figs e 50% pig feaces	100% Banana peels	50% banana peels, jackfruit, cocoyam, chayote leaves; 50% pig feces

Table 2. Selected solutions for the different communities.

Community	Construction	Number of digesters	Processing capacity (kg/day)	Generated biogas (m^3/day)
Mendes da Silva	Local	1	220	11
Santa Jenny	Pre-fabricated	1	94	5.1
Novo Destino	Pre-fabricated	1 + 2	20 + 60	1.1 + 3.4

Table 3. Fuel use per day, type and community.

Community	Oil (m³/day)	Fuelwood (t/day)
Santa Jenny	1.5	48
Mendes da Silva	7.8	138
Novo Destino	6.8	108

At this phase of the project, although all families will contribute by sending their waste to the digester, biogas production will be small and only some families will use it. This is due to the fact that initially, bacterial communities need to develop and get established and thus, as the amount of microorganisms is small, waste is slowly degraded and biogas generation is also small. Later on, biogas production will be approximately stable. The families that will use the biogas were selected and, after collecting the necessary data, the Ecovisão and DGA teams initiated the installation of the digesters and the gas grid with help from the population. Pre-fabricated units were installed on *Santa Jenny* and *Novo Destino*, whereas in *Mendes da Silva*, due to the population's high motivation, a unit was built locally. *Novo Destino* received 2 extra units, because of the exclusion of *Dona August* community and to benefit as many families as possible. Table 2 shows the selected solutions. Daily biogas production was based on each waste type potential for biogas production according to Teodorita El Seadi (2008) and its mean composition.

3 METHODOLOGY FOR THE ANALYSIS OF THE IMPACT OF THE PILOT PROJECT ON GHG EMISSIONS

The Intergovernmental Panel on Climate Change's (IPCC) methodology for the construction of national GHG inventories was used for the determination of GHG emissions before and after the implementation of the project, per its good practices.

First, the following key categories were identified: Energy; Agriculture, Forest, and Other Land Use; Waste. Then, the Tier 1 methodology was chosen, due to the quality and quantity of the data available for the analysis of the pilot project. Tier 1 was also used by STP for the determination of their national inventories. Results were expressed in Mg to ensure transparency, and their uncertainties were determined by partially applying IPCC's Approach 1, using IPPC's standard uncertainties when the real ones were unknown. Uncertainties were only determined for the original scenario (where only some families from the 3 communities used biogas for cooking and domestic human waste was not used to produce biogas).

For the determination of waste from agricultural practices, Food and Agriculture Organization of the United Nations (FAO) productivity levels for STP and surrounding areas were used. The percentage of crops that would be discarded was determined from FAO's statistics. Agricultural area size was not measured; therefore, it was considered that it was, percentage-wise, directly proportional to the percentage of the country's agricultural area. Fuelwood use was estimated by weighting some families' fuelwood use per meal. Oil consumption was considered to be 1 L/meal as stated by some families (previously referred). Table 3 sums up fuel use.

4 RESULTS

Table 4 shows the results for estimates of GHG emissions before and after the project, considering: without using human waste (use of all organic waste collected by the community, except for human waste); using human waste; and using all available waste (i.e. considering those used for other activities, and not just the waste the families collected) so all families could use biogas for cooking. It should be emphasized that, with exception of the energy sector, different "before" conditions exist between scenarios since in the initial scenario, the contribution of human waste, was not considered.

The analysis of Table 4 shows that the project will have a positive impact in terms of GHG emission reduction. Considering the present situation, its full implementation has the potential to reduce emissions in about 126.79 Mg of CO_2 equivalent yearly (approximately 20% of the communities' total emissions for the evaluated sectors). Considering a production of 1.224 kg /person/ day of human waste (Polprasert C., 2007), which was either disposed on unmanaged dump sites (open defecation on the forest) or pit latrines, it can be concluded that if this waste type was considered, the emissions would increase in about 5.7%; however, because a total of 12.8 m^3 of biogas could be produced and used by 14 additional families (considering a production 0.028 m^3 biogas/ kg human waste (Buxton, 2010)), the reduction of GHG emissions could be 58%.

To enable biogas use by all the families, all the waste, even that which the families didn't send for the digester because an additional use was given, would need to be collected, meaning that the emissions of GHG from waste would be null. The necessary amount of biogas was estimated based on the biogas used per family in each community. It was considered that all families would use solely biogas to cook. The results are on Table 4 and show that if the project was extended to all families, the GHG emission for the analyzed sectors could be reduced by about 97% – 651.26 Mg of CO_2 equivalent. The remaining emission of 16.91 Mg CO_2 equivalent is the result of the combustion of biogas and use of the digestate as a fertilizer.

If the project was extended to all the communities with no access to the water and power network in São Tomé and Príncipe, under a similar success rate, a meaningfully reduction of the country's emissions could be achieved. Assuming that the other communities from the same district have a similar behavior in terms of GHG emissions and biogas production potential (for example, the community of *Mendes da Silva*, from the *Mé-zóchi* district is representative for the GHG per capita of this district, and the total districts' emission is obtained by extrapolation), the reduction of emissions was estimated, too. The results are presented in Table 5.

The extension of the project to these communities could imply a reduction of GHG emissions of 49.44 Gg equivalent CO_2, corresponding to 98% of the evaluated sectors emissions.

Table 4. GHG emissions inventory per community, considering the original scenario, the use of human waste and the use of all available waste.

Equivalent CO_2 (Mg)

Key Sector\Community	Without using human waste		Using human waste		Using all available waste	
	Energy					
	Before	After	Before	After	Before	After
Mendes da Silva	37.42	31.36	37.42	29.91	37.42	0.14
Novo Destino	31.20	24.23	31.20	16.49	31.20	0.68
Santa Jenny	9.56	8.65	9.56	7.83	9.56	0.06
Total	78.18	64.24	78.18	54.23	78.18	0.88
Agriculture, Forest and Other Land Use						
Mendes da Silva	252.30	212.48	252.30	183.33	252.30	7.22
Novo Destino	198.36	153	198.36	91.34	198.36	5.07
Santa Jenny	89.80	75.84	89.80	58.66	89.80	3.74
Total	540.45	44.32	540.45	333.33	540.45	16.03
Waste						
Mendes da Silva	6.98	0	23.99	0	23.99	0
Novo Destino	3.23	0	20.68	0	20.68	0
Santa Jenny	3.50	0	4.86	0	4.86	0
Total	13.72	0	49.54	0	49.54	0
Total (3 categories)	632.35 ± 235.06	505.56 ± 189.54	668.17	387.56	668.17	16.91

Table 5. Estimates on GHG emissions before and after project implementation on all communities with no access to the energy grid.

District	Communities (No.)	Population	Gg CO_2 equivalent, Before	Gg CO_2 equivalent, After
Água Grande	1	61	0.11	2.92×10^{-3}
Cantagalo	26	11 168	23.83	0.56
Caué	27	4 976	8.75	0.24
Lembá	17	1 007	1.59	0.06
Mé-Zóchi	42	10 477	16.39	0.38
Total			50.67	1.23

Total national GHG emissions on 2005 were 196.63 Gg equivalent CO_2, which means that such reduction is equivalent to about 25% of their national inventory.

5 CONCLUSIONS

The Bioenergy project has shown a high potential not only as a tool to reduce GHG emissions in São Tomé and Príncipe, but also to improve its population's social and health status, capacitating them with tools to become more resilient to the effects of climate change. The implementation of the pilot phase (considering the limitation of not using human waste to generate biogas for cooking) can allow for a reduction of 126.79 Mg of equivalent CO_2 per year (20% of the total, for the sectors that were analyzed). If human waste was used, the estimated reduction would be of 58%, to which human waste contributes with about 280.61 Mg of CO_2 equivalent. The expansion of the project to the remaining families from these communities, using all waste available and not only that which the population has collected at the project, could diminish their emissions by 97% - a reduction of 651.26 Mg of CO_2 equivalent. Expansion of the project to other communities which have no access to the energy grid could allow for a reduction of 25% of the GHG emissions of STP, corresponding to 49.44 Gg of CO_2 equivalent.

ACKNOWLEDGMENTS

The project is framed, in terms of Climate Change, by the Portuguese Initiative for Immediate Implementation ('Fast Start') and is funded by the Portuguese Carbon Fund. The Camões Institute for Cooperation and Language (Camões, I.P.) and the Portuguese Environment Agency were responsible for the technical monitoring of the plan's execution. J.M. Dias thanks Project UID/EQU/00511/2013-LEPABE (FEDER,COMPETE2020) and national FCT funds.

REFERENCES

Buxton, D. 2010. From Small Steps to Giant Leaps...putting research into practice. *EWB-UK National Research Conference 2010*. The Royal Academy of Engineering.
DGA 2013. *Apresentação de Propostas de Programas, Projetos ou Acções de Cooperação em Geral. Proposta de Financiamento de Projecto—Bioenergia em São Tomé e Príncipe (STP): Aproveitamento energético de biogás*. Direção Geral do Ambiente de São Tomé e Príncipe.
Goodwin, N.R. 1997. *The Consumer Society*. 269–299.
Polprasert, C. 2007. *Organic Waste Recycling*. London: IWA Publishing.
RDSTP 2012. *Deuxième Communication Nationale. Convention-Cadre des Nations Unies sur les Changements Climatiques*. São Tomé and Príncipe.
RDSTP 2015. *São Tome and Principe-Intended Nationally Determined Contribution*. São Tomé e Príncipe: Democratic Republic of São Tomé e Príncipe.
Seadi, T.A. 2008. *Biogas Handbook*. Denmark.

WASTES – Solutions, Treatments and Opportunities II – Vilarinho, Castro & Lopes (Eds)
© 2018 Taylor & Francis Group, London, ISBN 978-1-138-19669-8

Garden waste quantification using home composting on a model garden

T. Machado, B. Chaves & L. Campos
Lipor—Serviço Intermunicipalizado de Gestão de Resíduos do Grande Porto, Porto, Portugal

D. Bessa
Escola Superior de Tecnologias da Saúde do Porto, Porto, Portugal

ABSTRACT: Garden waste produced at households can be treated by home composting, but very little information can be found about this subject. On this study conducted by Lipor a model of a house garden was used to assess the amount of garden waste produced while also measuring a compost bins capacity to treat it. The results can then be used to assess the impact of home composting on the deviation of organic waste from municipal collection systems. The study concluded that a garden can produce from 0.95 kg/m^2 to 3.24 kg/m^2, while a compost bin can admit an input of 204.68 kg of garden waste per year.

1 INTRODUCTION

Household composting can be a method of preventing waste production as organic matter is used on the same place it is produced (Lipor, 2013). It is estimated that this process can accommodate up to 50% of municipal organic waste in Europe (Vásquez & Soto, 2017). By treating organic matter (kitchen and garden waste) as a separate waste flow, a rich compost can be produced and used directly on gardens and planters has a soil ammender and fertilizer (Tatáno, et al., 2014).

In northern Portugal a waste management system composed of 8 municipalities was able to divert more than 3000 t of organic waste through a household composting scheme with 9300 compost bins distributed in as many households (Lipor, 2013). Currently, the kitchen waste production can be assesd by monitoring visits done by composting masters to households—were they weight the amount of kitchen waste produced weekly—or by volunteer communication of values by the composting scheme participants. Due to the special characteristics of garden waste, the weighting of this matter was difficult to make and was never assessed correctly. This issue raised a problem as it was impossible to gather reliable data to be used in the waste prevention indicators calculation Furthermore, little information can be found regarding the amount or compositon of garden waste (Boldrin & Christensen, 2010). In order to collect information that mirrored the household reality and clearly assess the amount of waste produced by a garden Lipor conducted a study in it's waste management facility, particullary at Lipor school for household composting—the Horta da Formiga. The aim of this study was to quantify the production of garden waste from a model household garden as well as to acess the capacity of a commercial 300 L compost bin to manage the garden waste.

2 METHOD

2.1 Defining a model

To quantify the waste production on house gardens we needed to model the same system used at citizen's houses and apply the same technics used by them. The study of this model had to be conducted for at least one year to accommodate the garden waste production fluctuations.

Table 1. Study areas characterization.

Garden	Area (m²)	Characterization
1 – small garden	45	Lawn, bush hedge and a tree (6 m)
2 – medium garden	200	Lawn, bush hedge, tree (6 m) and rose garden
3 – big garden	379	Lawn, bush hedge at all perimeter, 2 trees (6 m)

For the model design, we analyzed the data collected from the different monitoring visits to the households of composting scheme participants. From this data 3 different types of gardens and average garden areas could be determined—corresponding to small, medium and big gardens -, thus representing the main types of gardens that exist on this region.

2.2 *Defining the areas of study*

At Lipor's Horta da Formiga, a green area of 1,5 hectares, the areas meant to be used as a model for the house gardens were defined as follows:

All the organic waste produced in those gardens was collected and deposited on an assigned site for garden waste deposition. The materials where then prepared to be deposited on a 300 L compost bin. This preparation consisted on shredding or aerating some of the materials and weighting the garden waste to assess its production for each garden type and measuring the amount of garden waste admitted by the compost bins. It should be noted that the composting and garden waste deposit area was located on a site where all good composting conditions were provided.

To determine the maximum capacity of a commercial compost bin a fourth composter was also used and kept in full capacity for an entire year—using garden waste from Horta da Formiga. To assure the quality of the process, and mimic the technics and characteristics of the products used at household composting, kitchen waste was added every week to each of the 4 compost bins.

The process of collecting and composting the organic waste produced on the gardens under study occurred for 12 months, from April 2015 to March 2016, and all the corrections on humidity and aeration of the composters were done by a team of expert composters.

3 RESULTS

3.1 *Results on garden waste production model gardens*

Table 2. Garden waste production results by garden type.

Garden	Garden waste production (kg)	Area	Garden waste production (kg/m²/yr)
1	146.01	45	3.24
2	205.23	200	1.03
3	358.77	379	0.95

3.2 *Results on garden waste composting capacity of a 300L compost bin*

Table 3. Amount of organic waste deposited on compost bins.

Composter	Garden waste	Kitchen waste	Total
1	163.74	165.41	329.15
2	218,34	122	340.34
3	286.79	113.9	400.69
4	204.68	182.92	387.6

4 DISCUSSION

4.1 *Garden waste production on a model house garden*

The results of the study confirm the differences on garden waste production by square meter between gardens with different dimensions. Garden 3 produced only 2.4 times more garden waste then garden 1. Nevertheless, garden 3 is 8 times bigger than garden 1. This correlation between garden size and waste production per square meter is not linear and does not rise with the garden size as expected. A bigger garden tends to have a lower value for $kg/m^2/year$ than a small garden. This might be explained by analysing the area in study where we can observe the same kind of plant species on the same gardens. On a small garden the waste concentrates because we have a small area, while in a big garden it is diluted on the area value.

We also observed that a space with 379 m^2 produces over 300 kg of garden waste per year, reaching a value of 0.95 $kg/m^2/year$.

4.2 *Garden waste composting capacity of a 300L compost bin*

When analyzing the data collected from the compost bin 4 it was verified that it admitted 387.6 kg of waste (garden and kitchen) over a year. However, when the values of composter 3 (which admitted residues from Garden 3) were observed it was found that the latter admitted 399.98 kg of waste (garden and kitchen). It can be can said that the maximum value allowed in a 300 L compost bin is 400 kg/year of organic waste, when a well-managed process is applied.

The difference between composter 4 (which has always been maintained at maximum capacity) and composter 3 may be because composter 3 has admitted large amounts of grass, while composter 4 received less. This could speed up the composting process in composter 3, thus allowing greater waste disposal. Also, the greater weight of grass (due to the high-water content) may have contributed to this difference.

5 CONCLUSIONS AND FUTURE RECOMMENDATIONS

5.1 *Conclusions*

This work allowed to clarify some critical points in the evaluation of the production of garden waste. It is concluded that:

- There are differences between the production of garden waste in small gardens (<45 m^2) and large gardens (>150 m^2). Smaller gardens have a waste production value per square meter ($kg/m^2/year$) higher than larger gardens;
- A garden with an area of over 300 m^2 produces more than 300 kg of garden waste per year;
- A 300 L compost bin serves up to a 300 m^2 garden;
- A 300 L compost bin requires approximately 150 kg/year of garden waste for its correct operation;
- A compost bin of 300 L has a great potential for deviation of organic waste from the waste management circuit and, consequently, from incineration or landfill.

This study allowed to assess a value for garden waste production for a model house garden and the capacity of a compost bin to treat and manage the garden waste produced. With the value found for garden waste prevention indicators can be better calculated, offering a closer view to the impact of home composting to the prevention of waste production at households in northern Portugal.

Finally, we must also mention that the gardens analysed (in the Horta da Formiga) are spaces that follow the philosophy of sustainable gardening—spaces where the production of garden waste is minimized—so the value of the actual production of green waste in the Lipor region may be slightly larger. This does not invalidate the study and provides an information base on the minimum production of garden waste in the Lipor intervention area.

5.2 *Future recommendations*

The values found with this study may be of great use to other waste management systems, municipalities or companies that need to assess the impact and waste deviation that is possible through composting, reporting upgraded prevention indicators.

For a correct evaluation of composting impact, we should correlate the size of the garden the compost bin is placed on with the production of garden waste by creating a scale of garden waste production and garden size as follows:

- <45 m^2 – 3 kg/m^2/year;
- >45 up to 200 – 1 kg/m^2/year;
- >200 up to 1000 – 0.95 kg/m^2/year;
- >1000 – 0.6 kg/m^2/year.

REFERENCES

Boldrin, A. & Christensen, T.H. 2010. Seasonal generation and composition of garden waste in Aarhus (Denmark). *Waste Management*: 30.
Lipor 2013. Sustainablility report. Available in http://www.lipor.pt/en/libraries/.
Tatáno, F., Pagliaro, G., Floriani, E. & Mangani, F. 2014. Biowaste home composting: Experimental process monitoring and quality control. *Waste Management* 38.
Vásquez, M.A. & Soto, M. 2017. The efficiency of home composting programmes and compost quality. *Waste Management*.

Acid esterification vs glycerolysis of acid oil soapstock for FFA reduction

E. Costa, M. Cruz, C. Alvim-Ferraz, M.F. Almeida & J.M. Dias
LEPABE, Faculty of Engineering, Oporto University, Porto, Portugal

ABSTRACT: In this study, the efficiency of esterification vs glycerolysis for the reduction of Free Fatty Acid (FFA) content of acid soapstock oil (128.8 mg KOH g^{-1}) was studied. Further, alkaline transesterification of the pre-treated oil mixed with sunflower oil (to achieve FFA content <2%) was conducted to maximize Fatty Acid Methyl Esters (FAME) production. Homogeneous acid esterification was conducted at 65°C, using 2.0 wt% H_2SO_4 as catalyst and 9:1 molar ratio of methanol to oil. Glycerolysis was performed at 200°C using a 6:1 molar ratio of glycerol:oil. Both reactions lasted 6 h. Final acid value of the product was around 10 mg KOH g^{-1} for both processes, showing negligible differences between them. After transesterification of the mixtures it was possible to obtain biodiesel with around 90% FAME content, thus showing potential for the recovery of such waste material.

1 INTRODUCTION

For humanity to continue economic growth and maintain the high standard of living, energy has become a crucial factor. Globally, the awareness of the energy and environmental problems associated with burning fossil fuels along with their scarceness encourages many researchers to consider alternative energy sources (Atabani *et al.*, 2012). A special attention has been given to biofuels, with particular reference to biodiesel, since it has many technical and environmental benefits over conventional fossil diesel. Main reported biodiesel advantages, are: i) high biodegradability and minimal toxicity; ii) greater lubrication capacity; iii) higher flash point; iv) requires minor modifications to current systems; and v) residual emissions of sulphur and aromatic compounds and other chemicals that afect the environment (Atabani *et al.*, 2012; Nie *et al.*, 2015).

In the production of biodiesel, more than 95% of feedstocks are still edible oils (Borugadda & Goud, 2012), however more than 300 alternative raw materials have been identified which could be used to produce biodiesel, including non-food oils, acid waste oils, used cooking oils and animal fats (Shahid & Jamal, 2011). Taking into account economic indicators, Piloto-Rodriguez *et al.* (Piloto-Rodríguez *et al.* 2014) indicates that more than 75% of the production costs of biodiesel are due to raw materials cost.

The soapstock is a waste from the chemical process used in the refining of vegetable oils, resulting from the neutralization of FFA. Usually, it is acidified with a strong mineral acid, such as sulfuric or hydrochloric acid to produce an acid oil that consists essentially of FFA (Basheer & Watanabe, 2016). The use of soapstock for biodiesel production is therefore a very good alternative for cost reduction of biofuel production and to solve environmental problems such as the competition of land use for food production against energy production (Piloto-Rodríguez *et al.*, 2014).

Biodiesel is conventionally a mixture of fatty acid methyl esters obtained by the transesterification reaction of triglycerides through alkaline catalysis. However, the alkaline catalysis cannot be applied to a material with high FFA (as the acid oil from soapstock) content due to soap formation; the conventional reported limit is 1 wt.% FFA content (Dias et al., 2009). When the aim is to lower the FFA content, two processes are highlighted: acid esterification and glycerolysis. Homogeneous acid esterification by which the FFA present in the oil react with al alcohol to produce esters and water, is highly implemented for the reduction of FFA content of highly acidic raw materials, with the main variables being temperature, catalyst concentration and type, and alcohol: oil molar ratio (Piloto-Rodríguez et al., 2014). The drawback of this process is the use of a highly corrosive reagent (mostly sulfuric acid) and the need to refine the products to separate/recover the catalyst. Glycerolysis allows the esterification by reaction of FFA to obtain glycerides (MG, DG and TG) as well as water. This process implies high temperatures, but can be conducted without homogeneous catalyst addition, thus avoiding expensive purification stages.

The present work compared the acid esterification vs glycerolysis for the pretreatment of an acid oil from soapstock aiming at biodiesel production. The main objectives were: (i) to evaluate homogeneous acid esterification and glycerolysis effectiveness for FFA reduction; (ii) to maximize FAME production after alkaline transesterification of a mixture of the pre-treated acid oils with sunflower oil.

2 EXPERIMENT

2.1 *Materials*

The acid oil from soapstock of vegetable oil refining (mixture of seeds) was provided by the company Nature Light, S. A. Commercial sunflower oil was used. Methanol (Fischer Scientific ≥99%) and glycerol were used as the acyl acceptor. The sulfuric acid used as catalyst had 98% of concentration. All reagents were of analytical grade.

2.2 *Analytical methods*

The physicochemical properties determined in the raw material were acid value and water content. The acid value was determined by volumetric titration, according to NP EN ISO 660 (2009) and it was also determined to evaluate the progression of the esterification reactions. Taking into account the expected values for water content, moisture content was determined by weight loss at $T = 105°C \pm 2°C$, until constant weight, according to EN 12880 (2000).

The methyl ester content in the final product (after transesterification) was determined according to EN 14103 (2003) by gas chromatography, using a Dani Master GC with a DN-WAX capillary column of 30 m, 0.25 mm internal diameter and 0.25 μm of film thickness. The temperature program used was as follows: 120°C was initially selected as the starting temperature, followed by a temperature rise at 4°C per minute, up to 220°C, held for 10 min.

2.3 *Esterification*

All esterification experiments were performed in duplicate and the variation of the results is expressed as the relative percentage difference to the mean (RPD).

2.3.1 *Homogeneous acid esterification*

For the acid esterification reactions, the following reference conditions were selected according to the literature (Dias *et al.*, 2009): 65°C, 2.0 wt% H_2SO_4 catalyst, 6 h and 6:1 molar ratio of methanol to acid oil. In addition, a 9:1 molar ratio of methanol:oil and a 4.0% of acid concentrations were studied, maintaining the other variables.

The setup of the batch reactor and the monitoring of the reaction (0.5 mL of sample from the reaction mixture was collected at different time intervals) was made according to Dias et al. (2009). The amount of sample (acid oil) used for each experiment was 50 g. At the end of the esterification reactions the pre-treated oils were washed to remove excess catalyst and dried for determination of the final real acid value of the product and fatty acid methyl ester (FAME) content.

2.3.2 *Glycerolysis*

The best conditions and the installation of Costa *et al.* (2015) were used (T = 200°C, 3:1 molar ratio of glycerol:oil, 6 h). The use of a 6:1 glycerol:oil molar ratio was also evaluated, maintaining the same temperature. The glycerolysis was also carried out using 50 g of acid oil and 0.5 mL samples were also collected to evaluate the evolution of the reaction.

2.4 *Transesterification*

Synthesis of biodiesel was made with the best result in terms of reduction of acidity obtained by the esterification reactions studied. When conducting a transesterification reaction by homogeneous alkaline catalysis, the acidity limit of 2 wt.% was considered. Thus, the reaction products were mixed with sunflower oil to obtain a raw material with such maximum acidity.

The alkaline transesterification was performed in a batch reactor at 65°C during 1 h (Dias *et al.*, 2009). The amount of catalyst was 1 wt.% NaOH and the methanol:oil molar ratio was 6:1. Acid/water washings and drying procedures also agreed with the mentioned study.

3 RESULTS AND DISCUSSION

3.1 *Raw material properties*

The properties of the acid waste oil are presented in Table 1.

The acid value obtained agrees with those found in the literature, which ranged from 81.9 mg KOH g^{-1} to 129.8 mg KOH g^{-1} (Dias et al. 2009; Shah et al. 2014; Soares, et al. 2013; Su & Wei, 2014). Similar water content was reported by Pérez-Bonilla (2011) (2.2 wt.%) for soapstock oil; however, Su & Wei (2014) studied an acid oil from soapstock with much less water content, of 0.185 wt.%.

3.2 *Monitoring of esterification*

3.2.1 *Homonegeous acid esterification*
Figure 1 shows the results from the acid esterification of the soapstock oil, under the different reaction conditions studied. It should be highlighted that the acid value was determined in the reaction mixtures and the acid concentration and methanol amount will affect such value; for that reason, initial acid values differ and normalized values to the highest initial acid value (difference between initial acidities corrected for all sampling points) are plotted.

The reaction progresses fast until 120 min and then tends to stabilize. The normalized values show negligible differences under the different reaction conditions. Considering the real values, the highest acidity reduction as (Acidity 0 h – Acidity 6 h)/Acidity 0 h 100, in percentage was achieved when the highest methanol to oil molar ratio was used, being around 77%, close to that obtained using 6:1 methanol:oil molar ratio and 2 wt.% catalyst (74%). The lowest reduction, of around 64 wt.% was obtained for the highest concentration studied. Such results differ slightly from what is observed for the normalized values. Taking into account the results variations (RPD <7%), differences are not considered meaningful.

Dias et al. (2009) selected as the best reaction conditions for acid lard acid esterification (initial acid value of 26 mg KOH g^{-1}) 65°C, 2.0 wt% H2SO4, 6:1 molar ratio of methanol to lard and 5 h. In the present study, despite the use of a raw material with a much higher acidity, the results are in agreement.

The FAME content of the final products was very similar in all cases, being between 68.3 wt.% and 69.3 wt.%. Table 2 presents the acid values of the esterification product, after

Table 1. Acid waste oil physicochemical properties determined.

	Acid value (mg KOH g^{-1})	Water content (wt.%)
Result	129.8	2.7
RPD* (%)	0.4	3.5

* Relative Percentage Difference to the mean.

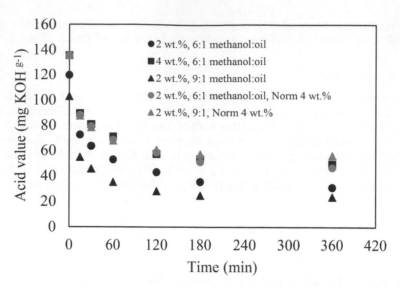

Figure 1. Evolution of the acid value during the esterification of the acid oil (65°C), considering real values and normalized.

Table 2. Acid value of pre-treated oils after acid esterification at the different reaction conditions (H₂SO₄, 65°C).

Reaction conditions	Acid value (mg KOH g⁻¹)	RPD (%)
2 wt.%, 6:1 methanol:oil molar ratio	13.86	2.6
4 wt.%, 6:1 methanol:oil molar ratio	20.71	4.5
2 wt.%, 9:1 methanol:oil molar ratio	11.46	7.4

purification. The lowest acid value was in fact obtained when using the highest methanol to oil molar ratio; for this reason, this oil was used for biodiesel production by alkaline transesterification in mixture with sunflower oil.

3.2.2 *Glycerolysis*

Figure 2 presents the results from the glycerolysis of the acid oil. The glycerol dilution effect affects the acid value of the mixtures but reaction progresses very fast at the beginning leading to high variations for the measurements at time zero; the curves show different behavior, compared to what was observed under different reaction conditions in section 3.2.1, being more accentuated in the case of using the of 3:1 methanol:oil molar ratio. At the end of the reaction period, differences were minor. Overall, taking into account the initial acid values (zero time of reaction), the FFA reductions were 87% and 83%, using 3:1 and 6:1 methanol:oil molar ratios.

The lowest acid value of the product was around 10 mg KOH g⁻¹. It should be refered that Costa et al. (2015) achieved in the first 15 min of glycerolysis reaction a reduction in the acid value of 92% using an oil obtained from an urban wastewater sludge as raw material with 140 mg KOH g1 of initial acid value reaching a 8.21 mg KOH g⁻¹ in the final product under the same conditions evaluated in the present work.

3.2.3 *Alkaline transesterification of oil mixtures*

The biodiesel production was performed using the mixture of sunflower oil with the pretreated oils from both reactions studied. With the objective to obtain one mixture with an

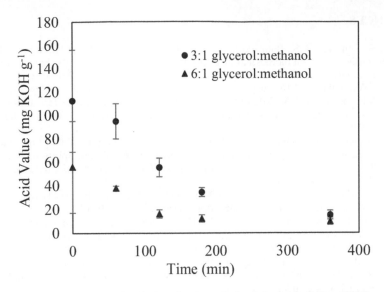

Figure 2. Evolution of the acid value during the glycerolysis of the acid oil (T = 200°C).

Table 3. Fatty acid methyl ester content of biodiesel produced using different raw materials.

Raw-materials	FAME (wt.%)
Sunflower oil	97.6
Acid Esterification Product + sunflower oil	88.2
Glycerolysis Product + sunflower oil	89.2

acid value suitable for biodiesel production (max. 2 wt.%), the following equation was used to determine the amount of each component:

$$\text{Acid Value}_{mix} = \left(\text{Acid Value}_1 * X\right) + \left(\text{Acid Value}_2 * \left(1 - X\right)\right) \tag{1}$$

The acid value of commercial sunflower oil was 0.214 mg KOH g^{-1} and the value used for the Acid Value$_{mix}$ was 4 mg KOH g^{-1} due to reasons explained. FAME content of the final products is presented in Table 3.

According to the values in Table 3, the differences between using the two products are negligible, which agrees with the efficiencies previously observed and the final similar acid values of both products. The obtained FAME content is clearly below the requirements for use as automotive fuel (> 96.5 wt. required).

4 CONCLUSION

In the present study, homogeneous acid esterification and glycerolysis of a highly acidic soapstock oil (128.8 mg KOH g^{-1}) showed potential towards free fatty acid reduction, with a final acid value of the product being close to 10 mg KOH g^{-1} using both processes.

After alkaline transesterification of a mixture of the products with sunflower oil, biodiesel with 90 wt.% fatty acid methyl ester content was obtained. Although FAME content is still below the requirements imposed by EN 14214, the processes show clear potential for the recovery of this alternative waste raw material.

REFERENCES

Atabani, A. E., Silitonga, A.S., Badruddin, I.A., Mahlia, T.M.I., Masjuki, H.H., & Mekhilef, S. 2012. A comprehensive review on biodiesel as an alternative energy resource and its characteristics. *Renewable and Sustainable Energy Reviews* 16(4): 2070–2093. doi: 10.1016/j.rser.2012.01.003.

Basheer, S., & Watanabe, Y. 2016. Enzymatic conversi on of acid oils to biodiesel. *Lipid Technology* 28.

Borugadda, V.B., & Goud, V.V. 2012. Biodiesel production from renewable feedstocks: Status and opportunities. *Renewable and Sustainable Energy Reviews.* 16(7): 4763–4784. doi: 10.1016/j.rser.2012.04.010.

Costa, E.T., Almeida, M.F., Dias, J.M., & Matos, A. 2015. Glycerolysis of two high free fatty acid waste materials for biodiesel production. Paper presented at the Wastes: Solutions, Treatments and Opportunities—*Selected Papers from the 3rd Edition of the International Conference on Wastes: Solutions, Treatments and Opportunities, 2015.*

Dias, J.M., Alvim-Ferraz, M.C.M., & Almeida, M.F. 2009. Production of biodiesel from acid waste lard. *Bioresource Technology* 100(24): 6355–6361. doi: http://dx.doi.org/10.1016/j.biortech.2009.07.025.

Lam, M.K., Lee, K.T., & Mohamed, A.R. 2010. Homogeneous, heterogeneous and enzymatic catalysis for transesterification of high free fatty acid oil (waste cooking oil) to biodiesel: A review. *Biotechnology Advances* 28(4): 500–518. doi: http://dx.doi.org/10.1016/j.biotechadv.2010.03.002.

Nie, K., Wang, M., Zhang, X., Hu, W., Liu, L., Wang, F., Tan, T. 2015. Additives improve the enzymatic synthesis of biodiesel from waste oil in a solvent free system. *Fuel* 146: 13–19. doi: http://dx.doi.org/10.1016/j.fuel.2014.12.076.

Pérez-Bonilla, A., Frikha, M., Mirzaie, S., García, J., & Mateos, G.G. 2011. Effects of the main cereal and type of fat of the diet on productive performance and egg quality of brown-egg laying hens from 22 to 54 weeks of age. *Poultry Science* 90: 2801–2810.

Piloto-Rodríguez, R., Melo, E.A., Goyos-Pérez, L., & Verhelst, S. 2014. Conversion of by-products from the vegetable oil industry into biodiesel and its use in internal combustion engines: A review. *Brazilian Journal of Chemical Engineering* 31(2): 287–301. doi: 10.1590/0104–6632.20140312s00002763.

Shahid, E.M., & Jamal, Y. 2011. Production of biodiesel: A technical review. *Renewable and Sustainable Energy Reviews* 15(9): 4732–4745. doi: 10.1016/j.rser.2011.07.079.

Su, E., & Wei, D. 2014. Improvement in biodiesel production from soapstock oil by one-stage lipase catalyzed methanolysis. *Energy Conversion and Management* 88: 60–65. doi: 10.1016/j.enconman.2014.08.041.

Sweet potato bioethanol purification using glycerol

J.O.V. Silva, M.F. Almeida & J.M. Dias
LEPABE, Department of Metallurgical and Materials Engineering, Faculty of Engineering,
University of Porto, Porto, Portugal

M.C. Alvim-Ferraz
LEPABE, Department of Chemical Engineering, Faculty of Engineering, University of Porto,
Porto, Portugal

ABSTRACT: Bioethanol is obtained from biomass sources (maize, sugar cane and beet) and sweet potato has been presented as a promising alternative for this purpose. Anhydrous ethanol can be obtained by extractive distillation of hydrated alcohol with chemical agents and it is usable for biodiesel production, from which a glycerol byproduct is obtained. The present work evaluates the use of glycerol as dehydrating agent to obtain anhydrous ethanol. With this objective, two grades of sweet potato bioethanol, 88.0 and 94.0°GL respectively, were used in the experiments with three volumetric ratios of 0.25, 0.50 and 0.75 V/V of glycerol/bioethanol and the results compared with those obtained using ethylene glycol under the same conditions. The highest ethanol grade obtained was 96.9°GL using glycerol and 96.6°GL using ethylene glycol, both using a 0.75 V/V ratio. The purity of bioethanol increased with the amount of drag agent used, whereas the yield decreased.

1 INTRODUCTION

The use of biomass based fuels in the global energy matrix to minimize environmental impacts on the planet and gradually replace the use of fossil fuels (coal, oil and natural gas) presents relevant positive aspects (Alper and Oguz 2016). Biofuels (biodiesel and bioethanol) account for a large proportion of these clean energy sources currently produced and increasing investment prospects are expected until 2030 (IEA 2016).

Biodiesel is a mixture of esters that can be produced by various processes (Haas et al. 2006). One of the most used is the transesterification where triglyceride molecules react with an alcohol (methanol or ethanol) in the presence of generally a alkaline homogenous catalyst (NaOH, KOH, their methoxides, among others), generating esters of fatty acids (biodiesel) and glycerol (Dias et al. 2008). According Ardi et al. (2015), for every 100 lbs of biodiesel produced, about 11 lbs of glycerol is generated. Taking into account that, in 2015, the world production of biodiesel was 30.1 Mm^3, it can be stated that relevant amounts of glycerol were generated as a by-product of the biodiesel industry. Therefore, many studies have been developed to add value to this by-product of the biodiesel industry (REN 2016).

The world production of ethanol in 2015 was about 98 Mm^3, with the United States being the largest producer with about 56 Mm^3, followed by Brazil with around 30 Mm^3 (REN 2016). This biofuel is mainly obtained as a first generation product from fermentable sugars sources (sugar cane, sorghum, sugar beet), as well as from starchy sources (corn, wheat, rice and cassava). Sweet potatoes production with yields of 10467 L/ha (Silveira et al. 2007), compared to corn production with 4250 L/ha (Dunn et al. 2013), and to sugarcane with 7224 L/ha (Urquiaga et al. 2005), is looked as a very promising complementary biomass for the production of ethanol.

According to the Brazilian standard no. 7/2011 (ANP 2011), ethanol is classified as Ethanol Hydrated Fuel (EHC), with an alcoholic content between 95.1 to 96.0°GL (or Vol.%) and as Commercial Anhydrous Ethanol (EAC) with a minimum of 99.6°GL. EHC can be easily obtained by fractional distillation, but the maximum purity achieved is 95.6°GL due to the azeotropic mixture between ethanol and water (Silva and Campos 2013). In order to obtain EAC, it is first necessary to submit EHC to a dehydration process to remove residual water.

Among the best known dehydration processes are extraction distillation, vacuum distillation, azeotropic distillation, molecular sieves, adsorption dehydration and diffusion distillation (Kumar et al. 2010). Extraction distillation is one of the most used methods for the dehydration of ethanol. One of the agents used in this process is benzene, one of the first to be used, that due to its carcinogenic effect, promoted the search for new alternatives (Wasylkiewicz et al. 2003), and was soon replaced by others such as toluene and hexane (both obtaining purity of ethanol ≥ 99.6°GL); the latter is the most used at the moment by the industries, although with the disadvantage of its high flammability, that makes its use quite dangerous (Gomis et al. 2005). Ethylene glycol has also been studied for this purpose, mainly because of the ease of reuse, low cost and final purity of ethanol (99.4°GL), but with the disadvantage of needing part of the required energy in the form of high pressure steam (Navarrete-Contreras et al. 2014).

Some studies present glycerol as a promising alternative as a dragging agent for ethanol dehydration, mainly due to its low cost (by-product of biodiesel), and the high ethanol purity obtained (98.9°GL) (Souza et al. 2013). However, no study was found on the integration of biodiesel and bioethanol processes, aiming at using glycerol as a dehydrating agent for the production of anhydrous ethanol. Thus, it is extremely important for the sustainable growth of biofuel production to develop studies that present perspectives of integration of these two processes for obtaining renewable fuels.

The aim of the present work was the use of glycerol as a dehydrating agent in the process of extractive distillation of sweet potato bioethanol with 88.0 and 94.0°GL purity as well as the comparison with ethylene glycol (under the same operating conditions).

2 MATERIALS AND METHODS

2.1 Bioethanol of sweet potato

The alcohol was obtained in a small bioethanol production plant of sweet potato with capacity of 3000 L/day, located in Palmas-Tocantins (Brazil), according to previously established operational conditions (Silveira et al. 2007, Silva 2012), from which an average alcohol yield of 161.4 L/t and purity of 90.0°GL was obtained.

Bioethanol with a lower alcohol content of 88.0°GL was subjected to fractional distillation (Figure 1) directly on a rotary evaporator (Heidolph, Laborota 4000 efficient), under prior

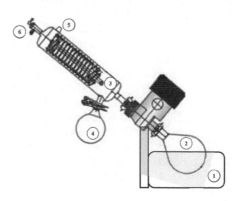

Figure 1. Schematic diagram for the extractive distillation, performed using a Rotary Evaporator under reduced pressure: (3) condenser, (4) final distillate, (5) air outlet with pressure control and (6) exhaust valve.

optimized conditions of 76°C, 600 mbar, 30 rpm and 1 h. The resulting 94.0°GL bioethanol was one of the two alcoholic degrees studied in this work.

The bioethanol grade (°GL) was determined according to NBR5992 (ABNT 2008) using a densitometer (Gay-Lussac/ALLA alcohol, 0–100°GL).

2.2 *Extractive distillation process*

Ethylene glycol with 99.5% purity (CARLO ERBA) and glycerol with 99.5% purity (LABCHEM) were used as drag reagents, considering the work by Souza et al. (2013).

Each experiment was carried out with 150 mL of sweet potato ethanol of 88.0 or 94.0°GL purity and blends with dehydrating agent/bioethanol at volumetric ratios of 0.25, 0.50 and 0.75 V/V of glycerol/bioethanol. Fractional distillation followed the operating conditions described in section 2.1.

3 RESULTS AND DISCUSSION

3.1 *Purification with glycerol*

Figure 2 shows the purity of bioethanol and yield obtained using glycerol as the drag agent. The highest ratio of 0.75 V/V gave the maximum purity values of 96.1 ± 0.1°GL and 96.9 ± 0.1°GL for both 88.0°GL and 94.0°GL grades of bioethanol, as well as the minimum yields of 55.83 ± 0.50 Vol.% and 60.67 ± 0.67 Vol.%, respectively. This tendency of increasing purity with blends richer in glycerol was reported by Souza et al. (2013).

Taking into account the Brazilian standard no. 7/2011 (ANP 2011), the best of the purities achieved was $95.1_{EHC} < 96.9 < 99.6_{EAC}$°GL, thus slightly higher (about 2%) than hydrated ethanol fuel and slightly lower (about 3%) than commercial anhydrous ethanol. In two other studies found (Pla-Franco et al. 2013, Souza et al. 2013), the purities of 96.3 to 98.9°GL were obtained with reflux control, thus indicating the possibility of improving the process. In the present study with no reflux, the results obtained were considered satisfactory.

3.2 *Purification with ethylene glycol*

Figure 3 shows the purity of bioethanol and yields obtained using ethylene glycol as the dragging agent. The maximum purity of bioethanol and minimum yield were obtained for

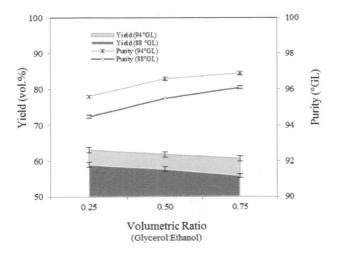

Figure 2.   Mean yield and purity of sweet potato bioethanol (88.0 and 94.0°GL of initial purity) using extractive distillation with glycerol at different volumetric ratios (0.25, 0.50 e 0.75 V/V).

the volumetric ratio of 0.75 V/V being of 95.1 ± 0.1°GL and 96.6 ± 0.1° GL purity as well as 57.33 ± 0.67 vol.% and 62.17 ± 0.83 vol.% yield, for 88.0°GL and 94.0°GL respectively. The tendency of increasing purity with blends richer in ethylene glycol, as in case of glycerol, can be explained by the elimination of the azeotrope with increasing drag reagent, similar behavior was also reported by Navarrete-Contreras et al. (2014).

According the Brazilian standard no. 7/2011 (ANP 2011), the higher purities were also between hydrated and anhydrous fuel requirements, as shown for glycerol (95.1$_{EHC}$ < 96.6 < 99.6$_{EAC}$). Results of 95.3°GL and 98.4°GL were found in two other studies (Navarrete-Contreras et al. 2014, Pla-Franco et al. 2013). In these studies the experiments were done either with reflux control, or using a distillation column with five stages, explaining the higher value in the range obtained.

Table 1 presents the maximum mean purity values obtained in this work as well as other information from different sources. As seen, purification of two grades of bioethanol with 88.0 and 99.4°GL using glycerol is slightly better than using ethylene glycol, since the results indicate values of achievable purity above, respectively + 1.0 and + 0.3°GL. The values are in the range of those found in literature, although below the best ones, probably because the different operational conditions may be further improved (ex. reflux control).

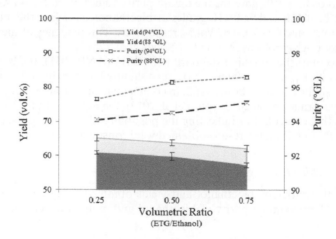

Figure 3.   Mean yield and purity of sweet potato bioethanol (88.0 and 94.0°GL of initial purity) using extractive distillation with ethylene glycol at different volumetric ratios (0.25, 0.50 e 0.75 V/V).

Table 1.   Maximum purity obtained for bioethanol in the present work and values in literature and standards.

Source	Maximum purity (bioethanol)		Reference value
	(88.0°GL)	(94.0°GL)	
Purification with glycerol	96.1 ± 0.1	96.9 ± 0.1	96.3, 98.9*
Purification with ethylene glycol	95.1 ± 0.1	96.6 ± 0.1	95.3, 99.8**
Brazilian Standard (2011)	NA	NA	>99.6***
American standard (2007)	NA	NA	>99.0***
European Standard (2007)	NA	NA	>99.8***

Note: All values listed in the table are expressed in °GL.
*Literature values for glycerol (Pla-Franco et al. 2013, Souza et al. 2013).
**Literature values for ethylene glycol (Navarrete-Contreras et al. 2014, Pla-Franco et al. 2013).
***Anhydrous ethanol, commercial.

The results obtained clear indicate dependency of purity achievable on the grade of bioethanol in the mixture, thus suggesting the use of better bioethanol than 88.0°GL for producing commercial anhydrous ethanol usable in biodiesel production.

Also, yield was clearly affected by the volumetric ratio of bioethanol/dragging agent, increasing with less glycerol or ethylene glycol in the mixtures, therefore with worst results for the purity obtained. Since studies focus only on the ethanol grade, no studies were found concerning such yield variations for comparison.

4 CONCLUSIONS

Glycerol and ethylene glycol were found to be viable dragging agents for the production of bioethanol with high purity, by minimizing the effects of the azeotrope in the extractive distillation process. The results in the present work seem to indicate glycerol as having a slightly better performance than ethylene glycol. Also, a clear dependency of purity achievable and the grade of bioethanol to be purified were found, thus suggesting the use of less hydrated bioethanol for the production of higher alcoholic degrees. Bioethanol yield and purity were clearly affected by the volumetric ratio of bioethanol/dragging agent with purity increasing by increase of dragging agent concentration and yield showing the inverse trend.

The values of purity required by the standards ANP no.7/2011 (Brazil) (2011), ASTM D 4806 (EUA) (2007) and EN 15376 (UE) (2007) for commercial anhydrous ethanol were not achieved in the present study and still are significantly far. However, with improvements in the extractive distillation process, including reflux control and some stages of distillation, this goal seems feasible and it is essential, since glycerol being the main by-product of biodiesel production, its use for obtaining commercial anhydrous ethanol usable in biodiesel production will be a step forward in the integration of bioethanol and biodiesel production systems.

ACKNOWLEDGEMENTS

This work was supported by the project UED/EQU/00511/2013-LEPABE (Laboratory of Process, Environment, Biotechnology and Energy Engineering—EQU/00511) by ERDF funds through the Operational Program for Competitiveness and Internationalization – COMPETE 2020 and national funding of FCT – Foundation for Science and Technology. Joab Silva thanks the IBRASIL Project-Erasmus Mundus Program (ibrasil@univ-lille3.fr), for the doctorate scholarship (IB-14DF/0138). Joana Dias thanks FCT for SFRH/BPD/112485/2015.

REFERENCES

ABNT 2008. NBR-5992: *Ethanol and its mistures and with water—Determination of the Density an Alcoholic Strenght—Glass Densimeter Method.*

Alper, A. & Oguz, O. 2016. The role of renewable energy consumption in economic growth: Evidence from asymmetric causality. *Renewable and Sustainable Energy Reviews* 60: 953–959.

ANP 2011. ANP n° 7: Resolução—Diário Oficial da União, 10 de fev/2011. *In:* AGÊNCIA NACIONAL DO PETRÓLEO, G.N.E.B. (ed.). Resolução—Diário Oficial da União, 10 de fev.2011.

Ardi, M., Aroua, M. & Hashim, N.A. 2015. Progress, prospect and challenges in glycerol purification process: a review. *Renewable and Sustainable Energy Reviews* 42: 1164–1173.

Dias, J.M., Alvim-Ferraz, M.C.M. & Almeida, M.F. 2008. Comparison of the performance of different homogeneous alkali catalysts during transesterification of waste and virgin oils and evaluation of biodiesel quality. *Fuel* 87: 3572–3578.

Dunn, J.B., Mueller, S., Kwon, H.-y. & Wang, M.Q. 2013. Land-use change and greenhouse gas emissions from corn and cellulosic ethanol. *Biotechnology for Biofuels* 6: 1–13.

Force, T.T. 2007. White paper on internationally compatible biofuel standards. *Tripartite Task Force Brazil, European Union & United State of America.*

Gomis, V., Font, A., Pedraza, R. & Saquete, M.D. 2005. Isobaric vapor–liquid and vapor–liquid–liquid equilibrium data for the system water + ethanol + cyclohexane. *Fluid Phase Equilibria* 235: 7–10.

Haas, M.J., McAloon, A.J., Yee, W.C. & Foglia, T.A. 2006. A process model to estimate biodiesel production costs. *Bioresource Technology* 97: 671–678.

IEA 2016. World Energy Outlook: Energy and Air Pollution. *Special Report 2016.* Paris, France.

Kumar, S., Singh, N. & Prasad, R. 2010. Anhydrous ethanol: A renewable source of energy. *Renewable and Sustainable Energy Reviews* 14: 1830–1844.

Navarrete-Contreras, S., Sánchez-Ibarra, M., Barroso-Muñoz, F.O., Hernández, S. & Castro-Montoya, A.J. 2014. Use of glycerol as entrainer in the dehydration of bioethanol using extractive batch distillation: Simulation and experimental studies. *Chemical Engineering and Processing: Process Intensification* 77: 38–41.

Pla-Franco, J., Lladosa, E., Loras, S. & Montón, J.B. 2013. Phase equilibria for the ternary systems ethanol, water + ethylene glycol or + glycerol at 101.3 kPa. *Fluid Phase Equilibria* 341: 54–60.

REN 2016. Renewables 2016 Global Status Report. *In:* SECRETARIAT, R. (ed.) *Ren21.* France: Ren21.

Silva, J.O.V. 2012. Comparação entre metodologias visando obtenção de maior rendimento de etanol a partir da batata-doce [*Ipomoea batatas (L.) Lam.*], em Palmas-TO. *In:* MARIANO, W.S. (ed.) *Library of UFT.* Master's in Agro-energy/UFT: Pedro & João Editores.

Silva, W.A. & Campos, V.R. 2013. Métodos de Preparação Industrial de Solventes e Reagentes Químicos. *Revista virtual de quimica* 5: 1007–1021.

Silveira, M., André, C., Alvim, T., Dias, L., Tavares, I., Santana, W. & Souza, F. 2007. A cultura da batata-doce como fonte de matéria-prima para a produção de etanol. Palmas, Universidade Federal do Tocantins. 45 p. *Boletim Técnico.*

Souza, W.L.R., Silva, C.S., Meleiro, L.A.C. & Mendes, M.F. 2013. Vapor–liquid equilibrium of the (water + ethanol + glycerol) system: Experimental and modelling data at normal pressure. *The Journal of Chemical Thermodynamics* 67: 106–111.

Urquiaga, S., Alves, B.J.R. & Boodey, R.M. 2005. Produção de biocombustíveis A questão do balanço energético. *Revista de política agrícola.*

Wasylkiewicz, S.K., Kobylka, L.C. & Castillo, F.J. 2003. Synthesis and design of heterogeneous separation systems with recycle streams. *Chemical Engineering Journal* 92: 201–208.

WASTES – Solutions, Treatments and Opportunities II – Vilarinho, Castro & Lopes (Eds)
© 2018 Taylor & Francis Group, London, ISBN 978-1-138-19669-8

Methodology for the assessment of non-hazardous waste treatment areas

V. Amant, A. Denot & L. Eisenlohr
Cerema, Bron, France

ABSTRACT: During a natural hazard, sea or river pollution or a nuclear accident, large amounts of waste may be generated. Finding new treatment sites for non-hazardous waste requires the synthesis of much information and data about the area. The purpose of this article is to propose a mapping method for prior identification of waste treatment sites. The methodology, developed with a Geographical Information System (GIS), can be used to select and prioritize potential sites in an area, taking into account environmental, economic and societal issues. Initially, the areas in which regulations or technical imperatives make it impossible to create a waste treatment site are excluded. Secondly, land that is not affected by an exclusion order is listed in order of priority. This method is a decision-making tool to help with choosing the location of waste treatment sites.

1 INTRODUCTION

The European directive on waste management (Directive 2008/98/EC) sets the principles and obligations in terms of waste management. In particular, it ensures that the environment is managed respectfully and provides information on location criteria for identifying waste management sites.

In France, facilities for treatment prior to recovery or disposal operations are adapted to the type of waste. There is currently no operational method available to take into account the context of the area, the technical imperatives associated with it, and the environmental, economic and societal issues.

The methodology presented in this article concerns sites for temporary processing of waste from arising, for example, from natural disasters, sea and river pollution or a nuclear accident. It was developed with state departments in France and tested across a region of nearly 8 million inhabitants.

2 MATERIAL AND METHODS

2.1 Available data

Data collection is the preliminary stage and is essential for setting up the methodology. It is based on an exhaustive inventory of existing map data available in databases usually administered by a public authority. It takes into account environmental, economic and societal, and technical imperatives.

Table 1 shows the type of data to be integrated into the method.

2.2 Environmental issues

Areas of environmental interest have been taken into account on the basis of the French National Inventory of Natural Heritage (*Inventaire National du Patrimoine Naturel* – INPN). The French National Natural History Museum (*Muséum National d'Histoire Naturelle*—MNHN)

Table 1. Inventory of available data to be collected.

Nature	Interest	Map layers
Physical context	Quick geographic referencing	Administrative boundaries (national, regional, departmental, communal), Roads, Towns
Reference data (such as IGN)	Identification, use of data sets, data processing	Mapping standards (raster or vector): SCAN25©, Topo© Data-Base, Carto© Data-Base, Alti© Data—Base
Hydrographic and hydrogeology data	Take into account ground and surface waters	Groundwater vulnerability
Territorial planning documents	Integrate urban planning rules related to planning regulations	Zoning plans for the prevention of natural and technological hazards: floods, flooding by the sea, coastal, mining, fire, earthquake and technological
Land use and land use planning	Consideration of urban areas, agricultural areas, vegetation cover	Zoning regulations around housing, historic, heritage and architectural property. Consideration of land use and agricultural areas
Natural environment	Integration of data with an environmental impact	Data pertaining to plant and animal species and their habitats

takes scientific responsibility for inventories carried out in this context. The INPN, drawn up for the entire French national territory, – land, rivers and sea—by the French Environmental Code, groups together inland and marine animal and plant species, natural habitats, protected areas and geological heritage within an information system.

For each type of set the database shows the relevant protection: regulatory, contractual or land protection, or designation under European or international agreements.

The entire database has been used in the context of the methodology.

2.3 Economic issues

The economic value of the land was taken into account in the methodology through urban planning projects. Also, all of the selection criteria (land ownership, land use, heritage and architecture protection zoning) were kept at plot level for potential sites identified by the methodology for the temporary treatment of waste.

2.4 Societal issues

In the case of a natural or technological disaster, one of the priority goals for local authorities is to regain the territories as quickly as possible after the disaster. Setting up temporary waste treatment sites helps towards achieving this goal.

The decree of 30 July 2012 relating to facilities for waste from marine and river pollution accidents or natural disasters (Decree of 30/07/12), and that of 3 December 2014 on waste from a nuclear or radiological accident (Decree of 03/12/14) specify the minimum distances of treatment sites from dwellings. The greater distance, common to the two decrees, was selected for this methodology.

Also, the data on drinking water resources (drinking water catchment areas, protection boundaries, volumes of groundwater identified as strategic resources for drinking water) and agricultural zoning were taken into consideration in the methodology.

2.5 The GIS tool and data processing

The mapping layers are all integrated into a Geographical Information System (GIS). They correspond to point, line or polygon features. The point and line features are processed by buffer zone to produce a polygon that can be used with the methodology.

Point features concerned by the methodology are related to drinking water catchment areas. These are associated with protective boundaries which are included in their entirety and generate a polygon. For catchments whose perimeters are not defined or were not declared as being of public utility a 100-meter buffer zone was taken into account.

Once the data have been collected, the different layers are processed to excluded certain areas and prioritize the remaining ones.

3   RESULTS AND DISCUSSION

Figure 1 shows the flowchart of the site identification method. It is composed of five steps with two cartographic data processing phases.

3.1   *Collection and classification of institutional data*

This step is based on the identification of environmental, economic and societal issues. The regulations related to these issues are also sought to allow discrimination of the data in the exclusion and prioritization steps. Technical imperatives, induced for example by the topography, were also taken into account.

All data may not be available in a usable digital format, or the mapping work may not have been carried out on the study area. The available data are then used. For example, in

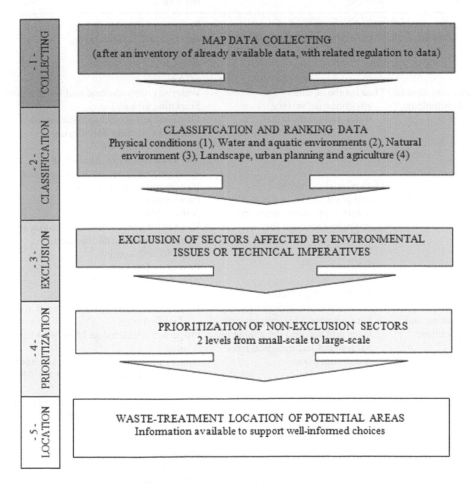

Figure 1.   Flowchart showing the steps of the methodology.

the absence of a flood risk prevention plan, non-statutory data from the atlas of flood zones are used.

To make cartographic analysis easier, all the data are grouped into four themes: physical conditions, water and aquatic environment, natural environment, landscape, urban planning and agriculture. The data collected and their ranking are presented in Table 2.

3.2 *The regulatory and technical exclusion phase*

This is the first step in cartographic data processing. It rules out sites where environmental issues and imperatives do not allow a temporary waste treatment site to be set up.

It is performed on all the data that are impacted by: regulatory protection, contractual protection, protection by land management, a convention, a technical imperative. Table 3 shows the data taken into account for the exclusion phase.

The data map layers to be excluded were integrated into the GIS. Synthesis of the surfaces to be excluded is used to retain only those territories of the study area that are not affected by exclusion (e.g. land sloping by more than 10%, drinking water catchment protection boundaries).

Table 2. Thematic grouping.

Theme	Interest	Map layers
Physical conditions	This data set is used as base map throughout the methodology.	Perimeter of the study area, Administrative boundaries, Plots, Roads, SCAN25©.
Water and aquatic environment	Data on the aquatic environment, surface water and groundwater.	Waterway, flow channel and flood channel, Spawning areas, Drinking water catchment areas and their protection boundaries, RAMSAR wetlands, The vulnerability of groundwater, Strategic resources for drinking water.
Natural environment	Data related to animal and plant species, natural habitats, protected areas and geological heritage sites.	Biotope protection orders, Parks and reserves, Natura 2000 zoning, Land belonging to the conservatory of natural spaces and the coastal protection agency, Protective forests, Areas of geological interest.
Landscape, urban planning and agriculture	Data related to protection of architectural or agricultural heritage sites. This data set also takes into account the zoning of natural and technological risk prevention plans	100-meter zones around houses, Protective zones around registered and classified sites, Properties registered as UNESCO heritage sites, Areas enhancing architecture and heritage, Discrimination of slopes from the Alti © database, Town and village urban development plans, Protected designation of origin agricultural areas, Land use, Zoning of risk prevention plans: flooding, earthquake, landslide, mining, shoreline, fire and technology.

Table 3.   Data used by the method during the exclusion phase.

Theme	Map layers
Physical conditions	This data set is used as base map throughout the methodology.
Water and aquatic environment	Waterway, flow channel and flood channel, Spawning areas, Drinking water catchment areas and their protection boundaries, RAMSAR wetlands.
Natural environment	Biotope protection orders, National Parks, Regional Natural Parks, Integral biological reserves, Managed biological reserves, Biosphere reserves, National and regional nature reserves, Natura 2000 network zoning, Land belonging to the conservatory of natural spaces and the coastal protection agency, Protective forests and classified wooded areas, Areas of geological interest.
Landscape, urban planning and agriculture	100-meter zones around houses, Protective zones around registered and classified sites, Properties registered as UNESCO heritage sites, Areas enhancing architecture and heritage, Slopes greater than 10%, Heavily urbanized municipalities, Urbanized areas of town and village urban development plans, Controlled areas of risk prevention plans: flooding, earthquake, landslide, mining, shoreline, fire and technology.

In the absence of a prescribed or approved risk prevention plan (RPP) for the study area, areas known for the hazard under consideration were excluded (for example areas subject to subsidence in the absence of a landslide RPP or a mining RPP). Also, local studies were taken into account. For example, flood zone atlases have no regulatory status but are a reference to be considered in the absence of a flooding RPP.

3.3   *The prioritization phase*

The second step of the method concerns prioritization of territories identified as potential. It takes into account all the environmental issues and technical imperatives that have no exclusion criteria.

Prioritization is based on the territories remaining after the exclusion phase. It is performed at two levels:

– Prioritization at municipal level defining 3 types of treatment sites: the least suitable, intermediate, the most suitable, taking into account the vulnerability of groundwater, for example.
– Prioritization at plot level taking into account the economic, technical and urban issues of potential sites with, for example, areas reserved for future urbanization.
Figure 2 shows the cartographic processing phases.

3.4   *Summary*

The exclusion and prioritization phase is used to locate potential sites for waste treatment. These sites are presented at plot scale with the associated characteristics (distance from a road, land ownership, land use).

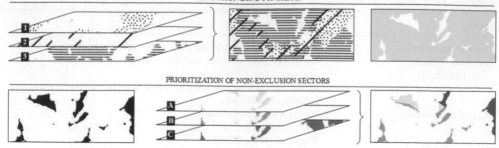

PRIORITIZATION OF NON-EXCLUSION SECTORS

Figure 2. Cartographic processing: Exclusion and prioritization phase (1: Water and aquatic environment, 2: Natural environment, 3: Landscape, urban planning and agriculture – A: Least suitable area, B: Intermediate area, C: Most suitable area).

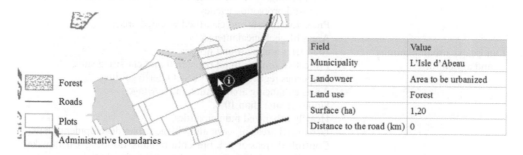

Field	Value
Municipality	L'Isle d'Abeau
Landowner	Area to be urbanized
Land use	Forest
Surface (ha)	1,20
Distance to the road (km)	0

Figure 3. Identification of potential sites for waste treatment and information about the plots concerned.

The mapping of plots is illustrated in Figure 3.

Application of the methodology gives land that meets the regulatory criteria for setting up waste treatment sites. The data related to the land is used to prioritize it. The final results obtained in this way should be used as assistance for decision-making following sea or river pollution, a natural disaster or a nuclear incident. Also, this method can be adapted to finding sites for treatment of all types of waste.

REFERENCES

Decree of 30/07/12 on general requirements for classified facilities subject to declaration under heading No. 2719 (Temporary facility for transit of waste from marine and river pollution accidents or waste from natural disasters).

Decree of 03/12/14 on general instructions applicable to facilities classified for the protection of the environment subject to declaration under heading 2798 (Temporary facility of for transit of radioactive waste from a nuclear or radiological accident).

Directive 2008/98/EC of the European Parliament and of the Council of 19 November 2008 on waste.

MAESTRI efficiency framework as a support tool for industrial symbiosis implementation

A.J. Baptista & E.J. Lourenço
INEGI—Instituto de Ciência e Inovação em Engenharia Mecânica e Engenharia Industrial, Porto, Portugal

P. Peças
IDMEC, Instituto Superior Técnico, Universidade de Lisboa, Lisboa, Portugal

E.J. Silva & M.A. Estrela
Instituto de Soldadura e Qualidade, Oeiras, Portugal

M. Holgado, M. Benedetti & S. Evans
Department of Engineering, Institute for Manufacturing, University of Cambridge, Cambridge, UK

ABSTRACT: Industrial Symbiosis (IS) envisages a collaborative approach to resource efficiency, encouraging companies to recover, reprocess and reuse waste within the industrial network. Several challenges regarding the effective application of IS continue to limit a broader implementation of this area of Industrial Ecology. The MAESTRI project encompasses an Industrial Symbiosis approach within the scope of sustainable manufacturing for process industries that fosters the sharing of resources (energy, water, residues, etc.) between different processes of a single company or between multiple companies. The Industrial Symbiosis approach is integrated with Efficiency Framework in the so-called MAESTRI Total Efficiency Framework. Efficiency Framework is devoted to the combination of eco-efficiency (via ecoPROSYS) and the efficiency assessment (via MSM – Multi-Layer Stream Mapping). In this manuscript the benefit of the combination of the Efficiency Framework as an facilitator to a more effective application of Industrial Symbiosis, within or outside the company's boundaries, is explored.

1 INTRODUCTION

Industrial Symbiosis, within the field of Industrial Ecology, can be defined as a concept engaging "traditionally separate entities in a collective approach to competitive advantage involving physical exchange of materials, energy, water" (Chertow, 2000). In practice, it means that manufacturers can make better use of all inputs to their processes through exchanges of waste, by-products and energy with other companies/sectors (Manufacturing Commission, 2015).

Symbiotic exchanges can occur as a one-off material waste exchange between two parties or as a more continuous flow exchanged between entities characterised by a certain geographic proximity (Chertow, 2000). It has been highlighted in literature that IS opportunities usually arise at process level (Lombardi and Laybourn, 2012), thus, entities taking part in symbiotic exchanges can be both companies and factories. Therefore, the IS concept covers both the cases in which IS opportunities are realised by a single company (intra-firm IS) and those realised in partnership with other companies (inter-firm IS) (Holgado et al., 2016). In addition, symbiotic exchanges can be self-organised, i.e. resulting from the serendipitous creation of networks through random social processes, facilitated, i.e. initiated by brokers that engage individual firms in the collaboration, or planned, i.e. characterised by the presence of an administrative member which aligns objectives and coordinate activities (Paquin and Howard-Grenville, 2009).

It is important to highlight that, in any of the above mentioned configurations, the success and effectiveness of IS planning and implementation is highly dependent on context specific characteristics (e.g. company size, production process, geographical and regulatory landscape) that, influencing each individual case, will inevitably end up shaping the scope and opportunities for IS (Holgado et al., 2016). Thus, tools and methods to facilitate IS implementation that take contextualisation issues into account are highly needed. Furthermore, IS opportunities should always be compared to other resource efficiency improvement opportunities (deriving for example from eco-efficiency actions), as IS implementation might not be the optimal mechanism to be implemented in each and every case (Holgado et al., 2016). This is a crucial issue to be addressed for an efficient, other than effective, IS implementation, that could also lead to a better integration of IS practices within companies' operations.

The integration of IS methods within the Total Efficiency Framework developed in MAESTRI project, and presented in the followings, represents a first attempt to embed Industrial Symbiosis in a broader industrial management system, also taking efficiency and eco-efficiency into account, as well as contextual factors.

2 GOAL AND MAIN FOCUS

The main aim of this work is to explore the benefit of the MAESTRI Efficiency Framework as an operational facilitator for a more effective and efficient application of Industrial Symbiosis, in intra-firm or inter-firm context. In fact, the IS approach is integrated within MAESTRI in the so called Total Efficiency Framework, which for one of its pillar and contains the Efficiency Framework that is devoted to the combination of eco-efficiency assessment, via the ecoPROSYS methodology (Baptista, 2016), and the overall efficiency and value/waste assessment, via MSM – Multi-Layer Stream Mapping (Lourenço, 2013). With the integrated results and parametrization of both MSM and ecoPROSYS it is possible to deliver a significant combination of information (waste quantification and characterization based on Lean Principles, environmental impact assessment and environmental performance assessment, value and cost analysis) to better support the analysis of IS opportunities and their implementation.

3 MAESTRI EFFICIENCY FRAMEWORK IN SUPPORT TO INDUSTRIAL SYMBIOSIS

3.1 *The MAESTRI efficiency framework*

The proposed Efficiency Framework, developed under the H2020 SPIRE Project MAESTRI, consists in the integration of two methodologies, namely eco-efficiency and efficiency assessment methods, and Information and Communication Technology (ICT) tools (MAESTRI Project, 2015). The eco-efficiency reference method within the project is oriented to the evaluation and assessment of eco-efficiency performance, via ecoPROSYS, while the lean based efficiency method, MSM, is applied to assess overall efficiency performance, waste quantification and disaggregated cost analysis.

The eco-efficiency approach ecoPROSYS aims to promote continuous improvement and underpin the efficient use of resources and energy, by providing a set of indicators easy to understand/analyse. The goal is to assess eco-efficiency performance in order to support decision making and enable the maximization of product/processes value creation while minimizing environmental burdens (Baptista 2016). The common expression for eco-efficiency is the ratio between value and environmental influence.

The resource efficiency assessment methodology, MSM, takes into account the base design elements from the Value Stream Mapping. Namely, by considering the value streams, in order to identify and quantify, at each stage of the production system, the "value added" (VA) and "non-value added" (NVA) actions, i.e. all types of waste and inefficiencies along the production system

(Lourenço 2013). Therefore, the basic principle relates to Lean Principles via clear definition of value and waste (in the Lean Principles context). The goal is to assess the overall performance, by the calculus of efficiency ratios for all parameter or variable (Equation 1), e.g. time, energy, water, raw material, associated to one or more processing stages (Figure 1).

Consequently, the approach provides both a high resolution efficiency analysis for each process step (see equation 1) and variable, and an efficiency aggregation assessment for the hierarchical production system (machine > line > section > factory), which can support the decision making process and helps to prioritize the implementation of improvement actions by identifying inefficiencies (wastes) in a very direct manner.

$$Efficiency = \frac{Value\,added\,(VA)}{Value\,added\,(VA) + Non-value\,added\,(NVA)} \tag{1}$$

The outline of the Efficiency Framework consists in the integration of eco-efficiency and efficiency methods through the mutual exchange of information and results, which corresponds to the central objective of the Efficiency Framework. Such integration strategy enables to obtain, besides the efficiency and eco-efficiency stand-alone, results to support decisions and new integrated results, namely the Total Efficiency Index (TEI)—obtained by combining normalized eco-efficiency results with efficiency results (Figure 2).

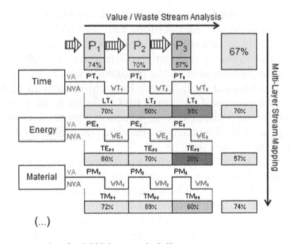

Figure 1.   Conceptual example of a MSM expanded diagram.

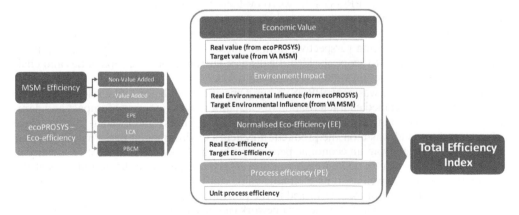

Figure 2.   Total efficiency index diagram calculus structure.

307

Regarding TEI, this index is calculated for each unit process (subsection of the processing stage, corresponding to a single action) of the production system under analysis. In quantitative terms, the TEI is obtained by multiplying the normalized eco-efficiency and the efficiency assessment results. The logic behind this index is to combine two fundamental efficiency aspects, namely eco-efficiency, which considers the ecology and economy aspects, and resource and operational efficiency, which considers the NVA and VA activities aligned with the Lean Principles from Multi-layer Stream Mapping (MSM). Consequently, TEI main outcome relates with providing the ability of evaluating if eco-efficiency performance variation is due to higher or lower environmental influence, or due to higher or lower economic value.

3.2 *MAESTRI efficiency framework as an operational facilitator for industrial symbiosis implementation*

In order to effectively and efficiently support the implementation of IS in manufacturing and process industry, MAESTRI project includes the development of a toolkit (T4IS – Toolkit for Industrial Symbiosis), which is based on four guiding questions:

- How to see waste?
- How to characterise waste?
- How to value waste?
- How to exploit waste?

Considering the design of the Efficiency Framework approach, relevant information that may be used to address these guiding questions has been identified. This information, as detailed in the followings, is particularly useful to support IS implementation, as it allows to quantify waste materials and energy in the production system, as well as the expected impact of IS practices/actions implementation on the production system and the environment. Furthermore, the interaction between the Efficiency Framework tools and T4IS is clearly a complementary and iterative process. This means that different tools provide complementary outcomes and can inform the decision making process in a more effective and comprehensive way. In the followings, identified interactions between the Efficiency Framework and T4IS are detailed for each guiding question.

How to see waste? The production system value and waste mapping (in MSM) identifies all wastes occurring in each unit process with a lean principles basis (either for material or energy flows). This provides a clear understanding and quantification of what type of residues and wastes are generated in the processes and a comprehensive list of potential resources to take into consideration during the definition of improvement opportunities. In addition, the process mapping also enables a first characterisation of waste materials (that are classified according to the European Waste Catalogue), current quantities of these residues and wastes (per unit process), as well as associated costs considering current disposal or treatment practices. In addition, ecoPROSYS allows to evaluate the production system's performance in economic and environmental terms, taking company's specific needs and concerns into account. This results in a first prioritisation and recognition of resources to be evaluated from an IS implementation perspective. Both of these outcomes are clearly useful inputs to the T4IS, in particular to scope the identification of current practices, capabilities and challenges regarding production system's residues and wastes.

How to characterise waste? ecoPROSYS has a specific module for Life Cycle Assessment, which performs a detailed characterisation of each unit process in terms of impact and damage to the environment. This information can be used to assess and evaluate the influence of each residue and waste on the production system's environmental performance. Similar results are provided from an economic perspective by the Process Based Cost Modelling in ecoPROSYS. Information on how each unit process is contributing to the overall costs of the production system as well as to its economic value can be retrieved. This may also improve the prioritisation of resources performed in the previous step.

How to value waste? The Efficiency Framework can be used to understand what is the contribution of each residue and waste to the production system's efficiency and eco-efficiency performance. Considering that residues and wastes are clearly related to a loss of material or energy, i.e. an increment of non-value added activities, unit processes with lower efficiency or eco-efficiency tend to present higher rates of wastes' and residues' generation. Therefore, these unit processes might have higher improvement potentials. It is possible to associate a precise amount of waste to each unit process, quantifying this loss, and allowing a first evaluation of IS potential. Moreover, even when these losses are first tackled with an eco-efficiency approach, focusing on the reduction of related non-value added activities, i.e. reducing the generation of residues and wastes, the full removal of residues and waste may not be feasible in all cases or it may require substantial economic investments or technological shifts not in line with company's targets and goals. IS represents a distinct but complementary approach, which allows the creation/capture of value related to the inevitable generation of these residues and wastes. In this respect, the T4IS aims to support the identification and implementation of this IS opportunities and can therefore supply useful information to the Efficiency Framework simulation module, where different improvement opportunities are combined and assessed in different scenarios.

How to exploit waste? Exploiting the complementarity between the tools, participants can also use the Efficiency Framework, in particular its simulation module, to jointly assess the identified/selected IS and, eco-efficiency and efficiency solutions. This module aims to present the expected impacts of these solutions, in terms of production system efficiency, environmental, cost and economic value performance, by generating alternative scenarios. Moreover, the solutions can be assessed in both joint or standalone perspective, which could help participants on the definition of priorities by selecting those, for instance, with higher economic value generation, higher increase of efficiency or representing a higher reduction of costs or environmental influence. This simulation functionality gives the opportunity to choose the most profitable solution among several potential improvement options, which were identified using, both or either, the T4IS and the Efficiency Framework.

Both the simulation module and the T4IS will bring additional information to companies regarding the improvement/symbiotic opportunities, in order to allow them to make better decisions related to their residues and wastes. As a consequence, and targeting the overall aims of the different tools, participants are able to take more informed decisions, by understanding not only their specific requirements but also by knowing in advance their potential consequences.

As a simplification of the above mentioned interactions and information flows occurring between the different tools, Figure 3 presents how this complementarity can be achieved.

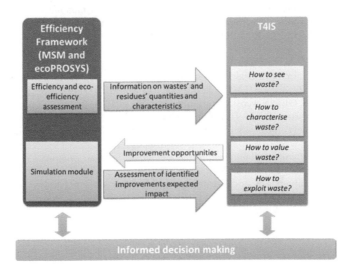

Figure 3.   Interactions and information flows between efficiency framework and T4IS.

4 FINAL REMARKS

The complementarity between information resulting from the Efficiency Framework and the T4IS is evident, and the outcomes may be used to support decision making considering different IS perspectives. Moreover, it represents, in conjunction with other MAESTRI methods and tools, a clear connection with efficiency assessment approaches (as MSM) and eco-efficiency assessment (as ecoPROSYS) to better support the IS opportunities evaluation and their implementation. Thus, it may represent a trigger/catalyst for IS implementation, which, in its absolute concept, aims to increase resource and energy efficiency in overall terms either for intra or inter-firm approach.

Within this perspective, it is clear that both the Efficiency Framework and T4IS share the same base goals and, in order to achieve higher efficient and valuable production systems, interactions between the developed tools should be explored as far as possible at several implementation levels. This means that, apart from using the information resulting directly from the Efficiency Framework implementation, T4IS should be also complementarily used by Efficiency Framework features. This is particularly important considering the potential of T4IS of providing solutions to specific problems occurring in the production system, namely on the identification of improvement opportunities. All in all, the interactions between the Efficiency Framework and T4IS arises as an high potential integrated approach to facilitate IS implementation.

ACKNOWLEDGMENTS

This work was supported by the European Union's Horizon 2020 research and innovation program through the MAESTRI project (grant no. 680570).

REFERENCES

A.J. Baptista, E.J. Lourenço, J.P. Pereira, F. Cunha, E.J. Silva, P. Peças 2016. ecoPROSYS: An Eco-efficiency Framework Applied to a Medium Density Fiberboard Finishing Line. *Procedia CIRP* 48: 170–175. ISSN 2212-8271. http://dx.doi.org/10.1016/j.procir.2016.04.061.

Chertow, M.R. 2000. Industrial symbiosis: literature and taxonomy. *Annual Review of Energy and the Environment* 25(1): 313–337.

Holgado, M., Morgan, D., Evans, S. 2016. Exploring the scope of industrial symbiosis: implications for practitioners. In: Setchi R, Howlett R, Liu Y, Theobald P (eds) *Sustainable, Design and Manufacturing 2016*. Smart Innovation, Systems and Technologies 52. Springer, Cham.

Lombardi, D.R., Laybourn, P. 2012. Redefining industrial symbiosis. *Journal of Industrial Ecology* 16(5): 28–37.

Lourenço, E.J., Baptista, A.J., Pereira, J.P., Dias-Ferreira, C. 2013. Multi-Layer Stream Mapping as a Combined Approach for Industrial Processes Eco-efficiency Assessment. *Re-engineering Manufacturing for Sustainability*: 427–433. http://dx.doi.org/10.1007/978-981-4451-48-2_70.

MAESTRI Project 2015. Official website http://maestri-spire.eu/.

Manufacturing Commission 2015. *Industrial Evolution: Making British Manufacturing Sustainable.* http://www.policyconnect.org.uk/apmg/sites/site_apmg/files/industrial_evolution_final_single-paged.pdf.

Paquin, R., Howard-Grenville, J. 2009. *Facilitating regional industrial symbiosis: Network growth in the UK's National Industrial Symbiosis Programme*. The Social Embeddedness of Industrial Ecology, Edward Elgar, Cheltenham, UK and Northampton, MA, 103–127.

WASTES – Solutions, Treatments and Opportunities II – Vilarinho, Castro & Lopes (Eds)
© 2018 Taylor & Francis Group, London, ISBN 978-1-138-19669-8

Activated carbons from Angolan wood wastes for the adsorption of MCPA pesticide

E.F. Tchikuala
Departamento de Química, Escola de Ciências e Tecnologia, Centro de Química de Évora,
Instituto de Investigação e Formação Avançada, Universidade de Évora, Évora, Portugal
Departamento de Ciências Exactas, Universidade Katyavala Bwila, Benguela, Angola

P.A.M. Mourão & J.M.V. Nabais
Departamento de Química, Escola de Ciências e Tecnologia, Centro de Química de Évora,
Instituto de Investigação e Formação Avançada, Universidade de Évora, Évora, Portugal

ABSTRACT: The work now reported presents the activated carbons production from Angolan woods wastes, namely Candeia, Hama, Njiliti, Nuati and Tchitiotioli. The physical activation with carbon dioxide produced materials with apparent surface area between 603 and 801 m^2/g, pore volume from 0.26 to 0.36 cm^3/g, mean pore width from 0.68 to 0.98 nm, and low external surface areas, less than 47 m^2/g for all samples. All samples present a basic nature with point of zero charge in the range 8.58 to 11.90. Selected samples were tested for the adsorption of a problematic pesticide, MCPA (4-chloro-2-methylphenoxyacetic acid) from aqueous solutions. The maximum adsorption was between 85 and 295 mg/g after 24h of equilibrium. With this work a time window of potential applications for these precursors is open, with a not negligible economic impact for the country.

1 INTRODUCTION

The valorisation of wastes without commercial value by its use as precursors for the production of materials with added value is increasingly relevant. One example is the use of biomass wastes for the activated carbons (ACs) production (Bansal et al. 2005, Mourão et al. 2011). The work now reported can have a significant positive impact as we have used wood wastes from Angolan trees for the production of activated carbons by activation with carbon dioxide. These wastes are usually burnt by the population to produce heat, so its use for the production of materials with a significant potential are of interest for Angola a developing country (FOSA—Forestry Outlook Study for Africa—Country Report—Angola).

Activated carbons are used in a wide range of applications performed in gas and liquid medium that include, among others, medicinal uses, gas storage and on environmental issues as pollutants removal from water streams (Bansal et al. 2005, Marsh et al. 2006). The pollutants can be of diverse nature such as emerging pollutants (e.g. drugs, pharmaceuticals), metals (e.g. cadmium, mercury) and herbicides and pesticides (Rathore et al. 2012).

Amongst the processes available to remove pollutants from water the adsorption on activated carbons is one of the most efficient methods as it allows a high adsorption capacity and a good selectivity in particular when they are used with other physical processes such as filtration or coagulation. The uptake of an adsorbate from aqueous solutions by ACs is complex, it depends on various factors, which include the type of precursor, the physical nature (surface area, pore size, pore volume, ash content, particle size) and functional groups present on the adsorbent, the nature of adsorbate (pKa, polarity, molecular weight, size, solubility) and the solution conditions (pH, temperature) (Belo et al. 2016, Dias et al. 2007, Marsh et al. 2006, Rufford et al. 2014).

2 EXPERIMENT

2.1 *Materials*

We have used five type of wood wastes, namely from Candeia, Hama, Njiliti, Nuati and Tchitiotioli trees, collected in Angola, region of Benguela. The precursors were crushed into pieces of size up to 3 mm in one dimension and dried before its pre-washing with an acid aqueous solution of 20% in H_2SO_4 by a period of 24 h. The suspension was filtrated and the solid material washed with distilled water, until the pH doesn't change, and completely dry at 110°C.

The production of activated carbons were done in a horizontal tubular furnace. The first step was the carbonisation of the precursor, by heating approximately 5 g of precursor under a flux of dry nitrogen at 800°C for 30 minutes. The gas was then switched to CO_2 for the activation step at 800°C for different times namely 180, 240, 300, 360 and 360 minutes, for Njiliti, Nuati, Hama, Tchitiotioli and Candeia, respectively. Samples were cooled to room temperature under an inert atmosphere and then removed from the furnace. The samples designations uses the following nomenclature (Wood Name)–(% of burn-off).

2.2 *Methods*

The determination of wood precursors content in cellulose and lignin was made by *Agroleico* (Porto Salvo, Portugal) using Portuguese Standards NP2029 and ME-414, respectively. Helium density was determined by pycnometry in an Accupyc 1330 from Micromeritics. Prior to the analysis, the equipment was calibrated against a standard volume and samples were dried. For each sample 10 analytical runs were made before the average value was defined.

Activated Carbon elemental analyses of C, H, N and S were carried out using a EuroEA elemental analyser, from Eurovector, the O amount was determined by difference.

The surface functional groups were characterized by FTIR, spectra recorded using a Perkin Elmer Spectrum Two FTIR Spectrophotometer by the KBr disc method, with a resolution of 4 cm^{-1} and 20 scans between 4000 and 450 cm^{-1}.

Prior to nitrogen adsorption measurements at −196.15°C, all the activated carbon samples were outgassed at 300°C for a period of 3 h on a Master Prep unit from Quantrachrome Instruments. The isotherms were measured on a Quadrasorb gas adsorption manometric equipment from Quantachrome Instruments, using nitrogen of 99.999% purity supplied by Air Liquide. The 4-chloro-2-methylphenoxyacetic acid (MCPA) was purchased from Sigma-Aldrich with a purity of HPLC-Grade 95%.

The study of the MCPA adsorption from liquid-phase on activated carbon samples was carried out at 298.15°C under pH = 3 by using a fixed amount of AC that was added to an

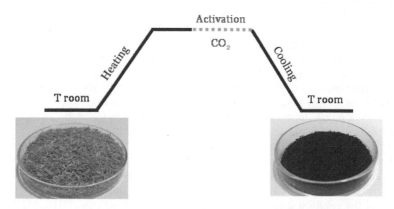

Figure 1. Scheme of activated carbon production, and an example of precursor and AC from Candeia.

aqueous solution of the pesticide, with initial specific concentrations, on Erlenmeyer flasks hermetically closed, under agitation on a thermostated shaker bath. After the equilibrium time, the AC suspensions were filtered, and the residual pesticide concentration determined by UV/Visible spectrophotometry, using a PerkinElmer Lambda 850 spectrophotometer, at a wavelength of 279 nm.

3   RESULTS AND DISCUSSION

3.1   *Precursors and activated carbons*

The wood wastes used as precursors in the production of the ACs have some differences, which are reflected not only on physical properties like density, but also, on the composition, namely the cellulose, hemicellulose and lignin content (Table 1). Whereas the first is an indicator of the softness or hardness of the wood, the second set of characteristics could influence, not only the structural, but also the chemical surface properties of the ACs. As can be seen in Table 1, Nuati present the higher content in cellulose and lignin, respectively 46.4% and 24.6%, whereas the lowest value of cellulose is 39.7% for Hama wood. The Tchitiotioli wastes have the smaller content in Lignin, 18.8%.

The analysis of the structural data from Table 2 and the profile of nitrogen isotherms on Figure 2 indicate that all the activated carbons, prepared by physical activation with CO_2, are essential microporous. The samples presents apparent surface areas between 603 and 801 m²/g, pore volume from 0.26 to 0.36 cm³/g, mean pore width from 0.68 to 0.98 nm, and low external surface areas, less than 47 m²/g. All samples have similar burn-off but dissimilar porous development. Sample Candeia-AC, with a burn-off of 47%, shows the highest apparent surface area (838 m²/g) and sample Njiliti-AC, with a burn-off of 40%, the smallest (603 m²/g).

All activated carbons samples were of basic nature as pointed out by the analysis of FTIR spectra and the point of zero charge (pzc). These features are typical of those type of carbon adsorbents that are prepared by physical activation with CO_2 in particular from the majority

Table 1.   Wood precursor characteristics.

Wood	Density* (g/cm⁻³)	Cellulose (wt%)	Hemicellulose (wt%)	Lignin (wt%)
Nuati	1.4550	46.4	13.9	24.6
Candeia	1.4538	45.9	16.5	21.8
Tchitiotioli	1.4534	42.9	15.2	18.8
Njiliti	1.3609	40.4	11.6	19.6
Hama	1.4858	39.7	14.4	21.8

*Density determined by Helium pycnometry.

Table 2.   Textural characteristics of the activated carbons.

Sample	Burn-off (%)	A_{BET} (m²/g)	A_S (m²/g)	V_S (cm³/g)	V_0 (cm³/g)	L_0 (nm)
Nuati-AC	33	801	22	0.36	0.31	0.74
Candeia-AC	47	838	24	0.36	0.33	0.71
Tchitiotioli-AC	55	741	47	0.32	0.30	0.98
Njiliti-AC	40	603	27	0.26	0.23	0.68
Hama-AC	57	657	36	0.28	0.26	0.81

Burn-off – percentage of mass loss due to the activation step, A_{BET} – apparent surface area, A_S – external surface area, V_S – pore volume from α_S plot, V_0 – pore volume from DR plot, L_0 – mean pore width.

313

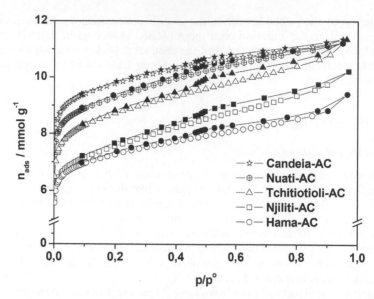

Figure 2. Adsorption isotherms determined on samples prepared by physical activation with CO_2 at 800°C (open symbols represent adsorption, closed symbols represent desorption).

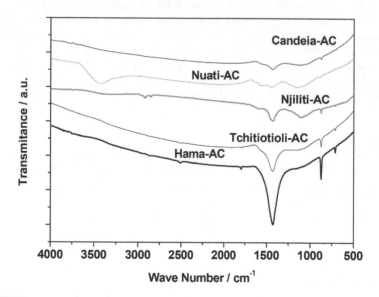

Figure 3. FTIR spectra of samples prepared by physical activation with CO_2 at 800°C.

of lignocellulosic precursors. Figure 3 shows the FTIR spectra of the ACs. It is possible to see that all samples have more or less the same absorption bands but with different intensities between samples. It is possible to identify characteristic bands of the following surface functional groups: alcohol (R-CH_2-OH, HO-R-OH, at 3429 and 1442 cm^{-1}), alkane and alkene (R-CH_2-R, at 2923, 2851, 875 cm^{-1}) and amide (NH, R-CO-NH_2, at 720, 1800 cm^{-1}). The intense absorption band around 1430 cm^{-1}, present in all samples, could be assigned to CH_3, CH_2 and CH groups and to C = C bonds associated with the typical aromatic structure of the ACs. All this assignments are in agreement with the content of carbon and oxygen, and also the lower values of hydrogen, nitrogen and sulphur of all the samples. The small amount of sulphur probably can justify the sulfone bands, C-SO_2-OH, of low intensity in the interval 1165–1150 cm^{-1}.

An analysis of data on Table 3, shows that all the samples are of basic nature, with values of pzc in the range between 8.85 and 11.90. Samples Njiliti-40 and Hama-57 show the highest pzc value, 9.89 and 11.90, respectively.

The characteristics of the ACs produced prove the suitability of the precursors tested to produce carbon materials with potential to be used as adsorbent for pollutants like MCPA pesticide, which is very common in rural areas, in particular in Angola.

3.2   *Adsorption of MCPA pesticide*

The equilibrium isotherms presented in Figure 4 show that the maximum adsorption capacity varies with different samples, it range from 85 to 295 mg/g, respectively for Njiliti-AC and Candeia-AC.

Taking in account the structural characteristics and the profile of isotherms, in particular the maximum adsorption capacity, it can be seen that the MCPA adsorption is proportional with the porous development of the ACs. Higher values of apparent surface area and pore volume are related with higher values for the maximum adsorption capacity. It is worthwhile to mention that the presence of small micropores seems to promote the MCPA adsorption. Also, when we compare the maximum adsorption capacity of samples Tchitiotioli-AC and Nuati-AC, with similar micropore volume (V_0, 0.30 and 0.31 cm³/g respectively), and despite the reduction of the external surface area (A_S, 47 to 22 m²/g), sample Nuati-33 shows a higher adsorption (245 mg/g) than sample Tchitiotioli-55 (220 mg/g).

Table 3.   Elemental composition of the activated carbons and pzc.

Sample	C (%)	H (%)	N (%)	S (%)	O* (%)	pzc
Nuati-AC	88.45	0.33	0.19	0.29	10.74	8.75
Candeia-AC	80.11	0.49	0.42	0	18.98	8.58
Tchitiotioli-AC	83.70	0.02	0.34	0.83	15.11	8.83
Njiliti-AC	82.29	0.60	0.25	0	16.86	9.89
Hama-AC	73.08	0.01	0.19	2.72	24.00	11.90

*Estimated by difference.

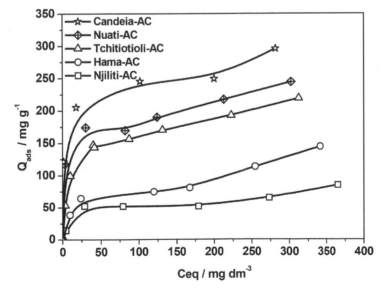

Figure 4.   Adsorption isotherms of MCPA onto the ACs produced.

Taking into account the pH of the solution MCPA is predominantly dissociated in the acid form. However, in our study the increment of sample's basicity was not sufficient to counterbalance the reduction of apparent surface area or micropore pore volume, as example for Tchitiotioli-AC and Hama-AC.

4 CONCLUSIONS

This work demonstrated the viability to produce ACs from wastes of woods from Angola with potential applications for adsorption not only in gas-phase, but in particular, for the removal of MCPA from aqueous liquid-phase. Also, according to our knowledge, is the first work were various Angolan natural renewable biomass resources are compared for this purpose.

The samples prepared by physical activation with CO_2 present well developed porous structures, with mean pore size below to 0.98 nm and, up to, an apparent surface area of 838 m²/g and a pore volume of 0.36 cm³/g. All samples are strongly enriched in carbon and oxygen, with the typical surface chemical groups that provided its basic nature, with pcz higher than 8.5.The MCPA adsorption achieved was quite interesting reaching values almost 300 mg of MCPA per gram of AC (Candeia-AC). The correlation of these values with the structural and chemical characteristics of the adsorbents allows us to infer that the porosity is the primary factor to influence the adsorption.

Finally, through this study, we intend to contribute decisively to the valorization of the wastes tested by the production of an added value product, with a large range of applications.

ACKNOWLEDGEMENTS

The authors gratefully acknowledge the funding by Fundação para a Ciência e a Tecnologia (project UID/QUI/0619/2016) with National (OE) funds. Emílio Tchikuala is thankful to Fundação Calouste Gulbenkian (Doctoral Grant n° 126541-"Bolsas de apoio à investigação para estudantes de doutoramento dos PALOP").

REFERENCES

Bansal, R.C., Meenakshi Goyal 2005. *Activated Carbon Adsorption*. Boca Raton: CRC Press.
Belo, C.R., Cansado, I.P.P., Mourão, P.A.M. 2016. Synthetic polymers blend used in the production of high activated carbon for pesticides removals from liquid phase, *Environmental Technology* 38 (3): 285–296.
Dias, J.M., Alvim-Ferraza, M.C.M., Almeida, M.F., Rivera-Utrilla, J., Polo, M.S. 2007. Waste materials for activated carbon preparation and its use in aqueous-phase treatment: A review. *Journal of Environmental Management* 85: 833–846.
FOSA—Forestry Outlook Study for Africa—Country Report—Angola. FOA Food and Agriculture Organization of the United Nation (accessed on April 2017 at http://www.fao.org/docrep/003/X6772E/X6772E00.htm#TOC).
Marsh, H., Rodriguez-Reinoso, F. 2006. Activated Carbon. *Elsevier Ltd.*
Mourão, P.A.M., Laginhas, C., Custódio, F., Nabais, J.M.V., Carrott, P.J.M., Ribeiro Carrott, M.M.L. 2011. Influence of oxidation process on the adsorption capacity of activated carbons from lignocellulosic precursors. *Fuel Processing Technology* 92: 241–246.
Rathore H.S., Nollet, L.M.L. 2012. *Pesticides: Evaluation of Environmental Pollution*. Boca Raton: CRC Press.
Rufford, T.E., Zhu, J., Hulicova-Jurcakova, D. 2014, *Green Carbon Materials: Advances and Applications*. Boca Raton: CRC Press.

Separate collection of packaging waste: Characterization and impacts

V. Oliveira
CERNAS, College of Agriculture, Polytechnic Institute of Coimbra, Coimbra, Portugal

J.M. Vaz
ECOGESTUS Lda, Waste Management Consulting, Figueira da Foz, Portugal

V. Sousa
CERIS, Department of Civil Engineering, Architecture and GeoResources, Tecnico Lisboa—IST, Lisbon, Portugal

C. Dias-Ferreira
CERNAS, College of Agriculture, Polytechnic Institute of Coimbra, Coimbra, Portugal
Materials and Ceramic Engineering Department, CICECO, University of Aveiro, Aveiro, Portugal

ABSTRACT: This study evaluates separate collection rate of packaging waste in the coastal area of the "Centro" region of Portugal, where collection is based on road-side waste containers. The area comprises 42 municipalities and the collection system of two waste management companies were studied: ERSUC and VALORLIS. In 2015, the average separate collection rate was 7.6% (30.7 kg person^{-1} year^{-1}), in ERSUC, and 7.7% (29.2 kg person^{-1} year^{-1}), in VALORLIS. Furthermore, separate collection rates varied significantly between municipalities, ranging from 4.5% to 12.6% of the total waste generated. Recyclables in unsorted waste represent a significant economic and environmental burden, and municipalities could save 8.1 million euros per year and avoid the emission of 86 thousand tonnes of CO_2 equivalent. The implementation of door-to-door collection and adoption of variable waste tariffs are further efforts required to increase separate collection rate, in order to achieve national targets.

1 INTRODUCTION

Separate collection of packaging waste allows the recovery of recyclables with higher quality that could be used as secondary raw material, contributing to a circular economy. The increase of separate collection performance leads to less energy consumption (e.g. glass and steel industry) and more economic development (Eriksson et al. 2005).

Currently there are 23 inter- and multimunicipal associations for municipal solid waste (MSW) recovery and treatment in Portugal. These associations are responsible for collection and treatment of source separated waste discarded by households. However, more than half of these associations only separately collect less than 10% of total MSW generated (APA 2016). Since separate collection of packaging waste is one of the objectives of the Portuguese Strategic Plan for Municipal Solid Waste Management (PERSU 2020), approved in 2014, which established a national target of 47 kg person^{-1} year^{-1} (MAOTE 2014), the majority of the municipal associations face a significant challenge in the near future.

Although separate collection of waste has been thoroughly implemented in Portugal since the mid-90's, the number of studies found in the literature about this subject is scarce. These mainly focus the environmental, economic and social costs and benefits in the recycling system (Ferrão et al. 2014, Da Cruz et al. 2012, Da Cruz et al. 2014, Ferreira et al. 2014), economies of density and size in MSW recycling (Carvalho & Marques 2014), economic viability of packaging recycling (Marques et al. 2014) and viability of implementing separate collection of bio-waste in restaurants and canteens (Rodrigues et al. 2015). However, other than Oliveira et al. (2017), who modelled source separation rates based on socio-economic and waste collection indicators, none of the references targeted specifically the issue of low separate collection rates in Portugal.

The present research is a contribution to fill this gap by assessing and discussing separate collection of packaging waste in a specific region, the coastal area of the "Centro" of Portugal, in which the separate collection rate is 7.6%. This work evaluates the waste management performance over the last 5 years in terms of amounts of MSW collected, separate collection rates and economic and environmental impacts. Finally, an overview of the different strategies/solutions adopted by municipalities to increase separate collection of packaging waste is presented.

2 STUDY CASE—COASTAL AREA OF THE "CENTRO" REGION OF PORTUGAL

Portugal is a country in the Iberian Peninsula located in the southwest of Europe. The coastal area of the "Centro" region of Portugal includes three districts: "Aveiro", "Coimbra" and "Leiria" divided into 42 municipalities. The study area is bordered by Oporto in the North and Lisbon in the South, has 1245241 inhabitants and an area of 8848 km^2, which represents 12% of the Portuguese population and 9.6% of total territory.

2.1 Municipal solid waste: Management and generation

The collection and treatment of unsorted MSW is the responsibility of local authorities (municipalities). Collected waste is delivered at three Integrated Centers for Treatment and Recovery of MSW (located in "Aveiro", "Coimbra", and "Leiria"). These facilities comprise a mechanical and a biological treatment (MBT) unit for the treatment of unsorted waste which is composed by i) an automated screening station for treatment of recyclable waste; ii) a unit for the preparation of refuse derived fuel for the fraction with calorific value; iii) an anaerobic digester and composted station for biological treatment of separated organic fraction; iv) a unit for energy recovery from the biogas produced by anaerobic process and; v) a refuse landfill. The treated and matured organic fraction of unsorted MSW is considered an organic amendment, which is sold or offered to the local authorities or individual citizens. When MTB units are not able to sort and treat all unsorted waste discarded by municipalities the waste is directly landfilled.

"EGF – Empresa Geral do Fomento, S.A." is the main player in the coastal area of "Centro" Region regarding the collection, treatment and recovery of separately collected packaging waste. EGF is currently responsible for 52% of all packaging materials sent for recycling to "Sociedade Ponto Verde", Portugal's non-profit organisation for recycling packaging waste. Two EGF companies carry out separate collection of packaging waste in the study area, "ERSUC – Resíduos Sólidos do Centro S.A." (hereafter referred as ERSUC) and "VALORLIS – Valorização e Tratamento de Resíduos Sólidos S.A." (hereafter referred as VALORLIS).

In 2015, almost 500 thousand tons of MSW were generated by households in the study area, representing about 10.5% of all MSW discarded in Portugal. From these, about 37 thousand tons were separately collected, representing a separate collection rate of 7.6% (29.9 kg person^{-1} year^{-1}).

2.2 Source separated waste collection system

Separate collection is based on road-side containers or bring-banks (95.7%) and civic amenity sites (4.3%). In 2015, ERSUC had in place 12494 containers for collection of packaging waste, which represents about 2 containers per km^2 and one container for every 77 inhabitants. During this year roughly 14264 t of glass, 7814 t of paper/cardboard and 6683 t of plastic/metal were separately collected. (ERSUC S.A. 2015a). Each "ERSUC resident" discarded at the bring-banks 15.20 kg year^{-1} of glass, 8.33 kg year^{-1} of paper/cardboard and 7.12 kg year^{-1} of plastic/metal, totalling 30.7 kg of packaging waste separately collected per person per year (ERSUC S.A. 2015b).

In the same period, VALORLIS had 2 containers per km^2 and one container for every 85 inhabitants, with 3607 containers for packaging waste in place (VALORLIS S.A. 2015a). The amount collected was 3965 t of glass, 2936 t of paper/cardboard and 2060 t of plastic/metal. The amount of packaging waste separately collected was of 29.2 kg person^{-1} year^{-1} distributed by 12.9 kg person^{-1} year^{-1} for glass, 6.7 kg person^{-1} year^{-1} for plastic/metal and 9.6 kg person^{-1} year^{-1} for paper/cardboard (VALORLIS S.A. 2015b).

3 RESULTS AND DISCUSSION

3.1 *Evolution of MSW collected by ERSUC and VALORLIS system*

The evolution of the amount (kg person^{-1} year^{-1}) of unsorted and source separated waste collected from 2010 to 2015 was negative in both ERSUC and VALORLIS (Table 1).

Overall, the company ERSUC registered a decrease of 7.7% in the amount of unsorted waste and a decrease of 14.8% in the amount of source separated waste between 2010 and 2013, while in 2014 this trend began to reverse and an increase on the amount of unsorted (3.0%) and source separated waste (5.5%) was noted. The company VALORLIS also registered a decrease in the collection of unsorted waste (12.2%) between 2010 and 2013. A slight increase took place in 2014 (3.5%). The amount of packaging waste collected by VALORLIS registered a gradual decrease until 2014 (20.2%) and a small increase was noted in 2015 (1.3%).

During the period 2010 to 2015, separate collection growth rate of packaging waste was negative: −3.8% for ERSUC and −10.5% for VALORLIS. In 2015, separate collection rate waste was 7.6% (30.7 kg person^{-1} year^{-1}) in ERSUC and 7.7% (29.2 kg person^{-1} year^{-1}) in VALORLIS, being far below of the 2020 target in PERSU2020, which refers a separate collection of packaging waste of 46 kg person^{-1} year^{-1} for ERSUC and 42 kg person^{-1} year^{-1} for VALORLIS.

3.2 *Comparison of separate collection rate of packaging waste across municipalities*

In 2015, separate collection rate varied across municipalities in the study area between 4.5% and 12.6% (Fig. 1). Municipality of "Sever do Vouga" (pop. 12000) had the highest rate: 12.6% followed by "Cantanhede" (12.5%), "Mealhada" (10.9%) and "Castanheira de Pêra" (10.8%). Separate collection rate were lowest in "Pampilhosa da Serra" (4.5%) and "Marinha Grande" (5.4%). In some cases, small and rural municipalities have higher separate collection rates than the most urbanized district capitals, "Coimbra" (8.9%; pop. 130000), "Aveiro" (6.4%; pop. 77229) and "Leiria" (7.7%; pop. 126897). The municipality of "Sever do Vouga" installed in the village center seven ecologic isles (set of waste containers for recyclables and unsorted waste, placed in an organized manner) and added more containers for the separate collection of glass in commercial areas and restaurants. In addition, the municipality also began a weekly door-to-door collection system of paper/cardboard in commercial and strategic areas and equipped all schools of the municipality with eco-points (set of containers for collection of recyclables). These actions were carried out together with awareness programs involving the citizens and companies. These strategies might explain the higher separate collection rate achieved by "Sever do Vouga" within the study area.

Table 1. Unsorted and source separated waste (kg person^{-1} year^{-1}) collected by ERSUC and VALORLIS and separate collection rates (%) between 2010 and 2015.

		2010	2011	2012	2013	2014	2015
ERSUC	Unsorted waste (kg person^{-1} year^{-1})	394.3	391.7	372.5	363.9	374.7	373.8
	Source separated waste (kg person^{-1} year^{-1})	33.8	34.4	30.8	28.8	30.4	30.7
	Separate collection rate (%)	7.9	8.1	7.6	7.3	7.5	7.6
VALORLIS	Unsorted waste (kg person^{-1} year^{-1})	385.4	363.3	338.7	338.3	350.2	349.6
	Source separated waste (kg person^{-1} year^{-1})	36.1	34.6	31.0	29.6	28.8	29.2
	Separate collection rate (%)	8.6	8.7	8.4	8.1	7.6	7.7

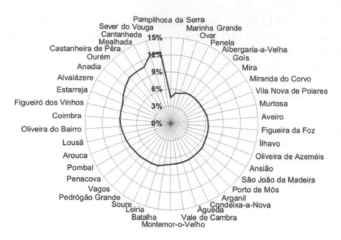

Figure 1.   Separate collection rate (%) in municipalities of the coastal area of the "Centro" region of Portugal.

3.3   *Impacts arising from increasing separate collection of packaging waste*

Current composition of unsorted waste includes 9–14% of plastic/metal, 7%–13% of paper/cardboard and 5%–7% of glass. The remaining fractions are 36%–39% of putrescible matter and 34–35% of fines (<20 mm) and others. This confirms that there are relevant amounts of packaging waste that are discarded together with unsorted waste and that would be valued if diverted to separate collection.

The total amount of packaging waste currently separately collected is shown in Figure 2. The collection potential for packaging waste, which considers the current amount plus 50% of the recyclables present in unsorted waste is also shown in Figure 2, for comparison. In 2015, 28.8 thousand tons of packaging waste was collected by ERSUC and 9 thousand tons by VALORLIS. If only 50% of recyclable waste currently deposited by households into unsorted waste was diverted to separate collection system, ERSUC would collect 74.4 thousand tons (79 kg person^{-1} year^{-1}) and VALORLIS 25.6 thousand tons (83 kg person^{-1} year^{-1}). This would represent an increase of almost 3 times the current amounts, allowing both ERSUC and VALORLIS to comply with the targets defined in PERSU 2020.

The replacement of virgin material by recyclables in the process of manufacturing of new products allows a reduction on the emission of CO_2 equivalent (CO_2 eq): 354 kg of CO_2 eq per tonne of paper/cardboard, 202 kg CO_2 eq per tonne of glass and 1763 kg of CO_2 eq per tonne of plastic/metal (Eunomia 2015). The estimated CO_2 eq emissions avoided for currently collected packaging waste and the potential according the rationale above are presented in Figure 2. The avoided CO_2 eq emissions in manufacturing of new products could reach 3 to 4 times the current values.

Management of unsorted waste discarded by households is a responsibility of the local authorities (municipalities). If the amount of packaging waste separately collected increases, costs associated with unsorted waste management will decrease for municipalities. According to Rodrigues et al (2015), transportation, collection, treatment and disposal of one ton of unsorted waste costs 81.12 € in one of the ERSUC municipalities ("Aveiro"). Adopting this value as a reference for the study area, the savings for the municipality of not handling the currently separately collected packaging waste was calculated and is presented in Figure 2. Additionally, the potential savings in case 50% of the packaging waste currently discarded in unsorted waste became separately collected was also calculated and is shown in Figure 2. Findings are that the 42 municipalities in the study area would save more than 8.1 million euros per year, instead of the currently 3 million euros per year. It should be noted that the effective cost reduction might be smaller depending on the proportion of fixed costs with the existing unsorted waste collection (that will remain even with the reduction of waste to be collected) and the eventual cost increase with the separate waste collection (e.g., need for additional equipment and/or more frequent collection).

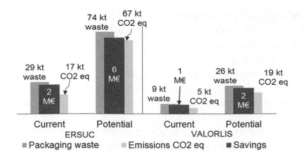

Figure 2. Current and full impacts arising increase of separate collection of packaging waste in 2015: avoided CO_2 eq emissions (thousand tons) and savings (million €) and packaging waste (thousand tons).

3.4 *Examples of good practices for increase separate collection of packaging waste*

In Portugal, the PERSU 2020 referred the implementation of door-to-door collection and the adoption of variable tariff system as the main strategies to increase separate collection of packaging waste. These solutions already started to be implemented at several municipalities, together with existing solutions.

One such example is the CSI programme, promoted by ERSUC and the municipality of "Aveiro", in which door-to-door collection of recyclables is carried out in 85 commercial facilities of the city center using plastic bags distributed free of charge. The same programme is also implemented in "Coimbra" and there is a plan to extend it to other city centers. The implementation of variable waste tariff is being currently tested in "Aveiro", "Condeixa" and Lisbon under the framework the European LIFE Programme (Project "LIFE PAYT – Tool to reduce waste in South of Europe"). The awareness campaigns promoted by VALORLIS, for example in schools and charity institutions stimulating permutation of recyclable materials by equipment and school supplies, are also examples of initiatives for increasing separate waste collection.

The municipality of "Maia", located in the north of Portugal (outside of the study area), is a success case because, during the last 10 years, separate collection rate increased 130%, from 14% to 33%. With a territorial area of 83 km² and about 136000 inhabitants, the strategy focused on a project called "Ecopoint at home" where bins and containers were distributed to households and a door-to-door collection scheme was implemented. In addition, in the municipality there are 5 civic amenity depots and awareness campaigns are carried out in schools, through the installation of outdoors, flashmobs and the creation of games. Another strategy was also implementation of a pilot project based on PAYT tariffs, in the second semester of 2014.

Another case of success is the town of "São João da Madeira" (about 8 km²), in which separate collection rates rose from 1% to 10% between 2002 and 2014 as a result of the construction of a civic amenity depot and consecutive awareness campaigns. The implementation of door-to-door collection system of recyclables in commercial waste producers with the distribution of containers or bags was also one of the strategies implemented in this municipality, resulting in an increase of the separate collection rate by 160% between 2007 and 2014.

Finally, between 2003 and 2014, Lisbon municipality tripled its separate collection rate from 6% to 22% due to the implementation of a door-to-door collection system in 2003. Currently about 61% of population is served by door-to-door collection system followed by road containers (23%) and ecological isles (15%).

4 CONCLUSIONS

This study assessed separate collection of packaging waste (glass, paper/cardboard and plastic/metal) in the coastal area of the "Centro" region of Portugal.

In 2015, separate collection rate of packaging waste across municipalities in the study area was between 4.5% and 12.6%, well below national targets. Additionally, significant amounts of packaging waste are still discarded as unsorted waste. Improving source-segregation to separately collect 50% of such packaging waste could lead to: i) savings of more than 8 million euros per year for the 42 municipalities at the study area and ii) a reduction of more than 86 thousand tons in the emissions of CO_2 eq if recyclable materials were used instead virgin materials in the manufacturing of new products.

Currently, in Portugal, separate collection of packaging waste relies substantially on the citizens' altruism, goodwill and sense of civic duty. However, this model show signs of stagnation and changes are required to boost separate waste collection rates. Implementation of different collection system (such as door-to-door) and the creation of PAYT tariff seem to be main strategies/solutions adopted by waste management companies and municipalities in Portugal.

This work shows that there is a huge potential for improvement in the Portuguese separate collection system and points out some alternative/complementary solutions based on success cases.

ACKNOWLEDGEMENTS

C. Dias-Ferreira and V. Oliveira gratefully acknowledge FCT – Fundação para a Ciência e para a Tecnologia (SFRH/BPD/100717/2014; SFRH/BD/115312/2016) and project Life PAYT – "Tool to Reduce Waste in South Europe" (LIFE15 ENV/PT/000609) for financial support.

REFERENCES

APA 2016. *RESÍDUOS URBANOS—Relatório anual 2015*, Departamento de Resíduos, Agência Portuguesa do Ambiente I.P., Amadora, Portugal. pp 41 (In Portuguese).

Carvalho, P. & Marques, R.C. 2014. Economies of size and density in municipal solid waste recycling in Portugal. *Waste Manag.* 34: 12–20.

Da Cruz, N.F., Ferreira, S., Cabral, M., Simões, P. & Marques, R.C. 2014. Packaging waste recycling in Europe: Is the industry paying for it? *Waste Manag.* 34: 298–308.

Da Cruz, N.F., Simões, P. & Marques, R.C. 2012. Economic cost recovery in the recycling of packaging waste: the case of Portugal. *J. Clean. Prod.* 37: 8–18.

Eriksson, O., Reich, M.C., Frostell, B., Björklund, A., Assefa, G., Sundqvist, J.O., Granath, J., Baky, A. & Thyselius, L. 2005. Municipal solid waste management from a systems perspective. *J. Clean. Prod.* 13: 241–252.

ERSUC S.A. 2015a. *Plano de Ação do PERSU 2020 - Sistema Multimunicipal de Valorização e Tratamento de Resíduos Urbanos do Litoral Centro*. Portugal. pp 31 (In Portuguese).

ERSUC S.A. 2015b. *Relatório e Contas 2014*. Portugal. pp 117 (In Portuguese).

Eunomia 2015. *Recycling Carbon Index: England, Wales & Northern Ireland*. Local Authorities 2013/14.

Ferrão, P., Ribeiro, P., Rodrigues, J., Marques, A., Preto, M., Amaral, M., Domingos, T., Lopes, A. & Costa, E.I. 2014. Environmental, economic and social costs and benefits of a packaging waste management system: A Portuguese case study. *Resour. Conserv. Recycl.* 85: 67–78.

Ferreira, S., Cabral, M., da Cruz, N.F. & Marques, R.C. 2014. Economic and environmental impacts of the recycling system in Portugal. *J. Clean. Prod.* 79: 219–230.

MAOTE 2014. *Portaria nº 187-A/2014: Plano Estratégico para os Resíduos Urbanos (PERSU 2020), para Portugal Continental*. Ministério do Ambiente, Ordenamento do Território e Energia. Portugal. pp 87.

Marques, R.C., Da Cruz, N.F., Simões, P., Faria Ferreira, S., Pereira, M.C. & De Jaeger, S. 2014. Economic viability of packaging waste recycling systems: A comparison between Belgium and Portugal. *Resour. Conserv. Recycl.* 85: 22–33.

Oliveira, V., Sousa, V., Vaz, J.M. & Dias-Ferreira, C. 2017. Model for the separate collection of packaging waste in Portuguese low-performing recycling regions. *J. Environ. Manage.* (Accepted).

Rodrigues, J., Oliveira, V., Lopes, P. & Dias-Ferreira, C. 2015. Door-to-Door Collection of Food and Kitchen Waste in City Centers Under the Framework of Multimunicipal Waste Management Systems in Portugal: The Case Study of Aveiro. *Waste and Biomass Valorization* 6: 647–656.

Valorlis S.A. 2015a. *Plano de Ação do PERSU 2020 - Sistema Multimunicipal de Valorização e Tratamento de Resíduos Sólidos Urbanos da Alta Estremadura*. Portugal. pp 31 (In Portuguese).

Valorlis S.A. 2015b. *Relatório & contas 2014*. Portugal. pp 126 (In Portuguese).

WASTES – Solutions, Treatments and Opportunities II – Vilarinho, Castro & Lopes (Eds)
© 2018 Taylor & Francis Group, London, ISBN 978-1-138-19669-8

Improvement of a clayey soil with alkaline activation of wastes

M. Corrêa-Silva, T. Miranda, N. Araújo & J. Coelho
University of Minho, Guimarães, Portugal

N. Cristelo
University of Trás-os-Montes e Alto Douro, Vila Real, Portugal

A. Topa Gomes
University of Porto, Porto, Portugal

ABSTRACT: A clayey soil was improved by means of addition of cement and lime and with an alternative approach using the alkaline activation of an industrial waste (fly ash). A set of laboratory tests, namely UCS, CBR and seismic P-waves, were carried in mixtures with different binder content and curing ages. The results showed that the alkali mixtures have comparable or even better mechanical performances than the best results obtained with the traditional binders.

1 INTRODUCTION

During the construction of transport infrastructures, it is common to deal with underperforming soils. The cost and technical difficulties associated with the substitution of the local soil are responsible for other solutions, aiming the improvement of the original, on site materials. Stabilisation with cement and/or lime is probably the most effective technique. However, the economical and, especially, the environmental impact associated with the production of cement is increasing the worldwide awareness of such environmental problems. Currently, the production of 1 ton of cement produces approximately 0.8 ton of CO_2-eq. Therefore, it is imperative to find new, more sustainable binders, which can substitute cement without losses in mechanical effectiveness. Alkaline activation constitutes a very promising alternative regarding the development of these sustainable binders (Palomo *et al.*, 1999; Davidovits, 2002; Torgal & Jalali, 2009). It consists on the activation, using strongly alkaline solutions, of raw materials (usually residues like fly ash or blast furnace slag) with a high level of amorphisation, and rich in Si and Al and, even, in Ca. The present work was developed aiming the possible stabilization of a clayey soil using alkali activated fly ash, having in mind the construction of road bases and sub-bases.

The experimental campaign started with the mechanical characterization of the initial soil, which was then mixed with the precursor (fly ash), in different proportions, and the resulting solids were activated with a combination of sodium hydroxide and sodium silicate. A similar study was simultaneously developed using traditional binders, i.e. cement and lime, during which the dry power content (cement or lime) was the only variable considered.

The mechanical behaviour of the mixtures was assessed using uniaxial compressive strength (UCS) tests, performed after different curing periods. The most effective mixture of each of the three binders considered was then further studied using CBR tests and seismic wave velocities.

2 EXPERIMENTAL PROGRAM

2.1 *Materials*

The soil used throughout the present study was collected in the North of Portugal, near the city of Porto, and was classified as a lean clay (CL), according to the Unified Soil

Classification System (ASTM D 2487, 2006). Based on the American Association of State Highway and Transportation Officials (AASHTO, 2004), it is included in the group A-4 (3). Table 1 presents the main geotechnical properties of the soil.

Fly ash from the Portuguese thermo-electric central of Pego was used as precursor. This fly ash is, according to the ASTM C 618 (2012) standard, classified as class F, due mainly to its low calcium content. The semi-quantitative evaluation of the fly ash, determined by electronic dispersive spectroscopy (EDS) is included in Table 2. Portland cement CEM II /B-L 32.5 N and air lime (containing 93% of calcium hydroxide) were also used. The liquid phase of the mixtures was constituted by tap water, sodium hydroxide (originally in pellets which were mixed with water to form a 10 m concentration solution) and sodium silicate (already in solution form, with a Na2O/SiO2 ratio of 0.5). The activator was thus obtained by adding two parts of silicate and one part of hydroxide.

2.2 Specimen preparation and testing

Table 3 characterises all the mixtures considered in the first phase of the experimental work. The density and liquid content values were from the results of Proctor tests (LNEC E 197,

Table 1. Main geotechnical properties of the original soil.

Tests				
Proctor	LNEC E 197, 1966	Water content	W_{opt}	14.4%
		Dry density	$\rho_{d,}$ máx	1.81 g/cm^3
California Bearing Ratio	LNEC E 198, 1997	Normal	CBR_{Normal}	48%
		After 24 hours immersion	$CBR_{embedded}$	14%
			Expansion	1.20%
Uniaxial Compression Test	ASTM D 2166, 2000	Maximum stress	R_c	0.4 MPa
		Yielding strain	ε_r	12‰
		Tangential deformability modulus	E_{tang}	32.3 MPa

*E_{tang} was determined in the linear phase of stress-strain diagram using values of extensions normally comprised between 0% and 0.5%.

Table 2. Chemical composition of the fly ash in percentage (%).

Si	Al	Na	Mg	P	S	K	Ca	Ti	Fe
48.81	21.77	1.31	1.56	0.58	1.17	4.42	3.85	1.79	14.74

*The fly ash consists mainly of silicon and alumina and the amount of material available to be activated is approximately 71%.

Table 3. Identification and characterisation of all the mixtures tested.

Mixtures	Soil (%)	Lime (%)	Cement (%)	Fly ash (%)	Water content (%)	Alkaline activator content (%)	Density (g/cm^3)
Soil	100				14.4		1.81
LIM_5	95	5			14.2		1.80
LIM_7.5	92.5	7.5			14.2		1.80
LIM_10	90	10			14.2		1.80
CEM_5	95		5		14.5		1.84
CEM_7.5	92.5		7.5		14.5		1.84
CEM_10	90		10		14.5		1.84
FA_10	90			10		13	1.80
FA_15	85			15		13	1.80
FA_20	80			20		13	1.80

1966) performed on the mixtures LIM_7.5, CEM_7.5 and FA_15. Comparing the results from the Proctor tests it is possible to observe that the ideal conditions for compaction are not substantially different, and thus it was decided not to estimate the compaction properties based on interpolations, but instead to use the values obtained for each of the Proctor tests performed.

To fabricate the specimens, the dry soil was first mixed with the binding powder (fly ash, lime or cement) until a homogeneous mixture was obtained. The liquid phase (activator or tap water) was then added to the solids, and further mixing was applied. The resulting paste was compacted in three layers inside a cylindrical stainless steel mold with 70 mm of diameter and 140 mm height in order to obtain the desired unit weight. The mould + specimen was then stored in a humid chamber, at 20°C and 95% of relative humidity. After 48 h the specimen was removed from the mold and wrapped in cling film before being stored again in the humid chamber. Curing periods of 7, 14, 28 and 90 days were considered, after which the UCS tests were performed based on the content of the ASTM D2166 (2000) and ASTM D1633 (2000) standards. For reproducibility reasons, each UCS result is the average of three tested specimens. The tests were carried out under monotonic displacement control, at a rate of 0.18 mm/min.

The second phase of the experimental program comprised California Bearing ratio (CBR) tests (after 28 days curing) and seismic wave velocity measurements (up to 189 days curing). The ultrasound equipment used allowed only the measurement of P-waves.

3 RESULTS AND DISCUSSION

3.1 *Uniaxial compressive strength tests—first stage of the experimental program*

The average failure values, as well as the respective strain at failure and the resulting deformability modulus, are presented in Tables 4 and 5 (for the soil-cement and soil-lime mixtures) and in Tables 6 (for the soil-ash mixtures). The deformability moduli were determined considering the initial linear phase of the stress-strain curves, between the 0% and 0.5% strain values. The relatively low values of the standard deviation and variation coefficient of the cement and lime mixtures show an acceptable consistency of the results. The ash mixture produced higher variation coefficients, which is probably explained by the higher difficulty to guarantee a good homogenization of the mixtures. The highest coefficient of variation were observed in the FA_20 mixtures at 90 days which presented a coefficient of variation for stress and strain rupture and tangential deformability modulus of 13.6%, 33.6% and 39.1%.

Table 4. Results obtained for the soil-lime mixtures at different curing times (days).

	7			14			28			90		
Mixtures	R_c (MPa)	ε_r (%)	E_{tang} (MPa)	R_c (MPa)	ε_r (%)	E_{tang} (MPa)	R_c (MPa)	ε_r (%)	E_{tang} (MPa)	R_c (MPa)	ε_r (%)	E_{tang} (MPa)
LIM_5	0.8	0.9	72.6	1.0	1.0	101.8	0.9	0.8	137.6	1.3	0.7	259.3
LIM_7.5	0.9	0.9	87.5	0.9	0.8	96.4	1.1	0.8	159.5	2.2	0.7	481.3
LIM_10	1.5	0.9	161.7	1.0	1.0	107.6	1.1	0.9	159.5	2.4	0.7	503.9

Table 5. Results obtained for the soil-cement mixtures at different times of curing (days).

	7			14			28			90		
Mixtures	R_c (MPa)	ε_r (%)	E_{tang} (MPa)	R_c (MPa)	ε_r (%)	E_{tang} (MPa)	R_c (MPa)	ε_r (%)	E_{tang} (MPa)	R_c (MPa)	ε_r (%)	E_{tang} (MPa)
CEM_5	2.2	0.8	455.7	2.5	0.9	442.5	3.0	0.7	600.9	3.6	0.8	700.9
CEM_7.5	2.4	0.8	498.5	3.8	0.8	732.7	3.8	0.7	849.2	5.3	0.7	1170.8
CEM_10	3.4	0.8	682.1	4.5	0.8	895.1	6.0	0.8	1314.3	7.2	0.7	1317.3

Table 6. Results obtained for the soil-ash mixtures at different times of curing (days).

Mixtures	7			14			28			90		
	R_c (MPa)	ε_r (%)	E_{tang} (MPa)	R_c (MPa)	ε_r (%)	E_{tang} (MPa)	R_c (MPa)	ε_r (%)	E_{tang} (MPa)	R_c (MPa)	ε_r (%)	E_{tang} (MPa)
FA_10	1.0	1.4	136.4	1.2	1.0	188.5	1.5	0.5	447.4	2.9	0.5	893.5
FA_15	1.1	1.1	197.0	1.4	0.9	300.8	2.4	0.6	629.1	4.5	0.6	1166.4
FA_20	1.1	1.2	158.7	1.4	0.7	307.0	3.2	0.6	766.7	8.6	0.5	2459.8

The mixture of 5% lime more than doubled the original soil strength, while 10% lime produced 4x that. The rapid strength increase, after only 7 days curing, is a consequence of the well-known flocculation of the soil, and is not related with the long-term strength gain, which is due to the pozzolanic reactions between the clay minerals and the calcium. In general, the most significant increase in strength was observed between the 28 and 90 days, while practically no increase was registered during the first 28 days. This is a consequence of an 'induction' period, occurring after the flocculation and before the full development of the pozzolanic reactions, during which the dissolution of the Si and Al from the soil is taking place.

Regarding the soil-cement mixtures, a different trend was observed, with more than 80% of the total strength gains appearing during the first 28 days, and the remaining increase between the 28th and the 90th curing day. The higher cement content clearly produced the higher strength (100% and 37% increase between 5% and 10% and between 7.5% and 10%, respectively), contrary to what was observed with the lime, in which case the strength difference between the 5% and 10% contents and between the 7.5% and 10% was of 76% and 7%, respectively.

Similarly to the soil-lime mixtures, the soil-ash mixtures also presented a substantial strength increase between the 28th and the 90th curing day. Nevertheless, and contrary to the soil-lime material, in which the slow strength development is due to the late surge of the pozzolanic reactions, in the soil-ash material the low Ca content results in the formation of a N-A-S-H type gel, with a slower development than the C-S-H gel usually obtained in the cement-based matrixes. The fly ash content had a significant influence in the UCS at every curing stage, with the maximum percentage tested (20%) showing a compressive strength 190% and 90% higher than the 10% and 15% mixtures, respectively, indicating that the binder quantity has a more significant influence on the mechanical behavior than the remaining binders tested. Although this seems a fairly obvious conclusion—i.e. an increase in binder results in more performing materials, is it important to remember that such trend isn't always observed (e.g. the lime content has an optimum value, after which the carbonation of lime consumed in the reactions starts to decrease the compressive strength). Figure 1 compares the strength and stiffness values obtained with each type of binder, allowing to conclude that FA 20 was the mixture presenting the best mechanical performance at 90 days of curing, mostly regarding stiffness. In fact, FA 20 presented a stiffness of 2459.8 MPa, approximately double the one presented by CEM_10 at the same time.

3.2 Second stage of the experimental program

Based on the mechanical behavior, mixtures LIM_10, CEM_10, FA_15 and FA_20 were selected to perform additional tests, namely seismic wave velocity (to assess the evolution, with curing time, of the dynamic deformability modulus) and CBR tests.

3.2.1 CBR tests

This is an essential test in terms of pavement design. In this case, the data produced allowed to quantify the improvement rate obtained with each binder, relatively to the original soil. The tests were performed according with the Portuguese standard LNEC E 198 (1967), which establishes that one of the specimens is tested as it is after molding, while a second one is tested after submerged in water for 96 hours (allowing an additional result, which is the

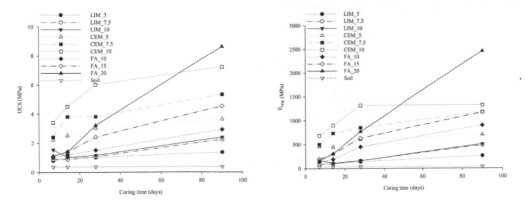

Figure 1.   Comparison between the UCS obtained with each mixture, at every curing stage considered.

Table 7.   CBR results after 28 days curing.

Mixtures	Normal	Embedded
Soil	48	14
LIM_10	92	79
CEM_10	318	283
FA_20	269	248

estimation of expansion due to water absorption). Contrary to what was observed in the original soil, the stabilized soil specimens didn't present any expansion while submerged, which is always relevant when dealing with clayey soils. The CBR results are presented in Table 7. It is clear the effect of the stabilization, and particularly the relative effect of each binder. Percentage increases of 92%, 562% and 460% were registered with the addition of lime, cement and ash, respectively, for the non-submerged specimens. The submersion in water resulted in a significant CBR decrease in the original soil (70%), which was substantially mitigated with the addition of the binders, with reductions of just 14% (lime), 11% (cement) and 7.8% (ash). Nevertheless, it appears that the ash-based binder was less susceptible to water than the lime and cement binders, which is an important advantage. Therefore, the submersion produced major effects on the relative CBR increases relatively to the original soil, with values of 464%, 1921% and 1671%.

3.2.2   Seismic P-wave analysis

The measurement of the velocity of the P-waves was performed throughout a curing period of up to 189 days, in order to estimate the dynamic modulus of deformability (E) of the mixtures. Based on the exposed in ASTM C597 (2002), the E value (in MPa) can be obtained using Equation 1.

$$V = \sqrt{\frac{E(1-\mu)}{\rho(1+\mu)(1-2\mu)}} \tag{1}$$

where V = wave velocity (m/s); ρ = density of the specimen (kg/m^3); μ = dynamic Poisson's coefficient. The velocity measurements were performed twice a day during the first 7 days of curing, once a day between 7 and 28 days, twice a week between 28 and 100 days, and once a month until 189 days of curing. The dynamic Poisson's coefficient considered was 0.25, obtained after (Amaral et al., 2011). The results are presented in Figure 2.

Immediately after the molding process, the cement-based mixture showed a deformability modulus approximately 4x higher than the remaining mixtures. Up until the 7th day of

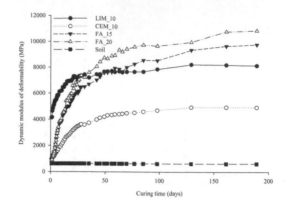

Figure 2. Evolution of the dynamic modulus of deformability with curing time.

curing, both fly ash mixtures presented similar E values but, after that, the effect of different ash contents started to show, especially after the 13th day. After 28 days the E value of the cement and 20% FA mixtures was very similar (7332 MPa and 7354 MPa, respectively) followed by the 15% FA (6475 kPa) and lime mixtures (3617 kPa). There is a similar trend to that registered for the UCS values. i.e. the lime mixture was clearly the least effective, the evolution rate of the cement mixture significantly decreased after 28 days, while both ash mixtures presented a significant E value increase up until day 189. After day 35 (20% FA) and 60 (10% FA), the E value of the ash mixtures became higher than of the cement.

At the final day of P-wave measurement, the following E values were obtained: 8241 MPa (for the cement mixture, representing a 12% increase relatively to the 28th day); 10871 MPa (for the 20% FA, 48% increase); 9789.8 MPa (for the 15% FA, 51% increase); 4988.6 MPa (for the lime mixture, 38% increase). These results clearly show the effect of a C-H-S type gel in the soil-cement mixture, which is known for its rapid development, and the effect of secondary chemical reactions in the remaining mixtures, namely the crystallization of the initial zeolitic gel (ash-based mixtures) and pozzolanic reactions (lime-based mixtures), which are responsible for a slower curing process.

4 CONCLUSIONS

The laboratory tests carried out in the binder soil mixtures at different curing times, allowed to conclude that the mechanical evolution of resistance and stiffness evolved in quite different forms over time. In general, soil-cement mixtures showed a more pronounced evolution of these parameters until 28 days of age, in contrast to the soil-lime and activated ash-soil mixtures where this evolution occurred over a longer period of time. Nonetheless, the alkali mixtures reached better results in the long term. The CBR tests allowed to verify that the addition of a binder to the soil, not only enhances the mechanical behavior, but also significantly reduces its sensitivity to water action. Finally, it is possible to conclude that the application of the technique of alkaline activation of fly ash in the improvement of soils led very good results, comparable or even superior to the results obtained with the addition of cement and lime to the soil. This technique revealed very interesting results from a mechanical point of view, superior to the results observed in the soil-cement mixtures at 90 days of age and in the soil-lime mixtures at all curing time under analysis.

REFERENCES

AASHTO 2004. *Classification of soils and soil-aggregate mixtures for highway construction purposes.* M 145–91. Georgia.

Amaral, M., Viana da Fonseca, A., Carvalho, J., Consoli, N. 2011. Dynamic Poisson ratio analysis. Analyse du coefficient de Poisson dynamique. *Personal Communication.*

ASTM C 597 – 02. 2002. *Standard Test Method for Pulse Velocity Through Concrete.* ASTM International.

ASTM C618 – 08. 2012. *Standard Specification for Coal Fly Ash and Raw or Calcined Natural Pozzolan for Use in Concrete.* ASTM International.

ASTM D 2487 – 06. 2006. *Standard Practice for Classification of Soils for Engineering Purposes (Unified Soil Classification System).*

ASTM D1633 – 00. 2000. *Standard Test Methods for Compressive Strength of Molded Soil-Cement Cylinders.* ASTM International.

ASTM D2166 – 00. 2000. *Standard Test Method for Unconfined Compressive Strength of Cohesive Soil1.* ASTM International.

Davidovits, J. 2002. Environmentally Driven Geopolymer Cement Applications. *Geopolymer Conference,* October 28–29, Melbourne, Austrália, 1–9.

LNEC E 197 1966. *Solos. Ensaio de Compactação.* Especificação do Laboratório Nacional de Engenharia Civil, Lisboa.

LNEC E 198 1967. *Solos. Determinação do CBR.* Especificação do Laboratório Nacional de Engenharia Civil, Lisboa.

Palomo, A., Grutzeck, M., Blanco, M. 1999. Alkalini—activated fly ashes. *A cement for the future. Cement and Concrete Research* 29: 1323–1329.

Torgal, F. & Jalali, S. 2009. *Ligantes obtidos por ativação alcalina.* 1º edição, Edição TecMinho. 153 p.

Gasification of RDF from MSW—an overview

F.V. Barbosa & J.C.F. Teixeira
Department of Mechanical Engineering, University of Minho, Guimarães, Portugal

M.C.L.G. Vilarinho
CVR—Center for Waste Valorization, Mechanical Engineering Department,
University of Minho, Guimarães, Portugal

J.M.M.G. Araújo
CVR—Center for Waste Valorization, University of Minho, Guimarães, Portugal

ABSTRACT: European legislation is very restrictive in relation to the deposition of Municipal Solid Wastes (MSW) in landfills and so, the waste energy recovery plays a key role in society. Several technologies ensure the conversion of wastes into energy, being incineration the most implemented. However, this process has high impacts on emissions of hazardous gases and production of harmful wastes. To mitigate such problems, thermal treatments such as pyrolysis and gasification emerged as alternative solutions. Several studies mention that gasification presents great advantages, however, to improve the process, it is important to maximize the energy potential of MSW (non-homogeneous), and this is possible through their conversion into refuse derived fuels, RDF (homogeneous). Thus, this paper aims to demonstrate that gasification of RDF from MSW is a good solution for waste energy recovery in Portugal as well as the potential of this country for the production of RDF.

1 INTRODUCTION

The growth of society's consumer habits has led to a significant increase in the production of waste per capita over the last decades. The most convenient and quick solution found for the management of these wastes was landfill deposition. However, the EU realized that this solution was not sustainable from an environmental point of view. To solve this problem, the EU started to implement action programs for waste management in all member states, with landfill being penalized. According to the hierarchy principle of the EU 2008/98/CE directive, the management of municipal solid wastes (MSW) should be focused on: reduction of waste production; ensure that the use of a good succeeds a new use or if not feasible, to recycle or convert it; the waste elimination is the last option; the waste responsibility is shared between the producer and the consumer. Inside the waste management, the waste thermal treatment is highlighted since it allows the energy recovery of wastes. Technologies for energy recovery from MSW play a vital role in mitigating the environmental pollution, reducing waste quantities that require final disposal and in the recovery of a significant amount of energy (Begum et al., 2012). Within the processes that allow the waste energy recovery, incineration has been the most commonly used technology. However, due to environmental concerns, waste incinerators have been required to install sophisticated exhaust gas cleaning equipment which is large and expensive (Morris & Waldheim, 1998). As an alternative to this technology, waste gasification seems to be an interesting and viable solution. The main benefits of gasification are the combination of high fuel flexibility, good environmental performance (lower generation of pollutants like dioxins, furans and NOx) and essentially a high net power production efficiency, which can be well above 31% using live steam with up to 540°C (Bolhàr-nordenkampf & Isaksson, 2014). A study presented by

Consonni & Viganò (2012) also mention that the syngas resultant from gasification is easier to control, meter and handle, and it can be used, after proper treatment, in gas turbines, combine cycle and Otto engines, or to generate high-quality fuels (Consonni & Viganò, 2012). Panepinto et al. (2015) presents some data about electrical and thermal efficiency in incineration and gasification processes for MSW treatment. The study mentions that the power generation efficiency is 7% higher in gasification (38% against 31%) and regarding the overall generation efficiency, 2% higher (27% against 25%). The study also reports that the main advantage of gasification (combined with gas engine and gas turbine) compared to incineration is the generation of large amounts of thermal energy which can be self-consumed or consumed by third users without compromising the power production, ensuring a fuel utilization higher than 50%. Gasification process allows a thermal generation of 45% and a thermal energy efficiency of 32% using a gas engine, and in the case of the gas turbine these values amount to 42% and 29% respectively (Panepinto et al., 2015). Relative to the emissions, Zaman (2013) mentions that the realise of hazardous gases to the atmosphere is about 2 times higher in incineration. In addition, Ionescu et al. (2013) explained that the sub-stoichiometric atmosphere of gasification process limits the formation of dioxins and large quantities of SO_x and NO_x (Ionescu et al., 2013).

To improve the energy recovery from MSW, refuse derived fuel (RDF) are a good alternative since they have a significant quantity of energy, which can be recovered in a fuel (Arena, 2012) (Malkow, 2004) (Panepinto et al., 2015). Even if the use of RDF in waste-to-energy plants is still limited (Agency, 2016) (Panepinto et al., 2015), the importance of the energy recovery from wastes has been growing, becoming a good alternative to the conventional energy sources. A recent trend shows that RDF and waste-to-energy facilities may be integrated together to deal with the situation of continuous landfill closing and to improve the quality of energy recovery processes (Chang & Chen, 2010). Considering the advantages that the combination of gasification with RDF can bring to the society, this work aims to demonstrate the benefits of this technology for MSW treatment and energy recovery, allowing to improve the waste management in Portugal. Nowadays in Portugal, the MSW management is performed by 23 waste management systems, 12 multimunicipal (of which 11 are part of Empresa Geral do Fomento, EGF, and BRAVAL) and 11 intermunicipal. The EGF manage 64% of the overall MSW produced in the country. The waste treatment processes used in these 23 plants are essentially landfill (50% of the total infrastructures), mechanical treatment (9.4%), mechanical and biological treatment (23.4%), organic recovery (7.8%), incineration (3.1%) and RDF preparation unit (6.3%) (Marçal et al., 2015). Considering that the country started taking the first steps in the production of RDF, it is now time to evaluate the potential of this fuel for energy recovery and to estimate the possibility to introduce the gasification process in waste treatment plants.

2 RDF FROM MSW

2.1 RDF characterization

Refuse derived fuel (RDF) from MSW are produced by shredding and dehydrating these urban wastes, after the separation of inert material, such as glass and metals, consisting mainly of organic components, essentially plastics and biodegradable wastes (Molino et al., 2013). The production of RDF through MSW is possible by following a suitable sequence of operations, composed of primary and secondary shredding, grading, wind sifting and screening, magnetic and eddy-current separation. The combinations of such operations (Fig. 1) may convert MSW, packaging, wood, paper and plastics into better manageable and storable fuel with more predictable characteristics and specifications (Buekens, 2013). During the production of RDF, particular attention must be paid to the combustion unit. For example, in order to facilitate the handling, storage, and transportation, it may be necessary to produce a densified fuel (i.e., a pelletized fuel) that meets the required specifications. RDF can be defined according to the final shape or its origin. According to the shape, they are distinguished between pellets—RDF produced by agglomeration of loose material in cub, disc or cylinder, whose diameter or its equivalent is generally less than 25 mm; fluff—loose and low density material that can be carried by air; and briquette—block or cylinder produced by agglomeration of loose material whose diameter in generally greater than 25 mm. Regarding the origin of the RDF, they can be PDF (Plastic Derived Fuel) or PPF (Paper and Plastic Fuel) (Martins-Dias et al., 2006). The quality of the RDF, according to technical specification CEN/TS 15359:2011, is evaluated through the analysis of three fundamental parameters: the net calorific value (NCV) and the content in chloride (Cl) and mercury (Hg). The RDF features should be able to maximize the thermal efficiency of the combustion, to minimize the corrosion effect in the boiler/gasifier and to minimize the emission levels (Hg) (Carvalho, 2011). Thus, the classification system is based on the limit values of these three parameters, being divided in five classes, as presented in Table 1. In reference to the composition of RFD, Fu et al. (2005) refer that this fuel usually have a volatile content (product of pyrolysis) of about 72.6–76.3% and a water content less than 10%. Regarding the additive of calcium hydroxide the same study mentions that a content of 2–4% is essential in the process of shaping RDF. Considering that the RDF approached in this work results from MSW, Vounatsos et al. (2012) refer that this fuel is classified as Class 4 in terms of NCV and Class 2 in terms of Cl and Hg contents.

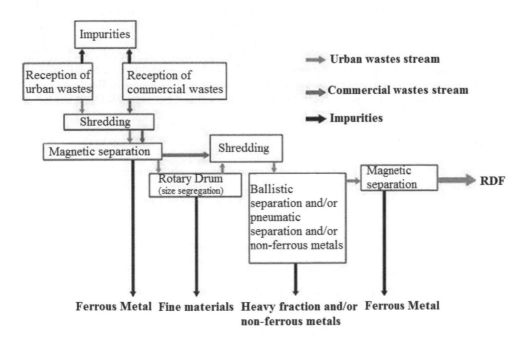

Figure 1. Representation of the RDF production process (adapted from Martins-Dias et al. (2006)).

Table 1. Classification system of RFD according to EN 15359:2011 (CEN, 2011).

Parameter	Statistic measure	Unit	Class 1	2	3	4	5
NCV	Mean	MJ/kg (as received)	≥ 25	≥ 20	≥ 15	≥ 10	≥ 3
Content in Cl	Mean	% (dry base)	≤ 0.2	≤ 0.6	≤ 1.0	≤ 1.5	≤ 3
Content in Hg	Median	Mg/MJ (as received)	≤ 0.02	≤ 0.03	≤ 0.08	≤ 0.15	≤ 0.50
	Percentile 80	Mg/MJ (as received)	≤ 0.04	≤ 0.06	≤ 0.16	≤ 0.30	≤ 1.00

Table 2. Amount of RDF and feedstock for RDF production in tonnes (t) (based on Marçal et al., 2015).

Processes	2011	2012	2013	2014
Screening Stations	–	–	120	757
Mechanical Treatment Plants (TM)	–	5709	11484	69996
Mechanical and Biological Treatment Plant (TMB)	5412	27153	20498	34198
Total	5412	32862	32102	104951

2.2 Production and use of RDF in Portugal

In Portugal, before 2011, the production of RDF from MSW did not exist. The units for waste management were only four mechanical and biological MSW treatment: the AMAVE composting plant located at Vale do Ave; the AZC composting plant located at Fundão; the AMTRES composting plant located in Oeiras/Cascais; and AMARSUL composting plant located at Setúbal. These waste treatment units aimed to valorize the organic fraction of wastes, existing in parallel some efforts for the recycling in the region as a pre-treatment of wastes (Marçal et al., 2015). The processes used in these units are similar to those for production of RDF, represented in Figure 1, with a previous biological stabilization. The rejected fractions obtained in each processes are mostly sent to landfill, however, if correctly used, they constitute the base product for future RDF. In 2011, the production of RDF from MSW was registered for the first time in Portugal, increasing significantly in 2012. Although production remained relatively constant in 2013, in 2014 it increased again with the implementation of two urban waste management system of Tratolixo and Resitejo. Table 2 presents the amount in tonnes of RDF production in Portugal between 2011 and 2014.

3 GASIFICATION

Gasification is a thermo-chemical process that converts a solid or liquid combustible raw material into a partially oxidized gas called syngas. This product gas contains a mixture of CO, H_2, CH_4 and some other inert gases (Consonni & Viganò, 2012). Gasification process of RDF (Fig. 2) can be divided in three steps: first, the RDF is dried inside the gasifier at a temperature between 100°C and 150°C; it follows the pyrolysis, which consists in the separation of CO, H_2, CO_2 (non-condensable gases), water vapor and organic liquids from the fixed carbon (solid carbon) and ash. The product that results from this first conversion step comprises mostly polyromantic hydrocarbons (PAHs) and tar which need to undergo a second gasification step. Through several reactions (as presented in Figure 2), a final syngas is obtained (Materazzi et al., 2016). The endothermic gasification reactions need heat which is generated by the combustion of part of the fuel, char, or gases (Boerrigter & Rauch, 2006).

In RDF gasification it is important to notice that the char resulting from the process is porous in nature leading to an incomplete chemical control of the reaction which limits the conversion of the fuel into a noble syngas (Materazzi et al., 2016). On the other hand, due to the high content of Cl, heavy metal and incombustible wastes associated with this fuel, significant fouling, corrosion and corrosion-erosion in the gasifier can occur if adequate measures are not taken during the design (Bolhàr-nordenkampf & Isaksson, 2014). Through gasification process, a syngas can be obtained from the transformation of the energy content of RDF, which can be re-used as chemical feedstock (Gendebien et al., 2003) or to produce energy.

Typically, a gasification system is made up of three fundamental elements: the gasifier, to produce the combustible gas; the gas clean-up system, necessary to remove harmful compounds from the combustible gas; and the energy recovery system. It is important to note that the gasification technology is selected on the basis of feedstocks properties and quality, gasifier operation, the desired product gas and its quality (Luque & Speight, 2015). In reference to the

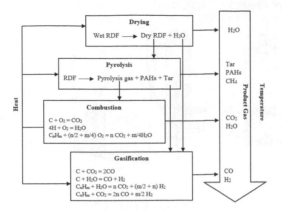

Figure 2. Gasification steps (based of Minteer (2011) and Materazzi et al. (2016).

energy conversion efficiency (defined as the ratio of the energy obtained within the product gas resulting from gasification to the energy content of the fuel added to the system) obtained from RDF gasification, this varies between 73% and 90% depending on the gasification process (Materazzi et al., 2016).

4 GASIFIER

The gasifier is the component in which the conversion of a feedstock into fuel gas takes place. Its efficiency depends on the phenomena of pyrolysis, partial oxidation of gaseous products, char gasification and conversion of tar and lower hydrocarbons (Barba et al., 2016). There are different types of gasifiers such as fixed bed, fluidized bed, entrained bed and plasma reactor. A key factor of gasifiers is the capacity to produce a gas with low tar content since a high concentration causes a wide range of problems to energy recovery systems due to its corrosive characteristics (Belgiorno, De Feo, Della Rocca, & Napoli, 2003). The basis for a successful design of a gasifier is to understand the properties and thermal behaviour of the fuel which feeds the gasifier (Minteer, 2011).

In reference to the fixed bed gasifier there are two types, the updraft (countercurrent) and downdraft (concurrent), shown in Figure 3 (a) and (b) respectively. In the first one the feed is introduced from the top and moves downwards while gasifying agents (air, steam, etc.) are introduced at the bottom of the grate moving the product gas upward. In a downdraft gasifier, both feed and product gas moves downward. The disadvantage of the updraft gasifier is the large amount of tars in product gas as opposed to the downdraft where most of tars are consumed (Kumar, Jones, & Hanna, 2009). Fluidized bed gasifiers (Fig. 3 (c)) are applied in various RDF energy recovery plants in the world due essentially to their high power production efficiency, high proven availability and reliability, capability to co-fire a wide range of different fuels, excellent environmental performance with low gas emissions and its low maintenance cost (Arena, 2012). In this technology the feed is introduced at the bottom and fluidized using air, nitrogen and/or steam, moving the product gas upward. These gasifiers can be divided in: bubbling, in which the flow rate of the fluidizing agent is compared to the minimum fluidizing velocity[1], or circulating which has a higher gasifying agent flow that increases the heat transfer and conversion rate of the feedstock (Belgiorno et al., 2003). However, when the main objective is to operate at high pressures ($\approx$ 25 bar), for example to treat coal and to refine residues and mixed plastic wastes, the entrained flow gasifier (Fig. 3 (d)) is the most appropriate. The main features of this technology is the possibility to make solid fuel feeding at high pressures and the propitious energy content to sustain the gasification reaction (Arena, 2012). The process consists in a mixture of pulverized solid fuel and

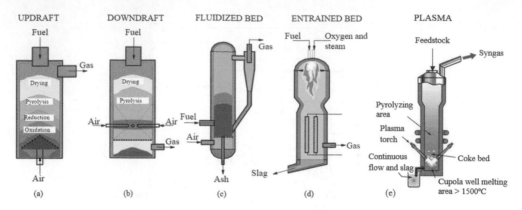

Figure 3. Types of gasifiers (based on Chevron & Gutsol, 2010).

steam/oxygen stream subsequently converted in a turbulent dust flame at high temperatures (> 1200°C) in a few seconds, which originates an almost tar-free syngas and a leach-resistant molten slag (Henrich & Weirich, 2004). On the other hand, plasma technology (Fig. 3 (e)), generates significant heat across the system originating an ionised gas stream, or plasma. In this process, plasma torches (located at the bottom of the gasifier) fire the MSW, breaking the feedstock and all hazardous and toxic components into their elementary constituents. A vitrified slag, highly resistant to leaching, is obtained from the melting of residual materials of inorganic elements of the feedstock (e.g. heavy metals) (Speight, 2014).

In Europe, RDF gasification has already been implemented essentially in Italy, France, UK, Switzerland, Finland, Norway, Germany and Belgium (Panepinto et al., 2015) and the gasifiers applied in these plants are fluidized bed (Heermann et al., 2001) (Martins-Dias et al., 2006). This technology is the most promising due essentially to the operating flexibility in the feedstock types used in the process, the nearly constant temperature along the gasifier and the great flow mixing between the reactants (Materazzi et al., 2016).

5 CONCLUSIONS

The principal challenge related to the MSW is its heterogeneity, limiting the amount of energy recovery that can be obtained directly from these wastes. An interesting solution is the production of RDF from MSW allowing to obtain a homogeneous fuel with controlled characteristics. Portugal has implemented some systems of production of feedstock that can be the base of RDF. Considering the energy potential of RDF from MSW, this work shows that a good way to recover energy from this fuel is through a gasification process. The advantages of this waste thermal treatment in comparison with incineration, which is a widespread technology in Portugal, were presented in this work. The main advantages mentioned were the low emissions to the atmosphere, which results essentially from the controlled air, and the production of a controlled syngas that ensures the production of higher amount of heat and electricity. According to the studies presented along this work, the thermal generation efficiency using RDF gasification is about 29–30% (no thermal generation results from incineration process) and the power generation efficiency is about 38% (against 30% obtained through combustion), considering a fuel utilization higher than 50% (in the case of combustion this value is less than 25%). A first step was made to improve the waste treatment and management in Portugal, however, at this moment, it is essential to implement RDF gasification plants. This plants could be integrated with preexisting industrial and thermoelectric plants due to their flexibility and compactness. This overview shows that the RDF gasification seems to be a great solution to improve the waste management and treatment in Portugal, but also as a good step to help the country to achieve the goals stablished in the Paris Agreement.

REFERENCES

Agency, E.E. 2016. Municipal waste management across European Country. *Eurostat*, 1–7.

Arena, U. 2012. Process and technological aspects of municipal solid waste gasification. A review. *Waste Management* 32(4): 625–639. https://doi.org/10.1016/j.wasman.2011.09.025.

Barba, D., Capocelli, M., Cornacchia, G., & Matera, D.A. 2016. Theoretical and experimental procedure for scaling-up RDF gasifiers: The Gibbs Gradient Method. *Fuel* 179: 60–70. https://doi.org/10.1016/j.fuel.2016.03.014.

Begum, S., Rasul, M.G., & Akbar, D. 2012. An Investigation on Thermo Chemical Conversions of Solid Waste for Energy Recovery, 6(2): 624–630.

Belgiorno, V., De Feo, G., Della Rocca, C., & Napoli, R.M.A. 200). Energy from gasification of solid wastes. *Waste Management* 23(1): 1–15. https://doi.org/10.1016/S0956–053X(02)00149–6.

Boerrigter, H., & Rauch, R. 2006. Review of applications of gases from biomass gasification. In H.A.. Knoef (Ed.), *Handbook Biomass Gasification* (p. 33). Netherland: Biomass Technology Group.

Bolhàr-nordenkampf, M., & Isaksson, J. 2014. Refuse Derived Fuel Gasification Technologies for High Efficient Energy Production, 379–388.

Buekens, A. 2013. Incineration Technologies. *Springer Briefs in Applied Sciences and Technology* (Vol. 1). USA: Springer. https://doi.org/10.1017/CBO9781107415324.004.

Carvalho, I. dos R.G. de. 2011. *CDR,* um Resíduo ou um Produto, e a sua Viabilidade Técnico-Económica: Análise do Estudo Caso. Universidade Nova de Lisboa.

CEN. EN 15359:2011–11: Solid recovered fuels—specifications and classes (2011). Belgium.

Chang, Y.H., & Chen, W.C. 2010. Evaluation of heat value and its prediction for refuse-derived fuel. *Renewable and Sustainable Energy Reviews* 44(1–3): 1522–1527. https://doi.org/10.1016/j.energy.2012.01.014.

Chevron, A.F.G., & Gutsol, A.F. 2010. Warm Discharges for Fuel Conversion. In *Handbook of Combustion.* WILEY-VCH. https://doi.org/10.1002/9783527628148.hoc085.

Consonni, S., & Viganò, F. 2012. Waste gasification vs. conventional Waste-To-Energy: A comparative evaluation of two commercial technologies. *Waste Management* 32(4): 653–666. https://doi.org/10.1016/j.wasman.2011.12.019.

Fu, Z.M., Li, X.R., & Koseki, H. 2005. Heat generation of refuse derived fuel with water. *Journal of Loss Prevention in the Process Industries* 18(1): 27–33. https://doi.org/10.1016/j.jlp.2004.09.001.

Gendebien, A., Leavens, A., Blackmore, K., Godley, A., Lewin, K., Whiting, K.J., & Davis, R. 2003. *Refuse Direved Fuel, Current Practice and Perspectives (B4-3040/2000/306517/MAR/E3).*

Heermann, C., Heermann, C., Schwager, F.J., Schwager, F.J., Whiting, K.J., & Whiting, K.J. 2001. *Pyrolysis & Gasification of Waste: A Worldwide Technology & Business Review.* (Vol. 2).

Henrich, E., & Weirich, F. 2004. Pressurized Entrained Flow Gasifiers for Biomass. *Environmental Engineering Science* 21(1).

Ionescu, G., Cristina, E., Ragazzi, M., Ma, C., Badea, A., & Apostol, T. 2013. Integrated municipal solid waste scenario model using advanced pretreatment and waste to energy processes 76: 1083–1092. https://doi.org/10.1016/j.enconman.2013.08.049.

Kumar, A., Jones, D.D., & Hanna, M.A. 2009. Thermochemical biomass gasification: A review of the current status of the technology. *Energies* 2(3): 556–581. https://doi.org/10.3390/en20300556.

Luque, R., & Speight, J.G. 2015. *Gasification for Synthetic Fuel Production.* (Elsevier, Ed.). UK: Woodhead Publishing Series in Energy. https://doi.org/10.1016/B978–0-85709–802–3.00010–2.

Malkow, T. 2004. Novel and innovative pyrolysis and gasification technologies for energy efficient and environmentally sound MSW disposal, 24: 53–79. https://doi.org/10.1016/S0956–053X(03)00038–2.

Marçal, A., Mateus, I., Silva, F. 2015. *Resíduos Urbanos—Relatório Anual 2014. Âgencia Portuguesa do Ambiente, I.P.* Amadora. https://doi.org/10.1017/CBO9781107415324.004.

Martins-Dias, S., Silva, R.B., Barreiro, F., & Costa, M. 2006. Avaliação do Potencial de Produção e Utilização de CDR em Portugal Continental. *Instituto Superior Técnico—Centro de Engenharia Biológica E Química.*

Materazzi, M., Lettieri, P., Taylor, R., & Chapman, C. 2016. Performance analysis of RDF gasification in a two stage fluidized bed—plasma process. *Waste Management* 47: 256–266. https://doi.org/10.1016/j.wasman.2015.06.016.

Minteer, S.D. 2011. *Handbook of Biofuels Production. Handbook of Biofuels Production.* https://doi.org/10.1533/9780857090492.2.258.

Molino, A., Iovane, P., Donatelli, A., Braccio, G., Chianese, S., & Musmarra, D. 2013. Steam gasification of refuse-derived fuel in a rotary kiln pilot plant: Experimental tests. *Chemical Engineering Transactions* 32(1992): 337–342. https://doi.org/10.3303/CET1332057.

Morris, M., & Waldheim, L. 1998. Energy recovery from solid waste fuels using advanced gasification technology. *Waste Management* 18(6–8): 557–564. https://doi.org/10.1016/S0956–053X(98)00146–9.

Panepinto, D., Tedesco, V., Brizio, E., & Genon, G. 2015. Environmental Performances and Energy Efficiency for MSW Gasification Treatment. *Waste Biomass Valor* 6: 123–135. https://doi.org/10.1007/s12649–014–9322–7.

Speight, J.G. 2014. *Gasification of Unconventional Feedstocks. Gasification of Unconventional Feedstocks.* https://doi.org/10.1016/B978–0-12–799911–1.00006–6.

Vounatsos, P., Koufodimos, G., Agraniotis, M., Roufos, K., & Eleftheriadis, C. 2012. Characterization and classification of Refuse Derived Fuel in the Materials Recovery Facility of EPANA S.A. In *Proceedings of ATHENS 2012 Conference.*

Zaman, A.U. 2013. *Life cycle assessment of pyrolysis—gasification as an emerging municipal solid waste treatment technology*, 1029–1038. https://doi.org/10.1007/s13762–013–0230–3.

Extraction of copper from dumps and tails of leaching by hydrochloric acid

K.K. Mamyrbayeva, V.A. Luganov, Y.S. Merkibayev, Zh. Yesken & S.D. Orazymbetova
Kazakh National Research Technical University, Almaty, Kazakhstan

ABSTRACT: The article presents the results of studies on hydrometallurgical processing of dumps and tails of leaching from Balkhash region, Kazakhstan by leaching with hydrochloric acid both with and without an oxidizer, followed by refining and concentrating copper by liquid extraction. The extraction of copper from the dumps into the solution by acid leaching without an oxidizer is 34%, from the tails –29%. The extraction of copper during leaching of the samples in the presence of an oxidizer (hydrogen peroxide) amounted to 75% from the dumps, and 47% from the leaching tails. The refining and concentration of copper from the productive hydrochloric acid solutions was carried out by extraction using LIX9858 NSC and ACORGA ORT5510 extractants followed by sulfuric acid re-extraction. The throughout extraction of copper from the dumps and tails of leaching into the production solution was from 74 to 80%.

1   INTRODUCTION

The copper industry is one of the key branches of Kazakhstan. Kazakhstan has significant copper ore resources and as estimated by Kazakhstan geologists, there are about 40 million tons of copper within subsoil assets of the country, which allows the country, developing 17 deposits, to be one of the world's largest producers and exporters of refined copper.

The issues of involvement of these deposits in development are primarily related to the solution of technological matters that allow for profitable mining of low-grade ores.

Supportability of mining enterprises by copper reserves ready for exploitation is short and is equal approximately to 25–30 years. The matters of rational and comprehensive use of the existing mineral resources base presume the involvement of accumulated off-balance ores, dumps, etc., containing millions of tons of non-ferrous, rare and noble metals, into the processing.

To extract copper from such raw materials, the methods of leaching (Davenport, G.W. et al., 2002) by various reagents, that successfully used by operating enterprises all over the world. For oxidized minerals—heap and underground leaching with sulfuric acid (Petersen J. 2016, Mamyrbayeva K.K. 2015, Sinclaira L., 2015), and for sulphide minerals— leaching with ferrous sulfate $Fe_2(SO_4)_3$ with the regeneration with Thiobacilius ferooxidans bacteria during the process (Roberto A., 2017), autoclave leaching (Seepon copper process, Mt. Gordon process, Biocop process) (Dreisinger D., 2003).

Kazakhstan's first commercial copper-leach solvent extraction-electrowinning plant, Central Asia Metals' Kounrad project, has been in operation since April 2012. The 10,000 tpa of copper cathode plant recovers copper from solutions derived from dump leaching of low grade stockpiles accumulated over nearly 80-year period of open pit mining. The commercial application of dump leach-SX-EW is not without its challenges during the sub-zero winter periods experienced at Kounrad. Nevertheless, the leach process and the plant have consistently met targets and produced high quality copper cathodes at the cost of production less than US$1/lb. Through successful expansion of copper leaching and recovery operation to the Western Dumps, plant productivity has been increased to 13,500 tpa.

The purpose of this study was to develop a hydrometallurgical technology for deep processing of substandard dumps, as well as tails of leaching for the Balkhash Mining and Processing Plant.

The paper presents the results of studies aimed at reducing the copper content in the leaching tails, increasing the extraction of copper into the productive solution.

To increase the extraction of copper from the ore, various acids and the combination of acids—sulfuric, nitric and hydrochloric—are used. Leaching of sulphide minerals often involves oxidative leaching.

Often, sulfuric acid is used as an oxidant and a solvent for metal sulfides. Some sulphides can be easily sulfatized from 50°C, with educing of hydrogen sulphide. Another group—chalcopyrite, chalcocite and covellite, begins to react with sulfuric acid at a higher temperature (150°C), with educing of sulfur dioxide into the gas phase. Almost completely the listed sulphides revealed with sulfuric acid when heated to 300°C.

The autoclave processes stand separately and occur at elevated pressures of oxygen and increased to 140–150°C temperature.

For the leaching of certain types of substandard raw materials, combinations of sulfate and chloride solutions are proposed (Zaiganov V.G., 2010).

In the paper were studied the nonoxidative and oxidative leaching of dumps and tails of leaching of dumps with using hydrochloric acid and a mixture of lactic acid and an oxidizer (hydrogen peroxide) as a solvent. Copper extracted from the obtained hydrochloric acid productive solutions by extraction, re-extraction is carried out with sulfuric acid.

2 RESEARCH AND RESULTS

2.1 Source materials

Samples of dump ore and tails from sulfuric acid leaching of the dumps of one of the enterprises of Kazakhstan were used. The copper content in the heap is 0.50%, in the tail, 0.33%. The main oxidized copper minerals in the samples are malachite, azurite, cuprite, sulphide minerals – chalcopyrite, covellite, chalcocite; iron minerals – siderite and pyrite.

X-ray spectroscopy revealed the presence in the ore of the following minerals of gangue: 80% Silicon Oxide – SiO_2, the rest: aluminum silicate – Al_2SiO_5, muscovite – $KAl_2(Si,Al)4O_{10}(OH)_2$, potassium hydrogen sulfate – $K(HSO_4)(H_2SO_4)$, aluminum sulfate – $Al_2(SO_4)_3$, kaolinite $Al_2Si_2O_5(Oh)_4$, andalusite – $Al_{1.920}Mn_{0.023}Fe_{0.057} (O(SiO_4))$, pyrophyllite – $Al_2Si_4O_{10}(OH)_2$, nacrite – $Al_2Si_2O_5(OH)_4$.

3 LEACHING

One-stage leaching of dump and tail samples was carried out with 100 g/l hydrochloric acid solution, as well as with a 100 g/l hydrochloric acid solution in a mixture with 15% hydrogen peroxide.

3.1 Nonoxidizing leaching

The influence of the duration of the process, the concentration of the initial hydrochloric acid and the solid-liquid (S:L) ratio on the extraction of copper in the solution was studied.

3.1.1 Effect of leaching duration

Experiments to study the effect of the duration on the extraction of copper from the samples were carried out at a ratio of S: L = 1: 2, a temperature of 20°C, and a duration of 0.5–5.0 hours with constant stirring. The weight of the initial samples was 65 g, the size was 0.1 mm.

The results show (Table 1) that increasing the leaching time from 0.5 to 4 hours leads to an increase in copper extraction from 18% to 34% from the dumps and from 18% to 29% from the leached tails.

Further increase in the duration of the process does not lead to an increase in copper extraction

3.1.2 Influences of acid concentration

Studies on the effect of the concentration of hydrochloric acid on the extraction of copper from samples were carried out with acid solutions of concentrations of 25, 50, 75, 100, 125 g/l, with a duration of 4 hours.

It is established (Figure 1) that when the concentration of hydrochloric acid rises from 25 to 100 g/l, copper extraction rises from 8% to 34% of the dumps, and from 10.0% to 29% from the leaching tails.

3.1.3 *The influence of the S:L ratio*

The studies were carried out at a temperature of −20°C, a hydrochloric acid concentration of 100 g/l, and a duration of 4 hours. The ratio of S:L varied from 1:1 to 1:5.

The results obtained (Table 2) show that S:L equal to 1:2 is the optimal ratio for extracting copper from dumps and tails of leaching.

Table 1. Extraction of copper from dump and tail samples, depending on the duration of leaching.

No.	t, h	Dumps leaching results			Tails leaching results		
		C^*, g/l	m^*, g	E^*, %	C^*, g/l	m^*, g	E^*, %
1	0.5	0.520	0.09	18	0.35	0.06	18
2	1.0	0.929	0.16	32	0.42	0.07	22
3	1.5	0.952	0.16	32	0.43	0.07	22
4	2.0	0.955	0.16	32	0.49	0.08	25
5	2.5	0.96	0.16	33	0.51	0.09	26
6	3.0	0.97	0.16	33	0.52	0.09	27
7	3.5	0.98	0.17	33	0.54	0.09	28
8	4.0	1.00	0.17	34	0.56	0.10	29
9	4.5	1.00	0.17	34	0.56	0.10	29
10	5.0	1.00	0.17	34	0.56	0.10	29

*C – *the content of copper in the productive solution.*
*m – *mass transferred from the initial sample into a solution of copper.*
*E – *copper extraction into solution.*

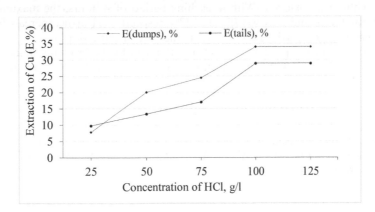

Figure 1. The effect of acid concentration on copper extraction into solution.

Table 2. Copper extraction depending on the S:L ratio.

S:L	Dumps leaching results			Tails leaching results		
	C_{Cu}, g/l	m_{Cu}, g	E, %	C_{Cu}, g/l	m_{Cu}, g	E, %
1:1	0.72	0.13	25.00	0.97	0.07	23
1:2	1.00	0.17	34.00	0.55	0.09	29
1:3	0.63	0.17	34.00	0.35	0.10	29
1:4	0.46	0.17	34.20	0.26	0.10	29
1:5	0.36	0.17	34.30	0.21	0.10	30

Table 3. Results of a leaching of dumps and tails with an oxidizer.

t, h	Dumps leaching results			Tails leaching results		
	C, g/l	m, g	E, %	C, g/l	m, g	E, %
0.5	1.5	0.15	47	0.65	0.072	34
1.0	2.03	0.203	63	0.71	0.078	37
1.5	2.10	0.210	66	0.73	0.080	38
2.0	2.25	0.225	70	0.83	0.091	43
2.5	2.30	0.230	72	0.84	0.092	44
3.0	2.26	0.226	71	0.85	0.094	45
3.5	2.35	0.235	73	0.86	0.095	45
4.0	2.41	0.241	75	0.90	0.099	47
4.5	2.41	0.240	75	0.90	0.099	47
5.0	2.41	0.240	75	0.90	0.099	47

3.2 Oxidative leaching

To study the influence of the oxidizer on the leaching of copper from the samples was used a solution of hydrochloric acid with a concentration of 100 g/l with adding an oxidizer—hydrogen peroxide to a concentration of 15%. The experiments were carried out under the following conditions: a ratio of S: L = 1: 2, a temperature of 20°C, a duration of 0.5–5.0 hours, a sample size of 0.1 mm.

The influence of the duration of the process on the extraction of copper was studied. The effect of acid concentration and the S:L ratio was not studied, but was taken on the basis of the results of the non-oxidative leaching.

The results (Table 3) show that the degree of copper extraction in the presence of the oxidizer, both from the dumps and from the tails, is higher compared to the degree of extraction in leaching without an oxidizer. With a leaching period of 4 hours, the maximum copper extraction from dumps and tails was 75% and 47%, respectively. Further increase in the duration of the leaching process does not lead to an increase in extraction.

The composition of the productive solutions obtained by the leaching of the dumps, g/l: Cu – 1.5–2.4, Fe – 1.0–1.5; and from the tails, g/l: Cu – 0.65–0.9, Fe – 0.5–0.8, respectively.

4   EXTRACTION OF COPPER FROM PRODUCTIVE SOLUTIONS

The resulting product solutions were sent for extraction refining and concentration.

We used extractants Lix 9858 NSC and Acorga 5510 to extract copper from chloride solutions obtained by leaching, and copper from dumps and tails.

4.1 Solvent extraction of copper

Extraction of copper from the productive solutions obtained by leaching the dumps of the composition, g/dm^3: Cu 2.4 g/dm^3, Fe–1.5 was carried out at pH of the medium: 1.5; 2.5; 3.5; 4. The ratio of organic and aqueous phases is 1: 1, the temperature is 20°C, the phase contact time is 300 sec, and the settling time is 300 sec. The extraction was carried out with a solution of sulfuric acid with a concentration of 200 g/dm^3. The extraction results were estimated by the extraction value E (%) of metals in the organic phase. The concentration of copper in the aqueous phase was determined by titrimetric analysis.

It can be seen from the obtained results (Figure 2) that both extractants allow selective extraction at a pH above 3. The degree of copper extraction in the pH range of 0.5–3 varies for Lix 9858 NSC from 5% to 92%, for the extractant Acorga 5510 from 3% to 89%.

At pH more than 3.0, the degree of copper extraction with both extractants is reduced. This can be explained by the loss of a certain amount of copper with the onset of hydrolysis

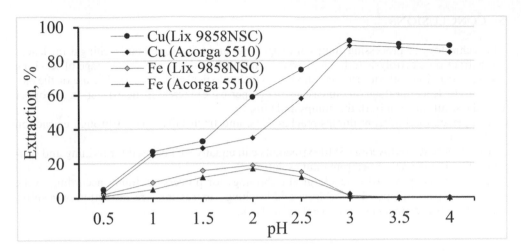

Figure 2. The extraction of copper and iron from productive solution.

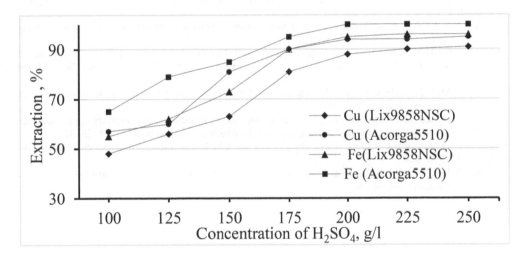

Figure 3. Stripping degree of copper and iron from loaded organic phase.

of copper chloride. The calculated copper and iron ß separation factors for Lix 9858 NSC at pH 3 are 733, for Acorga 5510–539, which allows for a deep separation of copper and iron.

4.2 *Stripping of copper from loaded organic phase*

The stripping of copper and iron from the organic phase with loaded copper was carried out with a solution of sulfuric acid with a concentration of 100–250 g/l at the following conditions: temperature –20°C, ratio of organic and aqueous phases: 1:1, stirring and settling for 5 minutes.

The obtained results show that the degree of copper extraction as well as the iron from the organic phase obtained using LIX9858 NSC is higher than using the extractant Acorga 5510. Optimal results for the extraction of copper and iron in the re-extract are obtained at the concentration of sulfuric acid in the solution going to the stripping 200 g/l (Figure 3).

Concentration and refining of copper was carried out according to the scheme 3E-1 W-2S, calculated according to the program Isocalk by BASF. As a result, a solution, g/l: 36–40 Cu, 0.1–0.2 Fe, is obtained, which suitable for copper production by electrolysis.

The results obtained correspond to the calculated parameters of the Isocalk program.

5 CONCLUSIONS

The technology of processing dumps of copper ores and tails from their leaching using leaching-extraction-electrolysis technology when using hydrochloric acid at the stage of leaching and re-extraction with sulfuric acid showed its technological effectiveness. The use of the oxidizer in the leaching stage (in our case hydrogen peroxide) increases the extraction of copper into the solution from both the dumps and tails.

Copper extraction from dumps reaches 75%, and from tails – 47% (through extraction – 74–80%).

Lix 9858 NSC and Acorga 5510 extractants can equally well be used for refining and concentrating copper solutions from dilute solutions.

The technology of hydrometallurgical technology for processing dumps according to the scheme—"spent dumps—hydrochloric acid leaching—extraction—re-extraction with sulfuric acid"— is recommended for an enlarged check.

REFERENCES

Davenport G.W., King J.M., Schlesinger E.M., Biswas A.K. 2002. *Extractive Metallurgy of Copper*. Oxford. Elsevier.

Dreisinger D. 2003. Copper leaching from primary sulfides: Options for biological and chemical extraction of copper. *Hydrometallurgy* 83(1):10–20.

Mamyrbayeva K.K., Luganov, V.A., Eshmoldayeva A.B. Processing of Aktogai (Kazakhstan) mixed copper ore. 2015. *The XVI Balkan Mineral Processing Congress*, Belgrad, Serbia, June 17–19: 713–718.

Petersen J. 2016. Heap leaching as a key technology for extraction of values from low-grade ores—A brief overview. *Hydrometallurgy* 165 Part 1: 206–212.

Roberto A. Bobadilla-Fazzini, Pérez A., Gautier V., Jordan H., Parada P. 2017. Primary copper sulfides bioleaching vs. chloride leaching: Advantages and drawbacks. *Hydrometallurgy* 168: 26–31.

Sinclaira L., Thompson J. 2015. In situ leaching of copper: Challenges and future prospect. *Hydrometallurgy* 157: 306–324.

Zaiganov V.G. 2012. Objective necessity of conversion of primary mineralization technology. *Mountain Journal of Kazakhstan*: 48–52.

WASTES – Solutions, Treatments and Opportunities II – Vilarinho, Castro & Lopes (Eds)
© 2018 Taylor & Francis Group, London, ISBN 978-1-138-19669-8

Construction wastes application for environmental protection

A.S. Sakharova, L.B. Svatovskaya, M.M. Baidarashvili & A.V. Petriaev
Emperor Alexander I St. Petersburg State Transport University (PGUPS), St. Petersburg, Russia

ABSTRACT: In this article the authors examine the construction wastes such as autoclave foam concrete and silicate brick. Studies have shown that these materials have geoecoprotective properties, because they can neutralize heavy metal ions. Therefore the authors suggest applying these materials in geoecoprotective technology of transport construction to reduce environment pollution. The determination of the geoecoprotective ability of the construction wastes and dependences between different geoecological parameters are being shown. The use the gabion structure as the sewage treatment plant near roads and railways is proposed. The construction wastes in the dispersion form are used as gabion geoecoprotective fillers.

1 INTRODUCTION

Environment pollution near road and railway structures results from unsafe transportation of dangerous loads. Heavy metal ions are the most dangerous pollutants from road and railway transport. As opposed to other pollutants heavy metal ions are kept in soils for a long time even when a pollution source is eliminated. Periods removal of heavy metals from soils are several thousand years. Heavy metals have the ability to migrate to the plant, into the lakes and rivers, groundwater and underground water and also they can accumulate in the food chain. Disorder of the cardiovascular system and the occurrence of severe allergies are the result of contact human with the pollutant. Most of the heavy metals have embryotropic and carcinogenic properties. They are genetic poisons that are accumulated in human body with a long-term effect (Davydova & Tarasov 2002).

Chemical composition of storm sewage from road pavement has showed that concentrations of some pollutants significantly exceed maximum allowable concentration (MAC), from 2 to 19 times.

Study (Kazantsev et al. 2007) have proved that the railway transport is one of the environment pollution sources. According to the research (Belkov 2013), the samples of crushed ballast contain copper at a concentration that exceeds the MAC 30–190 times and 2–6 times for lead. Possible sources of ballast contamination by copper are: line electric locomotives, railroad cars.

Thus, the main aim of this research is the study of construction wastes application for environmental protection through the use of geoecoprotective technology. This purpose can be achieved by using such construction wastes as autoclave foam concrete and silicate brick.

The main tasks of the research are to determine the geoecoprotective properties of listed materials and to develop technology in order to protect the environment from harmful effects of heavy metal ions.

The analysis of modern transport construction structures has shown that some of them can have geoecoprotective properties. Modernization of technological operations during the construction of such structures are based on the replacement of natural materials which are used in their constructional design for construction wastes having geoecoprotective properties. The use of these geoecoprotective transport structures will make it possible to solve two ecological problems. Firstly they minimize negative impacts on the environment near roads and railways. Secondly they contribute to the new application of construction wastes (Titova 2005).

2 DETERMINATION OF CONSTRUCTION WASTES GEOECOPROTECTIVE PROPERTIES

In that work geoecoprotective properties of construction wastes are considered to be absorbing properties against heavy metals ions. Scientists of the Engineering Chemistry and Natural Sciences Department of Emperor Alexander I St. Petersburg State Transport University have carried out such a research for the last 20 years (Svatovskaya et al. 2010). They have identified geoecoprotective properties of such materials as non-autoclave foam concrete, granulated blast-furnace slag, phosphogypsum, crushed shungite ballast, etc.

In our opinion, construction wastes such as autoclave foam concrete and silicate brick are mi-neral geoantidotes. Mineral geoantidotes («geo» – «earth», «antidotes» – «antitoxin») are solid difficultly soluble technogenic, artificial or natural substances, in dispersion form. They have the composition corresponding to natural composition of the crust (calcium and magnesium silicates and hydrosilicates). They can decontaminate of different pollutants, for example HMI, by forming of difficultly soluble substances. It is spontaneous reaction ($\Delta G^0_{298} < 0$). Such materials have geoecoprotective properties (Svatovskaya et. al. 2012a).

2.1 Laboratory experiment conditions

Four fractions of materials were selected for the research: 0.14–0.315 mm, 0.315–0.63 mm, 0.63–1.25 mm, 1.25–2.5 mm. First of all, the materials were cut into smaller pieces. Then they were placed in a concrete mixer. After that they were ground in a mortar and scattered through sieves with corresponding cell sizes.

Studies of construction wastes geoecoprotective properties were carried out with the standardized test solutions of heavy metal salts. Those solutions had concentration of 10^{-5}, 10^{-4} and 10^{-3} mol/l that exceeded maximum allowable concentration (MAC) 200 times and more. The following salts were used for the solutions: $Cd(NO3)2$ and $Pb(NO3)2$. Laboratory experiment conditions are presented in Table 1.

Determination of heavy metal ion concentration in a solution was performed with the help of an electronic analyzer «Expert- 001», using «ALICE» series ion selective electrodes, before and after the interaction of that solution with the tested materials. During the experiment the measuring flasks were filled with 100 ml of the standardized test solution which contained of heavy metal ions of different concentrations. Then 1 gram of various fraction tested materials was added in the flasks. The suspensions were stirred up every 5–10 minutes. The contact time was 1 hour that was specified by the adsorption-desorption equilibrium. As the time passed the materials were separated from the test solutions and put on a filter paper. The final concentration of heavy metal ions was determined in each sample. It was a self-reaction at the air temperature of 293 K.

2.2 Laboratory experiment results

The notion of geoecoprotective activity has been introduced by the authors to use quantitative characteristic of geoecoprotective properties of mineral geoantidotes. Geoecoprotective

Table 1. Laboratory experiment conditions.

Measurement units of the initial concentration of HMI	Initial concentration of heavy metal ions (HMI)					
	Cd(II)	Cd(II)	Cd(II)	Pb(II)	Pb(II)	Pb(II)
mol/l	10^{-5}	10^{-4}	10^{-3}	10^{-5}	10^{-4}	10^{-3}
mmol/l	0.01	0.1	1	0.01	0.1	1
mg/l	1.12	10.75	104.43	2.28	24.40	250.64
The excess ratio of maximum allowable concentration (MAC)	224	2150	20886	380	4067	41773

activity, A_{gep} is ability of mineral geoantidotes (construction wastes) to neutralize pollutants independently of the mechanism of the purification process. The specific geoecoprotective activity is the ratio of neutralized pollutants mass to unit mass of mineral geoantidote. Geoecoprotective activity is calculated by the formula (1):

$$A_{gep} = \frac{(C_i - C_r) \cdot V}{m} \qquad (1)$$

where A_{gep} = geoecoprotective activity of mineral geoantidotes, mg/g; C_i = initial concentration of heavy metal ions, *mg/l*; C_r = residual concentration of heavy metal ions, *mg/l*; V = solution volume, *l*; and m = mineral geoantidotes mass, g.

The mineral geoantidotes geoecoprotective activity values depending on various initial concentrations in solutions of HMI are given in Table 2. Thus, the value of Agep of the considered materials increases, when initial concentration of HMI is increased too. The values of geoecoprotective activity were calculated for the fractions of the construction waste 0.14–0.315 mm. At the same time calculations on formula (2) have showed that the treatment efficiency, Et, is decreased, when the metal cation concentration is increased in the initial solution.

Treatment efficiency was calculated by the formula (2):

$$E_t = \frac{C_i - C_r}{C_i} \cdot 100 \qquad (2)$$

where E_t = treatment efficiency,%; C_i = initial concentration of pollutant, mg/l; and C_r = residual concentration of pollutant, mg/l.

The laboratory experiment results have shown that the most effective purification of polluted water by hydrosilicates is achieved when the initial content of polluted cation is 0.1 mmol/l. It is significantly higher than the level of real cadmium pollution on railways.

The dependence of the solutions treatment efficiency on the construction wastes dispersion degree at an initial concentration of HMI 10^{-3} mol/l is presented in Table 3.

Table 2. The dependence of construction wastes geoecoprotective activity on initial concentration of HMI in solutions (fraction of materials is 0.14–0.315 mm).

Construction waste	Geoecoprotective activity, A_{gep} mg/g, to HMI					
	Pb (II)	Pb (II)	Pb (II)	Cd (II)	Cd (II)	Cd (II)
Autoclave foam concrete	0,227	2,48	22,07	0,105	1,05	5,95
Silicate brick	0,226	2,41	10,73	0,100	0,99	3,66
Initial concentration of HMI, mol/l	10^{-5}	10^{-4}	10^{-3}	10^{-5}	10^{-4}	10^{-3}

Table 3. The dependence of the solutions treatment efficiency, %, on the construction wastes dispersion degree at an initial concentration of HMI 10^{-3} mol/l.

Construction waste	Dispersion degree, mm							
	0,14–0,315		0,315–0,63		0,63–1,25		1,25–2,5	
	Cd (II)	Pb (II)	Cd (II)	Pb (II)	Cd (II)	Pb (II)	Cd (II)	Pb (II)
Autoclave foam concrete	97	100	98	100	99	100	98	100
Silicate brick	89	100	90	100	91	97	91	89

The data of Table 3 show that the solutions treatment efficiency is 90–100% with the sizes of the materials fractions from 0.14 to 2.5 mm. But the best result is achieved for autoclave foam concrete during lead ions solutions treatment with any degree of dispersion.

3 PRACTICAL APPLICATION OF CONSTRUCTION WASTES USING GABION STRUCTURE

Geoecoprotective technology can involve the use of welded gabion structure, which is filled by different stone placeholders in the process of construction and installation work.

It is proposed instead of the standard gabion filler to use the filler from investigated geoecoprotective material in dispersion form such as cement clinker (Figure 1) (Sakharova et. al. 2016, Baidarashvili et. al. 2017). Such structure is arranged to release surface runoff contaminated by heavy metal ions from the drainage gutter along roads and railways to the nearby river (Figure 2) (Sakharova et. al. 2016, Baidarashvili et. al. 2017).

Welded gabion will consist of three sections. The two outer sections are filled by standard stone filler. The inner section is filled by filler from investigated material having geoecoprotective properties towards heavy metal ions. The inner surface of gabion structure section is covered by nonwoven geotextile having high filtration properties. Then it filled by geoecoprotective material.

Taking into account the real level of pollution from the road and railway transport the fraction of construction waste of 1.25–2.5 mm was chosen from the studied fractions. It is optimum for characteristics such as the culvert ability and runoff treatment efficiency.

The use this construction technology will allow to reduce or to minimize environmental pollution by heavy metal ions accumulating on the railway track subgrade and roadbed. Polluted wastewater is purified before it gets into the water resources.

Technological operations schematic diagram for drainage gutter is shown in Figure 3.

This technology was introduced during the work of drainage gutters on the line section Korsakov-Nogliki of Far Eastern Railway of Sakhalin Island. The gabion is filled by cement clinker of 1.25–2.5 mm fraction. This allowed reducing the concentration of copper ions in surface sewage from 11 mg/l that exceeded MAC 11 times, to 1 mg/l. Thus treatment efficiency was 91%.

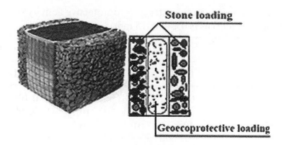

Figure 1.   Gabion structure with stone and geoecoprotective filler.

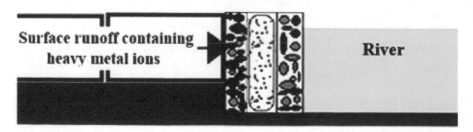

Figure 2.   Geoecoprotective gabion structure on release of drainage facility for surface runoff.

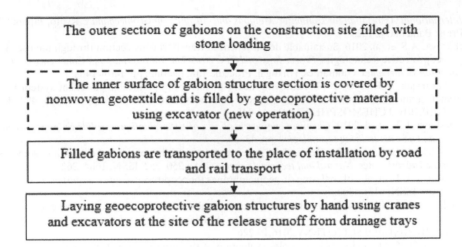

Figure 3. Technological operations schematic diagram for drainage gutter.

As the laboratory experiment results have shown that the most effective purification of polluted water by means hydrosilicates are achieved when the initial content of cation is 0.1 mmol/l that exceeded maximum allowable concentration (MAC) 2000 times for cadmium (II) and 4000 times for lead (II).

4 CONCLUSIONS

The dependence of the geoecoprotective activity of the investigated construction wastes on the initial concentration of the HMI in the solutions and the dispersion degree was established. The laboratory experiment results have shown that the most effective purification of polluted water by hydrosilicates is achieved when the initial content of polluted cation is 0.1 mmol/l. The best degree of dispersion for autoclave foam concrete is from 0.14 to 2.5 mm and for silicate brick is from 0.14 to 0.63 mm.

Investigations results showed that the examined construction wastes can be used as geoecoprotective filler in different road and railway structures in order to minimize negative impact of ions of heavy metals in the water resources. It is possible through the use of geoecoprotective gabion structure. Geoecoprotective gabion structure is sewage treatment plant (Sakharova et al. 2016). The use of gabion sewage treatment plant and construction wastes as a whole will allow to reduce or to minimize environmental pollution by heavy metal ions accumulating on the railway track subgrade and roadbed.

REFERENCES

Baidarashvili, M.M. et. al. 2017. The modern structure for storm sewage purification of roads. *Procedia Engineering* 189: 576–581. DOI.ORG/10.1016/J.PROENG.2017.05.091.
Belkov, V.M. 2013. *Land pollution of infrastructure*. Railway track and facilities, 7: 2–4.
Davydova, S.L. & Tarasov, V.I. 2002. Heavy metals as supertoxicants of XXI century. *Moscow: Publishing House Peoples Frendship University*. pp. 21–29.
Ecological substantiation of the scheme of disposal and storm water treatment with bridges and road painting, works head JP Bogdanov, Saint-Petersburg scientific research Institute of the Academy of municipal economy behalf of K.D. Pamfilova, 2001.
Kazantsev, I.Z. et. al. 2007. Influence of rolling stock on the content of heavy metals in soils and plants ROW railways. *Naturalistic series* 2 (52): 172–179.
Sakharova, A.S. et. al. 2014. Building wastes and cement clinker using in the geoecoprotective technologies in transport construction. *Proceedings of 14th IACMAG/The14th International Conference of the*

International Association for Computer Methods and Advances in Geomechanics. Kyoto, Japan: CRC Press: Balkema. p. 152.

Sakharova, A.S. et. al. 2016. Sustainable development in transport construction through the use of the geoecoprotective technologies. *Proceedings of ICTG 2016/The 3rd International Conference on Transportation Geotechnics*. Guimarães, Portugal: Procedia Engineering. pp. 1401–1408.

Sounthararajah, D.P. et. al. 2017. Removing heavy metals using permeable pavement system with a titanate nano-fibrous adsorbent column as a post treatment. *Chemosphere* 168: 467–473. DOI. ORG/10.1016/J.CHEMOSPHERE.2016.11.045.

Svatovskaya, L.B. et. al. 2010. *Chemical engineering bases of determination of solids geoprotective properties and new neutralizing technologies*. pp. 43–50.

Svatovskaya, L.B. et. al. 2012a. Research of geoecoprotective ability of cement clinker and some technogenic hydrosilicates. *Natural and technical science* 5(61): 250–252. ISSN: 1684–2626.

Svatovskaya, L.B. et. al. 2012b. Using of geomembranes for ecoprotective aims. *Transport construction* 8: 26–28. ISSN: 0131-4300.

Tedoldi, D. et. al. 2017. Spatial distribution of heavy metals in the surface soil of source-control stormwater infiltration devices—Inter-site comparison. *Science of The Total Environment* 579: 881–892. DOI.ORG/10.1016/J.SCITOTENV.2016.10.226.

Titova, T.S. 2005. Methodology of complex evaluation of the impact of new technologies on a geoecological situation. *Vniizht bulletin (railway research institute bulletin)* 5: 2. ISSN: 2223-9731.

Titova, T.S. & Potapov, A.I. 2010. *Solving ways of environmental problems of railroad transport*. Scientific, methodical. St. Petersburg: Gumanistika. pp. 546–547.

WASTES – Solutions, Treatments and Opportunities II – Vilarinho, Castro & Lopes (Eds)
© 2018 Taylor & Francis Group, London, ISBN 978-1-138-19669-8

Design of a laboratory scale circulating fluidized bed gasifier for residual biomass

D.A. Tibocha, D.C. Guío-Pérez & S.L. Rincón
Universidad Nacional de Colombia, Bogotá, Colombia

ABSTRACT: In order to take advantage of the solid waste generated by the Colombian agricultural industry, it is proposed to design an experimental unit at a laboratory scale that allows testing of different residual biomasses for gasification at different operating conditions. In the present work, the methodology used for the basic design of the different components of this unit is presented. The possibilities offered by the design in the use of different residual biomasses and the generation of expected thermal energy are analyzed. The following characteristics were previously selected for the design: gasification with air, semi-continous operation, circulating configuration, fluidization regimes from bubbling up to fast, thermal power of 10 kWth, possibility to gasify different types of residual biomass.

1 INTRODUCTION

Several strategies have been proposed during the last decades at the global level to reduce CO_2 and other greenhouse gas emissions. The use of renewable energies, the increment of efficiency in transformation processes are some of the main guidelines for achieving acceptable levels of pollutant emissions (IEA 2012). Among renewable energy sources, biomass opens up possibilities for reducing both emissions and dependence on fossil fuels (González-Salazar et al. 2014). In Colombia, due to its geographical location and variety of climates, residual biomass is suggested, on the one hand as a source with great potential for power generation and even for the synthesis of fuels and, on the other, as an alternative energy source in Non-Interconnected Zones (UPME 2010, UPME 2014). However, residual biomass generally presents high variability in properties, and in some cases, undesirable performance for use in energy conversion processes, such as high moisture, low ash melting temperature, and low heating value (Demirbas 2004).

Thermochemical transformation carried out in fluidized beds are characterized by being flexible to the type of fuel due to the high contact efficiency between the solid particles and the gaseous medium (Basu 2006), which appears to be advantageous for facing the problem of properties variability and allowing comparatively higher efficiencies. Circulating fluidized bed gasification processes combine the benefits of fluidized bed technology with the possibility for energy diversification offered by the gasification process, making possible the efficient transformation of biomass into thermal energy, power and a fuel gas of high added value (synthesis gas).

In the present work, the design of an experimental unit at laboratory scale is proposed aiming to investigate the performance of Colombian residual biomass in such process at different operating conditions. The design considerations are described together with the methodology used for the basic design of the different components. In the global design the following characteristics were previously selected:

- Gasification with air: the initial design considers only air gasification but should remain flexible to modifications in order to operate with different gasification agents.

– Semi-continous operation: as a laboratory facility, the unit should offer the possibility for steady state operation during some hours.
– Fluidization regimes from bubbling up to fast: operation with medium to high velocities of fluidization agent should be possible. A secondary air injection is considered so that fluidization velocity and air equivalence ratio can be controlled independently.
– Circulating configuration: the operation at higher velocities requires the circulation system. Operation at low velocities can be carried out at very low or no solids circulation rates.
– Thermal power of 10 kW$_{th}$: this scale is limited by the services installed capacity at the laboratory, but still offers the possibility of analyzing the whole process.
– Possibility of gasifying different types of residual biomass: the design should therefore consider that operation conditions may vary from one biomass to other.

2 METHODS

2.1 *Reactor*

The reactor has a circular section and an expansion at a middle height which compensates for the increase in the volume of the gases flowing inside. According to this, the reactor is divided into two zones: lower and upper (the latter of larger diameter). The design takes into account the properties of some types of residual biomass available in Colombia with high potential for thermochemical utilization (Guío-Pérez et al. 2016). The design process conditions (temperature, pressure, and composition of the gasifying medium) were selected based on a detailed literature review (750°C, 1 atm). The approximate composition of the product gas, process yield, and mass flows were calculated based on information extracted from literature and considering the composition and thermochemical properties of the biomasses selected (Table 1). For the calculations of the main flows and dimensions, the following methodology was used:

i. From the desired power output $(\dot{Q})$ and the lower heating value of the product gas $(LHVg)$, a theoretical volumetric flow of gas at normal conditions $(\dot{V}g)$ can be calculated, Equation 1. The LHVg depends on the gas composition, which can be determined experimentally or from equilibrium or kinetic models; for air-gasification of residual biomass in fluidized beds, calorific values between 3–7 MJ/Nm^3 are reported in literature (Lahijani et al. 2010, Tamer et al. 2016, Jin et al. 2016), these values were used for the design.

ii. The overall mass balance of the process relates the biomass $(\dot{m}_{bm})$ and the gasification medium $(\dot{m}_{gm})$ mass flows required to produce the desired product gas mass flow $(\dot{m}_g)$. The balance is given by the Equation 2. The gas mass flow can be the determined using the values for volumetric flow $(\dot{V}g)$ and the gas density (ρ_g), which can be calculated based on the composition and operating conditions (as reported in Basu, 2010). The air mass flow

Table 1. Proximate analysis on dry basis (percentages by weight).

Biomass	%C	%H	%N	%O	%S	HHV/kJ/kg
Rachis of palm (Lahijani et al. 2010)	43,52	5,72	1,2	48,9	0,66	15220
Sugar cane bagasse (Tillman et al. 2004)	48,64	5,87	0,16	42,82	0,04	19010
Coffee husks (Lugano et al. 2010)	49,4	6,1	0,81	41,2	0,07	18340
Rice husks (Wu et al, 2009)	39,78	4,97	0,46	40,02	0,2	14144
Poultry litter (Tillman et al. 2004)	38,1	5,6	3,5	34,9	0,6	14850
Acacia mangium (Pérez et al. 2014)	53,02	6,71	0,33	39,65	0,02	18759
Flooded gum (Pérez et al. 2014)	53,31	6,74	0,39	39,26	0,02	19083
Gmelina (Pérez et al. 2014)	52,73	6,96	0,47	39,12	0,02	18896
Patula pine (Pérez et al. 2014)	55,01	7,21	0,81	36,72	0,01	19154
Thinleaf pine (Pérez et al. 2014)	54,45	7,04	0,52	37,79	0,01	19031

(gasifying medium, $\dot{m}_{mg}$) required for the gasification can be calculated from Equation 3. Here, the stoichiometric ratio air-fuel for complete combustion (m_{th}) and the equivalence ratio (ER) are used. The recommended value of ER for biomass gasification is between 0.2 and 0.3 (Enden et al. 2004, Basu 2010), so that the amount of oxygen required is guaranteed. Combinig Equations 2 and 3, the mass flow of biomass is obtained as in Equation 4.

iii. The diameter of the gasifier is defined taking into account the flows to be handled (volumetric flows of gasifying medium and product gas) and the desired fluid dynamic behaviour (superficial velocity, U, corresponding to the selected regime). For this purpose, the velocity limits in the upper and lower zones of the reactor were located on the Grace diagram (Kunii et al. 1990) as shown in Figure 1a. The dimensionless particle diameter (dp*) and the dimensionless velocity (u*), given by Equations 5 and 6. The operating range for the lower zone of the reactor was fixed between the onset of the bubbling regime and upper velocity limit of the fast regime. For the upper zone the onset of the fast regime was used as reference. The mentioned areas are presented in Figure 1a.

iv. The cross-section area required for the reactor in the upper and lower zones can be determined using Equation 7, where the volumetric flux of product gas is used for the upper and the flux of gasification medium for the lower part. The mentioned limits were calculated for different mean particle sizes of the bed material (Fig. 1b). The diameters of the upper and lower sections of the gasifier were then selected making a compromise between the flow conditions in the two sections and the mean particle size. Silice sand was selected as bed material with a mean particle diameter (dp) of 200 μm and a particle density of 2600 kg/m³. According to this, the diameter selected for the lower zone was 6.35 cm (2 ½ in) and for the upper zone 7.62 cm (3 in). With these diameters the operating window is ensured for the expected flows and the flexibility to work in different fluid dynamic regimes is given.

v. Once the diameters have been defined, the actual volumetric flows of the product gas and power output can be calculated for the selected biomasses. Product gas volume flow of 6–100 Nm³/h and power output of 3–54 kW could theoretically be reached in the unit. The airflow required varies from 1.3 to 27.5 Nm³/h, and the biomass flux between 1.2 and 18 kg/h.

vi. The height of the lower section of the reactor was defined based on the assumption that at low velocity fluidization conditions the bottom zone of the reactor should contain the bed mass, while the upper section would correspond to the freeboard. The pressure drop at minimum fluidization velocity was calculated and the corresponding void fraction was measured experimentally for the selected bed material (at room conditions), obtaining $\varepsilon_{mf} = 0.58$. The pressure drop due to the bed material was then used to calculate the approximate height required for the bed, using the Equation 8 (Kunii et al. 1991). A height of 0.4 m was selected for the lower section. For the upper section, the residence time of the gas phase was taken into account, since the reactor should offer enough time for the gases to convert (Enden et al, 2004). A height of 0.8 m corresponding to an average residence time of the gases (τ) of 0.15–1.7 s was selected as appropriate. The average residence time was obtained from the free volume in the reactor (V) and the volumetric flow of the product gas $\left(\dot{V}\right)$, as in Equation 9.

2.2 Distributor

Among the main types of distributors (Kunii et al. 1991, Basu 2010), the perforated plate was selected because of its easy construction and operation, low cost and suitability for small units; and the calculation procedure suggested by (Basu 2010) was followed. The design of the distributor uses the relationship between the pressure drop in the distributor (ΔPd) and the bed (ΔPb) as the main parameter (Enden et al. 2004). The recommended values for this relationship are between 0.15 and 0.3 for bubbling beds (Kunii et al. 1991) and between 0.15 and 0.25 for circulating fluidized beds (Basu 2010). According to this, the pressure drop in the distributor is given by Equation 10. The gas velocity at the orifice (U_o) is given by the Equation 11, where the

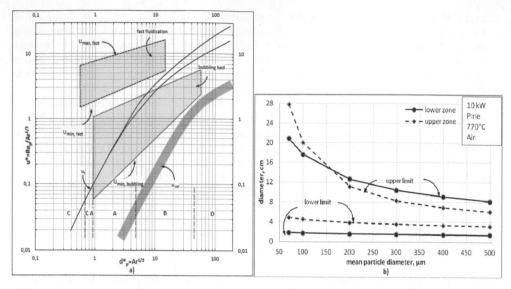

Figure 1. a) Grace diagram indicating the regions of operation of the reactor. b) Range of diameters suitable for the reactor using different particle sizes.

friction coefficient inside the hole, C_D, is given by Zenz (1981) as 0.8 and $\rho_{g,or}$ is the gas density passing through the hole. Orifice velocities for a circulating fluidized bed distributor between 30 and 90 m/s are considered adequate (Basu 2010). The number of holes in the distributor (N) is determined from the fraction of open area in the plate (Equation 12).

2.3   Solids separation system

In this work, the cyclone was selected as solids separation device. The most common design used in circulating fluidized bed gasifiers is the vertical axis cyclone with a tangential inlet and axial discharge (Basu 2010, Hoffman et al. 2002). There are multiple geometric relationships for this type of cyclone formulated by different authors (Stairmand 1951, Swift 1969, Lapple 1951). A high-efficiency cyclone design was chosen with the geometric relations presented by Swift (Swift 1969). Due to the comparatively higher load of solids typical of circulating beds, the separation efficiencies may be lower than expected. Therefore, some variations to the design recommended in literature (Hugi et al. 1998) were included: the vortex tube is eliminated and the inclination angle of the inlet duct is increased to 13° to prevent the stagnation of solids in the duct and direct the solids flux downwards.

2.4   Solid recirculation system

This system has the function of transporting the solids from a lower pressure section (cyclone) to a higher pressure one (bottom of the reactor). A proper operation is achieved by generating a pressure drop that counteracts this difference, so that the relation in Equation 13 is fulfilled; where ΔP_{cyc} = pressure drop in the cyclone, ΔP_b = pressure drop in the bed, ΔP_{sb} = pressure drop in the standpipe and ΔP_N = pressure drop in the N valve. In the present design, the N-type valve was selected. This design promises a comparatively lower pressure drop than in the U-type (traditional loop-seal design), thus reducing the required compression power (Peining et al. 2014), and improving at the same time the stability in solid recirculation. The geometries of the reactor and cyclone were used together with the properties of the bed material to propose the dimensions of the N-valve and to calculate the pressure drops in the circulating loop.

The calculation was repeated until the conditions for geometry and pressure drop were fulfilled. The pressure drop was obtained by a modification of the Ergun equation (Equation 14), using the relative velocity (Δu), the length of the valve (L) and the properties of the gas and the bed material.

2.5 Feeding systems

The feeding systems that transport the sand and the biomass into the reactor consist of hopper and screw sets. The design is based on recommendations from different authors (Bates 2000, Carson et al. 1992, Maynard et al. 2013, Marinelli et al. 1992). The screw is divided into three sections: the first (that connects with the hopper output) with variable diameter and constant pitch, the second with constant diameter and variable pitch, and the third with constant diameter and pitch. This design is intended to prevent stagnation of the material (Jenike 1964). The dimensioning of the hopper is done based on the Jenike procedure (Jenike 1964), a slot outlet, recommended when a screw feeder is used (Carson 1992), was selected. For biomass, it is suggested to make a design based on the mass flow pattern (Marinelli et al. 1992). A hopper angle of 70° was chosen in order to guarantee the flow pattern.

2.6 Heating system

The heating system provides the necessary energy for the reaction. The energy requirement is established by the overall energy balance of the process, Equation 15; where ΔH_e = enthalpy of the entrances, ΔH_p = enthalpy of products, ΔH_r = enthalpy of reaction. A cold-gas gasification efficiency of 0.8 was used according to Basu (2010). The general reaction for the balance is given by Equation 16.

Table 2. Equations.

$$\dot{V}_g = \left(\frac{\dot{Q}}{LHV_g} \right) \quad (1)$$

$$\dot{m}_g = \dot{m}_{bm} + \dot{m}_{gm} \quad (2)$$

$$\dot{m}_{air} = \dot{m}_{bm} \left(ER_{m\,th} \right) \quad (3)$$

$$\dot{m}_{bm} = \frac{\dot{m}_g}{\left(1 + ER_{m\,th}\right)} \quad (4)$$

$$dp^* = dp \left[\frac{\rho_g\left(\rho_s - \rho_g\right)g}{\mu^2} \right]^{\frac{1}{3}} \quad (5)$$

$$u^* = dp \left[\frac{\rho_g^2}{\mu\left(\rho_s - \rho_g\right)g} \right]^{\frac{1}{3}} \quad (6)$$

$$A = \frac{\dot{V}}{U} \quad (7)$$

$$\frac{\Delta Pb}{Lmf} = \left(1 - \varepsilon_{mf}\right)\left(\rho_s - \rho_g\right)\frac{g}{gc} \quad (8)$$

$$\tau = \frac{V}{\dot{V}} \quad (9)$$

$$\frac{\Delta Pd}{\Delta Pb} = 0.15 - 0.25 \quad (10)$$

$$U_o = C_D \left[\frac{2\Delta Pd}{\rho_{g,or}} \right]^{0.5} \quad (11)$$

$$N\frac{\pi}{4}d_o^2 = \frac{U}{U_o}\frac{\rho_g}{\rho_{g,or}} \quad (12)$$

$$\Delta P_b + \Delta P_{cyc} < \Delta P_N + \Delta P_{sp} \quad (13)$$

$$\frac{\Delta P}{L}g_c = 150\frac{\left(1 - \varepsilon_{mf}\right)^2}{\varepsilon_{mf}^3}\frac{\mu_g \Delta u}{\left(\phi_s d_p\right)^2} + 1.75\frac{1 - \varepsilon_{mf}}{\varepsilon_{mf}^3}\frac{\rho_g \left(\Delta u\right)^2}{\phi_s d_p} \quad (14)$$

$$\Sigma_{l=1}^{L}\dot{m}_e\Delta H_e = \Sigma_{l=1}^{L}\dot{m}_p\Delta H_p + \Delta H_r \quad (15)$$

$$\dot{n}_{bm}C_xH_yN_zO_w + \dot{n}_{air}\left(O_2 + 3.76N_2\right) \rightarrow$$

$$\dot{n}_{air}\left(x_{CO}CO + x_{CO_2}CO_2 + x_{CH_4}CH_4 + x_{H_2}H_2 + x_{N_2}N_2\right) + \dot{n}_{H_2O}H_2O \quad (16)$$

3 RESULTS AND DISCUSSION

The mass balance of the gasification process was obtained using the methodology described above. The necessary mass flows of biomass are calculated to meet the design energy requirement. The results obtained are presented in Figure 2b; all streams are presented as a range since biomasses of different characteristics were considered as fuel and the unit is designed to work at different fluidization conditions. In gasification of poultry litter, for example, significantly higher mass flow rates are required to fulfil the design parameters because of its comparatively lower stoichiometric requirement of air. Comparatively, the rachis palm is able to produce a larger output power, since the calorific value of the gas produced from this biomasss is higher; however, its bulk density is considerably lower, which demands higher feed rates. The design of the circulating fluidized bed experimental unit achieved using the methodology here presented is shown in Figure 2a.

Through a simple and systematic calculation procedure the basic design of a laboratory scale gasifier for residual biomass in a circulating fluidized bed was completed. Fundamental principles of fluidization, mass and energy balances and flow ratios were used. Information of product gas properties and recommended design ranges from specialized sources served as the basis for calculations. The experimental unit designed allows the use of different residual biomasses operating under different fluidization regimes from bubbling to fast. The design also offers flexibility in its construction, so that the geometry of each component can be changed for investigation purposes. The possibility of adapting the unit to gasification with air and steam mixtures is foreseen, for which minimum changes are necessary in the design. The results are important for the development of thermochemical processes in fluidized beds in Colombia, as they open the possibility for assesing the energy use of agricultural solid waste in such units.

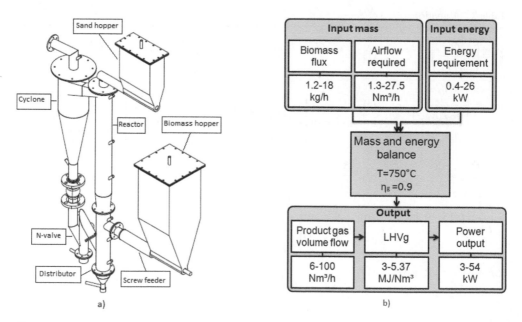

Figure 2. a) Experimental unit of circulating fluidized bed gasification. b) Overall balance of the process.

REFERENCES

Basu, P. 2006. *Combustion and Gasification in Fluidized Beds*. Boca Raton: Taylor & Francis.

Bates, L. 2000. *Guide to the Design, Selection, and Application of Screw Feeders*. Professional Engineering Publishing.

Carson, J.W. How to Design Efficient and Reliable Feeders for Bulk Solids. *Jenike & Johanson incorporated.*

Demirbas, A. 2004. Combustion characteristics of different biomass fuels. *Progress in Energy and Combustion Science* 30: 219–230.

Enden, P.J. van den. & Lora, E.S. 2004. Design approach for a biomass fed fluidized bed gasifier using the simulation software CSFB. *Biomass and Bioenergy* 26: 281–287.

Gonzalez-Salazar, M.A., Morini, M., Pinelli, M., Spina, P.R., Venturini, M., Finkenrath, M. & Poganietz, W. 2014. Methodology for biomass energy potential estimation: Projections of future potential in Colombia. *Renewable Energy* 69:488–505.

Guío-Pérez, D.C. et al. 2016. *Eur. Biomass Conf. And Exhib*. EUBCE. p. 9.

Hoffmann, A.C. & Stein, L.E. 2002. *Gas Cyclones and Swirl Tubes: Principles, Design and Operation*. Berlin: Springer-Verlag.

Hugi, E. & Reh, L. 1998. Design of Cyclones with High Solids Entrance Loads. *Chemical Enineering Technology* 21 (9): 716–719.

International Energy Agency (IEA) 2012. *Technology roadmap, bioenergy for heat and power*.

Jenike, A.W. 1964. *Storage and Flow of Solids*, Bul. 123, Utah Engineering Experiment Station.

Jin, W.K., Hee, M.C. & Bo, H.K. Gasification and tar removal characteristics of rice husk in a bubbling fluidized bed reactor. *Fuel* 181: 942–950.

Kunii, D. & Levenspiel, O. 1991. *Fluidization Engineering*. 2nd ed. London: Butterworth-Heinemann.

Lahijani, P. & Zainal, Z.A. 2010. Gasification of palm empty fruit bunch in a bubbling fluidized bed: A performance and agglomeration study. *Bioresource Technology*: 2068–2076.

Lapple, C.E. 1951. Process Use Many Collector Types. *Chemical Engineering* 58: 144–151.

Marinelli, J. & Carson, J.W. 1992. Solve Solids Flow Problmes in Bins, Hoppers and Feeders. *Chemical Engineering Progress*: 22–28.

Peining, W., Junfu, L., Wenchong, X., Hairui, Y. & Man, Z. 2014. Inpact of loop seal structure on gas solid flow in a CFB system. *Powder Technology* 264: 177–183.

Pérez, J., Ramos, S. & Barrera, R. 2014. Análisis energético y exergético como herramienta de selección de biomasa como materia prima para la producción de boisyngas para motores de combustión.

Rhodes, M. 2008. *Introduction to Particle Technology*. Wiley.

Schulze, D. 2008. *Powders and Bulk Solids Behavior, Characterization, Storage and Flow*. Berlin: Springer-Verlag.

Stairmand, C.J. 1951. The design and performance of cyclone separators. *Trans. Instn. Chem. Engrs*. 29: 356–383.

Swift, P. 1969. Dust Control in Industry. *Steam Heat Engineer* 38: 453–456.

Tamer, I., Abid, E., Monteiro. E. & Rouboa A. 2016. Eulerian-Eulerian CFD model on fluidized bed gasifier using coffee husks as fuel. *Applied Thermal Engineering: 1391*–1402.

Tillman, D.A. & Harding, N.S. 2004. *Fuels of Opportunity: Characteristics and Uses in Combustion Systems*.

Unidad de Planeación Minero Energética (UPME). 2010. Atlas de potencial energético de la Biomasa Residual en Colombia, Colombia.

Unidad de Planeación Minero Energética (UPME). 2014. Resumen ejecutivo. Plan indicativo de expansión de cobertura de Energía eléctrica 2013–2017, Colombia.

Wu, C., Yin, X.L., Ma, L., Zhou, Z. & Chen, H.. 2009. Operational characteristics of a 1.2 MW biomass gasification and power generation plant. *Biotechnology advances* 27: 588–592.

Zenz, F.A. 1981. Elements of Grid Design. *Gas Particle Industrial Symposium*, Engineering Society, Pittsburgh.

WASTES – Solutions, Treatments and Opportunities II – Vilarinho, Castro & Lopes (Eds)
© 2018 Taylor & Francis Group, London, ISBN 978-1-138-19669-8

Efficient activated carbons from chars of the co-pyrolysis of rice wastes

D. Dias, M. Miguel & N. Lapa
LAQV, REQUIMTE, Departamento de Ciências e Tecnologia da Biomassa, Faculdade de Ciências e Tecnologia, Universidade Nova de Lisboa, Lisboa, Portugal

M. Bernardo, I. Matos & I. Fonseca
LAQV, REQUIMTE, Departamento de Química, Faculdade de Ciências e Tecnologia, Universidade Nova de Lisboa, Lisboa, Portugal

F. Pinto
Laboratório Nacional de Energia e Geologia, Unidade de Bioenergia, Lisboa, Portugal

ABSTRACT: Chars obtained from the co-pyrolysis of rice husk (50% w/w) and polyethylene (50% w/w) were converted to activated carbons by physical activation. A fraction of the resulting activated carbon (PCPA) was submitted to chemical treatment with HNO_3 for functionalization purposes (PCAPCT). Both activated carbons were used in Cr(III) removal assays from liquid-phase. A commercial activated carbon, either submitted or not submitted to HNO_3 treatment (CACCT and CAC, respectively) was used for comparison purposes. During Cr(III) removal assays, two different solid/liquid ratios (S/L) were tested: 5.0 and 10.0 g L^{-1}. Concerning Cr(III) removal, PCPA was more efficient than CAC; with a S/L ratio of 10.0 g L^{-1}, PCPA was able to remove 99.9% of Cr(III) by precipitation. On the other hand, PCPACT performed slightly better with a S/L ratio of 5 g L^{-1} than CACCT, with 50.9% Cr(III) removed only by adsorption mechanism.

1 INTRODUCTION

Considered as an essential source of nourishment, rice is one of the most important cereals in the world. As one of the major producers in Europe and the biggest European consumer of rice, Portugal generates a large amount of wastes from the cultivation and harvesting of rice (e.g., rice straw and polyethylene), from processing (e.g., rice husk), and from packaging (e.g., different plastic polymers). These wastes are commonly reused in ways that disregard their energetic content (Alvarez et al., 2014; Mohammed et al., 2016).

The production of biochars through pyrolysis or gasification of agricultural wastes has been a recurring topic vastly studied in the last years (Ahmad et al., 2014; Inyang e Dickenson, 2015; Shackley et al., 2012). Biochars have been further tested in applications such as soil bioremediation and removal of contaminants from wastewaters through precipitation or adsorption (Tan et al., 2015).

In order to improve the ability of biochars to remove pollutants, physical or chemical activations can be applied in order to convert the biochars into activated carbons. These materials may present high porosity, with minerals or functional groups on their surfaces, according to the type of activation used. In the case of physically activated carbons, the resulting materials can be submitted to acid chemical treatment, further improving their adsorption capacity through ionic exchange mechanism, which combines well with the highly porous structure previously developed during the physical activation process (Ternero-Hidalgo et al., 2016; Yahya et al., 2015).

Even though several studies with chars obtained from rice wastes can be identified in literature, only few of them were dedicated to convert chars into activated carbons (Kalderis et al., 2008; Sugashini e Begum, 2015). Also, only rice husk is used as feedstock in these studies, instead of a blend of rice husk and polyethylene to improve their composition. Furthermore, none of these studies focus on Cr(III) removal only by adsorption mechanism. The majority of them considers both the adsorption and precipitation together, which makes difficult to understand whether precipitation is more important that adsorption, and vice-versa.

According to a report on raw materials in the European Union (EU), chromium is one of the 21 most critical raw materials for EU industrial sector: it has a high economic value and is largely used in tannery and metallurgic industries (European Commission, 2014). Therefore, its recovery from industrial wastewaters is a valuable asset in terms of economic and environmental sustainability. Cr(III) is commonly removed from industrial wastewaters by precipitation, resulting in a highly concentrated sludge with low economic value, due to contamination with other compounds. This makes difficult the Cr(III) recycling (Ma et al., 2016). Through adsorption, Cr(III) can be recovered from the activated carbons, which in turn can be regenerated and reused for a new adsorption process.

In this context, this work aims to study Cr(III) removal, from liquid effluents, using activated carbons produced from low-cost byproducts—chars obtained in the co-pyrolysis of wastes from rice production. Also, commercial activated carbon, with and without acid treatment, was used for comparison purposes.

2 MATERIALS AND METHODS

2.1 Physical activation and acid treatment assays

Details concerning the co-pyrolysis biochar production are referred in previous works (Dias et al., 2017; Pinto et al., 2016). Co-pyrolysis biochar was physically activated in a quartz reactor, which was heated in a vertically tubular oven at 800°C for 4 h, under a 150 cm^3 min^{-1} CO$_2$ flow. Part of the resulting activated carbon (PCPA) and the commercial activated carbon (CAC) (Norit GAC 1240) were submitted to a chemical treatment with HNO$_3$ (13 M), with a S/L ratio of 50.0 g L^{-1}, while stirred and heated in a silicone bath, at 90°C, for 6 h. The treated activated carbon (PCPACT) and the treated CAC (CACCT) were then washed with deionized water until stable pH. Both carbons were dried at 105°C until constant weight, and sieved to a particle size bellow 100 μm.

2.2 Characterization of materials

The obtained activated carbons (PCPA; PCPACT) and the commercial activated carbons (CAC; CACCT) were submitted to the following analyses:

i. Textural analysis—consisted in determining the surface area, pore volume and pore size distribution, through N$_2$ adsorption-desorption isotherm at 77 K, after sample degasification, under vacuum conditions, at 150°C;

ii. pH$_{pzc}$ – 0.1 M NaCl solutions with an initial pH (pH$_i$) between 1 and 12 were prepared. 0.05 g of activated carbon was added to 10 mL of each 0.1 M NaCl solution; the mixtures were shaken in a roller-table device, at 150 rpm, for 24 h; at the end of the shaking period, the final pH (pH$_f$) was measured. The pH$_{pzc}$ value corresponds to the point in which pH$_i$ = pH$_f$.

2.3 Cr(III) removal assays

Batch studies were performed with all activated carbons under the following conditions: Cr(III) initial concentration = 70 mg L^{-1}; Initial pH = 4.5; S/L ratios = 5.0 and 10.0 g L^{-1}; Mixing time = 24 h. A pH of 4.5 was used in order to prevent the precipitation of Cr(III), which begins to occur above 5.0. All the Cr(III) removal assays were conducted in a roller-table shaker, under constant mixing of 150 rpm. A standard Cr(NO$_3$)$_3$ solution of 1000 mg Cr(III) L^{-1} in 0.5 M of HNO$_3$ (Merck) was used for preparing a 70 mg L^{-1} Cr(III) solution. After the batch experiments, the samples were filtered through cellulose nitrate membranes, with 0.45 μm porosity. The pH was then measured and Cr(III) concentration quantified through ICP-AES.

The removal efficiency, $\eta(\%)$, and adsorbent uptake capacity, q_e (mg g^{-1}), were calculated by using equations 1 and 2, respectively:

$$\eta(\%) = \frac{\left(C_o - C_f\right)}{C_o} \times 100 \qquad (1)$$

$$q_e = \frac{\left(C_o - C_f\right)}{m} \times V \qquad (2)$$

where C_0 and C_f are Cr(III) concentrations (mg L^{-1}) before and after the batch assays, respectively, m is the adsorbent mass (g) and V is the Cr(III) solution volume (L).

3   RESULTS AND DISCUSSION

3.1   *Characterization of materials*

3.1.1   *Textural analysis*
According to Figure 1, all the activated carbons showed isotherms consisting of a mixture of types I and IV (IUPAC classification), indicating a porous structure mainly composed by micropores and mesopores. Additionally, all isotherms presented a type H4 hysteresis, associated to micro and mesoporous materials with narrow slit-shaped pores.

Table 1 shows the textural parameters of the activated carbons.

CAC and CACCT presented higher S_{BET} and V_{total} values than PCPA and PCPACT. Furthermore, the acid treatment performed over CACCT and PCPACT reduced the S_{BET} and V_{micro} values of the non-treated materials (CAC and PCPA), respectively, due to the collapse of micropores on their structure. The loss of microporosity can also be justified by the pore blocking due to the successful introduction of functional groups at the entrance.

3.1.2   *pH$_{pzc}$*
Figure 2 shows the pH$_{pzc}$ of the activated carbons. The non-treated activated carbons CAC and PCPA had similar pH$_{pzc}$ values (9.13 and 9.89, respectively). The alkaline nature of

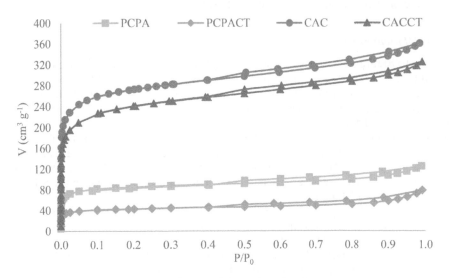

Figure 1.   N$_2$ adsorption-desorption isotherms of the activated carbons.

Table 1. Textural analysis of the activated carbons.

Parameter	PCPA	PCPACT	CAC	CACCT
S_{BET} (m^2 g^{-1})	325	164	1030	893
V_{micro} (cm^3 g^{-1})	0.100	0.051	0.304	0.248
V_{meso} (cm^3 g^{-1})	0.071	0.051	0.253	0.243
V_{total} (cm^3 g^{-1})	0.179	0.102	0.557	0.491

S_{BET}: BET surface area; V_{micro}: micropore volume; V_{meso}: mesopore volume; V_{total}: total pore volume.

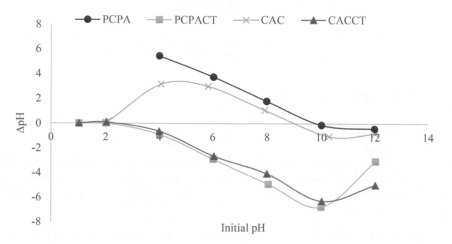

Figure 2. pH$_{pzc}$ of the activated carbons.

activated carbons, may induce an increase in the final pH values of Cr(III) solutions, causing chromium precipitation.

Also, CACCT and PCPACT showed similar pH$_{pzc}$ values (2.28 and 2.04, respectively). As these activated carbons presented a significant acidic nature, their use in Cr(III) adsorption assays may promote Cr(III) adsorption through electrostatic attraction and can prevent Cr(III) precipitation by lowering the initial pH value.

3.2 Cr(III) removal assays

According to Figure 3, Cr(III) was removed in higher percentages when the S/L ratio increased from 5.0 to 10.0 g L^{-1}. Due to final pH values above 5.0, CAC and PCPA removed Cr(III) almost through precipitation, at a S/L ratio of 10.0 g L^{-1}. For the S/L ratio of 5.0 g L^{-1}, Cr(III) precipitation might be less significant for these same activated carbons, as the pH values were around 5.0.

PCPACT showed similar Cr(III) removal percentages to CACCT (72.0 and 72.4%, respectively). The low final pH values, induced by the acidic nature of these activated carbons, ensures that Cr(III) was removed mainly by adsorption. However, the amount of Cr(III) removed was not twice as efficient in both activated carbons, when the S/L ratio was doubled from 5.0 to 10.0 g L^{-1}.

The Cr(III) uptake capacity and final pH values for each activated carbon are shown in Figure 4. For the highest S/L ratio (10.0 g L^{-1}), a decrease in the uptake capacity was registered for CACCT and PCPACT activated carbons.

For a S/L ratio of 5.0 g L^{-1}, PCPACT presented slightly better results than CACCT, with q_e values of 6.66 and 6.09 mg g^{-1}, respectively. High uptake capacity values for CAC and PCPA, with a S/L ratio of 10.0 g L^{-1}, are due to the occurrence of precipitation.

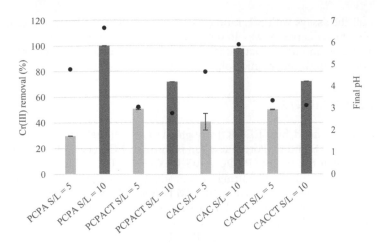

Figure 3. Cr(III) removal percentages for the activated carbons and final pH values (S/L ratio expressed in g L⁻¹).

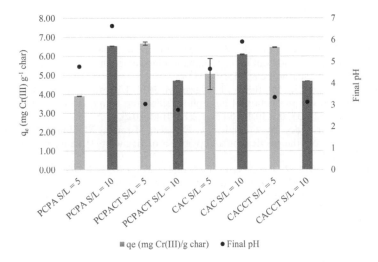

■ qe (mg Cr(III)/g char) ● Final pH

Figure 4. Cr(III) uptake capacities of the activated carbons and final pH values (S/L expressed in g L⁻¹).

4   CONCLUSIONS

Despite of the lower S_{BET} values, both PCPA and PCPACT removed Cr(III) similarly to or even better than CAC and CACCT, from the aqueous solutions. This suggests that the surface chemistry of activated carbons is more important than their surface area, regarding the removal of Cr(III) through adsorption.

The acidic nature of PCPACT makes it adequate for the goals of this work, as it prevents Cr(III) precipitation during the adsorption process. On the other hand, the alkaline nature of PCPA may induce Cr(III) precipitation, hindering the adsorption process but allowing removal percentages of almost 100%. This feature makes this an easily tunable material for Cr(III) removal.

PCPACT had similar Cr(III) removal and a slightly better uptake capacity than CACCT, meaning that an activated carbon from a low-cost biochar can perform just as well as a more expensive commercial activated carbon, even with the addition of an acid treatment.

PCPA and CAC removed nearly all Cr(III) through precipitation in the S/L ratio of 10.0 g L⁻¹, making them very efficient in the point of view of Cr(III) removal, but not very adequate for adsorption itself.

Although all the assays with S/L ratio of 10.0 g L^{-1} performed better in terms of Cr(III) removal, their uptake capacity was lower when the solid/liquid ratio was doubled, which means that using an amount of activated carbon above the necessary amount might not be the better approach in the economic point of view.

ACKNOWLEDGEMENTS

This research was funded by FEDER through the Operational Program for Competitive Factors of COMPETE and by Portuguese funds through FCT (Foundation for Science and Technology) through the project PTDC/AAG-REC/3477/2012—RICEVALOR "Energetic valorization of wastes obtained during rice production in Portugal", FCOMP-01-0124-FEDER-027827, a project sponsored by FCT/MTCES, QREN, COMPETE and FEDER.

The authors also acknowledge the Foundation for Science and Technology for funding Maria Bernardo's post-doc fellowship (SFRH/BPD/93407/2013), Diogo Dias's PhD fellowship (SFRH/BD/101751/2014), and LAQV/REQUIMTE through Portuguese funds (UID/QUI/50006/2013) and co-funds by the ERDF under the PT2020 Partnership Agreement (POCI-01-0145-FEDER-007265).

REFERENCES

Ahmad, M., Rajapaksha, A.U., Lim, J.E., Zhang, M., Bolan, N., Mohan, D., Vithanage, M., Lee, S.S., Ok, Y.S. 2014. Biochar as a sorbent for contaminant management in soil and water: A review. *Chemosphere*. doi:10.1016/j.chemosphere.2013.10.071.

Alvarez, J., Lopez, G., Amutio, M., Bilbao, J., Olazar, M. 2014. Upgrading the rice husk char obtained by flash pyrolysis for the production of amorphous silica and high quality activated carbon. *Bioresour. Technol*. 170: 132–137. doi:10.1016/j.biortech.2014.07.073.

Dias, D., Lapa, N., Bernardo, M., Godinho D., Fonseca, Miranda, M., Pinto, F., Lemos, F. 2017. Properties of chars from the gasification and pyrolysis of rice waste streams. *In press*. doi:10.1016/j.wasman.2017.04.011.

European Commission 2014. Report on critical raw materials for the EU, *DG Entreprises*.

Inyang, M., Dickenson, E. 2015. The potential role of biochar in the removal of organic and microbial contaminants from potable and reuse water: A review. *Chemosphere* 134: 232–240. doi:10.1016/j.chemosphere.2015.03.072.

Kalderis, D., Bethanis, S., Paraskeva, P., Diamadopoulos, E. 2008. Production of activated carbon from bagasse and rice husk by a single-stage chemical activation method at low retention times. *Bioresour. Technol*. 99: 6809–6816. doi:10.1016/j.biortech.2008.01.041.

Ma, H., Zhou, J., Hua, L., Cheng, F., Zhou, L., Qiao, X. 2016. Chromium recovery from tannery sludge by bioleaching and its reuse in tanning process. *J. Clean. Prod*. 142: 1–9. doi:10.1016/j.jclepro.2016.10.193.

Mohammed, I.Y., Lim, C.H., Kazi, F.K., Yusup, S., Lam, H.L., Abakr, Y.A. 2016. Co-pyrolysis of rice husk with underutilized biomass species: A sustainable route for production of precursors for fuels and valuable chemicals. *Waste and Biomass Valorization* 1–11. doi:10.1007/s12649-016-9599-9.

Pinto, F., André, R., Miranda, M., Neves, D., Varela, F., Santos, J. 2016. Effect of gasification agent on co-gasification of rice production wastes mixtures. *Fuel* 180: 407–416. doi:10.1016/j.fuel.2016.04.048.

Shackley, S., Carter, S., Knowles, T., Middelink, E., Haefele, S., Sohi, S., Cross, A., Haszeldine, S. 2012. Sustainable gasification-biochar systems? A case-study of rice-husk gasification in Cambodia, Part I: Context, chemical properties, environmental and health and safety issues. Energy Policy 42: 49–58. doi:10.1016/j.enpol.2011.11.026.

Sugashini, S., Begum, K.M.M.S. 2015. Preparation of activated carbon from carbonized rice husk by ozone activation for Cr(VI) removal. *New Carbon Mater*. 30: 252–261. doi:10.1016/S1872-5805(15)60190-1.

Tan, X., Liu, Y., Zeng, G., Wang, X., Hu, X., Gu, Y., Yang, Z. 2015. Application of biochar for the removal of pollutants from aqueous solutions. *Chemosphere*. doi:10.1016/j.chemosphere.2014.12.058.

Ternero-Hidalgo, J.J., Rosas, J.M., Palomo, J., Valero-Romero, M.J., Rodríguez-Mirasol, J., Cordero, T. 2016. Functionalization of activated carbons by HNO3 treatment: Influence of phosphorus surface groups. *Carbon N. Y*. 101: 409–419. doi:10.1016/j.carbon.2016.02.015.

Yahya, M.A., Al-Qodah, Z., Ngah, C.W.Z. 2015. Agricultural bio-waste materials as potential sustainable precursors used for activated carbon production: A review. Renew. Sustain. *Energy Rev.* 46: 218–235. doi:10.1016/j.rser.2015.02.051.

WASTES – Solutions, Treatments and Opportunities II – Vilarinho, Castro & Lopes (Eds)
© 2018 Taylor & Francis Group, London, ISBN 978-1-138-19669-8

Recovery of the polymer content of electrical cables for thermal and acoustic insulation

J. Bessa, C. Mota, F. Cunha & R. Fangueiro
2C2T—Center for Textile Science and Technology, University of Minho, Guimarães, Portugal

ABSTRACT: This article evaluates the potential of polymeric waste of electrical cables as a raw material for new products, identifying possible areas of application. In this study, several panels samples were produced, by compression moulding, using only waste of electric cables as raw material, with different weights and particle sizes, which were after mechanically, thermally and acoustically characterized. The increase of weight and the use of smaller particles sizes allows better mechanical properties, such as compressive strength, since it was verified the improvement of required force to reach 25% of the initial thickness to 1774.7 N. It was also verified an increase of thermal resistance with use of smaller particle size and of acoustic insulation for samples with greater weights. These results allow to conclude that the waste electrical cables can have potential to use on building's construction or rehabilitation, due to interesting properties of thermal and acoustic insulation.

1 INTRODUCTION

Currently, it has been experienced an increasing of environmental awareness, mainly imposed by European regulations such as the European Directives 2008/98/CE and 2012/19/EU. By the way, the purpose of the last Directive is to contribute to sustainable production and consumption by, as a first priority, the prevention of waste electrical and electronic equipment (WEEE) and, in addition, by the reuse, recycling and other forms of recovery of such wastes so as to reduce the disposal of waste and to contribute to the efficient use of resources and the retrieval of valuable secondary raw materials (2012). Besides that, according to this Directive, it is expected that from 2019, the minimum collection rate to be achieved annually shall be 65% of the average weight of electrical and electronic equipment (EEE) placed on the market in the three preceding years in the Member State concerned, or alternatively 85% of WEEE generated on the territory of that Member State (2012).

Since the consumption of electrical and electronic equipment continues to grow and innovation cycles are becoming even shorter, these materials are a source of waste in rapid growth (2012). Electrical cables residues, which falls into the category of electrical and electronic waste (REE), are a metallic conductor elements, like as copper or aluminum (Santos, 2005). These residues are obtained from separation of the insulating element, being a residue still little explored currently, due to its composition, since it is composed of different polymer matrices (Fanny Lambert, 2015) and it is troublesome to recycle in a sustainable way (Mehdi Sadat-Shojai, 2011).

The materials used in the production of insulators of conductive cables can be thermoplastic, thermosetting or elastomer polymers, being thermoplastic ones most commonly used, such as polyvinyl chloride (PVC) and polyethylene (PE). Other polymers that can be used are cross-linked polyethylene (PEX), which is a thermosetting one, and rubber-ethyl-propylene (EPM) or silicone rubber, which are elastomers (Santos, 2005).

The main motivation for the recycling of electrical cable waste is the high commercial value of the conductive metal, while the plastic that has lower value is often overlooked (Mats Zackrisson, 2014). However, as previously referred, the amount of electric cables waste tends to increase and should be implemented new methods and solutions for the recycling of metals and insulating polymers, to avoid serious constraints related with the sustainability of the planet and resources. On the other hand, the recycling of end-of-life electrical cables can also promote significant environmental savings. The disposal of polymer waste in landfill is no longer an option for the future and is theoretically banned since 2008 in the European Union, despite it still occurs (Annika Boss). For economic and environmental reasons, the recycling or recovery of electric cables residues is essential due to increasing scarcity of natural resources and the need to reduce environmental impact and the generation of greenhouse gases. In 2015 it was estimated a recycling of approximately 515 tons of PVC from electric cables, only on the European Union (EU)-28, Norway and Switzerland (Plus, 2016).

Separation techniques used actually allow effective separation of the different materials of the electric cables, and then the PVC or other polymers can be reprocessed into new products. The separation and recycling processes can be mechanical or chemical ones. Chemical processes have significant potential benefits, comparing with mechanical ones. Although mechanical recycling processes dominate the industry due of lower operating costs, chemical recycling allows neutralize dangerous additives such as heavy metal stabilizers (Network, 2015)

In this context, the purpose of this paper is to characterize samples produced from polymeric waste of end-of-life electric cables and identify potential areas of application for the same that justifies their recovery. The produced samples were obtained by a compression moulding process and subsequently their mechanical, thermal and acoustic properties were determined.

2 EXPERIMENT

2.1 *Composite materials production*

The samples produced for characterization were processed by compression moulding using electrical cables waste, varying the particle size, Figure 1, and amount of material used, in order to evaluate the effect of the volume and density of the material in mechanical, thermal and acoustic properties. Two samples were produced using material granules less than 1 mm, and two with particle size with size particles bigger than 2 mm, Figure 1. In both cases, the amount of material used in the production of the samples varied approximately 25% between each sample.

After the material selection, it was placed in a specific mould, with $25 \times 25 \times 0.4$ cm dimensions, Figure 2, which was then closed and the process was performed at 240°C and using

Figure 1. Waste with particle size bigger than 2 mm and smaller than 1 mm, respectively.

Table 1. Samples produced.

Sample	Particle size (mm)	Mass (g)
1	> 2	550
2	> 2	680
3	< 1	680
4	< 1	850

Figure 2.   Mould used for samples production.

Figure 3.   Acoustic insulation chamber.

10 ton pressure, for 15 minutes, in a heated plate press. The processing parameter of temperature was determined by differential scanning calorimetry (DSC), using an equipment from Perkin-Elmer, DSC-4 model, with a heat rate (φ) of 10°C min^{-1}, between 40 and 250°C.

After this heating step, the samples were cooled in the same equipment, turning on the cool water circulation to hot plates, and using 10 ton pressure, during 5 minutes. Finally, the final composites were extracted after the pressure withdrawal. In Table 1 it was summarized the size and mass characteristics of the samples produced.

2.2   *Characterization tests*

The obtained samples were subsequently subjected to several characterization tests. The mechanical tests were performed according to ASTM D575–91, in order to evaluate the performance in compression. The thermal tests were performed by a certified device Alambeta device with circular samples with radius of 55 mm, and the thermal conductivity was determined. The acoustic insulation properties of the material were obtained using an acoustic insulation chamber, Figure 3, according to E-90 standard, by measuring the reduction of noise in the interior zone, emitted by a sound source in the exterior zone.

3 RESULTS AND DISCUSSION

The samples present different appearances between them, with the clear differences, especially in cross-section, as visible in Figure 4 and Figure 5. In these figures, it can be observed an increased presence of air pockets in the samples produced with a higher particle size (samples 1 and 2), in comparison with samples made with a smaller particle size (samples 3 and 4), where higher degree of compaction is noticed.

The summary of results of several tests performed is presented in Table 2. From mechanical tests, namely compressive ones, it was verified an increase of required force to reach 25% of the initial thickness, when the mass of the sample is larger, for the same particle size. For instance, comparing samples 1 and 2, the required compression forces are 368.3 and 467.1 N, respectively. The same trend is also verified with samples 3 and 4, which registered a compression forces of 939.0 and 1774.7 N. Then, from these results, it can be concluded that increasing mass of the sample allow better mechanical resistance at compression, probably due of the better compaction and lowest presence of voids. On the other hand, comparing samples 2 and 3, which have the same sample's mass, it can be observed a bigger required force to reach 25% of the initial thickness, when a lower particle size is used. As presented in Table 2, the required compression forces were of 467.1 and 939.0 N, respectively. Thus, the use of smallest particle size can improve the compression resistance of the samples. This trend can be also justified for better compaction and lowest presence of voids with the use of smaller particles.

Regarding thermal properties, it can be observed that the samples produced with higher particle size allows better thermal resistance to the samples. Once again, this is due of the greater number of voids, and consequent air pockets, in the samples with this type of particles, namely samples 1 and 2, which have thermal resistances of 0.087 m^2 K W^{-1} and 0.093 m^2 K W^{-1}. On the other hand, the weight of samples, in this case, has little influence

Figure 4.   Front view of the samples 1, 2, 3 and 4, respectively.

Figure 5.   Cross-section view of the samples 1, 2, 3 and 4, respectively.

Table 2.   Characterization results of the several samples.

Sample	1	2	3	4
Particle size (mm)	> 2	> 2	< 1	< 1
Mass (g)	550	680	680	850
Compression force* (N)	368.3	467.1	939.0	1774.7
Thermal resistance (m^2 K W^{-1})	0.087	0.093	0.069	0.067
Noise reduction at 500 Hz (dB)	12.7	16.3	11.9	16.5

* Required force to reach 25% of the initial thickness.

on thermal properties, apparently, since there are few variations between both pairs of samples. For example, comparing samples 3 and 4, it was obtained thermal resistances of 0.069 m^2 K W^{-1} and 0.067 m^2 K W^{-1}.

The study of acoustic behavior of the produced samples allowed to determine sound absorption capacity of air sounds over the frequency spectrum between 31.5 Hz and 16 000 Hz. Then, from the obtained results for noise reduction at 500 Hz, it was verified a trend to increase this property with the use of greater mass and higher particle size in production of the samples. Comparing samples 1 and 2, which have the same particle size, it was observed a noise reduction at 500 Hz of 12.7 dB and 16.3 dB, respectively. Besides that, comparing the samples 2 and 3, which have the same sample's mass, it was observed a noise reduction at 500 Hz of 16.3 dB and 11.9 dB, respectively. As referred in the results of thermal tests, these results are explained by the formation of more voids and, consequently, air pockets, which contribute to the sound insulation.

4 CONCLUSIONS

The recycling and reuse of the materials plays an important role in protecting the environment, particularly with respect to electronic products. The volume of used electronic equipment is growing rapidly and now represents a significant part of the waste stream in Europe. So, that there is a growing need to find solutions to products of end-of-life. In this context, this study shows the potential of reuse waste electric cables. From this waste, it was produced compact samples, by a compression moulding process, and subsequently they are characterized. Mechanical compression tests showed an increase of at least 25% of the required compression force to reach 25% of the initial thickness, when the sample mass is increased by 25%, for the same particle size. On the other hand, it was verified an increase of approximately 100% on this required compression force, when is used the particles with size lower than 1 mm, for the same sample mass. Moreover, it was also demonstrated that the use of particle sizes greater than 2 mm, can improve the thermal resistance of the samples at least 26%, comparing to samples with particle size lower than 1 mm. In addition to this, it was also verified that the capacity of noise reduction of the samples is improved mainly by the use of greater mass in samples production, being verified an increase of at least 28% of this property, when an additional 25% of mass is used. These results are strongly affected by the voids and air pockets on the samples, which contributes to weaker mechanical properties, greater thermal insulation and interesting properties of acoustic insulation.

Thus, from this study, it is concluded that the recycling and reuse of electric cables can contributes to obtain new materials with interesting properties of thermal and acoustic properties. In this sense, these conclusions can arouse interest for some applications, namely in the buildings construction more sustainable or in the rehabilitation of older ones.

ACKNOWLEDGEMENTS

The authors gratefully acknowledge to project "*Fibrenamics Green—Platform for the development of innovative waste-based products*", code NORTE-01-0246-FEDER-000008, which is co-financed by the European Union, through the European Regional Development Fund (ERDF) under the NORTE 2020—North Portugal Regional Operational Program 2014–2020.

REFERENCES

Boss Annika [et al.]. *New technology for recycling of plastic from cable waste*. Sweden.
Lambert Fanny [et al.] 2015. Copper leaching from waste electric cables by biohydrometallurgy. *Minerals Engineering*. Belgium: Elsevier.

Network Healthy Building Post-Consumer Polyvinyl Chloride in Building Products 2015.

Parliament European Directive 2012/19/EU of the european parliament and of the council on wast electrical and electronic equipment (WEEE) 2012. [s.l.]. *Official Journal of the European Union.*

Plus Vinyl Progress Report 2016.

Sadat-Shojai Mehdi e Bakhshandeh Gholam-Reza 2011. Recycling of PVC wastes. *Polymer Degradation and Stability* 96 (4).

Santos J. Neves dos 2005. *Condutores e Cabos de Energia*. Faculdade de Engenharia da Universidade do Porto.

Zackrisson Mats, Jönsson Christina e Olsson Elisabeth 2014. Life Cycle Assessment and Life Cycle Cost of Waste Management—Plastic Cable Waste. *Advances in Chemical Engineering and Science*. Sweden: Scientific Research.

Recovery of wood dust in composite materials

J. Bessa, C. Mota, F. Cunha & R. Fangueiro
2C2T—Center for Textile Science and Technology, University of Minho, Guimarães, Portugal

ABSTRACT: This work compares the influence of different thermoplastic polymer/waste wood dust ratios on the mechanical and thermal properties of composite materials. Moreover, the effect of a coupling agent incorporation, namely Maleic Anhydride Polypropylene (MAPP), to promote better adhesion between both components, on these properties was evaluated. The composite materials were obtained by compression moulding process, and subsequently characterized through mechanical and thermal tests. The experimental results show that the waste wood dust has the capacity to reinforce the thermoplastic polymer used, PP (polypropylene), allowing better mechanical properties, such as Young's and flexural modulus, to 2808 MPa and 1739 MPa, respectively. The 2% MAPP incorporation leads to an even higher increase of the mechanical properties. These results allow to conclude that the increase of wood dust can increase the mechanical properties of the PP composites, and the MAPP incorporation contributes to the better compatibility between both materials, increasing also these properties.

1 INTRODUCTION

The recent world population growth has been one of the main factors for the technological development and innovation. In fact, exponential growth has taken place in the world population, being expected to reach 9 billion in 2042, compared to 3 billion in 1960.

This has led to the use and development of new materials and products, which can respond to these increasing needs and trends. Consequently, an increase in composite materials' use has been noticed mainly in construction and automotive areas, replacing conventional materials. In addition to the costs and weight reduction of components, these materials also allow to increase some mechanical properties and corrosion resistance.

On the other hand, this has also aroused concerns about environmental sustainability, related with the use of huge contents of synthetic materials, such as carbon and glass fibres, and the overexploitation of natural resources. For instance, wood is one of the most important renewable resources available, which leads to an efficient management need. Actually, the wood based panels industry had a fundamental position due to its innovation capacity in obtaining high value-added products. However, due to the great use of wood, a high generation of waste is generated, mainly resulting from the different stages of production and the end of their life cycle.

In this context, the recovery of this waste has become increasingly important, in order to contribute to a more sustainable and eco-friendly world. Several scientific groups and researchers have been looking for development of composites with natural materials, such as waste wood dust, as reinforcement. However, the combination of natural fibres with polymeric materials presents some constraints related to the compatibility and adhesion at the interface between both materials (Paulo Henrique Fernandes Pereira, 2005) (S. Mukhopadhyay, 2009). Therefore, the combination of hydrophilic natural fibrous materials with hydrophobic polymer matrices often requires the use of coupling agents to improve the adhesion between these materials at the interface (Quazi T.H. Shubhra, 2013).

This study intended to evaluate the combination between a thermoplastic polymer, polypropylene (PP), and waste wood dust, at different weight percentages. Besides that, the effect of coupling agent incorporation in the mixture, namely maleic anhydride polypropylene (MAPP) was investigated. The composite materials were produced by compression moulding and subsequently were characterized by mechanical tests, such as tensile, flexural and impact ones, and thermal tests.

2 EXPERIMENT

2.1 *Materials*

One of the most important factors to take into account on this study was the similarity of densities of the different materials, in order to guarantee a good and homogeneous mixture before the processing. All materials are presented in Figure 1. Then, the waste wood dust used is coming of mahogany tree, *Swietenia macrophylla*, which presents a density between 0.85–0.95 g cm^{-3}. The polymer material used was purchased from Resinex, Portugal, with a RXP 2101 commercial reference. This polymer has a melt flow index (MFI) of 100 g/10 min and a density of 0.9 g cm^{-3}, which is similar to that of mahogany wood.

The MAPP was supplied by BYK Additives & Instruments, with the commercial reference Scona TPPP 8112 FA. This polypropylene highly functionalized with maleic anhydride, in dust form, has a MFI higher than 80 g/10 min and a maleic anhydride content of 1.4% approximately.

2.2 *Composite materials production*

The composite materials were produced by compression, Figure 2. Firstly, the dust form of several compounds were manually mixed, according to mass fractions referred in Table 1. After the mixture, the processing occurs in a specific mould, to obtain composite materials with 25 × 25 cm dimensions. This process was performed at 210°C and using 20 ton pressure, for 20 minutes. After this heating step, the samples were cooled down in the same equipment,

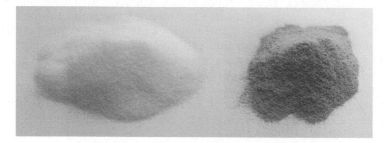

Figure 1.   PP and wood dust used.

Figure 2.   Moulding compression equipment and mould used.

Table 1. Composite materials produced.

Sample	PP/Wood ratio	% MAPP
3	70:30	0
5	60:40	0
7	50:50	0
15	50:50	2

Figure 3. Wood dust/PP composite.

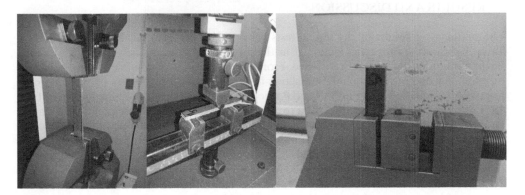

Figure 4. Specimen in a tensile, flexural and impact tests, respectively.

using cold water circulation to hot plates, and 20 ton pressure, during 5 minutes. Finally, the final composites, Figure 3, were extracted after the pressure withdrawal.

2.3 Characterization tests

The obtained samples were subsequently subjected to several characterization tests. The mechanical tests performed included tensile, flexural and impact tests, Figure 4. The tensile tests were performed according to EN ISO 527-2 standard, using 10 samples of 2 cm width and at least 150 cm overall length. The test parameters used were a crosshead speed of 2 mm min⁻¹ and 115 cm as initial distance between grips. The flexural tests were performed in accordance to EN ISO 178 standard, using 10 samples of 10 mm width, 80 mm length and 4 mm thickness. These specimens were tested at a crosshead speed of 5 mm min⁻¹. The impact resistance of the material composites obtained was measured according to ASTM D256–04 standard.

The thermal tests were performed by Alambeta device with 55 mm radius circular samples, and the thermal conductivity was determined. Finally, the moisture absorption test, with an

Figure 5. Alambeta device and analytical balance for thermal and moisture absorption tests, respectively.

analytical balance, allowed to measure the moisture content absorbed after 24 hours. The used equipments in both tests are presented in Figure 5.

3 RESULTS AND DISCUSSION

The summary of results of the several tests performed is presented in Table 2 and Table 3. Table 1 shows the properties obtained in the mechanical tests, namely Young's modulus, flexural modulus and impact resistance. From these results, it is possible to verify that the increase of wood dust in the mixture allows better mechanical properties of the composite materials. Comparing samples 3, 5 and 7, the tensile and flexural modulus increased from 1773 MPa to 2808 MPa and 1569 MPa to 1739 MPa, respectively. Consequently, the impact strength demonstrated slight variations. On the other hand, it is also possible to conclude that the coupling agent incorporation improved these mechanical properties, as well as the impact strength of composite materials. Comparing samples 7 and 15, it can be observed that an increase from 2808 MPa to 3320 MPa in Young's modulus and from 1739 MPa to 2022 MPa in flexural modulus. Besides that, the incorporation of MAPP causes also an increase of impact strength from 1.68 kJ m^{-2} to 1.76 kJ m^{-2}.

In terms of thermal conductivity and moisture absorption, Table 1, small variations between the samples were observed. A decreasing trend was verified with the increase of wood mass fraction, between samples 3, 5 and 7, by 0.145 W m^{-1} K^{-1} to 0.138 W m^{-1} K^{-1}. However, with MAPP incorporation, sample 7, this parameter increases again to 0.145 W m^{-1} K^{-1}.

Moreover, an increasing trend of moisture absorption with increasing wood dust ratio was clearly noticed. Comparing samples 3 and 7, it is verified and increased by 0.97% to 1.34% on moisture absorption. This fact is probably related to the wood's morphology, which is a porous structure. Then, despite of hydrophobic surface, wood has the capacity to absorb moisture through these pores, once the interior is constituted by hydrophilic compounds.

In Figure 6 is presented the influence of wood mass fraction on the several studied properties, namely mechanical, thermal and moisture absorption ones. As it can be observed, there is a trend to improve the mechanical properties of Young's and flexural modulus with increasing the wood mass fraction. Opposite to this, it is confirmed the trend to decrease the thermal conductivity of the samples, with the increase of wood mass fraction. In this context, both conclusions are positively important for the desired properties. However, the trend to moisture absorption is also verified with increasing of wood mass fraction, which is not so positive such as the previous properties.

Table 2. Mechanical properties of thermoplastic composites produced with wood dust.

Sample	PP/Wood ratio	% MAPP	Young's modulus (MPa)	Flexural modulus (MPa)	Impact strength (kJ m⁻²)
3	70:30	0	1773.91 ± 2.03	1569.60 ± 3.71	1.85 ± 0.33
5	60:40	0	2662.16 ± 4.06	1687.72 ± 8.63	1.45 ± 0.09
7	50:50	0	2808.48 ± 8.21	1739.61 ± 7.12	1.68 ± 0.26
15	50:50	2	3320.44 ± 7.42	2022.71 ± 11.24	1.76 ± 0.42

Table 3. Thermal and absorption properties of thermoplastic composites produced with wood dust.

Sample	PP/Wood ratio	% MAPP	Thermal conductivity (W m⁻¹ K⁻¹)	Moisture absorption (%)
3	70:30	0	0.145 ± 0.02	0.97 ± 0.13
5	60:40	0	0.141 ± 0.03	1.15 ± 0.17
7	50:50	0	0.138 ± 0.02	1.34 ± 0.19
15	50:50	2	0.145 ± 0.02	1.08 ± 0.15

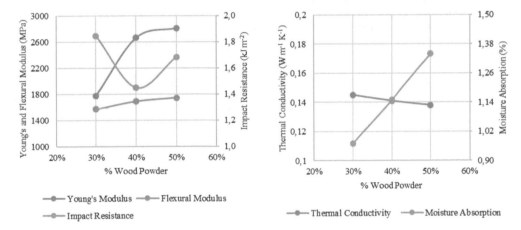

Figure 6. Influence of wood mass fraction on several studied properties of composite materials.

4 CONCLUSIONS

Nowadays, the recovery of natural materials in several sectors is a crucial trend in the development of innovation. The concerns related to sustainability of resources and life cycle of the materials has been the main factors for this trend's evolution. This study shows that the use of waste wood dust allows the reinforcement of thermoplastic polymers and increase of mechanical properties of composite materials. Tensile tests showed an increase of approximately 37% on Young's modulus, between samples with 70:30 and 50:50 ratios. Flexural tests showed an increase of approximately 10% on flexural modulus, between samples with 70:30 and 50:50 ratios. In addition to this, it is also demonstrated that the use of coupling agents, such as MAPP, can improve the adhesion at the interface between both components. With the use of MAPP, the Young's and flexural modulus increased approximately 15% and 14%, respectively, comparing to the sample without this coupling agent, in the same mass fractions.

Moreover, it was also demonstrated that the increase of wood's mass fraction can slightly improve the thermal properties, such as thermal conductivity, once this property decrease approximately 3%, between 70:30 and 50:50 ratios. Then, decreasing the thermal conductivity of the composite materials, the thermal insulation is improved.

Thus, from this study, it can be concluded that the increase of wood's mass fraction on the final composite materials can improve their mechanical properties and contributes to the slight increase of thermal insulation. Therefore, these conclusions are of paramount importance for some applications, namely in the automotive or building sectors.

ACKNOWLEDGEMENTS

The authors gratefully acknowledge to project *"Fibrenamics Green—Platform for the development of innovative waste-based products"*, code NORTE-01-0246-FEDER-000008, which is co-financed by the European Union, through the European Regional Development Fund (ERDF) under the NORTE 2020—North Portugal Regional Operational Program 2014–2020.

REFERENCES

Mukhopadhyay S. e Fangueiro Raul 2009. Physical Modification of Natural Fibers and Thermoplastic Films for Composites—A Review. *Journal of Thermoplastic Composite Materials*. 22: 135–162.
Pereira Paulo Henrique Fernandes [et al.] 2005. *Vegetal fibers in polymeric composites: a review*. P. 9–22.
Shubhra Quazi T.H., Alam A.K.M.M. e Quaiyyum M.A. 2013. Mechanical properties of polypropylene composites: a review. *Journal of Thermoplastic Composite Materials*. A de 2013: 362–391.

WASTES – Solutions, Treatments and Opportunities II – Vilarinho, Castro & Lopes (Eds)
© 2018 Taylor & Francis Group, London, ISBN 978-1-138-19669-8

Modified biological sorbents from waste for the removal of metal ions from the water system

L. Rozumová & P. Kůs
Centrum vyzkumu Rez, Husinec-Rez, Czech Republic

I. Šafařík
Department of Nanobiotechnology, Biology Centre, ISB, CAS, Ceske Budejovice, Czech Republic

ABSTRACT: The ability of orange peel and sawdust waste, a waste material derived from the commercial processing of orange and wood production, to remove Pb(II), Ni(II) and Cd(II) ions from aqueous solution was determined. Biosorption by raw waste materials could be a cost effective technique for removing metals ions from aqueous solutions. The performance of a new biosorbents system, consisting of biological matrix which was magnetically modified by iron oxide nanoparticles, was studied. The use of low-cost and eco-friendly adsorbents has been investigated as an ideal alternative to the current expensive methods. The sorption of metal ions was investigated by determination of adsorption isotherms. The binding of metal ions was found to be dependent on pH. Magnetic orange peel and magnetic sawdust were found to be an effective sorbent for metal ions in the range of low concentrations.

1 INTRODUCTION

Rapid urbanization and industrialization often lead to environmental contamination with heavy metals. Clean water is an essential and limited natural resource for the development of a series of living organisms in aquatic environments as well as for humans, all of which require its preservation. Nowadays, the inputs of the heavy metals in the environment are monitored and they are mostly faced by stringent regulations. The predominant applied methods are coagulation, flocculation, precipitation, electrochemical processes, ion exchange, extraction etc. Use of a suitable sorbent is another favorable way of removing pollutants from waste water. Price, availability, adsorption capacity and strong affinity to pollutants are limiting factors for sorbent application in waste water treatment hence new materials to be used as sorbents are evaluated constantly (Grohmann et al., 1992). Therefore, it is necessary to develop new environmentally friendly ways to clean up contaminants using low-cost methods. Among low-cost techniques, the adsorption is an efficient and economic method, based on simple and flexible operating conceptions and use of regenerative adsorbents, for the removal of organic or inorganic pollutants with a high efficiency in many cases. Biosorption processes, defined as the sorption of metal ions by biomass, are being employed as an alternative technique for the purification effluents and for the recovery of the retained metals (Tripodo et al., 2004; Wikandari et al., 2015). A convenient alternative for the classical treatment for the wastewater treatment is the usage of some non-conventional adsorbents (natural materials, biosorbents and waste materials from industry and agriculture) with lower cost and high efficiency (Rozumová et al., 2016; Rozumová et al., 2014).

Orange as the main citrus fruit is one of top fruit commodities that dominate the global production of market with fruits. More than half, about 50–60%, from global orange production ends up as waste including seed, peel, segment membrane and other by-products (Widmer et al., 2010). A large amount of this waste is still dumped every year (Pourbafrani et al., 2010), which causes both economic and environmental problems such as high transportation cost, lack of dumping site, and accumulation of high organic content material (López et al., 2010). Therefore, more effective and sustainable alternatives for using orange peel wastes such as an/the adsorbent are highly desirable. Agriculture industry produces many types of waste. There are many solid agricultural waste products and waste materials from forest industries as sawdust that are available in large quantities, are biodegradable and might be potential sorbents due to their physico-chemical characteristics.

Wastes of orange peel and sawdust were magnetically modified for using as adsorbents. Magnetic derivatization of adsorbents is a very important modification which has been proven to improve the manipulation and increase the adsorption capacity (Mockovciakova et al., 2010; Bourlinos et al., 2003; Vereš et al., 2009). In the present work, magnetically modified orange peels and sawdust were used for removing metal ions from water. This work deeply investigates the biosorption and desorption process of removing of heavy metals. The objective of the present study was to investigate the efficiency of adsorption for the removal of metal ions from different concentrations of solutions. Besides the influence of pH, the sorption biomass uptake has also been quantified by means of sorption and kinetic models.

2 EXPERIMENT

2.1 *Adsorbent and adsorbate*

Orange peels (OP) and sawdust (SD) were collected from locally available company and after drying, it was ground in a coffee mill. The particles with diameter less than 0.5 mm were prepared. Water-based ionic magnetic fluid stabilized with perchloric acid (analytical grade) was prepared using a standard procedure (Massart et al., 1991). The ferrofluid was composed of magnetic iron oxide nanoparticles with diameters ranging between 10 and 20 nm (measured by electron microscopy). The relative magnetic fluid concentration (25.2 mg.ml^{-1}) is given as the iron (II, III) oxide content determined by a colorimetric method. Three grams of powdered orange peels in a 50 ml polypropylene centrifuge tube were suspended in 40 ml of methanol and then it was added 6 ml of ferrofluid. The suspension was mixed in a rotary mixer (Dynal, Norway) for one hour. Subsequently, the sample was rinsed by methanol and dried at the laboratory temperature (Šafařík et al., 2010).

The sorption process of metal ions was conducted in a dynamic "bath" type. The variable process parameters were: concentration of metal ions, the pH of the solutions, the time. Model solutions of Pb, Ni and Cd ions were prepared from $Pb(NO_3)_2$ $Ni(NO_3)_2.6 H_2O$ and $Cd(NO_3)_2.4 H_2O$, analytical grade. Magnetically modified material (0.2 g) was suspended in 50 ml of solution with defined concentration of metal ions. The sorption process was carried out at the constant room temperature. The suspension was mixed over 5, 24, 48, 72 and 168 hours. The following step was the filtration by which the samples were separated from the suspension. The optimal value of pH was 5. Parameters of reaction time and pH value were chosen on the basis of the previous primary experiments (Rozumová et al., 2016*).

2.2 *Methods of characterization and data processing*

Morphological analyses of the samples were performed by using scanning electron microscopy (SEM LYRA3 TESCAN) equipped with energy-dispersive X-ray spectrometer (EDX). For the measurement was used backscattered electrons detector BSE.

The specific surface area was measured by the Quadrasorb EVO/SI and calculated by the QuadraWin software according to the BET isotherm. Results were obtained by means of pure liquid N_2 adsorption-desorption isotherms at 77 K. Prior to analysis, all samples were degassed under vacuum at 40 °C for 48 h The BET method was used to estimate the specific surface area (S_{BET}) of the samples, in which the conditions of linearity and considerations regarding the method were fulfilled.

Concentration of metals in aqueous leachates were determined by atomic absorption spectrometer AAS-FA (UNICAM SOLAAR M6).

The amount of adsorbed metal ions per gram of adsorbent ($mg.g^{-1}$) was calculated from the experimental data of metal ions concentration at equilibrium as follows in Eq. (1):

$$q = (c_0 - c)/m_{ads} \qquad (1)$$

where c_0 and c are the metal concentrations in liquid phase before and after adsorption experiments ($mg.l^{-1}$), respectively, and m_{ads} is mass of adsorbent in the solution (g).

The experimental data were fitted by the adsorption isotherms and the kinetic models to determine the isotherms coefficients. The isotherm/kinetic coefficients were calculated by regression statistics. The coefficient of determination (R^2) was used to describe the best corresponding model calculated by linear regression.

3 RESULTS AND DISCUSSION

3.1 Characterization of sorbents

The morphologic appearance of the magnetically modified orange peel and sawdust is shown in Fig. 1. The magnetic orange peel particles were different in shape. Fig. 1 shows the heterogeneity in the shapes and sizes of the different particles and formation of the clusters. SD sample has more regular structure of particles than OP. Different pore sizes may be observed. By mapping, magnetically modified material OP and SD was proven to have adsorbed particles of iron oxides on the surface material. Spectrum of EDS analysis is showing that among of iron is 40 Wt.% (OP) and 53 Wt.% (SD). Iron particles of ferrofluid are very well held on the surfaces of both samples.

Specific surface area before modification was 0.45 $m^2.g^{-1}$ (OP) and 0.44 $m^2.g^{-1}$ (SD), after magnetic modification it was increased to 1.60 $m^2.g^{-1}$ (OP) and 1.77 $m^2.g^{-1}$ (SD). OP and SD are formed from mezopores. Pore Volume was $1.35.10^{-3}$ $cm^3.g^{-1}$ (OP) and $1.54.10^{-3}$ $cm^3.g^{-1}$ (SD). Pore Size Radium was for both samples 1,585 nm. Powder diffraction proved the presence of the iron in the magnetic material, namely in the Fe_3O_4 form.

3.2 Sorption of metal ions on magnetically modified OP and SD

The uptake of metal ions onto magnetically modified OP and SD as a function of pH is shown in Fig. 2. The initial concentration of the followed metals was 10 $mg.dm^{-3}$ and the time was selected for 1 hour. The adsorbed amount is significantly different according (compared?) to the metal ions species in solution. This fact indicates that the pH value is important parameter in the process of sorption.

Metal removal efficiency by OP and SD was evaluated using one metal system of each followed metals. As can be observed in Fig. 3, metal ions were better removed from the prepared solution at lower concentration than at higher concentration.

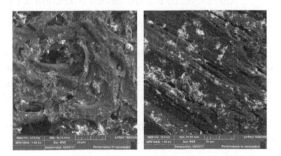

Figure 1.   SEM images of magnetically modified OP and SD before adsorption of metal.

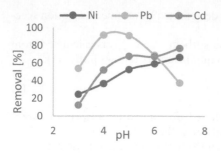
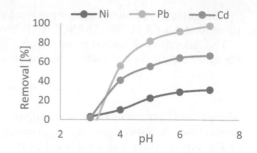

Figure 2. Metal ions removal efficiency as a function of pH on OP and SD samples.

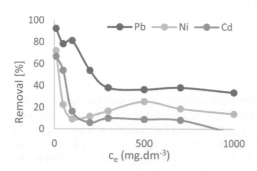
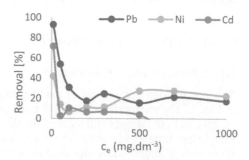

Figure 3. Metal ions removal efficiency as a function of initial concentration on OP and SD samples.

3.3 Adsorption isotherms

The experimental data of biosorption were processed through theoretical models. The presented data of biosorption of Pb(II) were processed by the Langmuir (L), Freundlich (F), Temkin (T) and Dubinin-Radushkevich (DR) isotherm models.

Determination coefficient of linear regression analysis (R^2) representing the proportion of the dependent variable expressed by the regression line was used to determine the best corresponding adsorption isotherm.

In the terms of R^2 values, the applicability of the above four models for present experimental data of approximately were followed in the order:

Magnetically modified OP: Pb: Freundlich > Langmuir > Temkin > Dubinin-Radushkevich
Ni: Freundlich > Temkin > Langmuir > Dubinin-Radushkevich
Cd: Langmuir > Freundlich > Dubinin-Radushkevich > Temkin

Magnetically modified SD: Pb: Freundlich > Temkin > Langmuir > Dubinin-Radushkevich
Ni: Freundlich > Temkin > Dubinin-Radushkevich > Langmuir
Cd: Dubinin-Radushkevich > Temkin > Freundlich > Langmuir

Isotherm parameters and coefficient of determination calculated by linear regression are presented in Table 1.

The R^2 indicates that for description of sorption process of Pb and Ni ions describe the best Freundlich isotherm model. The R^2 indicates that the Freundlich isotherm model describes the sorption process of Pb and Ni ions. The Freundlich equation accounts for a multisite adsorption isotherm for heterogeneous surfaces. The data show that the value is indicating the favorability of the metal ions sorption. The sorption of Cd ions is described Langmuir (OP) model and Dubinin-Radushkevich model (SD). Langmuir supposes the creation of monolayer. Dubinin-Radushkevich assumes the potential micropores in the material.

Table 1. Parameters of adsorption equation models for sample OP and SD for metal ions.

Isotherm models	Isotherm constant	Pb		Ni		Cd	
		OP	SD	OP	SD	OP	SD
L*	q_{max} (mg/g)	175.4	125.0	96.15	84.03	27.32	17.18
	K_L	0.016	0.003	0.003	0.001	0.009	0.002
	R_L	0.86–0.06	0.97–0.27	0.97–0.24	1.01–8.31	0.92–0.10	0.98–0.30
	R^2	0.991	0.876	0.737	0.483	**0.859**	0.172
F*	K_F (l/g)	2.103	1.349	1.132	0.195	1.571	0.222
	n	1.938	1.788	1.688	0.769	3.083	1.031
	R^2	**0.996**	**0.951**	**0.900**	**0.948**	0.769	0.696
T*	b_T	78	110	193	66	705	679
	a_T	0.139	0.045	0.211	0.019	0.739	0.0312
	R^2	0.876	0.891	0.867	0.832	0.688	0.822
DR*	a_D	93.74	51.66	34.25	56.12	16.15	9.751
	b_D	7.4E-07	1.5E-04	4.0E-06	7.2E-04	4.0E-06	0.001
	R^2	0.861	0.490	0.458	0.688	0.722	**0.962**

*L – Langmuir, F – Freundlich, T – Temkin, DR – Dubinin-Radushkevich.

Table 2. Parameters and coefficients of determination for the linearized pseudo-first and pseudo-second order models.

Kinetic models	Kinetic constant	Pb		Ni		Cd	
		OP	SD	OP	SD	OP	SD
Pseudo-first order	q_e (mg.g⁻¹)	1.778	2.649	2.542	1.834	0.683	1.202
	K_1	0.046	0.058	0.032	0.029	0.014	0.042
	R^2	0.951	0.978	0.975	0.918	0.085	0.944
Pseudo-second order	q_e (mg.g⁻¹)	4.950	5.003	3.741	2.197	3.096	3.322
	K_2	0.106	0.079	0.030	0.029	−0.256	0.158
	R^2	**0.999**	**0.999**	**0.996**	**0.974**	**0.991**	**0.999**

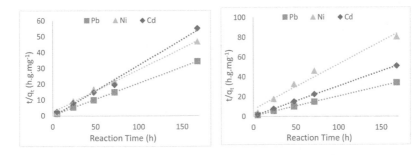

Figure 4. Pseudo-second order kinetic model of OP and SD samples.

3.4 Adsorption kinetics

Adsorption kinetics data are presented in Table 2. As can be seen from the results, the correlation coefficients were lower for pseudo-first order model than pseudo-second order model. As can be seen from the results, the correlation coefficients R^2 for pseudo-second order model were high (> 0.99) and it is suggesting that this model can be considered in the present adsorption system. Fig. 4 shows pseudo-second order kinetic model of magnetically modified OP and SD.

4 CONCLUSION

In our study, the metal ions were successfully removed by magnetically modified orange peels and sawdust. Biosorbents were found to be promising biosorbents for the removal of Pb(II), Ni(II) and Cd(II) ions from aqueous solution. The optimized experimental conditions, which led to higher metal removal, were initial pH in the range 5–6 and no control or adjustment of the pH during the biosorption process. The metal removal was shown to be strongly affected by changes in the solution pH. The removal efficiency of the sorption process of orange peels was more than 80% for lead for the concentration $\geq$100 mg.dm^{-3}. Nickel and cadmium ions reached the removal efficiency 65% for low concentration ($\geq$10 mg.dm^{-3}). The removal efficiency of sorption process of sawdust sample was only for low concentration in aqueous solution ($\geq$10 mg.dm^{-3}) for all followed metals.

ACKNOWLEDGEMENT

Authors thank for the financial support from the Ministry of Education, Youth and Sports of the Czech Republic; Project LQ1603 (Research for SUSEN). This work has been realized within the SUSEN Project (established in the framework of the European Regional Development Fund (ERDF) in project CZ.1.05/2.1.00/03.0108) and project TAČR—Gama project TG02010037.

REFERENCES

Bourlinos, A.B., Zboril, R., Petridis, D. 2003. A simple route towards magnetically modified zeolites. *Microporous & Mesoporous Materials.* 58: 155–162.

Grohmann, K., Baldwin E.A. 1992. Hydrolysis of orange peel with pectinase and cellulose enzymes. *Biotechnology Letters* 14: 1169–1174.

López, J.Á.S., Li, Q., Thompson, I.P. 2010. Biorefinery of waste orange peel. *Critical Reviews in Biotechnology* 30: 63–69.

Massart, R.A. 1981. Preparation of aqueous magnetic liquids in alkaline and acidic media. *IEEE Transactions on Magnetics* 17: 1247–1248.

Mockovciakova, A., Orolinova, Z., Skvarla, J. 2001. Enhacement of the bentonite sorption properties. *Journal of Hazardous Materials* 180: 274–281.

Pourbafrani, M., Forgacs, G., Sárvári Horváth, I., Niklasson, C., Taherzadeh, M.J. 2010. Production of biofuels, limonene and pectin from citrus wastes. *Bioresource Technology* 101: 4246–4250.

Rozumová, L., Životský, O., Seidlerová, J., Motyka, O., Šafařík, I., Šafaříková, M. 2016. Magnetically modified peanut husks as an effective sorbent of heavy metals. *Journal of Environmental Chemical Engineering* 4: 549–555.

Rozumová, L., Seidlerová, J., Šafařík, I., Šafaříková, M., Cihlářová, M., Gabor, R. 2014. Magnetically modified tea for lead sorption. *Advanced Science, Engineering and Medicine* 6: 473–476.

Rozumová, L., Kůs, P., Šafařík, I. 2016. *Sorption of metals on biological waste material.* SWWS Conference, in Press.

Tripodo, M.M. Lanuzza, F., Micali, G., Coppolino, R., Nucita, F. 2004. Citrus waste recovery: a new environmentally friendly procedure to obtain animal feed. *Bioresource Technology* 91: 111–115.

Veres, J., Orolinova, Z. 2009. Study of the treated and magnetically modified bentonite as possible sorbents of heavy metals. *Acta Montanistica Slovaca* 14: 152–155.

Widmer, W., Zhou, W., Grohmann, K. 2010. Pretreatment effects on orange processing waste for making ethanol by simultaneous saccharification and fermentation. *Bioresource Technology* 101: 5242–5249.

Wikandari, R., Nguyen, H., Millati, R., Niklasson, C., Taherzadeh, M.J. 2015. Improvement of Biogas Production from Orange Peel Waste by Leaching of Limonene. *Biomed Research International* 2015: 1–6.

WASTES – Solutions, Treatments and Opportunities II – Vilarinho, Castro & Lopes (Eds)
© 2018 Taylor & Francis Group, London, ISBN 978-1-138-19669-8

Phytoremediation of soils contaminated with lead by *Arundo donax* L.

S. Sidella & S.L. Cosentino
Dipartimento di Agricoltura, Alimentazione e Ambiente—Di3A, Università degli Studi di Catania, Catania, Italy

A.L. Fernando & J. Costa
Departamento de Ciências e Tecnologia da Biomassa/MEtRiCS, Faculdade de Ciências e Tecnologia, FCT, Universidade Nova de Lisboa, Caparica, Portugal

B. Barbosa
Universidade Federal do Oeste da Bahia, Barreiras, Brasil

ABSTRACT: *Arundo donax* L. is a high yielded perennial energy crop. Its cultivation for bioenergy may generate land-use conflicts which might be avoided through the establishment of energy crops on marginal land. In this context, this work aims to study the potentiality of *Arundo donax* in Pb contaminated soils (450 and 900 mg Pb kg^{-1}, dry matter) under a low irrigation regime (475 mm). Results showed that biomass productivity of giant reed was negatively affected by the contamination. Increased higher lead content in the biomass was obtained with increased lead contamination of the soil. Highest accumulation of lead was observed in the roots and rhizomes. Although the lead removal percentages by giant reed accumulation represent 3.2‰ maximum, after two consecutive years, towards the Pb soil bioavailable fraction, the establishment of a vegetative cover represent an approach to attenuate and stabilize contaminated sites with additional revenue to owners.

1 INTRODUCTION

Soil contaminated by heavy metals is a major problem causing vast areas of agricultural land to become derelict and non-arable and hazardous for both wildlife and human populations (Alloway, 1995). The area of soils contaminated with anthropogenic lead (Pb) is high, given its widespread deposition over the course of the last two centuries due to industrial waste, leaded paint, and automobile exhaust (McClintock, 2015). In the late 1970s, the anthropogenic deposition of Pb began to decline due to the gradual phase-out of leaded fuels and paints. Yet, lead historic deposition and Pb soil levels in certain spots (McClintock, 2015) still represent a source of pollution.

Giant reed (*Arundo donax* L.) is a C3 herbaceous perennial non-food crop characterized by relatively high yields (El Bassam, 2010). The plant belongs to the *Poaceae* family and to the *Arundineae* tribe, being the most common of the species of its genus (Mariani et al., 2010), growing spontaneously throughout the Mediterranean basin (Angelini et al., 2005; Cosentino et al., 2014). This perennial grass has been recognized as low-cost and low-maintenance crop and the high yielded biomass can be used for the production of energy (for both solid and second generation biofuels), paper pulp and biomaterials (Alexopoulou et al., 2011 and 2012).

Recent approaches utilize this crop in the phytoremediation of soils contaminated with heavy metals (e.g., Barbosa et al., 2015, Liu et al., 2017, Nsanganwimana et al., 2015, Sidella et al., 2016). Its robustness and physiological characteristics and its deep, dense and extensive root system, allows this grass to easily adapt to different types of soils and ecological conditions (Fernando et al. 2010), offering the possibility to associate soil decontamination and restoration with the production of biomass for bioenergy and biomaterials with additional revenue (Fernando et al., 2016). Additionally, when giant reed is cultivated in marginal/contaminated soils, land use conflicts with food crops are reduced (Dauber et al. 2012, Lewandowski, 2015), minimizing direct and indirect negative effects due to Land Use Change (LUC) (Fritsche et al. 2010). In this context, this research work aims to study the potentiality of giant reed production in Pb contaminated soils. Furthermore, considering the scarcity of water resources in the Mediterranean region, the study was conducted under a low irrigation regime, in order to reduce the impacts associated with irrigation.

2 MATERIALS AND METHODS

The aim of this work was to study the potential of giant reed in the phytoremediation of soils contaminated with lead under a low irrigation regime. The two-year pot experiment was conducted inside the *Campus* area of the Faculty of Sciences and Technology of the Universidade NOVA de Lisboa, from where the soil and rhizomes of *Arundo donax* L. were collected. The trial, established in April 2012, was run in pots containing 12 kg of soil (loam soil previously analyzed) artificially contaminated with a lead rich sludge (waste product derived from a battery manufacturing company, "Sociedade Portuguesa do Acumulador Tudor", located in Castanheira do Ribatejo, near Lisbon), containing 14% Pb (dry weight basis). Two concentrations of lead in contaminated soils were tested (450 and 900 mg Pb.kg^{-1} dry matter, corresponding to maximum allowable and to twice as maximum, respectively, Pb$_{450}$ and Pb$_{900}$) (Decreto-Lei n° 276/09, 2009). In each pot, two rhizomes were established (10 cm deep) (a pot for each level of contamination with replicates). After the establishment of the rhizomes, pots were fertilized: 3 g N/m^2 (urea, 46% N); 3 g N/m^2 (nitrolusal, mixture of NH$_4$ NO$_3$ + CaCO$_3$, 27% N); 17 g K$_2$O/m^2 (potassium sulphate, 51% K$_2$O); 23 g P$_2$O$_5$/m^2 (superphosphate, 18% P$_2$O$_5$). Simultaneously, a low irrigating regime was applied: 475 mm. The urea was applied when plants reached approximately 40–50 cm height. The same NK fertilization was applied in the 2nd year, when plants reached approximately 40–50 cm height, but not P once P fertilizer applied in the first year is enough for the growth of these perennial grasses for at least 10 years (El Bassam, 2010). Pots without plants were also prepared to investigate the influence of the soil-biomass system versus soil system in the remediation of the contamination.

At the end of each growing season (December–January), during two consecutive years (2012–2013), the plants were harvested and the aerial productivity and lead accumulation was monitored. Total below-ground dry weight and its lead accumulation were also determined in the second year. Lead content was determined by atomic absorption following calcination of biomass at 550°C for two hours, in a muffler furnace and nitric acid digestion of the ash material. Lead released by percolated waters was also evaluated along the two growing seasons. Bioavailable Pb in the soils at the beginning of the experiment was also evaluated by EDTA extraction (Iqbal et al. 2013).

The statistical interpretation of the results was performed using analysis of variance (one-way ANOVA) by means of CoStat software (version 6.0) and the means were separated according to the test of Student-Newman-Keuls (SNK) when ANOVA revealed significant differences (p ≤ 0.05).

3 RESULTS AND DISCUSSION

3.1 *Biomass productivity*

Figure 1 presents the aerial biomass productivity obtained in the trials, corresponding to stems leaves and litter produced during two consecutive years. According to the results obtained, there was an increase of productivity from the 1st to the 2nd year: this behavior was observed in all of the studied pots (contaminated and not contaminated). This reflects

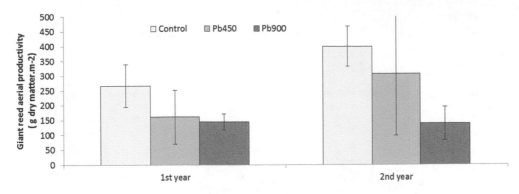

Figure 1. Giant reed aerial productivity (g.m^{-2} dry matter) during two growing cycles.

the energy spent by the plant, on the first years, to develop the extensive under-ground rhizome system (El Bassam, 2010). Leaves fraction represent the highest share in giant reed, ca. 53–58% of the total aerial biomass. Stems represent ca. 28–37% of the total aerial biomass. Litter represents ca. 5–19% of the total aerial biomass. Belowground biomass represents the highest share in the total biomass, ca. 89–92%. This is consistent with the common behavior of perennial crops, that spend more energy on the development and establishment of the belowground organs on the first years of cultivation (Fernando, 2005).

Lead contamination negatively affected biomass, both aerial biomass (Figure 1) and rhizomatous biomass. Roots and rhizome yields in Pb_{450} pots decreased by 37%, and roots and rhizome yields in Pb_{900} pots decreased by 60%, when compared with control (non-contaminated pots, with 3900 mg m^{-2}, dry matter). Yet, the reduction in yields observed in the pots contaminated with lead was not statistically significant (p > 0.05). Guo and Miao (2010) also reported phytotoxic effects on giant reed, but only when Pb concentration in soil was higher than 1000 mg kg^{-1}.

3.2 *Accumulation of Pb in the biomass*

Figure 2 present the Pb accumulated by the aerial biomass of giant reed in two growing seasons, and in the belowground biomass in the 2nd year.

Results show that biomass (aerial and belowground) obtained in Pb contaminated soils presented significantly higher lead accumulation then biomass from non-contaminated soils, thus showing phytoextraction capacity. The higher accumulation of Pb in the aerial fraction of the plant, from soils contaminated with Pb, shows the capacity of the plant to serve as a phytoextractor, once the metal will be uptaken and collected from the fields (Mirza et al., 2010). Yet, Pb transfer to the aerial fraction is limited and highest Pb remained accumulated in the belowground organs of the plant, which is consistent with other studies reported by Nsanganwimana et al. (2014). According to Kabata-Pendias (2001) the translocation of Pb from belowground organs to the aerial fraction is limited because Pb pyrophosphate bind to cell walls. This indicates that giant reed phytoremediation action is mostly due to its Pb immobilizing and stabilizing capacity in the roots and rhizomes.

3.3 *Pb remediation*

Figure 3 shows the Pb removal (%) by the aerial and belowground biomass of giant reed towards the bioavailable Pb content in the soil (Control, 7.5 mg kg^{-1}; Pb_{450}, 192 mg kg^{-1}; Pb_{900}, 515 mg kg^{-1}), after two consecutive years.

As it is observed by Figure 3, the removal percentage of bioavailable Pb in soil by the aerial biomass is very limited: 0.5‰ maximum and 0.3‰ average. Pb mostly accumulates in the belowground biomass: 2.7‰ maximum and 1.5‰ average. This means that if we consider the phytoextraction process, the results are not optimal, because to remove all the lead in the soil,

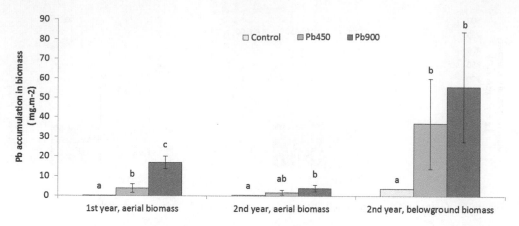

Figure 2. Lead accumulation in giant reed aerial biomass in two consecutive years and in below-ground biomass, on the 2nd year (mg Pb.m^{-2}). Different lower-case letters indicate statistical significance ($p < 0.05$, SNK test) between Pb treatments, for each type of biomass (aerial or belowground) and for each season (1st year or 2nd year).

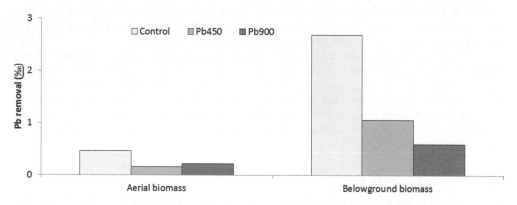

Figure 3. Giant reed biomass Pb removal (‰) from bioavailable lead in the soil, after two consecutive years.

several years would be needed. But, considering that the aerial biomass does not contain Pb in high amounts, than the aerial biomass can be economically valorized. Moreover, the plant showed tolerance to the contamination, and a vegetative crop can be established in this type of contaminated soil, with the environmental benefits associated with.

3.4 *Pb in percolated waters*

During the two years of trials, percolated waters were monitored. Results were always below 0.10 mg Pb.dm^{-3}, lower than the limit values for irrigation water (Pb, 5.0 mg.dm^{-3}, according to the Portuguese legislation, Decreto Lei n° 236/98, 1998). This means that the amount of Pb in the percolated waters do not represent a risk to ground/surface waters.

4 CONCLUSIONS

Giant reed showed lead phytoextraction and accumulation capacity. The higher the Pb contamination in the soil, the higher the accumulation of Pb in aerial and below ground biomass. The contamination with lead also affected negatively the production of giant reed, but not significantly. Results showed that giant reed have potential to simultaneously deliver high yields and restore soil properties. Due to the extensive radicular system, giant reed is associated with control of soil erosion, carbon sequestration and minimization of nutrient leaching as well as with the restoration of soil properties (fertility, structure, organic matter). The biomass being produced may contribute to a positive energy balance and to greenhouse gases emissions reduction. Yet, the removal of lead from the soil by giant reed is a slow process. Yet, it represents an opportunity to produce sustainable biomass in a resource constrained World.

Further studies are needed to assess the biomass quality obtained from those contaminated fields. The prospect of the valorization of the giant reed aerial biomass, for bioenergy or biomaterials, could lessen the financial costs of soil remediation, compared to the traditional physical—chemical processes with the associated revenue of environmental benefits. However, little is known about the fiber content, or the calorific value of giant reed produced in Pb contaminated soils, and if the contamination affects the contents, compromising the biomass valorization.

ACKNOWLEDGEMENTS

This work was supported by the European Union (Project Optimization of perennial grasses for biomass production (OPTIMA), Grant Agreement No: 289642, Collaborative project, FP7-KBBE-2011.3.1-02).

REFERENCES

Alexopoulou, E., Christou, M., Cosentino, S.L., Monti, A., Soldatos, P., Fernando, A.L., Zegada-Lizarazu, W., Scordia, D., Testa, G. 2011. Perennial Grasses and their Importance as Bioenergy Crops In EU27. In: Faulstich, M., Ossenbrink, H., Dallemand, J.F., Baxter, D., Grassi, A., Helm, P. (eds) Proceedings of the 19th European Biomass Conference and Exhibition, From Research to Industry and Markets. 6–10 June 2011, Berlin, Germany, Organized by ETA-Florence Renewable Energies and WIP-Renewable Energies, published by *ETA-Florence Renewable Energies*, pp 81–85.

Alexopoulou, E., Christou, M., Cosentino, S.L., Monti, A., Soldatos, P., Fernando, A.L., Zegada-Lizarazu, W., Scordia, D., Testa, G. 2012. Perennial Grasses: Important Biomass Source Feedstock for Bio-Based Products and Bioenergy, in: B. Krautkremer, H. Ossenbrink, D. Baxter, J.F. Dallemand, A. Grassi, P. Helm (Eds.), *Proceedings of the 20th European Biomass Conference and Exhibition, Milano*, 18–22 June 2012, 201–206.

Alloway, B.J. 1995. Heavy metals in soils, second ed. *Blackie Academic and Professional, Glasgow*, UK.

Angelini, L., Ceccarini, L., Bonari, E. 2005. Biomass yield and energy balance of giant reed (*Arundo donax* L.) cropped in central Italy as related to different management practices. *Eur. J. Agron.* 22: 375–389.

Barbosa, B., Boléo, S., Sidella, S., Costa, J., Duarte, M.P., Mendes, B., Cosentino, S.L., Fernando, A.L. 2015. Phytoremediation of Heavy Metal-Contaminated Soils Using the Perennial Energy Crops *Miscanthus* spp. and *Arundo donax* L., *BioEnergy Research* 8: 1500–1511.

Cosentino, S.L., Scordia, D., Sanzone, E., Testa, G., Copani, V., 2014. Response of giant reed (*Arundo donax* L.) to nitrogen fertilization and soil water availability in semi-arid Mediterranean environment. *Europ. J. Agronomy* 60: 22–32.

Dauber, J., Brown, C., Fernando, A.L., Finnan, J., Krasuska, E., Ponitka, J., Styles, D., Thrän, D., Groenigen, K.J.V., Weih, M., Zah, R. 2012. Bioenergy from "surplus" land: environmental and social-economic implications. *BioRisk* 7: 5–50.

Decreto-Lei nº 236/98, 1998. *Normas, critérios e objectivos de qualidade com a finalidade de proteger o meio aquático e melhorar a qualidade das águas em função dos seus principais usos*, Diário da República 176: 3676–3722 (in Portuguese).

Decreto-Lei nº 276/09, 2009. Anexo I, *Valores limite de concentração relativos a metais pesados, compostos orgânicos e dioxinas e microrganismos*, Diário da República, nº 192, I Série, 2 de Outubro 2009, 7154–7165 (in Portuguese).

El Bassam, N. 2010. Handbook of Bioenergy Crops. A complete reference to species, development and applications. *Earthscan Ltd.* London, UK.

Fernando, A.L., Barbosa, B., Costa, J., Papazoglou, E.G., 2016. Giant reed (*Arundo donax* L.): a multipurpose crop bridging phytoremediation with sustainable bio-economy. In: Prasad, M.N.V. (ed.) *Bioremediation and Bioeconomy, Elsevier Inc.*, UK, pp. 77–95.

Fernando, A.L., Duarte, M.P., Almeida, J., Boléo, S., Mendes, B. 2010. Environmental impact assessment of energy crops cultivation in Europe. *Biofuels, Bioproducts & Biorefining* 4: 594–604.

Fernando, A.L.A.C. 2005. Fitorremediação por *Miscanthus* x *giganteus* de solos contaminados com metais pesados, *PhD Thesis*, Universidade Nova de Lisboa (Lisbon, Portugal).

Fritsche, U.R., Sims, R.E.H., Monti, A. 2010. Direct and indirect land-use competition issues for energy crops and their sustainable production—an overview, *Biofuels, Bioprod. Biorefin* 4: 692–704.

Guo, Z.H., Miao, X.F. 2010. Growth changes and tissues anatomical characteristics of giant reed (Arundo donax L.) in soil contaminated with arsenic, cadmium and lead. *J Cent South Univ Technol* 17: 770–777.

Iqbal, M., Bermond, A., Lamy, I. 2013. Impact of miscanthus cultivation on trace metal availability in contaminated agricultural soils: Complementary insights from kinetic extraction and physical fractionation. *Chemosphere* 91: 287–294.

Kabata-Pendias, A. 2011. *Trace elements in soils and plants*. 4th edn. Boca Raton: CRC.

Lewandowski, I. 2015. Securing a sustainable biomass supply in a growing bioeconomy, *Global Food Sec* 6: 34–42.

Liu, Y.N., Guo, Z.H., Xiao, X.Y., Wang, S., Jiang, Z.C., Zeng, P. 2017. Phytostabilisation potential of giant reed for metals contaminated soil modified with complex organic fertiliser and fly ash: a field experiment. *Science of the Total Environment* 576: 292–302.

Mariani, C., Cabrini, R., Danin, A., Piffanelli, P., Fricano, A., Gomarasca, S., Dicandilo, M., Grassi, F., Soave, C. 2010. Origin, diffusion and reproduction of the giant reed (*Arundo donax* L.): a promising weedy energy crop. *Ann. Appl. Biol.* 157: 191–202.

McClintock, N. 2015. A critical physical geography of urban soil contamination. *Geoforum* 65: 69–85.

Mirza, N., Mahmood, Q., Pervez, A., Ahmad, R., Farooq, R., Shah, M.M., Azim, M.R. 2010. Phytoremediation potencial of *Arundo donax* in arsenic-contaminated synthetic wastewater. *Bioresour Technol* 101: 5815–5819.

Nsanganwimana, F., Marchland, L., Douay, F., Mench, M., 2014. *Arundo donax* L., a candidate for phytomanaging water and soils contaminated by trace elements and producing plant-based feedstock, a review. *Int J Phytorem* 16: 982–1017.

Sidella, S., Barbosa, B., Costa, J., Cosentino, S.L., Fernando, A.L. 2016. Screening of Giant Reed Clones for Phytoremediation of Lead Contaminated Soils. In: Barth, S., Murphy-Bokern, D., Kalinina, O., Taylor, G., Jones, M. (Eds.) Perennial Biomass Crops for a Resource Constrained World, *Springer International Publishing, Switzerland*, pp. 191–197.

WASTES – *Solutions, Treatments and Opportunities II – Vilarinho, Castro & Lopes (Eds)*
© *2018 Taylor & Francis Group, London, ISBN 978-1-138-19669-8*

Employment of industrial wastes as agents for inclusion modification in molten steels

F.A. Castro
University of Minho, Guimarães, Portugal
W2V, S.A., Guimarães, Portugal

J. Santos & P. Lacerda
Ferespe—Fundição de Ferro e Aço, Lda., Ribeirão, Portugal

R. Pacheco, T. Teixeira, A. Silva & E. Soares
W2V, S.A., Guimarães, Portugal

J. Machado & M. Abreu
University of Minho, Guimarães, Portugal

ABSTRACT: Chemical modification of sulphide and oxide inclusions is a normal practice to obtain more desirable mechanical properties in steels. For example, cryogenic resilience is a relevant property that may be affected relevantly by chemical and morphological modification of inclusions during liquid steel elaboration. This change in properties is done, in steelmaking, mostly during ladle treatment. In steel foundry practice, similar procedures are possible. In this work, it has been studied the employment of industrial residues, like calcium rich ones, for the purpose of treatment of molten steels. Results are evaluated and allowed to conclude for the usefulness of some of the wastes, enhancing better mechanical properties, especially at low temperatures.

1 INTRODUCTION

Calcium and calcium compounds and alloys play a relevant role in the treatment of steel melts for the purpose of metal desulphurization and dephosphorization, as well as for oxide and sulphide inclusions control and modification (Gatellier, 1991; Kiessling, 1978; Ghosh, 2001; Castro, 1987). For this purpose, metallic calcium and calcium compounds and alloys may be added by injection in steel melts, thus providing the desired chemical reactions. This is a normal treatment in steelmaking, normally done as a secondary metallurgy operation, in ladle furnaces. In foundry practices, similar procedures may be done, even if some difficulties arise due to the lower amounts of liquid metals, what implies lower metal-static pressures, thus providing more ability for calcium evaporation and consequent losses of this valuable metal.

As a matter of fact, at 1600°C and atmospheric pressure, calcium is present in the vapor phase. Thus, addition of calcium compounds or alloys leads to calcium vaporization. However, the pressure induced by liquid iron may reduce this effect, the deeper the insufflation of the material, the lower the tendency to evaporate. This is relevant because lesser tendency to evaporation allows more time of residence of calcium inside the liquid metal, thus enabling less loss of this expensive metal and greater efficiency for the chemical reactions. In foundry furnaces, the height of metal is normally in the order of 1 to 2 meters, while in steelmaking ladle furnaces that value may exceed 5 meters. This difference implies that, in steelmaking, the depth at which the insufflation of the material is done can control the evaporation of calcium, while in foundry practice it is more difficult to achieve. To avoid this effect, the adding of calcium oxide containing materials may be enough efficient, especially in the case of oxide inclusions modification. As a matter of fact, calcium oxide (CaO) tends to react with alumina

(Al_2O_3) inclusions that are formed after aluminum de-oxidation, in carbon steels, leading to the formation of calcium aluminates.

The relevant reaction is of the type:

$$CaO + Al_2O_3 \rightarrow CaO \cdot Al_2O_3 \tag{1}$$

Figure 1 presents the phase diagram of the $CaO - Al_2O_3$ system (Jerebtsov & Mikhailov, 2001), allowing to verify that in the central region there is a narrow range of compositions, approximately from 40% to 60% in mass of CaO, that corresponds to a liquid phase bellow 1600°C.

This is very relevant for inclusion modification. As a matter of fact, solid aluminum oxide inclusions, derived from aluminum de-oxidation, are small inclusions, but very hard and with irregular shapes (Rastogi & Cramb, 2001). This is not beneficial for some mechanical properties, as cryogenic temperatures resilience, for example. The formation of liquid inclusions, replacing Al_2O_3 ones, has a relevant effect on some of the mechanical properties, especially the above indicated impact resilience. Other benefit is related to the fact that liquid inclusions may easily coalesce, thus decanting more promptly, and being then absorbed by slag. This may result in a lower amount of inclusions in solid casts, with improvement of other mechanical properties, as tensile strength and tenacity, as referred by Rege, Szekeres & Forgeng (1970).

In this work, an industrial waste, containing high amounts of free lime, has been employed for the purpose of chemical modification of alumina inclusions in low carbon cast steels. The wastes whose results are here presented were industrial wastewater treatment sludge. In this case, liquid effluents from galvanic surface treatment operations are treated in the plant by lime neutralization. The resultant sludge contains high amounts of lime (normally more than 50%), present in the form of free lime or calcium compounds like sulphate, chloride and fluoride, thus being possible materials for the uptake of calcium to modify the oxide inclusions.

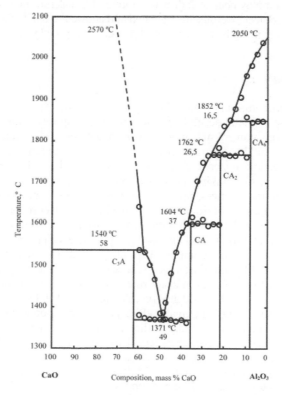

Figure 1. Phase diagram of the $CaO - Al_2O_3$ system (Jerebtsov & Mikhailov, 2001).

2 EXPERIMENTAL PROGRAM

2.1 *Employed wastes and preparation*

Wastes that have been employed in this work were a sludge from the physico-chemical treatment of wastewaters from the galvanic surface treatment of plastics, for chrome-platting. The industrial operation includes the platting of layers of copper, nickel and finally chromium. Wastewaters contain hence these three metals as well as phosphates and calcium sulphate. The sludge, originally containing 74% water, has been calcined at 1100°C for 6 hours, in an electric laboratory furnace, to eliminate water and decompose calcium sulphate to calcium oxide. The final composition of the waste after thermal treatment, obtained by X-ray Fluorescence Spectrometry is displayed in Table 1. Other elements were vestigial, like carbon, sulphur, tin and titanium, all present in contents lower than 1%.

A SiCaMn alloy with 68% Si, 18% Ca and 14% Mn, has been also employed as a chemical treatment precursor. Deoxidizing agent was aluminum scrap with more than 98% purity.

Experiments were conducted in an industrial induction furnace with capacity for 1 ton of steel. The alloy was a carbon steel with 0.14% C, 0.69% Mn, 0.47% Si, 0.12% Cr, 0.06% Ni, 0.020% S and 0.018% P. A mix of powderly materials, including waste and precursors, were injected through a low carbon steel lance inside the steel melt, as illustrated schematically in Figure 2, where the top cylinder is the powder container, the tube bellow it is the lance, the lower hole is the crucible top and the rest parts are the robotized arms. In green is indicated the tubes for argon conduction, which was employed as carrier gas to inject the powders.

Four experiments were done using 500 g of powder mixtures, with the following composition: 400 g of sludge with 100 g SiCaMn, this latter being grinded to less than 250 mm.

Table 1. Final composition of the waste, in % of weight.

Compound	CaO	P_2O_5	Cr_2O_3	NiO	CuO	SiO_2
Weight (%)	62	18	9	3	2	2

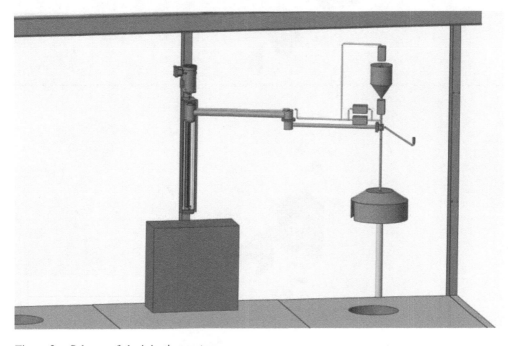

Figure 2. Scheme of the injection system.

In all experiments, a deoxidizing addition of 300 g of aluminum scrap has been done, just before the start of the injection of the powders.

The injection was done during around 90 seconds. Then, the liquid metal was poured in a ladle to produce castings. Metal samples were took some seconds before and after treatment, to evaluate the type of inclusions and low temperature resilience.

Charpy test samples were produced from collected samples, by machining. Results of tests, performed according to ASTM A370 standard, were compared with normal values obtained for the same type of steel.

2.2 Results

The metallographic observation of the samples obtained before the injection of waste + SiCaMn powder has been done by scanning electron microscopy and allowed to identify inclusions of pure alumina, as illustrated in Figure 3.

After treatment, the inclusions were completely modified, showing a round shape (see Fig. 4, obtained also by scanning electron microscopy), thus indicating that they were liquid at the treatment temperature.

The chemical analysis of the inclusions, obtained by Electron Dispersive Spectrometry (EDS) in the scanning electron microscope, allowed determine that they contain approximately the composition corresponding to the phases $Al_2O_3 \cdot CaO$ and $7\,Al_2O_3 \cdot 12CaO$. This is normally considered as beneficial for the mechanical properties of steels, including fatigue behavior (Juvonen, 2004).

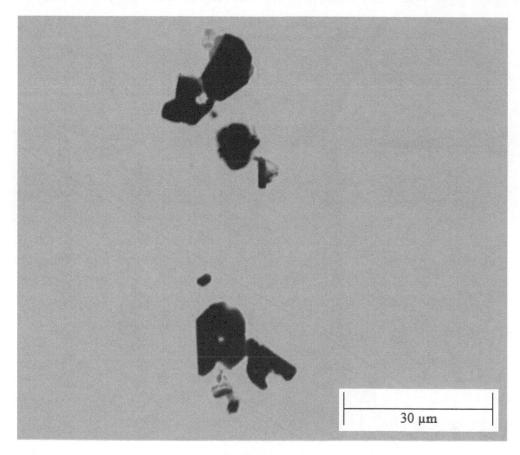

Figure 3.   Alumina inclusions present in steel before treatment.

392

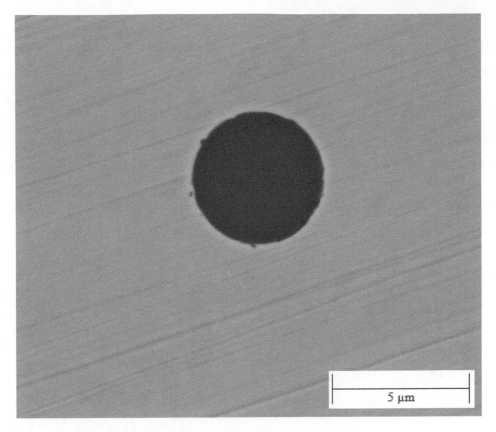

Figure 4. Globular calcium aluminate inclusion obtained after treatment.

Resilience determination at −50°C indicated, for the 4 samples determination, an average value of 76 J with a standard deviation of 22 J. This compare positively with the historical data obtained for more than 100 samples, for the same type of steel, of 47 J as average with a standard deviation of 18 J, when no treatment has been done.

3 CONCLUSIONS

The treatment of carbon steel foundry melts with calcium compounds and wastes allow the shape modification of oxide inclusions with the replacement of hard and irregular shaped alumina by round shaped calcium aluminate. This has a beneficial effect on low temperature resilience of the steel, measured at −50°C. The injection of mixed powders with deoxidizing and precursors for chemical modification of inclusions, combined with CaO containing wastes, is promising as a treatment of steel melts, for the enhancing of some mechanical properties in cast pieces, when it is induced by the types and morphology of inclusions present in the final product. Wastewater sludge are interesting waste materials providing CaO for the purpose of this chemical modification of oxide inclusions in cast steels, by means of injection procedures.

ACKNOWLEDGEMENTS

This work has been co-financed by COMPETE 2020, Portugal 2020 and the European Union through the European Regional Development Fund – FEDER within the scope of the project POCI-01-0247-FEDER-003330, RCS – Refined Cast Steel.

REFERENCES

Castro, F.A. 1987. *Modificação das Inclusões pelo Cálcio nos Aços de Alto Teor em Enxofre* (in Portuguese). PhD Thesis. University of Minho. Portugal: Braga.

Gatellier, C., Gaye, H., Lehmann, J., Pontoire, J.N. & Castro, F.A. 1991. Control of Complex Inclusions during Metallurgical Treatments on Liquid and Solid Steels. *Steelmaking Conference of The Iron and Steel Society of The American Institute of Metallurgical Engineers (ISS-AIME)*. Steelmaking Conference Proceedings' 91, Washington DC. ISSN 0932897630, Ed. ISS-AIME, p. 827–834.

Ghosh, A. 2001. *Secondary Steelmaking—Principles and Applications.* CRC Press.

Jerebtsov, D.A. & Mikhailov, G.G. 2001. Phase diagram of the CaO-Al$_2$O$_3$ system. *Ceramics International* 27(2): 25–28.

Juvonen, P. 2004. *Effects of non-metallic inclusions on fatigue properties of calcium treated steels.* PhD Thesis Helsinki University of Technology. Finland: Espoo.

Kiessling, R. 1978. *Non-Metallic Inclusions in Steel.* Parts I-IV. London: The Institute of Materials.

Rastogi, R.W. & Cramb, A.W. 2001. *Steelmaking Conference Proceedings.* ISS, Warrendale, PA, Vol. 84, pp. 789–829.

Rege, R.A., Szekeres, E.S. & Forgeng, W.D. 1970. Metallurgical Transactions, AIME, Vol.1, N°. 9, pp. 2652.

WASTES – Solutions, Treatments and Opportunities II – Vilarinho, Castro & Lopes (Eds)
© 2018 Taylor & Francis Group, London, ISBN 978-1-138-19669-8

Olive pomace phenolics extraction: Conventional *vs* emergent methodologies

M.A. Nunes, R.C. Alves, A.S.G. Costa & M.B.P.P. Oliveira
REQUIMTE/LAQV, Faculty of Pharmacy, University of Porto, Porto, Portugal

H. Puga
Centre for Micro-Electro Mechanical Systems, University of Minho, Guimarães, Portugal

ABSTRACT: Olive oil production has been steadily increasing. Along with the growth of this sector, a high rate of wastes is produced. Olive pomace presents high levels of phenolic compounds with potential applications in food, cosmetic and pharmaceutical industries. Sustainable and cost-effective technologies for bioactive compounds recovery are required. In this work, conventional extraction methodologies were compared with an ultrasound technique using a probe and only water as solvent. Total phenolics and antioxidant activity of the extracts were assessed. A conventional solid-liquid extraction (hydroethanolic solvent, 40°C, 60 min) was the best procedure to recover phenolics compounds from pomace (487 µg GAE/ml). However, in only 5 min, the sonication probe allowed significantly higher yields of recovery compared to aqueous conventional extraction (60 min) (402 and 257 µg GAE/ml, respectively). Therefore, it seems to be a very promising technique to obtain antioxidants from olive pomace in an organic solvent-free medium.

1 INTRODUCTION

In Europe, for waste management, the Waste Framework Directive (2008/98/EC) is the core legislative act. Accordingly, it is emergent to avoid waste generation and to improve the use of wastes as resources (Directive 2008/98/EC, 2008). Green practices are mandatory to allow a sustainable growth of the agro-industry sector. The conversion of wastes into added-value products is a challenging task that can have a positive socio-economic impact and simultaneously reduce the environmental burden.

Olive oil production has been increasing worldwide over the last years. European Union leads the international market, producing ≈80–84% of the total olive oil. Olive oil has a great economic importance in Mediterranean countries and in new producing countries of America, Africa, and Australia. Nevertheless, as other agro-industries, olive oil production generates a large quantity of wastes, namely olive pomace (Nunes et al., 2016). The resulting wastes depend on the technique applied for extracting olive oil. The common systems used are pressing mills, two-phase and three-phase processes. In the two-phase process, unlike the three-phase system, no water is used. This means that both olive oil and a humid semi-solid pomace are obtained. Despite resulting from an eco-friendly system, olive pomace has a high water content, which can difficult a proper storage. This semi-solid residue also includes skins, pulp and stones. After olive oil processing, olive pomace is discharged in large open-air containers (evaporation ponds) where it is stored through long periods until further transformation in olive pomace oil and a last final residue—extracted olive pomace (Fernández-Hernández et al., 2014). As the production of olive oil is increasing, the discharge of olive pomace is also escalating. Olive pomace oil industry partially solves this problem. Still, a final residue exists at the end of the processing that usually is used as fuel with a minimal added-value (Roselló-Soto et al., 2015).

In olive pomace, a huge amount of phenolic compounds and fatty acids still remains. These bioactive compounds are phytotoxic. Despite some countries allow land spreading, it must be done under a tight control (Di Bene et al., 2013). Although those pomace compounds are a hazard for the environment, they present benefic effects for human health (Nunes et al., 2016). This feature anticipates new opportunities to recover functional ingredients for further applications, namely for cosmetics and food industry.

The recovery of bioactive compounds from olive by-products, intending innovative applications and enhanced technological functions have been strongly explored over the last years. The added-value of these agro-industry by-products has been recognized, as well as the benefits of their bioactive compounds (Galanakis, 2012). The extraordinary content in phenolic compounds, such as phenolic acids and alcohols, secoiridoids, lignans and flavones, stands out olive pomace as a promissory valuable by-product (Mirabella et al., 2014).

Several conventional (*e.g.* solvents, heat and grinding) and non-conventional methods (*e.g.* supercritical fluid extraction, pulsed electric fields and microwaves) have been studied in order to achieve maximum recoveries of olive pomace bioactive compounds (Roselló-Soto et al., 2015). Nevertheless, an emergent question always rises: how to obtain, simultaneously, maximum yields of extraction, at low-cost, using eco-friendly technologies and in a minimum period of time? The development of sustainable methodologies for the recovery of bioactive compounds from agro-industrial by-products is emergent. Indeed, it is important to find simple, fast, cost-effective and efficient processes to achieve that aim. Extraction methodologies using solvents, microwave assisted or supercritical fluids have been used consistently for bioactive compounds extraction (Galanakis, 2012).

The useful effects of ultrasonic liquid processing are related to efficient transformation of electric energy (using an ultrasonic power supply or generator) into mechanical vibration by means of an ultrasonic transducer (operating in resonance), which is in the last step radiated to a medium in form of acoustic intensity. During this process, it is necessary to minimize thermal losses (in the ultrasonic generator, piezoelectric transducer and acoustic radiator) and maximize the real power load of acoustic energy in the water medium. Thus, with the application of the acoustic intensity, it is created kinetic energy in the liquid, which promotes the homogenization of the mixture. Moreover, the increasing of the medium temperature (related to bubbles collapse) promotes and accelerates the extraction.

This work aimed to explore and compare conventional methods of extraction (solid-liquid extraction with different solvents) with an emerging processing (an ultrasonic vibration technique using a probe and only water as extractor) in order to achieve an optimal extraction of olive pomace antioxidants.

2 MATERIALS AND METHODS

2.1 *Reagents and standards*

Gallic acid, heptahydrate ferrous sulfate, trolox, Folin-Ciocalteu's reagent, 2,2-diphenyl-1-picrylhydrazyl radical (DPPH), ferric chloride, 2,4,6-tripyridyl-striazine (TPTZ), and sodium acetate were all acquired from Sigma-Aldrich (St. Louis, USA). Anhydrous sodium carbonate and absolute ethanol were purchased from Merck (Darmstadt, Germany). All reagents were of analytical grade.

2.2 *Sample and sample preparation*

Olive pomace was obtained from an olive oil extraction unit in the Protected Designation of Origin Trás-Os-Montes (Portugal). In order to preserve olive pomace, samples were frozen (−80°C) and lyophilized (Telstar Cryodos-80 Terrassa, Barcelona).

2.3 *Nutritional analysis*

Moisture was determined using an infrared balance (Scaltec model SMO01, Scaltec Instruments, Heiligenstadt, Germany). All the analysis (protein, ash and fat contents) were performed according to AOAC (AOAC, 2012). Total carbohydrates were calculated by difference.

2.4 Extracts preparation

2.4.1 Conventional extraction

A classic solid-liquid extraction was performed using three different solvents: i) water, ii) water: ethanol (1:1, v/v), and (iii) ethanol. Extracts were prepared using a rigorously weighted amount (1 g) of lyophilized sample and 50 ml of solvent. The bioactive compounds were extracted with magnetic stirring at 40°C, for 1 h. Then, they were centrifuged (3000 rpm; 10 min) and filtered (Whatman No. 4).

2.4.2 Ultrasonic extraction

An ultrasound device (Figure 1(a)) was used for extraction, comprising a piezoelectric transducer, an acoustic radiator, a generator, a Compact DAQ for temperature acquisition, a type S thermocouple and a Windows compatible software to control processing. The whole ultrasonic device was previously optimized through Finite Element Model simulation (Fig. 1(b)) based in pre-established mathematical models. Parameters of extraction are described in Table 1.

2.5 Total phenolics content

The total amount of phenolic compounds was determined by spectrophotometry (BioTek Instruments, Inc) based on Costa et al. (2014), with some modifications. Briefly, 30 µl of each extract were mixed with 150 µl of Folin-Ciocalteu reagent (1:10) and 120 µl of a sodium carbonate solution (7.5 g/100 ml). The mixture was incubated at 45°C, protected from light, during 15 min. After 30 min at room temperature, the absorbance was measured at 765 nm. A calibration curve prepared with gallic acid was used (5–100 µg/ml; R^2 = 0.9996). Total phenolics were expressed as µg of gallic acid equivalents (GAE)/ml of extract. The assay was performed in triplicate.

2.6 Ferric Reducing Antioxidant Power (FRAP)

The FRAP assay was carried out according to Benzie and Strain (Benzie and Strain, 1996) with minor modifications (Costa et al., in press). In short, an aliquot of 35 µl of extract were

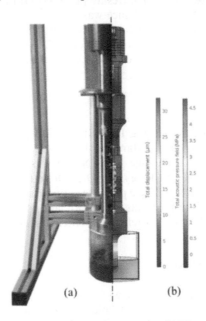

Figure 1. Ultrasonic extraction. (a) Experimental apparatus; (b) Numerical results of acoustic pressure in the sonicated medium (input electrical power of 160 W and f_0 = 20.01 ± 0.03 kHz).

Table 1. Total phenolics content and antioxidant activity of differently prepared extracts.

	Extract	Solvent	Frequency (kHz)	Input electric power (W)	Time (min)	Total phenolics µg GAE/ml	FRAP µmol/ml FSE	DPPH· inhibition µg TE/ml
Sonication	1	Water	20.05 ± 0.03	90	10	309.3 ± 17.3[d]	5.78 ± 0.41[c]	955.4 ± 93.2[c]
	2	Water	20.00 ± 0.03	160	10	370.2 ± 13.7[c]	6.68 ± 0.36[b]	989.1 ± 34.3[c]
	3	Water	20.01 ± 0.03	160	5	401.9 ± 20.2[b]	7.26 ± 0.82[b]	1162.8 ± 181.5[b]
Conventional	4	Water	n/a	n/a	60	256.4 ± 17.8[e]	3.57 ± 0.31[d]	490.0 ± 44.8[d]
	5	Water: Ethanol (1:1)	n/a	n/a	60	487.2 ± 22.8[a]	9.40 ± 0.38[a]	1403.1 ± 110.7[a]
	6	Ethanol	n/a	n/a	60	307.4 ± 13.2[d]	5.57 ± 0.35[c]	1070.6 ± 102.6[bc]

Different letters within each column represent significant differences at $p < 0.05$.
n/a, not applicable; GAE, gallic acid equivalents; FRAP, ferric reducing antioxidant power; FSE, ferrous sulfate equivalents; DPPH·, (2,2-diphenyl-1-picrylhydrazyl); TE, trolox equivalents.

mixed with 265 µl of the FRAP solution (containing 0.3 M acetate buffer, 10 mM TPTZ solution and 20 mM of ferric chloride). The mixture was kept for 30 min at 37°C protected from light. Absorbance was measured at 595 nm. A calibration curve was prepared with ferrous sulfate (25–500 µmol/l; $R^2 = 1$), and the ferric reducing antioxidant power was expressed as µmol of ferrous sulfate equivalents (FSE)/ml of extract. The assay was performed in triplicate.

2.7 DPPH· (2,2-diphenyl-1-picrylhydrazyl) scavenging activity

The radical scavenging ability of the different extracts was evaluated according to Costa et al. (2014), with slight modifications. Briefly, the reaction was initiated by transferring 30 µl of a diluted extract to 270 µl of a freshly prepared DPPH· solution (6.0×10^{-5} mol/l in ethanol). The decrease of the absorbance was measured in equal time intervals (2 min) at 515 nm, in order to observe the kinetic reaction. The reaction endpoint was attained in 30 min. A calibration curve was prepared with trolox (2.5–175 µg/ml; $R^2 = 0.9984$) and the scavenging activity was expressed as µg of trolox equivalents (TE)/ml of extract. The assay was performed in triplicate.

2.8 Statistical analysis

Statistical analysis was performed using IBM SPSS v. 20 (IBM Corp., Armonk, 241 NY, USA). Data were expressed as mean ± standard deviation ($n = 9$). The Shapiro-Wilk test was used to evaluate data normality. The one-way ANOVA was used to assess significant differences between samples, followed by Tukey's HSD or Dunett T3 post-hoc test (selected based on the equality of the variances) to make pairwise comparisons between means. The level of significance for all hypothesis tests (p) was 0.05. Correlations among the different parameters analyzed were achieved by Pearson correlation coefficients (r) at a significance level of 99%.

3 RESULTS AND DISCUSSION

The olive pomace analyzed in this study was composed, in a dry weight basis, by 3% of moisture, 7% of protein, 1% of ash, 6% of lipids and 83% of total carbohydrates (in which total fiber is included). Conventional solid-liquid extractions using different solvents were performed and compared with aqueous extraction using a sonication probe (Table 1).

The efficacy of extraction was assessed by analyzing the total phenolics content of the extracts, as well as their antioxidant activity. This was determined using two complementary methods. In the DPPH$^{\bullet}$ scavenging assay, the radical can be neutralized by both direct reduction (via electron transfer) or by radical quenching (via H atom transfer). In the FRAP assay, the reduction of Fe^{3+}-TPTZ to a colored product occurs, detecting natural compounds with redox potentials lower than 0.7 V (that of Fe^{3+}-TPTZ), but not compounds that act by radical quenching (H transfer) since only an electron transfer mechanism occurs. Nevertheless, both assays are reasonable screening methods to evaluate the potential ability of natural compounds to maintain the redox status in cells or tissues (Costa et al., in press). Table 1 presents the antioxidant profile of the olive pomace extracts obtained with the different experimental conditions.

As it can be clearly observed in Table 1, the conventional solid-liquid extraction using a hydroethanolic solvent (1:1), at 40°C, for 60 min, was the best procedure to recover antioxidants from olive pomace. Indeed, significantly higher values of total phenolics and antioxidant activity ($p < 0.05$) were achieved, compared to all the remaining tested procedures. This is certainly due to the type of phenolic compounds present in olive pomace (such as hydroxytyrosol, tyrosol, oleuropein, apigenin, luteolin, among others) (Araújo et al., 2015), which together are more extractable with an intermediate polarity solvent.

Nevertheless, comparing the aqueous extracts, in only 5 min (using an input electric power of 160 W), the sonication probe allowed a significantly higher ($p < 0.05$) yield of extraction ($\approx$more 60% of total phenolics) than that obtained with the aqueous conventional method (60 min). An extension of this extraction to 10 min showed to be detrimental probably due to the temperature increase of the medium (Fig. 2), which can lead to the degradation of the compounds. In fact, Uribe et al. (2013) also reported a loss of total phenolic compounds, especially above 70°C, in olive pomace submitted at air-drying temperature. These authors concluded that air drying temperatures affected antioxidant activity (Uribe et al., 2013) and our results are in agreement to those conclusions. Based on that, less extraction time and a higher input electric power showed to be more adequate to extract the natural antioxidants of this matrix.

The composition of the extracts varied widely according to extraction procedure used. Nevertheless, significantly high correlations ($p < 0.01$) were obtained between the different parameters analyzed (Table 2), namely between 0.828 and 0.963, showing clearly that phenolic compounds of olive pomace are the chemical constituents responsible for the sample antioxidant activity.

Several factors have impact on the olive pomace phenolic compounds. Primarily, olive pomace depends on the phenolic profile of olives, which, in turn, varies greatly with their variety, stage of ripeness, storage time, geographical origin and edaphoclimatic conditions; secondly, the olive oil processing technique will also influence the pomace composition. In a last instance, sample preparation and extraction methodologies (*e.g.* time, temperature,

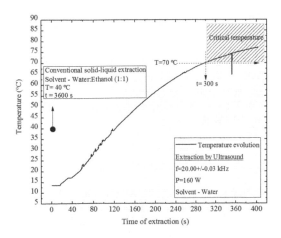

Figure 2.   Temperature increase during ultrasonic olive pomace antioxidants extraction.

Table 2. Pearson correlations coefficients (r) between the analyzed parameters.

	Total phenolics	FRAP	DPPH⁺ inhibition
Total phenolics	1	0.963**	0.828**
FRAP	0.963**	1	0.892**
DPPH⁺ inhibition	0.828**	0.892**	1

**Significant correlation at $p < 0.01$.
FRAP, ferric reducing antioxidant power; DPPH⁺, 2,2-diphenyl-1-picrylhydrazyl.

solvent) can lead to the great range of results found in scientific literature (Uribe et al., 2013). Nonetheless, a review from Dermeche et al. (2013) showed that total phenolics varied between 0.4 and 2.43%, which is in total agreement with the results of this work. In addition, an optimized methodology for phenolics recovery which studied the effect of a multistep extraction, including the time-temperature binomial, and a solvents sequence is also in line with the results here presented (Alu'datt et al., 2010).

Ultrasonic extraction seems, thus, to be a very promising technique to extract antioxidants in an organic solvent-free medium although, in this case, optimization should be further attempted in order to achieve the extraction yields obtained with the hydroalcoholic solvent.

4 CONCLUSIONS

Due to the olive pomace extracts richness in bioactive compounds, its valorization should be promoted. This study was conducted in order to evaluate the efficacy of different procedures to extract phenolic compounds from this by-product, contributing to its valorization and recycling (in a circular economy perspective), foreseeing health benefits since its extracts can be further used in food or cosmetic industries. New approaches to recover and stabilize bioactive compounds from extracts as well as new industrial applications are also promising.

High amounts of phenolic compounds were quantified in olive pomace, namely between 1.3 and 2.4 g GAE/100 g of pomace, depending on the extraction method used. Moreover, this work highlights the potential of ultrasound as a green extraction technique that requires less energy and time, without using organic solvents. The results of this work show that this type of technology can be of high interest to recover antioxidants from natural products, in a clean and efficient way. Nevertheless, optimization can still be performed in order to achieve the highest extraction yield obtained with the hydroalcoholic solvent.

ACKNOWLEDGEMENTS

The authors from REQUIMTE/LAQV thank the financial support to the project Operação NORTE-01-0145-FEDER-000011—Qualidade e Segurança Alimentar—uma abordagem (nano) tecnológica. This work was also supported by the project UID/QUI/50006/2013 – POCI/01/0145/FEDER/007265 with financial support from FCT/MEC through national funds and co-financed by FEDER.

REFERENCES

Alu'datt, M.H., Alli, I., Ereifej, K., Alhamad, M., Al-Tawaha, A.R., & Rababah, T. 2010. Optimisation, characterisation and quantification of phenolic compounds in olive cake. *Food Chemistry* 123: 117–122.
AOAC, 2012. Official methods of analysis. Association of Official Analytical Chemists, *Arlington VA*, USA.

Araújo, M., Pimentel, F.B., Alves, R.C. & Oliveira, M.B.P.P. 2015. Phenolic compounds from olive mill wastes: Health effects, analytical approach and application as food antioxidants. *Trends in Food Science & Technology* 45: 200–211.

Benzie, I.F.F. & Strain, J.J. 1996. The Ferric Reducing Ability of Plasma (FRAP) as a Measure of "Antioxidant Power": The FRAP Assay. *Analytical Biochemistry* 239: 70–76.

Costa, A.S.G., Alves, R.C., Vinha, A.F., Barreira, S.V.P., Nunes, M.A., Cunha, L.M. & Oliveira, M.B.P.P. 2014. Optimization of antioxidants extraction from coffee silverskin, a roasting by-product, having in view a sustainable process. *Industrial Crops and Products* 53: 350–357.

Costa, A.S.G., Alves, R.C., Vinha, A.F., Costa, E., Costa, C.S.G., Nunes, M.A., Almeida, A.A., Santos-Silva, A. & Oliveira, M.B.P.P. Nutritional, chemical and antioxidant/pro-oxidant profiles of silverskin, a coffee roasting by-product. *Food Chemistry*. doi: 10.1016/j.foodchem.2017.03.106

Dermeche, S., Nadour, M., Larroche, C., Moulti-Mati, F., & Michaud, P. 2013. Olive mill wastes: Biochemical characterizations and valorization strategies. *Process Biochemistry* 48: 1532–1552.

Di Bene, C., Pellegrino, E., Debolini, M., Silvestri, N. & Bonari, E. 2013. Short—and long-term effects of olive mill wastewater land spreading on soil chemical and biological properties. *Soil Biology and Biochemistry* 56: 21–30.

Directive 2008/98/EC of the European Parliament and of the Council of 19 November 2008 on waste and repealing certain Directives. *Official Journal of the European Union* 312: 1–9.

Fernández-Hernández, A., Roig, A., Serramiá, N., Civantos, C.G.-O. & Sánchez-Monedero, M.A. 2014. Application of compost of two-phase olive mill waste on olive grove: Effects on soil, olive fruit and olive oil quality. *Waste Management* 34: 1139–1147.

Galanakis, C.M. 2012. Recovery of high added-value components from food wastes: Conventional, emerging technologies and commercialized applications. *Trends in Food Science & Technology* 26: 68–87.

Mirabella, N., Catellani, V. & Sala, S. 2014. Current options for the valorization of food manufacturing waste: a review. *Journal of Cleaner Production* 65: 28–41.

Nunes, M.A., Pimentel, F.B., Costa, A.S.G., Alves, R.C. & Oliveira, M.B.P.P. 2016. Olive by-products for functional and food applications: Challenging opportunities to face environmental constraints. *Innovative Food Science & Emerging Technologies* 35: 139–148.

Rosselló-Soto, E., Koubaa, M., Moubarik, A., Lopes, R.P., Saraiva, J.A., Bousseta, N., Grimi, N. & Barba, F.J. 2015. Emerging opportunities for the effective valorization of wastes and by-products generated during olive oil production process: Non-conventional methods for the recovery of high-added value compounds. *Trends in Food Science & Technology* 45: 296–310.

Uribe, E., Lemus-Mondaca, R., Vega-Gálvez, A., López, L.A., Pereira, K., López, J., Ah-Hen, K. & Di Scala, K. 2013. Quality Characterization of Waste Olive Cake During Hot Air Drying: Nutritional Aspects and Antioxidant Activity. *Food and Bioprocess Technology* 6: 1207–1217.

Incorporation of metallurgical wastes as inorganic fillers in resins

A. Oliveira & C.I. Martins
Institute for Polymer and Composites/I3N, University of Minho, Guimarães, Portugal

F. Castro
Mechanical Engineering Department, University of Minho, Guimarães, Portugal

ABSTRACT: This work aims at incorporating industrial wastes arising from the metallurgical sector, as fillers into polymeric resins. The waste materials that were investigated are: brass foundry sand, ferrous slag, inert steel aggregate for construction, white slag from steelmaking ladle secondary treatment, electric arc furnace dusts and aluminum anodizing wastewaters treatment sludge. Two different types of resins were used: a thermoset phenolic resin normally used for the manufacture of foundry molds and a thermoplastic aqueous dispersion vinyl-based resin generally used for soils stabilization and retaining dust. The selection of the most suitable resin for each residue depends on the characteristics of the resins and the waste particle size: for powder residues, it was used a vinyl-based resin, whereas for larger particle size residues a phenolic resin has been employed. Samples were prepared by compression molding of parts at room temperature and characterized for mechanical properties, water absorption and leaching.

1 INTRODUCTION

Industrial activity generates large amounts of waste that are usually landfilled. This renders into a both environmental and economic problem that can be minimized with waste recovery processes (Waste Framework Directive, 2008). Some of these residues are in the form of sludge, sand and dust, with different characteristics, such as color, texture, electrical and magnetic properties, among others, that make them interesting materials for the replacement of natural material as fillers for polymers, providing new and improved characteristics to plastic composites. The residues can be incorporated into materials of different natures, creating hence innovative products.

Polymers are versatile materials in which fillers and reinforcements can be incorporated into composites (Biron, 2004). Thereby it is possible to incorporate materials of other natures and to obtain products with differentiated characteristics. The composites may be formed from a thermoplastic or thermosetting matrix. The processing depends on the type of resin used (Harper *et al.* 2003).

Some of these residues are already valued in the production of non-polymer composites used in the civil construction sector (Castro *et al.* 2009, Castro 2014 and Almeida *et al*, 2000), for example in the production of concrete (Dash *et al.* 2016). However, their incorporation into polymers is not yet used industrially, although research studies have been carried out in this area (Sousa *et al.* 2014 and Oliveira, 2016).

2 EXPERIMENTAL WORK

The resins used in this work were: phenol-formaldehyde resin RES3007 from Foresa, Spain, with 200–325 MPa.s viscosity, 1242–1252 kg/m^3 density, 12–13 pH and 4.5–6.5 min of gel time that is used with a catalyst CAT3010 from Foresa, with <35 MPa.s viscosity, 1100–1200 kg/m^3

density and 3–5 pH; and a vinyl based aqueous dispersion Greenfor Dust Plus (GDP) from Foresa, with 1–3 GPa.s viscosity, 1000–1100 kg/m^3 density and a pH of 3.5–5.5.

The metallurgical wastes used in this work were brass foundry sand (from non-ferrous alloy casting industry), ferrous slag (from casting of ferrous alloys industry), electric arc furnace steelmaking slag (from the melting of steel scrap), ladle white slag (derived from liquid steel treatment in ladle metallurgical operations), electric arc furnace dusts and aluminum anodizing sludge (from the treatment of wastewaters derived from surface treatment of aluminum profiles). The main characteristics of each residue are depicted in Table 1.

The resin to be incorporated with each residue was selected based on the waste granulometry, suggested by the resins supplier, due to the usual applicability of each resin. The phenolic resin RES3007 is used in the manufacture of casting molds (binding the sand) and the vinyl based resin GFD is used as binder in soil stabilization, dust containment and erosion control of sloping surfaces. Therefore, powder materials of particle size lower than 0.5 mm used GDP and those whose particle size lied between 0.5 and 2 mm used RES3007. Therefore, brass sand, ferrous slag and steelmaking slag were mixed with 5, 10 and 15wt% of RES3007 together with 5% of catalyst CAT3010. White slag, EAF dust and aluminum anodizing sludge were mixed with different portions of GDP, according to the following procedure: GDP resin was diluted in water and used in different concentrations, respectively: 0, 10, 20, 50, 100, 200 and 400 ml/l. The quantity of diluted resin to be used in each residue was optimized for its moisture content, using an adaptation of the Proctor compaction test, as presented in Table 1.

The materials were mechanically mixed and then laid in the mold. Two mold types were used to produce parts by compression molding: a cylindrical mold with 36 mm of diameter to produce specimens to determine compressive strength and a parallelepiped mold with 127 mm of length and 13 mm of width to produce samples to test flexural strength. The height of both types of samples depends of the amount of material used, ranging from 30 to 40 mm.

A compression force of 10 kN to cylindrical specimens and 50 kN to parallelepiped specimens was applied in the mold for about 1 minute, after which the parts were demolded. Some of the samples with phenolic resin were heat treated in an oven for 1 hour at 100°C; whereas the remaining samples were cured at room temperature, 20 to 22°C. This procedure was made

Table 1. Main characteristics of the residues and composites.

Residue information					Composite production			
Residue	LER Code	Moisture content [%]	Density [g/cm³]	Granulo-metry	Resin	Optimized Moisture [%]	Quantity of resin [wt%]	Dilution of GDP in water [ml/l]
Brass sand	10 10 08	0.17	1.99	≤1 mm	Phenolic		5, 10, 15	–
Ferrous slag	10 09 03	0.05	1.03	[0.5,2] mm	resin	–		
Steelmaking Slag	–		0.34	1.17	[0.5,2] mm	RES3007		
Ladle white slag	10 02 02	0.34	1.17	<500 μm	Vinyl based aqueous dispersion	17	0.17, 0.34, 0.85, 1.7, 3.4, 6.8, 17	
EAF dust	10 02 07*	1.42	2.11	<100 μm	GDP	9	0.09, 0.18, 0.45, 0.9, 1.8, 3.6, 9	0, 10, 20, 50, 100, 200, 400
Aluminum anodizing sludge	19 08 14	71.78	0.66	<100 μm		42	0.42, 0.84, 2.1, 4.2, 8.4, 16.8, 42	

to evaluate the curing characteristics of the resins and its influence on the overall properties of the parts, as phenolic resin RES3007 can quickly cure either by a catalyst or at temperatures between 100–150°C.

The samples were tested mechanically at Lloyd Instruments LRK Plus equipment. The cylindric specimens were tested to determine the compressive strength of the samples at a compression velocity of 0.010 mm/s. The parallelepiped specimens were tested to determine the flexural strength in three points bending at a velocity of 0.02 mm/s with a span of 100 mm.

To determine the water absorption rate, the samples were immersed in water for 24 h in a solid/liquid ratio of 1 kg/10l.

To characterize the leachate of the samples pH and salt leachability were measured by atomic absorption spectrometry and X-ray fluorescence spectrometry, this latter in the case of the analysis of the resulted evaporation crystals.

3   RESULTS

Figure 1 shows an example of each type of molded composite prepared with the phenolic resin RES3007 and the vinyl based resin Greenfor Dust Plus. The parts present different textures and colors according to the resin and waste material used. A compact structure has been obtained for each compound, indicating that it is possible to use these residues to obtain solid composites. Nevertheless, some of them have a brittle behavior or release particles easily when handling. Samples produced with GDP resin were prone to release particles more easily than those prepared with RES3007; also, samples prepared with aluminum anodizing sludge

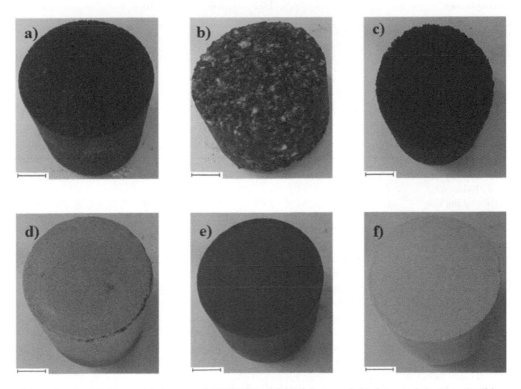

Figure 1.   Composite molded parts of RES3007 with: a) brass sand, b) ferrous slag, c) steelmaking slag and GDP with: d) ladle white slag, e) EAF dust and f) aluminum anodizing sludge. Scale bar corresponds to 10 mm.

Table 2. Properties of RES3007 composites.

Sample	Thermal treatment	Compressive strength [MPa]	Flexural strength [MPa]	Water absorption [%]	pH	Solubilized matter [g/L]	Leachate elements [mg/kg]
Brass sand +5%Res3007	none	1.3	0.42	18	9	2.5	Cr-31 Cu-30 Zn-7
Brass sand +10%Res3007		3.5	1.11	17	10	0.9	Cu-10 Zn-7
Brass sand +15%Res3007		3.9	1.10	14	10	2.1	Cu-5
Brass sand +5%Res3007	1h at 100°C	1.7	0.26	19	9	0.7	Cr-6 Cu-17 Zn-3
Brass sand +10%Res3007		3.9	1.11	16	9	1.2	Cu-29 Zn-4
Brass sand +15%Res3007		6.8	1.08	16	10	2	Cu-66 Zn-7
Ferrous slag +5%Res3007	none	2.2	0.38	22	9	0.7	Cu-2
Ferrous slag +10%Res3007		4.2	0.87	18	10	1.3	–
Ferrous slag +15%Res3007		4.4	1.44	17	9	1.8	Cu-5
Ferrous slag +5%Res3007	1h at 100°C	2.2	0.78	19	9	0.5	Cu-1
Ferrous slag +10%Res3007		4.2	1.72	14	10	1	–
Ferrous slag +15%Res3007		8.4	3.70	9	10	1.7	Cu-7
Steelmaking slag +5%Res3007	none	8.2	2.77	13	10	1.2	–
Steelmaking slag +10%Res3007		12.3	1.19	10	10	1.3	Cu-4
Steelmaking slag +15%Res3007		12.8	1.72	4	9	1.8	–
Steelmaking slag +5%Res3007	1h at 100°C	8.8	1.52	9	10	0.5	Cu-5
Steelmaking slag +10%Res3007		9.9	1.34	9	10	1.9	Cu-6
Steelmaking slag +15%Res3007		11.9	2.69	5	10	2	Cu-9

were more brittle. In the case of composites with RES3007 only those prepared with 5 wt% of incorporated resin were releasing particles.

Table 2 and 3 show the properties of RES3007 and GDP composites, respectively.

In general, for RES3007 composites, the increase of amount of incorporated resin improves the mechanical strength of the samples. The results are showing that 10 wt% of phenolic resin is enough to obtain an interesting compressive strength. Moreover, the thermal treatment also helps to increase the compressive strength of brass sand and ferrous slag based composites, especially for 15 wt% of resin used. In the case of steelmaking slag, no significant effect was observed.

The ability to absorb water lowers with the increase of resin incorporated and with the heat treatment applied. Still most of the samples absorb a large amount of water because the phenolic resin is quite hydrophilic. No significant relationship can be found between the

Table 3. Properties of Greenfor Dust Plus composites.

Residue	Concentration of GDP in H$_2$O for the optimize moisture [ml/l]	Compressive strength [MPa]	Flexural strength [MPa]	Water absorption [%]	pH	Solubilized matter [g/L]	Leachate elements [mg/kg]
Ladle White slag	10	4.9	2.66	18	12	2.5	–
	20	5.6	2.55	19	11	2.2	Cr-4 Zn-2
	50	7.7	1.47	18	10	1.9	Cr-15 Cu-3 Zn-1
	100	9.4	1.05	20	10	2.2	Cu-11
	200	10.1	1.43	19	10	2.1	Cr-5 Cu-4 Zn-1
	400	13.3	2.68	16	9	2.0	Cu-3
	Not diluted	14.0	2.69	9	10	2.1	Cu-3
EAF Dust	10	6.5	1.79	11	8	4.4	Cu-7 Zn-9
	20	7.7	1.34	20	7	3.9	Cd-19 Zn-37
	50	7.1	1.06	19	7	2.9	Cd-15 Cu-8 Zn-65
	100	9.0	1.36	17	7	3.9	Cd-6 Cu-6 Zn-60
	200	8.0	1.63	15	7	2.7	Cd-7 Cu-4 Zn-26
	400	8.6	1.13	15	7	4.9	Cd-11 Cu-5 Zn-75
	Not diluted	6.1	1.25	21	7	4.3	Cd-7 Cu-4 Zn-51
Aluminum Anodizing Sludge	10	0.7	0.11	19	8	2.5	Cu-6
	20	0.9	0.16	52	7	4.8	Cu-9
	50	1.4	0.17	39	7	3.1	Cu-11
	100	1.7	0.17	47	7	5.9	Cu-15
	200	2.1	0.16	53	7	4.7	Cu-10
	400	2.6	0.15	34	7	5.7	Cu-10

concentration amount of leached elements and the amount of resin incorporated, nor with the heat treatment.

The composites with better overall properties are those with steelmaking slag.

The properties of the composites prepared with the vinyl resin Greenfor Dust Plus vary with the amount of added resin, however this variation does not follow an explicit trend. The composites with concentrations in water between 200 ml/l and 400 ml/l of GDP are those that have better characteristics, such as higher mechanical strength and lower water absorption rate.

The composites that present better overall properties are those produced with Ladle White Slag. The composites with aluminum anodizing sludge present low compression resistance, being these composites most difficult to mold. This can be due to the viscosity of the sludge, therefore making it more difficult to produce these types of composites.

These composites, prepared with GDP, leach elements that are harmful for the environment, but in rather limited amounts.

4 CONCLUSIONS

This study shows that the incorporation of some residues of the metallurgical industry into resins is a promising method for employing the wastes as substitutes for natural charges and fillers. It was possible to produce composites with the incorporation of different residues in different quantities of resin, obtaining samples with different textures and properties. However, some composites are critically brittle, especially for lower resin contents.

Based on mortar requirements for pavement blocks (EN1338), one of the main foreseen applications for these composites, the compressive strength requirement of more than 3.6 MPa is generally fulfilled. However, water absorption requirements are not completely fulfilled in some of the composites.

Considering the environmental perspective, the composites with Greenfor Dust Plus resin are those that may cause greater environmental impact, due to a lower degree of inertization for some of the elements more critical for environmental protection purposes. The composites that showed lower environmental impact were those with RES3007 and Ferrous slag, and those with Greenfor Dust Plus and aluminum anodizing sludge.

The mechanical properties of the composites are higher, as expected, the higher the amount of resin. Those properties increase also with heat treatment. This is valid for composites made with both types of tested resins.

ACKNOWLEDGMENTS

Authors would like to acknowledge FORESA, Spain, for the supplying of resins and catalysts used in this work.

REFERENCES

2008/98/CE Directive. *Waste Framework Directive*. 19 November 2008.
Almeida, D.A., *et al.* 2000. Mechanical Behavior of Portland cement mortars with incorporation of Al-containing salt slags. *Cement and Concrete Research* 30: 1131–1138.
Biron, M. 2004. *Thermosets and Composites*. s.l.: Elsevier Advanced Technology. ISBN 1856174115.
Castro, F., Teixeira, T. 2014. Incorporation of industrial wastes from thermal processes in cement mortars, *CONAMET-IBEROMAT*, Santa Fe, 2014.
Castro, F., Vilarinho, C., Trancoso, D., Ferreira, P., Nunes, F., & Miragaia, A. 2009. Utilisation of pulp and paper industry wastes as raw materials in cement clinker produtions. *International Jounal of Materials Engineering Innovation* 1: 74–90.
Dash, M.K., Patro, S.K. and Rath, A.K. 2016. Sustainable use of industrial-waste as partial replacement of fine aggregate for preparation of concrete. *International Journal of Sustainable Build Environment* 5: 484–516
Harper, C.A. and Petrie, E.M. 2003. *Plastic Materials and Process—A Concise Encyclopedia*. s.l.: Jonh Wiley & Sons, Inc. ISBN0471456039.
Oliveira, A.C. 2016. *Incorporation of metallurgical industry waste in resins*. Master's thesis in Polymer Engineer. University of Minho.
Sousa, A.C.A., *et al.* 2014. Incorporation of Marble and Granite Rejects in Polymer Matrices. *Brazilian Congress of Engineering and Materials Science*. 09–13 November, Cuiabá, MT, Brasil.

Sustainability and circular economy through PBL: Engineering students' perceptions

A.C. Alves, F. Moreira & C.P. Leão
School of Engineering, ALGORITMI Center, University of Minho, Guimarães, Portugal

M.A. Carvalho
School of Science, Chemistry Center, University of Minho, Braga, Portugal

ABSTRACT: The learning on sustainability and circular economy, within an interdisciplinary Project-Based Learning methodology, on the first year of the Masters' degree program in Industrial Engineering and Management is investigated. By the end of semester, the perceptions about the project were collected and an analysis of the deliverables of each team was performed. The responses to the questionnaire unveils that the students recognize that the project is a stimulating challenge, while enabling to reflect and learn on sustainability issues. Meanwhile, it also shows that the articulation between the courses of the semester still holds some space for improvement. The analysis of the teams' deliverables revealed a variety of solutions that were based on sustainability concepts, namely on the choice of raw materials, digital and physical project deliverables, such as the ones detailing the products, representing the shop floor or making proof-of-concepts, and the production systems design and product prototypes.

1 INTRODUCTION

Industrial Engineering and Management (IEM) is an engineering program often considered not to be directly related, or having low impact, on sustainability, at least when compared to other engineering programs, such as Environmental, Mechanical, Civil or Chemical Engineering. These perceptions were conveyed by Colombo et al. (2015), whose review specifically addressing IEM Education revealed a very small number of papers on Engineering Education for Sustainability. Another study introduced the hypothesis that the IEM program had a reduced number of courses related to sustainability (Colombo & Alves, 2017). Nevertheless, IEM Engineers are responsible for the management of production systems that transform raw materials, using a number of resources, such as machines, energy, people and information, into products. This activity, along with the products themselves, should preferably be eco-friendly, by minimizing all wastes and resource use (Alves et al., 2014), and exploiting its full potential to reenter the economy upon their end-of-life, following the European Commission directives:

> "*The European Commission aims to ensure coherence between industrial, environmental, climate and energy policy to create an optimal business environment for sustainable growth, job creation and innovation. To support this, the Commission has established an ambitious agenda to transform EU economy into a circular one, where the value of products and materials is maintained for as long as possible, bringing major economic benefits.*" (EC, n.d.).

So, it is imperative that IEM students recognize the importance and develop awareness for environmental issues, and that of the triple bottom line, i.e. Planet, People and Profit (3P), to guide their professional activity. In order to achieve this goal, first year IEM students of the University of Minho, have been developing semester wide interdisciplinary projects akin to sustainability. These projects are developed under a PBL methodology. Herein, the present study describes how and if PBL methodology acts as a facilitator supporting the students' awareness and learning on sustainability and that of the circular economy in an active way.

1.1 *Sustainability and circular economy*

Sustainability is considered to be a difficult topic to define and teach due to its subjectivity (Christie et al., 2013). It is a multidisciplinary topic embracing several and distinct dimensions, such as environmental, sociocultural, scientific and technological, among others. Historically, engineering has a fundamental role in the development of societies, so it is expected that engineers are aware and qualified in all these dimensions (Alves et al., 2015). In the early 2000s the number of works regarding "education for sustainability" started to increase exponentially till 2013. This growth led to innovative approaches, as well as new concepts and the "circular economy" concept emerged (Côrtes & Rodrigues 2016). So, circular economy could be defined as a new economic model, encouraging effectiveness of the processes by reducing waste, and improving reuse and recycling. These concepts were considered important in the highly qualified training of IEM students and were introduced through an interdisciplinary project, which was based on the PBL methodology. In order to increase productivity, eco-efficiency, and reduce costs through wastes elimination, it is important that companies not only adopt a Lean Production approach, but also to establish the link with Green Production. This can be done by designing and operating the production systems under a Lean-Green setting (Moreira et al., 2010; Alves et al., 2016; Abreu et al., 2017).

1.2 *PBL methodology*

PBL can by defined as an approach that uses a project as a central teaching-learning methodology. Students work in groups to solve a real world problem, over a semester. Furthermore, students train and develop different skills, such as literature searching, oral and writing communication, team work, conflicts managing, core values for pursuing an engineering profession where innovation and entrepreneurship are pillars (Harmer, 2014).

The experience and the results of the group of teachers and tutors that has been implementing and follow the development of the PBL projects, particularly, the Project-Led Engineering Education model of Powell & Weenk (2003), are well documented (Lima et al. 2007, Alves et al. 2012; 2015; 2016 and Fernandes et al. 2014). The PBL methodology was firstly applied in the 2004–05 academic year on the first year of the Master Degree on IEM program (Lima et al., 2007), and dwells on more than 13 years of experience. Students work in teams, normally of eight to nine members each, and are accompanied by tutors. In spite of some identified implementation difficulties (Alves et al. 2016), PBL is widely regarded as a valid learning methodology in higher education, with several studies supporting its success: group-processing skills, and increased motivation and interest in learning (Andrew et al. 2016, Alves et al. 2012, Thomas 2009).

The IEM program of the first semester, first year has six courses, each holding five ECTS, and all contributing as Project Supporting Courses (PSCs) for the development of the project using PBL approach. PSCs of Engineering School are: Algorithms and Programming (AP), Integrated Project on Industrial Engineering and Management 1 (IPIEM1), and Introduction to Industrial Engineering and Management (IIEM). The PSCs of Sciences School are Calculus (CC), Linear Algebra (LA) and General Chemistry (GQ).

Since the first edition, the project has persistently focused on sustainability issues (Moreira et al., 2011; Colombo et al., 2014). The 2016–17 PBL edition is no exception, it addressed the theme of "Remanufacture/Recycling of fashion accessories or footwear made of synthetic material". This paper attempts to point out how and if students see PBL as a methodology to raise their sustainability awareness.

This paper is divided in four sections. After this first section, which makes an introduction on PBL, sustainability and circular economy, section two presents the research methodology.

The third section presents the results and the discussion of the students' perceptions and PBL teams outcomes. Finally, section four draws the conclusions.

2 MATERIAL AND METHODS

A questionnaire was used to gather the IEM students' perceptions that participate in the 2016–17 PBL edition, regarding sustainability and circular economy. The full online questionnaire contains six sections related to the project itself, learning and skills acquired, teamwork, teachers role, assessment model and, PBL as learning methodology, with a total of fifty-five questions. Each question was scored based on a 5-point Likert-type scale ('1' corresponds to "strongly disagree" and '5' to "totally agree"). From the 48 enrolled students, 66.7% (corresponding to 32 students) answered the questionnaire that was available online for two weeks, after the end of the 2016–17 first semester (Alves et al., 2017). In this paper only seven closed questions, directly related to the topic, will be presented and discussed, as shown in Table 1.

In Colombo et al. (2014; 2015) the authors identified the students' sustainability competences acquired during the PBL project. Other related competencies included, the application of eco-efficiency and eco-design as well as Life-Cycle Assessment (LCA).

Throughout the PBL 2016–17 edition, students accomplished several important milestones and deliverables as reported in Alves et al. (2015; 2016; 2017). The learning outcomes were in accordance with the project main objective, as well as with each PSC contents. Focusing the main objective of this paper, the PBL teams' outcomes are described regarding two PSCs: IIEM and GC. One of the IIEM outcomes was the introduction of actions for promotion of sustainable development (e.g. products and processes). Therefore, the teams were required to deliver a short report on the actions to promote greener products and low environmental impacts deriving from the industrial activity. This included a life-cycle perspective on products and activities that transcended the walls of the company, to protect people and promote social equity. A template of a diagram for eco-design strategies was provided, encompassing eight dimensions, namely a dimension on new product/service concepts. A significant part of the GC contents were applied in the project (Remanufacture/Recycling of fashion accessories or footwear made of synthetic material). In GC, students identified the polymer to be used in the remanufacturing, its chemical structure description and physical properties. The students also applied concepts of thermodynamic to the production process.

Table 1. List of questions to be used in the analyses and discussion.

Section	Code	Question
I. Project "Remanufacture/Recycling of fashion accessories or footwear made of synthetic material"	Q2	I consider that the subject of the project was interesting and motivating.
	Q3	The fact that the project is opened (with several solutions) was a stimulating challenge.
	Q4	For the learning of the courses of this semester, I consider that the proposed project was adequate
	Q5	The articulation between the courses of the semester was well achieved with the project.
	Q7	I believe that this project has made me think more about issues related to sustainability.
II. Learning and Skills acquired	Q8	The project allowed me to better understand the relevance of the contents of the courses.
	Q9	Through the project, it was possible to see the application of the contents in real situations.

3 RESULTS AND DISCUSSION

During the PBL project the students generally looked motivated and interested (with minor exceptions). The teams developed their work in a room specifically prepared for that purpose, where each team have their specific space, organized according to their own preferences. This allows the establishment of their specific work plan, improve their awareness on the distribution of the several individual tasks and roles, and visually follow each element tasks progress and reminders of team meetings. Students were encouraged to use Gantt diagrams, tables and coloured post-its with different shapes and content to schedule and monitor work progress.

3.1 *Students' perceptions*

The distributions of the student's answers to questions Q2, Q3, Q4, Q5, Q7, Q8 and Q9 are illustrated in Figure 1, ordered by the decreasing mean value. In average, the answers show a positive perception regarding the topics under study, since the values obtained were higher than 3.7.

Students, in average, agree that through the project, it was possible to see the application of the contents in real situations (Q9, mean = 4.28) allowing them to better understand the relevance of the contents of the courses (Q8, mean = 4.21). The study also reveals that the openness of the project is considered to be a stimulating challenge (Q3, mean = 4.21). Regarding the results on Q7, the students also agree that the project enables to reflect about issues related to sustainability (Q7, mean = 4.03). Furthermore they, consider that the proposed project was adequate for the learning of the courses contents (Q4, mean = 3.79). The less positive agreement was on Q5 (the articulation between the courses was well achieved with the project), mean = 3.72, and Q2 (I consider that the subject of the project was interesting and motivating) with 3.72.

3.2 *PBL teams outcomes*

During the project, the students needed to meet important milestones, provide deliverables and conduct many other activities, namely a final report and an oral presentation and discussion, where the main results and conclusions of the project were presented. The teams were also stimulated on delivering prototypes and other digital and physical project deliverables detailing the products, representing the shop floor or making proof-of-concepts, which they normally do.

Table 2 summarizes the multiple outcomes of each team: company logo, main raw material, the production system design, the product(s) prototypes developed, and blogs created to register their work progress throughout the project development stage. Due to the fact that

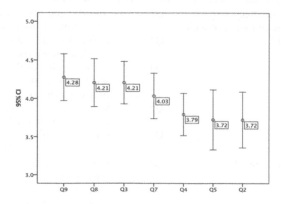

Figure 1.   Students' answers distribution for questions Q2, Q3, Q4, Q5, Q7, Q8 and Q9.

Table 2. Deliverables and others activities of the IPIEM1 teams.

Team	G1	G2	G3	G4	G5	G6
Company logo	phoenix	we camp	WooChair	NewLife4U	biif	GreenBl ck
Raw materials + containers prototypes	PVC, used in synthetic leather	Polyester (PET)	(LDPE) Wootex	Polyamide Socks containers	Polyurethane	Polyurethane PVC; Synthetic leather
Production system	3D represent.	2D + 3D represent.	2D+3D represent.	3D represent.	3D represent.	3D represent.
Products prototypes	Cover	Tent	Chair	Racket strings	Spuff	Bricks
Blog (in Portuguese)	http://phoenixpiegi.wixsite.com/phoenixblog/processos	http://piegi1617.wixsite.com/miegi2/conceito	http://g3piegi.wixsite.com/woochair/equipa-1	http://newlf4u.wixsite.com/newlife4u	http://biif-piegi.blogspot.pt/	http://grupo6miegi.wixsite.com/grenblock

the project is opened, several solutions with different raw materials were developed, adding value to the project and promoting the students' creativity. Another materials created by the teams included some information and promotional videos appealing for recycling socks, flyers about the companies, identification cards, production system simulation videos, etc.

Regarding the IIEM-related competences development, during the project, students' proposals tackled most phases of the products life-cycle, i.e. from eco-design of the product (decision on the type of waste materials to be used on the new products and the respective provenience; estimating impacts of realization, i.e. on bringing the product to market and on how to make them durable and valuable, to prevent early obsolescence), to sustainable production systems (which incorporated a number of concerns on emissions, residues, water, energy or materials consumption, transportation need within the factory, followed by proposals for mitigating them). In this way, the concept of circular economy was accomplished.

The issue of deeply linked sections within the factory was brought forward by one of the teams, so that lower requirements on transports were needed, thereby saving time, space, money, energy and GHG emissions. Natural and efficient lighting, along with the selection of energy efficient machines was also brought forward by a number of teams. Some teams were also concerned on how to implement a return to factory strategy, focused on promoting the remanufacturing of end-of-life products and rewarding the costumers promoting behaviour changes (Cappuyns & Stough, 2016).

4 CONCLUSIONS

The **PBL** methodology was applied to promote motivation and the learning on sustainability and on circular economy in the first year of the Masters' on Industrial Engineering and Management, of the University of Minho.

The results of the students' questionnaire, issued at the end of the semester, show that the PBL methodology stimulates the learning of the contents of the courses and, simultaneously, endorses and promotes sustainability actions during problem solving. Moreover, the specific deliverables of each team reinforce that PBL is an efficient strategy to acquire and apply the concepts of sustainability and circular economy. These results included prototypes, digital representations of the shop floor and products, distinct choices on raw materials, durability of the products, strategies to reuse and recycle end-of-life products, design of production systems to minimize the use of materials, energy, water, and emissions of pollutants and consequent human exposure, as well as impacts on the ecosystems. The students recognize, at least to some extent, the importance of acquiring competences on circular economy through PBL, aiming for improved professional practice. The authors consider that there is still a lot to be done to promote these concerning matters. Future work should focus on extending this approach to the following academic years of the IEM program, which at present times is poorly achieved.

ACKNOWLEDGEMENTS

This work has been supported by COMPETE: POCI-01-0145-FEDER-007043 and FCT— Fundação para a Ciência e Tecnologia within the Project Scope: UID/CEC/00319/2013. The authors are also grateful to the students that participated in the PBL project.

REFERENCES

Abreu, F., Alves, A.C., Moreira, F. 2017. Comparing Lean-Green models for eco-efficient and sustainable production. *Energy*. DOI: 10.1016/j.energy.2017.04.016 (In Press).

Alves, A., Sousa, R., Moreira, F., Carvalho, M.A., Cardoso, E., Pimenta, P., Malheiro, M.T., Brito, I., Fernandes, S. & Mesquita, D. 2016. Managing PBL difficulties in an industrial engineering and management program, *Journal of Industrial Engineering and Management* 9(3): 586–611.

Alves, A.C., Kahlen, F.-J. Flumerfelt, S., Manalang, A.B.S. 2014. Fostering Sustainable Development Thinking Through Lean Engineering Education. *In Proc. of the ASME-IMECE 2014*. DOI: 10.1115/IMECE2014-38192.

Alves, A.C., Moreira, F., Carvalho, M.A., Oliveira, S., Malheiro, T., Brito, I., Leão, C.P. & Teixeira, S. 2017. Integrating STEM contents through PBL methodology in Industrial Engineering and Management first year program. *European Journal of Engineering Education*, (Submitted 2017. 03.14, *in revision*)

Alves, A.C., Moreira, F., Lima, R. Sousa, R., Dinis-Carvalho, J., Mesquita, D., Fernandes, S. & van Hattum-Janssen, N. 2012. Project Based Learning in First Year, First Semester of Industrial Engineering and Management: some results. In *Volume 5: Education and Globalization; General Topics*, p. 111; *ASME-IMECE2012*, EUA.

Alves, A.C., Sousa, R., Fernandes, S., Cardoso, E., Carvalho, M.A., Figueiredo, J. & Pereira, R.M.S. 2015. Teacher's experiences in PBL: implications for practice, *European Journal of Engineering Education* 41(2):1–19.

Alves AC, Moreira F, Abreu F, Colombo CR. 2016. Sustainability, Lean and Eco-efficiency symbiosis. In Triple Helix Interactions for Sustainable Competitiveness (Innovation, Technology, and Knowledge Management). *Springer International Publishing*, 2016. ISBN: 978-3-319-29675-3 DOI: 10.1007/978-3-319-29677-7. http://link.springer.com/chapter/10.1007/978-3-319-29677-7_7.

Andrew, K., Richards, R. & Ressler, J.D. 2016. Engaging Preservice Teachers in Context-based, Action-oriented Curriculum Development, *Journal of Physical Education, Recreation & Dance* 87(3):36–43.

Cappuyns, V. & Stough, T. 2016. Dealing with societal challenges of a circular economy in engineering education. In *Engineering Education for Sustainable Development, 4–7 Sept. 2016*, Bruges, Belgium.

Christie, B.A., Miller, K.K, Cooke, R. & White, J.G. 2013. Environmental sustainability in higher education: how do academics teach?, *Environmental Education Research* 19(3):385–414.

Colombo, C. Alves, A.C., Moreira, F. y van Hattum-Janssen, N. 2015. *A study on impact of the UN Decade of Education for Sustainable Development on Industrial Engineering Education*. Dirección y Organización, DYO—56 (July 2015), pp. 4–9. ISSN: 2171-6323.

Colombo, C.R., & Alves, A.C. 2017. *Sustainability in engineering programs in a Portuguese Public University. Production*, 27 (spe), e20162214. http://dx.doi.org/10.1590/0103-6513.221416.

Colombo, C.R., Alves, A.C., van Hattum-Janssen, N. and Moreira, F. 2014. Active learning based sustainability education: a case study. In *Proceedings of the Sixth International Symposium on*

Project Approaches (PAEE2014), Medellin, Colombia, 27–28 July, ID55.1–55.9. http://hdl.handle.net/1822/30173.

Colombo, R.C., Moreira, F., Alves, A.C. 2015. Sustainability Education in PBL Education: the case study of IEM—UMINHO. In *Proceedings of the Seventh International Symposium on Project Approaches in Engineering Education (PAEE'2015),* June 2015, 221–228, San Sebastian, Spain.

Côrtes, P.L. & Rodrigues, R. 2016. A bibliometric study on "education for sustainability". *Brazilian Journal of Science and Technology* 3(8).

EC (n.d.) *Sustainability and Circular Economy.* Available at: https://ec.europa.eu/growth/industry/sustainability_pt [Accessed 24 May 2017].

Fernandes, S., Mesquita, D., Flores, M.A., & Lima, R.M. 2014. Engaging students in learning: Findings from a study of project-led education. *European Journal of Engineering Education* 39(1): 55–67.

Harmer, N. 2014. Project-based learning Literature review. Available at: https://www.plymouth.ac.uk/uploads/production/document/path/2/2733/Literature_review_Project-based_learning.pdf [Accessed 24 May 2017].

Lima, R.M., Carvalho, D., Flores, A. & van Hattum-Janssen, N. 2007. A case study on project led education in engineering: students' and teachers' perceptions. *European Journal Eng. Edu.* 32(3): 337–347.

Moreira, F., Alves, A.C. & Sousa, R.M. 2010. Towards Eco-efficient Lean Production Systems. IFIP Advances in Information and Communication Technology, Volume 322, *Balanced Automation Systems for Future Manufacturing Networks*, Pages 100–108, ISBN-13: 978-3-642-14340-3, Springer, Available from: http://www.springerlink.com/content/v6181026252 × 1025/.

Moreira, F., Mesquita, D., van Hattum-Janssen, N. 2011. The importance of the Project Theme in Project Based Learning: a Study of Student and Teachers Perceptions. *In: Proceedings of the 3rd Ibero American Symposium on Project Approaches in Engineering Education (PAEE'2011)*, Lisboa-Portugal, (CD-ROM). ISBN: 978-989-8525-05-5, pp. 65–71.

Powell, P.C., & Weenk, W. 2003. Project-Led Engineering Education. *Utrecht: Lemma.*

Thomas, I. 2009. Critical Thinking, Transformative Learning, Sustainable Education, and Problem-Based Learning in Universities. *Journal of Transformative Education* 7(3): 245–264.

WASTES – Solutions, Treatments and Opportunities II – Vilarinho, Castro & Lopes (Eds)
© 2018 Taylor & Francis Group, London, ISBN 978-1-138-19669-8

Suitability of agroindustrial residues for cellulose-based materials production

D.J.C. Araújo
Institute for Polymers and Composites/I3N and CVR—Centre for Waste Valorization, University of Minho, Guimarães, Portugal

M.C.L.G. Vilarinho
Mechanical Engineering and Resources Sustainability Centre, University of Minho, Guimarães, Portugal

A.V. Machado
Institute for Polymers and Composites/I3N, University of Minho, Guimarães, Portugal

ABSTRACT: The depletion of fossil resources and negative environmental impact related to conventional polymeric materials life cycle have fostered the search for renewable raw materials suitable for their manufacturing. This work aims to present a methodology for the selection of agroindustrial residues with potential to be used as feedstock in cellulose-based materials production. The suitability of main residues identified was calculated taking into consideration their intrinsic characteristics, namely the cellulose content, cellulose-to-lignin ratio and availability. The selection and generation estimates of residues were based on the reality of Portugal's agriculture sector. The results indicate a range of residues with potential to be used as raw materials. In addition to residues generated in harvest fields, the processing industries can also be considered a potential source of byproducts suitable for the application concerned.

1 INTRODUCTION

Over the last few decades, environmental pressures and concerns about the limited availability of fossil resources have fostered the search for technological alternatives and feedstocks less harmful to the environment. The fast growing of the polymeric industry in the last 30 years has significantly boosted the development of society, but it has also triggered many environmental problems (Barnes et al 2009). Only in 2014, the world plastic production reached a value close to 311 million tons (Statista 2016). From this total, a main portion is derived from fossil resources and has the characteristic of being non-biodegradable (Tokiwa et al. 2009), which makes its accumulation potentially adverse to terrestrial and aquatic ecosystems.

Therefore, the search of renewable raw materials suitable to produce biodegradable polymeric and bio-based materials is imperative. Biodegradable polymers are a specific type of polymer that can be naturally degraded by the action of weathering agents and microorganisms (Pillai 2014). Depending on environmental conditions and type of biopolymer high degradation rates can be reached (Song et al. 2009). In recent years, the group of bioplastics derived from renewable and biodegradable resources, such as cellulose, lignin and starch, has received greater interest. Indeed, bioplastics have been mentioned as a lead market by the European Commission, with global production capacity set to grow 350% by 2019 (European bioplastic, 2016).

Among available sources of renewable feedstock, the lignocellulosic biomass stands out in the global scenario. Lignocellulosic biomass is mainly composed of three natural organic polymers, cellulose, hemicellulose and lignin, and smaller amounts of proteins, pectin and extractives (Monlau et al. 2013). Cellulose is labeled as the most abundant organic polymer in the earth, and due to some of its properties, such as mechanical, thermal, biodegradable, renewable and low cost, its application to the development of new bio-based products is increasingly being explored (Wang et al 2016, Chaker et al. 2014). In general, lignocellulosic feedstock may result from agriculture, forestry activities, energy crops and municipal and industrial wastes (Lee et al. 2014). The set of activities associated with the vegetable agribusiness sector excels as the one of the main residues generators. It is estimated that the production of agricultural and forestry residues in the Europe Union exceed 200 million tons/year (Searle & Malins 2013, Scarlat et al 2010).

Taking into account the growing interest of this research topic, as well as the wide variety of residues available, this paper proposes the development of an evaluation methodology to support the selection of agroindustrial waste suitable to be used as raw material in the production of cellulose-based materials. The developed methodology was applied based on the agroindustrial market of Portugal continental and insular.

2 METHODOLOGY

2.1 *Suitability of agroindustrial residues for bioplastic production*

The first step to analyze the feasibility of agroindustrial waste valorization (S), regarding its application to produce polymeric products, consists in the identification of waste types that besides to have high or low content of cellulose and lignin, respectively, are also generated in significant quantities. Thus, three main parameters were taken into account in order to obtain an output that reflects the suitability of different agroindustrial residues, namely availability of residues (AR), cellulose content (CC) and cellulose to lignin ratio ($R_{C/L}$) (Equation 1).

$$S = f(AR; CC; R_{C/L}) \tag{1}$$

The parameters included in equation 1 are positively correlated to the final value of S, and the importance of each one on the final selection criteria is also considered. This can be accomplished by weighting each parameter as presented in equation 2. The constant β (45), of greater influence, is associated with CC, while γ (35) and α (20) are related to AR e $R_{C/L}$, respectively. Greater importance was given to CC as it deals with the raw material required to produce bio-based materials. Less importance has been devoted to AR and $R_{C/L}$, since limitations attributed to these parameters can be circumvented. Since the range of values of raw data varies widely and the parameters included in the formulation (1) encompass different measuring units, it was necessary to adjust values to a notionally common scale by applying normalization. Thus, equation 1 was rewritten as shown in equation 2.

$$S'_i = \gamma AR'_i + \beta CC'_i + \alpha R'_{C/Li}; \quad \text{Lignin content} > 0 \tag{2}$$

where,

$$AR'_i = a + \left[\frac{(AR_i - AR_{min})(b-a)}{(AR_{max} - AR_{min})}\right]; \tag{3}$$

$$CC'_i = a + \left[\frac{(CC_i - CC_{min})(b-a)}{(CC_{max} - CC_{min})}\right]; \tag{4}$$

$$R'_{C/Li} = a + \left[\frac{(R_{C/Li} - R_{C/Lmin})(b-a)}{(R_{C/Lmax} - R_{C/Lmin})}\right] \tag{5}$$

In equations 3, 4 and 5 the constants a and b correspond to arbitrary points, equivalent to 0.1 and 1, respectively, used to restrict the range of values in the dataset. In the same equations, the parameters designated by maximum and minimum subscripts are associated to residues that present the highest and lowest values of the parameter concerned, respectively.

Thus, for each residue, the S_i (suitability index) might vary from 10 to 100 and will be expressed as dimensionless.

2.2 *Agroindustrial residues availability*

The assessments of agroindustrial residues availability took into account the average annual production of crops (AAP), residue generation rate (RGR), sustainable removal rate (SSR) and other competitive uses. Average annual production of crops, as well as the selection of main crops, were obtained from FAO database with a temporal coverage from 2004 to 2013 (FAOSTAT, 2014). The sustainable removal rates of 40% for wheat straw, maize residues and rice straw and husk was adopted (EPE 2014). Besides, based on studies conducted by Scarlat et al (2010) and Searle and Marlins (2013), it was assumed that 30% of residues can have other use (Competitive uses—CU), such as consumption for livestock (animal bedding and feeding), cosmetics industries and mushroom production. It should be noticed that cellulose and lignin content were established based on data available in literature.

Exclusively for those industry-driven crops, residues generation took into consideration the portion of production directed to industrial processing (IP). According to data from Statistics Portugal (2014), approximately 97.3% of olive production are intended for olive oil production; 97.9% of grape production are intended for wine production; and 91.8% of tomatoes production are intended for industrial processing. Besides, it was assumed that 23% of apple (USDA 2011), 7% of orange (Euromedcitrusnet 2007) and 10% of potato (Commission of the European Communities 2007) national production are allocated to industries processing. By applying these information, as well as the average annual production of crops and the residues generation rates (Table 1), it is possible to obtain the average availability of crop residues (AR_{CR}) (equation 6) and industry-driven crops (AR_{IDC}) (equation 7).

$$AR_{CR} = (AAP)(RGR)(SRR)(1-CU); \tag{6}$$

$$AR_{IDC} = (AAP)(IP)(RGR)(1-CU); \tag{7}$$

3 RESULTS

Table 1 provides the main information necessary to identify relevant residues to be used as feedstock for the production of cellulose-based materials. Although it is possible to identify

Table 1. Chemical characterization of lignocellulosic biomass and residues availability. RGR—residue generation rate; AR—availability of residue; $R_{C/L}$ – Cellulose-to-lignin ratio.

Residue	Cellulose (%)	Lignin (%)	RGR ($t_{residue}/t_{crop}$)	AR (kt)
Olive pomace	19.27 [1]	11.32 [1]	0.35 [2]	91.69
Orange bagasse	24.52 [3]	7.51 [3]	0.50 [4]	5.17
Apple pomace	7.2 [5]	23.5 [5]	0.30 [6]	12.07
Grape pomace	27.9 [7]	63 [7]	0.25 [8]	149.90
Maize cob	31.2 [9]	15 [9]	0.33 [10]	63.63
Maize husk	62.07 [11]	14.6 [11]	0.22 [10]	42.42
Maize straw	40.8 [12]	22 [12]	1.96 [10]	377.90
Tomato pomace	29.1 [7]	57.4 [7]	0.04 [13]	32.86
Potato skin	10.5 [14]	4.0 [14]	0.27 [15]	10.21
Wheat straw	35.4 [16] [17]	18.75 [16] [17]	1.28 [10]	49.30
Rice husk	35 [18]	23 [18]	0.25 [10]	11.41
Rice straw	39.5 [19]	15.9 [19]	1.33 [10]	59.27

[1] Vlyssides et al 2004; [2] Brscic et al 2009; [3] Bicu & Mustafa 2011; [4] Foster-Carneiro et al 2013; [5] Dhillon et al 2012; [6] Dhillon et al 2013; [7] Chiou et al 2015; [8] Dwyer et al 2014; [9] Silvério et al 2013; [10] FAO 2014; [11] Ma et al 2015; [12] Chaker et al 2014; [13] Jiang et al 2015; [14] Rommi et al 2015; [15] Schieber et al 2001; [16] Kopania et al 2012; [17] Lee et al 2014; [18] Johar et al 2012; [19] Kim et al 2011.

the relevance of some residues by looking at the results showed in this table, the decision-making to select the most suitable is not an easy task.

Besides to cellulose content and residues availability, lignin amount should also be taken into account. The presence of lignin in lignocellulosic biomass can be considered as one of the major obstacles in pretreatment processes (Monlau et al 2013). Moreover, the presence of substantial amounts of non-cellulosic components in lignocellulosic fibers may negatively influence their biodegradability, crystallinity, density, tensile strength, modulus and moisture (Monlau et al 2013, Reddy & Yang 2005). Therefore, in this study, the influence of lignin over the selection of residues was taken into account by including the cellulose-to-lignin ratio in the formulation of suitability. The greater the ratio, the greater is the cellulose percentage compared to the lignin content, and thus more efficient would be the biomass treatment.

The ranking of residues with greater potential to be used as raw material (Figure 1) was obtained from the substitution of the calculated parameters AR_i', CC_i' e $R_{C/Li}'$ (Table 2) in equation 2. It can be noticed that mainly the parameters AR and $R_{C/L}$ have a significant influence on the residues ranking. The high cellulose content associated with maize husk resulted in a normalized cellulose-to-lignin ratio and cellulose content much higher than of other residues (Tables 1 and 2), which contributed significantly to the final value of suitability index. Among residues with the highest cellulose contents are maize husk, maize straw, rice straw and wheat straw (Tables 1 and 2). Considering only the individual contribution of residues available, the maize straw, grape pomace and olive pomace (Tables 1 and 2) account about 70% (619.5 kt) of the total available residues generation.

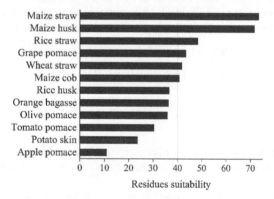

Figure 1.   Cellulose-based materials production suitability of agroindustrial residues.

Table 2.   RGA, $R_{C/L}$ and CC normalized values for each residue.

Residue	Normalized availability of residue (AR_i')	Normalized cellulose content (CC_i')	Normalized celulose-to-lignin ratio ($R_{C/Li}'$)
Olive pomace	0.40	0.298	0.418
Orange bagasse	0.1	0.384	0.775
Apple pomace	0.124	0.1	0.1
Grape pomace	0.602	0.439	0.131
Maize cob	0.241	0.493	0.504
Maize husk	0.190	1	1
Maize straw	1	0.651	0.453
Tomato pomace	0.196	0.459	0.145
Potato skin	0.117	0.154	0.629
Wheat straw	0.206	0.562	0.460
Rice husk	0.114	0.556	0.377
Rice straw	0.230	0.629	0.597

Among the major agroindustrial residues identified in Portugal, those with greater potential to be used as raw material are maize straw, maize husk, rice straw, grape pomace and wheat straw. On the other hand, lower suitability values were assigned to apple pomace and potato skin (Figure 1). It is important to emphasize that variables involved in the proposed methodology are directly influenced by climatic and soil conditions, as well as by agricultural business and farming practices. Consequently, the values assumed in this work tried to be as representative as possible of the reality of Portugal, thus the results obtained can only be applied to that country. Attempts to apply this methodology in other countries or regions should make use of a specific and reliable database.

According to the results, in the production and manufacturing cycle of agricultural products, the processing industries are also a potential source of suitable residues. About 40% of the total amount of waste available come from them. The parameters used to obtain the suitability index are also relevant for other bio-based applications, such as biofuels and bioenergy (Scarlet et al 2010). Hence, some residues identified in this work have already been used in other valorization routes (Arevalo-Gallegos et al 2017) besides cellulose-based materials production (Wang et al 2016), which supports and validates the obtained results. It should be noted that the results obtained do not neglect the possibility of using the other residues not highlighted in Figure 1. However, in such situations, the improvement and optimizing of treatments or pretreatments techniques must be carried out.

Additionally, the use of agricultural residues to the application concerned still need to encompass a number of issues besides the availability and chemical composition of residues, among which include: temporal variability of generation, technological alternatives available for valorization, perception and social impact on different farmers' category, in addition to logistical issues mainly influenced by the spatial variability of residues generation. For instance, in Portugal, the maize production takes place mainly in the Alentejo region, hence, the largest share of maize residues would be available in that region.

4 CONCLUSIONS

In this study, a methodology to select agroindustrial residues suitable for cellulose-based materials production has been conducted based on the agricultural sector of Portugal. The chemical composition and availability of residues were taken into account. Besides to residues generated during the crops harvesting, the processing industries are also a potential source of byproducts. Among the most suitable residues to the application concerned are maize straw, maize husk, rice straw and grape pomace. Mainly due to its easy application and wide availability of needed input data, this methodology can be used as an efficient management tool in aid for decision-making. However, seeing that residues generation is closely related to crops production, in order to improve this methodology, in-depth studies on economic issues and spatial and temporal variability of residues should be performed and taken into account over the proposed formulation.

REFERENCES

Arevalo-Gallegos, A., Ahmad, Z., Asgher, M., Parra-Saldivar, R., & Iqbal, H. M. 2017. Lignocellulose: A sustainable material to produce value-added products with a zero waste approach—A review. *International Journal of Biological Macromolecules* 99: 308–318.

Barnes, D.K.A., Galgani, F., Thompson, C.R., Barlaz, M. 2009. Accumulation and fragmentation of plastic debris in global environments. *Phil. Trans. R. Soc.* B 364, 1985–1998.

Bayer, I.S., Puyol, G.S., Guerrero, J.A.H., Ceseracciu, L., Pignatelli, F., Ruffilli, R., Cingolani, R., Athanassiou, A. 2014. Direct Transformation of Edible Vegetable Waste into Bioplastics. *Macromolecules* 47: 5135–5143.

Bicu, I. & Mustafa, F. 2011. Cellulose extraction from orange peel using sulfite digestion reagents. *Biosource technology* 102: 10013–10019.

Brščić, K., Poljuha, D., & Krapac, M. 2009. Olive Residues-Renewable Source of Energy. In: Management of Technology-Step to Sustainable Production, MOTSP 2009. *Sibenik* 10–12 June 2009. Croatia.

Chaker, A., Mutjé, P., Vilar, M.R., Boufi, S. 2014. Agriculture crop residues as a source for the production of nanofibrillated cellulose with low energy demand. *Cellulose* 21: 4247–4259.

Chiou, B.S., Valenzuela-Medina, D., Bilbao-Sainz, C., Klamczynski, A.K., Avena-Bustillos, R.J., Milczarek, R.R.,... & Orts, W.J. 2015. Torrefaction of pomaces and nut shells. *Bioresource technology* 177: 58–65.

Commission of the European Communities 2007. *The potato sector in the Europe Union.* Brussels, 118.

Dhillon, G.S., Kaur, S., & Brar, S.K. 2013. Perspective of apple processing wastes as low-cost substrates for bioproduction of high value products: A review. *Renew. and Sustain. Ener. Revie.* 27: 789–805.

Dhillon, G.S., Kaur, S., & Brar, S.K. 2013. Potential of apple pomace as a solid substrate for fungal cellulose and hemicellulose bioproduction through solid-state fermentation. *Industrial crops and products* 38: 6–13.

Dwyer, K., Hosseinian, F., & Rod, M. 2014. The market potential of grape waste alternatives. *Journal of Food Research* 3(2): 91–106.

EPE 2014. Série Recursos energéticos. *Nota técnica DEA 15/14: Inventário energético de resíduos rurais.* Rio de Janeiro, 51.

EUROMEDCITRUSNET 2007. "Safe and high quality supply chains and networks for the citrus industry between mediterranean partner countries and Europe". *Deliverable 9 - national citrus sector analysis: Portugal.* 37.

European bioplastic 2016. *Bioplastics: Facts and figures* [Online]. Available at: http://migre.me/vYb0k. Accessed at 02 January 2016.

FAO 2014. *Bioenergy and food security rapid appraisal manual. Natural resource module: user manual.* 41.

FAOSTAT 2016. *Domains—production/crops.* Available at: http://www.fao.org/faostat/en/#data/QC. Accessed at 20 March 2016.

Forster-Carneiro, T., Berni, M.D., Dorileo, I.L., & Rostagno, M.A. 2013. Biorefinery study of availability of agriculture residues and wastes for integrated biorefineries in Brazil. *Resour., Conserv. and Recyc.* 77: 78–88.

Jiang, F., Hsieh, Y.L. 2015. Cellulose nanocrystal isolation from tomato peels and assembled nanofibers. *Carbohydrate Polymers* 122: 60–68.

Johar, N., Ahmad, I., & Dufresne, A. 2012. Extraction, preparation and characterization of cellulose fibres and nanocrystals from rice husk. *Industrial Crops and Products* 37(1): 93–99.

Kim, J.W., Kim, K.S., Lee, J.S., Park, S.M., Cho, H.Y., Park, J.C., Kim, J.S. 2011. Two-stage pretreatment of rice straw using aqueous ammonia and dilute acid. *Bioresource Technology* 102: 8992–8999.

Kopania, E., Wietecha, J., & Ciechańska, D. 2012. Studies on isolation of cellulose fibres from waste plant biomass. *Fibres & Textiles in Eastern Europe* 167–172.

Lee, H.V., Hamid, S.B.A., Zain, S.K. 2014. Conversion of Lignocellulosic Biomass to Nanocellulose: Structure and Chemical Process. *The Scientific World Journal* 20.

Ma, Z., Pan, G., Xu, H., Huang, Y., & Yang, Y. 2015. Cellulosic fibers with high aspect ratio from cornhusks via controlled swelling and alkaline penetration. *Carbohydrate polymers* 124: 50–56.

Monlau, F., Barakat, A., Trably, E., Dumas, C., Steyer, J.P., Carrère, H. 2013. Lignocellulosic materials into biohydrogen and biomethane: impact of structural features and pretreatment. *Crit. Revi. in environ. scien. and techn.* 43(3): 260–322.

Pillai, C.K.S. 2014. Recent advances in biodegradable polymeric materials. Materials science and technology *in Polymer Science* 30(5): 558–566.

Reddy, N., Yang, Y. 2005. Biofibers from agricultural byproducts for industrial applications. *Trends in Biotechnology* 23(1): 22–27.

Rommi, K., Rahikainen, J., Vartiainen, J., Holopainen, U., Lahtinen, P., Honkapää, K., & Lantto, R. 2016. Potato peeling costreams as raw materials for biopolymer film preparation. *Journal of Applied Polymer Science* 133(5).

Schieber, A., Stintzing, F.C. & Carle, R. 2001. By-products of plant food processing as a source of functional compounds-recent developments. *Trends in food and technology* 12: 401–413.

Searle, S., & Malins, C. 2013. Availability of cellulosic residues and wastes in the EU. *International Council on Clean Transportation: Washington,* USA, 1–7.

Shah, A.A., Hasan, F., Hameed, A., Ahmed, S. 2008. Biological degradation of plastics: A comprehensive review. *Biotechnology advances* 26: 246–265.

Silverio, H.A., Neto, W.P.F., Dantas, N.O., Pasquini, D. 2013. Extraction and characterization of cellulose nanocrystals from corncob for application as reinforcing agent in nanocomposites. *Industrial Crops and Products* 44: 427–436.

Song, J.H., Murphy, R.J., Narayan, R., Davies, G.B.H. 2009. Biodegradable and compostable alternatives to conventional plastics. *Phil. Trans. R. Soc.* B 364: 2127–2139.

Statista 2016. *Production of plastics worldwide from 1950 to 2014* [On line]. Available at: http://migre.me/vYaY3. Accessed at 24 March 2016.

Statistic Portugal 2014. *Estatísticas agrícolas 2013*. Available at: http://migre.me/vYaU9. Accessed at 30 March 2016.

Tokiwa, T., Calabia, B.P., Ugwu, C.U., Aiba, S. 2009. Biodegradability of plastics. *Intern. Jorn. of molec. Scien.* 10: 3722–3742.

USDA 2011. EU-27 Fresh deciduous fruit annual: good prospects for EU-27 apple and pear production. *Global agricultural information network* 29.

Vlyssides, A.G., Loizides, M., Karlis, P.K. 2004. Integrated strategic approach for reusing olive oil extraction by-products. *Journal of Cleaner Production* 12(6): 603–611.

Wang, S., Lu, A., Zhang, L. 2016. Recent advances in regenerated cellulose materials. *Progress in Polymer Science* 53: 169–206.

WASTES – Solutions, Treatments and Opportunities II – Vilarinho, Castro & Lopes (Eds)
© 2018 Taylor & Francis Group, London, ISBN 978-1-138-19669-8

Pyrolysis of lipid wastes under different atmospheres: Vacuum, nitrogen and methane

L. Durão

Department of Sciences and Technology of Biomass, Faculty of Sciences and Technology, Mechanical Engineering and Resources Sustainability Center, New University of Lisbon, Caparica, Portugal
Polytechnic Institute of Portalegre, Portalegre, Portugal

M. Gonçalves, A. Oliveira, C. Nobre & B. Mendes

Department of Sciences and Technology of Biomass, Faculty of Sciences and Technology, Mechanical Engineering and Resources Sustainability Center, New University of Lisbon, Caparica, Portugal

T. Kolaitis & T. Tsoutsos

Renewable and Sustainable Energy Laboratory, Environmental Engineering Department, Technical University of Crete, Chania, Greece

ABSTRACT: Pyrolysis of high acidity olive oil, olive husk oil and animal fat was carried out under different atmospheres (vacuum, N_2 and CH_4), at constant temperature and different residence times. Bio-oils were the main pyrolysis products, with mass yields from 55.2% to 88.9%. The use of nitrogen or methane atmospheres led to an increase in the formation of gas products (14.1% to 26%), while increasing the residence time resulted in higher yields of gases and solids. The pyrolysis bio-oils contained 54.4% to 88.7% (w/w) of distillable liquids, collected as a light fraction and a heavy fraction. Both fractions were analysed by GC-MS to determine the hydrocarbon profiles and identification of functional groups. Vacuum pyrolysis promoted the formation of low molecular weight liquids while the use of a methane atmosphere increased the relative concentration of aliphatic components in the bio-oil.

1 INTRODUCTION

Low quality or waste lipids such as vegetable oils and animal fats are renewable carbon sources which can be converted to bio-oils with high calorific power that can be further upgraded for the production of drop-in-fuels. Lipids have elemental compositions and heating values more similar to those of petroleum-derived hydrocarbons than lignocellulosic biomass (Demirbas, 2005). Nevertheless, these waste lipids, specially animal fats, have high viscosities and poor cold flow properties so they are not adequate for direct use in internal combustion engines (Lappi & Alén, 2009).

About 95% of global olive oil production takes place in the Mediterranean basin and Portugal is one of the main producers, therefore products from this industry, such as high acidity olive oil and olive husk oil exist in very large quantities (Pinto et al., 2014). High acidity olive oils (OO) and olive husk oils (OHO) have high levels of free fatty acids, that cannot be esterified with basic catalysts (Poulli et al., 2005). Olive husk oil (OHO) is extracted with hexane or petroleum ether at very high temperatures leading to the formation of toxic compounds (Pinto et al., 2014). These characteristics make them unsuitable for direct consumption and for production of biodiesel by transesterification. Finding alternative valorisation pathways for these oils may contribute to improve the efficiency of the olive oil production industry through a circular economy approach. Another example of such strategy is the valorisation of animal wastes, produced in animal-processing plants. The production of such wastes has been increasing worldwide, at a rate that is proportional to the increase of the meat production sector. These wastes are heterogeneous and mostly solid at room temperature because of their high content of saturated fatty acids (Feddern, 2011). Pyrolysis is a thermochemical process in which the materials are subject to thermal cracking in oxygen-deficient conditions, followed by rearrangement of fragmented molecules, to give solid, gas and liquid products. Liquid products include an organic phase (bio-oil) and an aqueous phase, whose proportions and composition are highly dependent on the nature of the raw materials and on the conversion conditions (Bridgwater et al., 1999). This bio-oil can be used as a fuel for heat production, upgraded to yield a biofuel or valorised as raw material for the production of different chemicals. Pyrolysis conditions such temperature, residence time, catalyst and atmosphere strongly affect the yield and properties of products (Li et al., 2004). Most pyrolysis experiments are conducted under a modified atmosphere of N_2 (Zhang et al., 2011), but there are a few reports of performing pyrolysis under initial vacuum (Fan et al., 2014; Dewayanto et al., 2014).

In this work, we compared the yields and composition of the pyrolysis products obtained from vegetable oils and animal fat, at a temperature adequate for thermal cracking (420°C), at different residence times (10, 20 and 30 min) and using different modified atmospheres: initial vacuum, nitrogen and methane. The use of initial vacuum is interesting because it allows to decrease the concentrations of oxidant gases, especially oxygen, without requiring the injection of an inert gas. To the best of our knowledge this is the first example of the use of methane to modify the atmosphere during pyrolysis. Methane is a very stable compound but is not considered an inert gas such as nitrogen, therefore the purpose of these experiments is also to evaluate the influence of a methane atmosphere in the efficiency of lipid deoxygenation and in the quality of the bio-oils produced. Furthermore, CH_4 can be obtained from various renewable sources by processes of anaerobic digestion (Ge et al., 2014) or landfill gas collection (Niskanen et al., 2013).

2 MATERIAL AND METHODS

2.1 *Raw materials*

The high acidity olive oil (OO) and the crude olive husk oil (OHO) were supplied by OLICER, an olive mill from southern Portugal. The animal fat sample (AF) was a waste fat material from a poultry industry (AVIBOM) that had not sufficient quality for valorisation as feed or as biodiesel. The apparent density and acid value of the samples were determined according to the European Standard EN 14104. Samples were kept at room temperature (25°C), in the dark, since reception and during the experiments.

2.2 *Pyrolysis tests*

Pyrolysis tests were performed in a laboratory scale pyrolysis pilot plant Parr Instrument with a 1 L batch reactor connected to a PID programmable temperature controller. The experiments were carried out at 420°C, varying the residence time (10, 20 and 30 min) and the reaction atmosphere (N_2, CH_4 and vacuum).

After every experiment, the reactor was cooled to room temperature and the gas products were collected in a Tedlar bag. The reactor was weighed when empty, after loading with oil and after opening to determine the mass yields of solid, liquid and gas products. After opening the reactor, the liquids were separated from the solids through decantation and the solid residue was washed with a small volume of hexane to remove residual bio-oil, and air-dried before weighing and characterisation.

2.3 *Distillation and distillate characterization*

The bio-oils were fractionated by simple distillation and two distillable liquid fractions were collected: a light fraction (35–150°C) and a heavy fraction (150–250°C); a carbonaceous residue of non-distillable components was left in the distillation flask in some of the experiments. The distillates were characterised by gas chromatography and mass spectrometry (Focus GC,

PolarisQ, Thermo) to identify individual components and functional groups. The retention time of individual compounds was compared to a homologous series of hydrocarbons from C7 to C30 (C7 - C30 Saturated Alkanes, Ref.49451-U Supelco) to evaluate the carbon distribution of the bio-oils.

3   RESULTS AND DISCUSSION

3.1   *Raw material characterization*

The OO and OHO samples exhibit similar apparent densities at 25°C (0.948 and 0.951 g/mL, respectively), typical of vegetable oils (Noureddini et al., 1992) but higher than the apparent densities of conventional fuels, in the range of 0.77 to 0.87 g/mL (Demirel, Y., 2012). The AF sample had a lower density (0.847 g/mL) maybe due to the higher acidity of this lipid waste. On the other hand, the densities of the obtained bio-oils varied from 0.823 to 0.867 g/mL, values close to the density of diesel, showing that sensible transformations occurred during pyrolysis.

The acidity values of the raw materials were 5.8 mg KOH/g for OO, 15.0 mg KOH/g to OHO and 35.0 mg KOH/g for AF, values consistent to a high degree of triglyceride hydrolysis.

3.2   *Comparison of pyrolysis and distillation yields using different pyrolysis atmospheres*

The pyrolysis of OO, OHO and AF were performed, at the conditions indicated in Table 1, yielding bio-oil as main product, corresponding to 55.2 to 88.9% (w/w) of initial mass. Higher yields of liquid products were obtained with initial vacuum, while the use of nitrogen or methane atmospheres favored the formation of gas products. This behavior may result from the solvation of the primary pyrolysis products by nitrogen and methane, decreasing the probability of condensation reactions and therefore leading to the formation of lower molecular weight compounds.

The increase in residence time caused an increase in the yields of both solid and gas products because during a longer reaction time both thermal cleavage reactions and condensation reactions may occur, thus enhancing the yields of respectively gas products and solid products. The shielding effect of the nitrogen and methane atmospheres can be noticed by the lower formation of solid products, indicating that less condensation reactions took place in these modified atmospheres than when pyrolysis was performed in similar conditions but with initial vacuum.

Table 1.   Pyrolysis conditions at 420°C, pyrolysis yields and distillation yields of the bio-oils.

Raw materials	Time (min)	Pressure (bar)		Pyrolysis Yields (%w/w)			Distillation Yields (%w/w)		
				Gas	Liquid	Solid	Light Frac. (F_1)	Heavy Frac. (F_2)	F_1+F_2
OO	10	<0		11.2	88.8	0.0	33.6	20.8	54.4
OHO	10	<0		11.1	88.9	0.0	42.5	46.2	88.7
OO	20	<0		15.0	84.5	0.5	25.7	39.6	65.3
OHO	20	<0		17.4	82.3	0.2	48.1	32.4	80.5
OO	30	<0		16.1	82.0	1.9	33.7	34.4	68.1
OHO	30	<0		44.0	55.2	0.8	17.7	48.4	66.1
AF	30	<0		15.6	81.9	0.8	43.9	19.3	63.2
OO	10	Nitrogen	2.4	19.7	80.3	0.0	39.3	48.2	87.4
OHO	10	Nitrogen	3	18.1	81.9	0.0	12.7	73.1	85.9
AF	30	Nitrogen	3	26.0	74.0	0.0	18.3	58.0	76.3
OO	10	Methane	3	19.2	80.8	0.0	16.9	57.2	74.1
OHO	10	Methane	3	14.1	85.9	0.0	29.2	48.5	77.7
AF	10	Methane	3	18.7	81.3	0.0	18.1	40.4	58.5
AF	20	Methane	3	22.4	77.4	0.2	18.6	46.8	65.4

The obtained bio-oils were distilled in order to isolate the more volatile and low molecular weight fractions (Table 1). In comparable conditions of raw material and residence time the use of a nitrogen or methane atmosphere seems to favor the formation of distillable liquids and of the heavier fraction relatively to the light fraction; this behavior may result from limitations to the condensation reactions to form solids, leading to a stabilization of products that are liquid at room temperature. Since the use of methane or nitrogen also increased the yields of gas products, there were also less condensation reactions between the lighter primary products that remain in the gas phase therefore reducing the formation of liquids with low molecular weight.

The distillation of the liquid products is also essential to evaluate the "true" yield of bio-oils since the original raw materials and unreacted oils or fats are also liquids, but cannot be distilled in the same conditions.

3.3   *Comparison of the pyrolysis products obtained under different atmospheres*

The chromatographic analysis of the distilled fractions allowed to identify the main components present in the distilled bio-oils and to compare the chromatographic profile with a mixture of linear alkanes from heptane (C7) to triacontane (C30). The main functional groups identified in the distilled bio-oils were aliphatic and aromatic hydrocarbons, indicating that the deoxygenation of raw materials was efficient.

The carbon number distribution of the bio-oil components was evaluated by making the assumption that the compounds eluted between two adjacent linear hydrocarbons have carbon numbers identical to those hydrocarbons. Thus, their individual relative concentrations were grouped, and represented as a function of the corresponding carbon numbers, as seen in Figure 1 for both distilled fractions of the bio-oils obtained from the three raw materials. The carbon distribution of both distilled fractions shows clear differences for the different raw materials, showing as their nature is a major parameter influencing the bio-oil composition.

The light fraction shows higher relative concentrations of components with carbon numbers from C8 to C12, but different profiles for the two vegetable oils and the animal fats: olive oil (OO) and olive husk oil (OHO) had significantly higher concentrations of C8 to C10 hydrocarbons than those of components with higher carbon number while the light fraction of the bio-oil obtained from animal fat shows comparable concentrations for components from C8 to C16. These characteristics may result from different mechanisms of thermal decomposition acting in these different lipid materials. The vegetable oils are mainly composed by C18 fatty acids namely oleic acid and linoleic acid, that have double bonds in their hydrocarbon side chain; these double bonds act as reactive spots, favoring the decomposition of the side chain to yield C8 and C9 hydrocarbons. On the other hand, the animal fat presents much higher concentrations of saturated fatty acids, in particular stearic acid (C18) and palmitic acid (C16) that by decarboxylation yield C16 and C14 saturated hydrocarbons; these hydrocarbons may then suffer thermal cleavage but since they do not present double bonds,

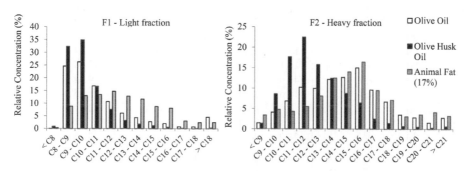

Figure 1.   Carbon number distribution of the distilled bio-oil components (F1 – light fraction and F2 – heavy fraction), from the pyrolysis of the various raw materials, under initial vacuum, at 420°C, during 30 min.

the probability of bond cleavage is equivalent for all bonds, leading to a regular distribution of concentrations for components with 16 carbons or less.

The composition of the heavy fraction (F2) also reflects these mechanisms, in particular the decarboxylation of C18 fatty acids to yield C16 hydrocarbons. Minor amounts of hydrocarbons with 19 carbons and higher may result from the combination of higher molecular weight primary products.as well as the decomposition of fatty acids with more than 18 carbons, present in the oils in residual concentrations. Olive husk oil (OHO) showed a higher proportion of lower carbon number components namely from C10 to C13 in the heavy fraction and C8 to C10 in the light fraction, a behavior that is consistent with the unsaturated nature and high acid value of this lipid material.

The carbon number distribution of the distilled fractions of OHO bio-oils obtained by pyrolysis under nitrogen or methane atmospheres are presented in Figure 2.

The use of CH_4 resulted in a stabilization of the lower molecular weight components of both fractions suggesting that besides solvating the pyrolysis primary products methane may also react with these products to yield higher molecular weight derivatives. In particular it is noticeable the increase in the concentration of components with more than 21 carbon numbers that were practically absent in the bio-oils obtained under initial vacuum.

When performing pyrolysis under vacuum or inert atmospheres the carbon-carbon bonds will be cleaved but there is not a source of hydrogen atoms to complete the four bonds required by carbon atoms in sp^3 hybridization. Therefore, the pyrolysis primary products tend to form double bonds and/or cyclic structures in order to yield more stable molecules. Total concentrations of aromatics and aliphatics in the bio-oils were evaluated as the sum of the relative concentrations of each individual component (Figure 3).

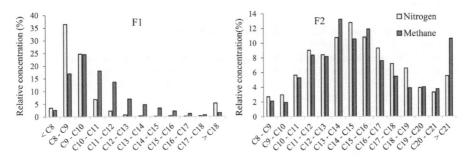

Figure 2. Carbon number distribution of the distilled fractions (F1 – light fraction and F2 – heavy fraction) of OHO bio-oils obtained by pyrolysis under nitrogen or methane atmospheres (initial pressure = 3 bar), at 420°C, during 10 min.

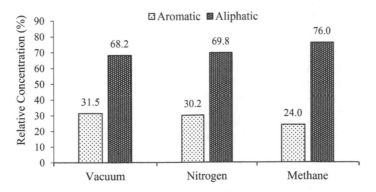

Figure 3. Relative concentrations of aromatic and aliphatic components present in the bio-oils obtained by pyrolysis of high acidity olive oil, during 10 min, at 420°C and atmospheres of initial vacuum, nitrogen or methane.

The relative concentrations of aromatic and aliphatic components were comparable when pyrolysis was performed under initial vacuum or a nitrogen atmosphere but a 6% decrease in total aromatics and an equivalent increase in total aliphatics was observed when nitrogen was replaced by methane; this observation is also an indication that methane may be acting not only as a non-oxidative atmosphere but also as a reactive gas, namely by stabilizing the pyrolysis primary products and constituting a source of hydrogen atoms.

4 CONCLUSIONS

Oils and fats with high contents of free fatty acids and water can be used as a source of carbon-rich bio-oils that could be further refined to yield different types of biofuels. The composition of the raw materials influences the thermochemical degradation pathways leading to different mixtures of products. The use of animal fats, long residence times and methane atmospheres favoured the formation of heavier components.

Performing pyrolysis at reduced pressure appears to allow deoxygenation and extensive decomposition of the lipid-rich materials without requiring consumption of an inert gas, therefore lowering the operational costs.

The ratio between the relative concentrations of aromatic and aliphatic components of the bio-oil is equivalent for pyrolysis under vacuum or nitrogen atmospheres but was reduced when the methane atmosphere was used indicating that methane may be acting both as a shielding atmosphere and as a reducing agent.

REFERENCES

Bridgwater, A.V., Meier, D., & Radlein, D. 1999. An overview of fast pyrolysis of biomass. *Organic Geochemistry* 30 (12): 1479–1493.

Demirbas, A. 2005. Biodiesel production from vegetable oils via catalytic and non-catalytic supercritical methanol transesterification methods. *Progress in Energy and Combustion Science* 31(5–6): 466–487.

Demirel, Y. 2012. Energy: Production, conversion, storage, conservation and coupling. In *Green Energy and Technology.* Springer-Verlag London Limited.

Dewayanto, N., Isha, R., & Nordin, M.R. 2014. Use of palm oil decanter cake as a new substrate for the production of bio-oil by vacuum pyrolysis. *Energy Conversion and Management* 86: 226–232.

Fan, Y., Cai, Y., Li, X., Yin, H., Yu, N., Zhang, R., & Zhao, W. 2014. Rape straw as a source of bio-oil via vacuum pyrolysis: Optimization of bio-oil yield using orthogonal design method and characterization of bio-oil. *Journal of Analytical and Applied Pyrolysis* 106: 63–70.

Feddern, V. 2011. Animal Fat Wastes for Biodiesel Production. *Biodiesel—Feedstocks and Processing Technologies*: 45–70.

Ge, X., Matsumoto, T., Keith, L., & Li, Y. 2014. Biogas energy production from tropical biomass wastes by anaerobic digestion. *Bioresource Technology* 169: 38–44.

Lappi, H., & Alén, R. 2009. Production of vegetable oil-based biofuels—Thermochemical behavior of fatty acid sodium salts during pyrolysis. *Journal of Analytical and Applied Pyrolysis* 86(2): 274–280.

Li, S., Xu, S., Liu, S., Yang, C., & Lu, Q. 2004. Fast pyrolysis of biomass in free-fall reactor for hydrogen-rich gas. *Fuel Processing Technology* 85(8–10): 1201–1211.

Niskanen, A., Värri, H., Havukainen, J., Uusitalo, V., & Horttanainen, M. 2013. Enhancing land fill gas recovery. *Journal of Cleaner Production* 55: 67–71.

Nouredinni, H., Teoh, B.C. & Clemens, L.D. 1992. Densities of vegetable oils and fatty acids. *Journal of the American Oil Chemists' Society* 69(12): 1184–1188.

Pinto, F., Varela, F.T., Gonçalves, M., Neto André, R., Costa, P., & Mendes, B. (2014). Production of bio-hydrocarbons by hydrotreating of pomace oil. *Fuel* 116: 84–93.

Poulli, K.I., Mousdis, G.A., & Georgiou, C.A. 2005. Classification of edible and lampante virgin olive oil based on synchronous fluorescence and total luminescence spectroscopy. *Analytica Chimica Acta* 542(2): 151–156.

Zhang, H., Xiao, R., Wang, D., He, G., Shao, S., Zhang, J., & Zhong, Z. 2011. Biomass fast pyrolysis in a fluidized bed reactor under N2, CO2, CO, CH4 and H2 atmospheres. *Bioresource Technology* 102(5): 4258–64.

Effect of temperature in RDF pyrolysis

A. Ribeiro & J. Carvalho
CVR—Centre for Waste Valorisation, Guimarães, Portugal

C. Vilarinho
*Mechanical Engineering Department, Mechanical Engineering and Resources Sustainability Center,
University of Minho, Guimarães, Portugal*

ABSTRACT: Refuse Derived Fuel (RDF) is a solid fuel made after basic processing steps or techniques that increase the calorific value of Municipal Solid Waste (MSW). Therefore, energy production from RDF can provide economic and environmental benefits, as reduces the amount of wastes sent to landfill and allows the energy recovery from a renewable source. In this work, it was studied the effect of temperature in RDF pyrolysis collected in a Portuguese company in gas production, gas composition and mass conversion. The effect of reaction temperature was studied at 450, 600 and 750°C. Results showed that, for the same operational conditions, pyrolysis was more efficient at 750°C. At this temperature, it was obtained a syngas of 11.2 MJ/m^3 and a specific gas production of 0.43 m^3 syngas/kg RDF. Results also proved that the increasing of temperature in pyrolysis reaction enhanced methane production and decreased carbon dioxide concentration in syngas.

1 INTRODUCTION

The sustainable management of wastes, has maintained the attention of citizens, stakeholders and scientific community for the last decades. Around 3 billion tonnes of waste are generated in the European Union (EU) each year—over 6 tonnes for every European citizen. In Portugal, according to data from the National Statistical Institute (INE) in 2014, was generated 4.7 million tons of urban waste and 11.6 million tons of industrial wastes, particularly from manufacturing industries and trade and contribution services. This has a huge impact on the environment, causing pollution and greenhouse gas emissions that contribute to climate change (Werge, 2009). In fact, the most relevant environmental impact regarding to incorrect waste management policies reported in scientific research on global warming is the effect of emissions of greenhouse gases (most significantly by the release of methane) from municipal solid waste landfill. For instance, in Portugal, the deposition of waste in landfill originated in 2012, 2.8 million tons of CO_2 *eq* of GHG emissions, which represents 34.1% of emissions from the waste sector and 4.0% of total national GHG emissions estimated for that year (APA, 2014). As a result, alternatives to landfill for municipal solid waste are frequently viewed as having a positive effect on global warming (EEA, 2016). Therefore, good waste management policies can significantly reduce these impacts.

Regarding to waste management policies, carried out by EU Waste Framework Directive 2008/98/EC, EU approach is based on the waste hierarchy, which sets the following priority order: prevention, (preparing for) reuse, recycling, recovery and, as the least preferred option, disposal (which includes landfilling and incineration without energy recovery) (Pires et al, 2011). Nevertheless, this hierarchy also assumes that when wastes cannot be prevented or recycled, recovering its energy content by waste-to-energy technologies is in most cases preferable to landfilling it, in both environmental and economic aspects (Saveyn et al, 2016).

In order to achieve these targets, Waste Framework Directive 2008/98/EC and Portuguese Strategic Plan for Municipal Solid Waste (PERSU 20200) reinforced the importance of using

RDF as an alternative fuel for intensive industries (power generation facilities, co-generation plants and heat-demanding processes) (Ferrão et al, 2016).

RDF or Solid Recovered Fuels is defined by EN 15359/2011 as solid fuel made from non-hazardous waste used for energy recovery in incineration and co-incineration plants, that can be production by specific waste, municipal solid waste, industrial waste, commercial waste, construction and demolition waste and sewage sludge (EN 15359:2011). Nevertheless, RDF production is most related to municipal solid wastes (MSW) treatment. Nowadays, one of the well-stablish technologies to produce RDF from MSW are dry stabilization process and especially mechanical and biological pre-treatment (MBT) (Montejo et al, 2011). According to Edo-Alcón et al (2016), in EU the amount of RDF produced from MSW with a high is about 12 million tonnes per year (Edo-Alcón et al, 2016). The most active EU Member in RDF production are Austria, Germany, Netherlands, Denmark, Italy, etc., which is related to higher performances in MSW source separation and recycling (Heidenreich & Foscolo, 2015).

Therefore, the utilization of RDF as a robust endogenous resource can respond to a range of problems associated with waste management, making it possible to:

- Minimize the deposition of MSW in landfill
- Contribute to national self-sufficiency in energy production
- Contribute to the reduction of greenhouse gases through the reduction of CH^4 emission from landfill
- Diversify the sources of fuels (Strezov et al, 2007).

In general, pyrolysis represents a process of thermal degradation of the waste in the total absence of air that produces recyclable products, including char, oil/wax and combustible gases. Pyrolysis has been used to produce charcoal from biomass for thousands of years. When applied to waste management, RDF can be turned into fuel and safely disposable substances (char, metals, etc.), and the pyrolysis process conditions can be optimized to produce either a solid char, gas or liquid/oil product, namely, a pyrolysis reactor acts as an effective waste-to-energy convertor. According to Velghe et al (2011), pyrolysis temperature varied from 300 to 900°C, but the typical running temperature is around 500–550°C with liquid products as major portion of products. At temperatures higher than 700°C, syngas is the vital product. Most of the researches paid more attention to the liquid and syngas products than the char due to the fact that the oil and syngas are more valuable; and the yields and composition of pyrolysis oil and syngas are mainly changing with temperature. The residence time of the materials in the reaction zone is another important parameter; which was reported to be in the range of a few seconds to 2 h. It was well recognized that longer residence time would enhance the tar cracking and result in higher gas yields; but at the same time longer residence time reduced water content and waxy material in the liquid products conduced to the improved quality (Velghe et al, 2011).

2 MATERIALS AND METHODS

2.1 Refused Derived Fuel (RDF) characterization

Refused Derived Fuel (RDF) used in the present paper were collected from a local company. According to the company, this RDF is a mixture of MSW from MBT plants and non-hazard industrial wastes. Physical-Chemical characterization of RDF was performed according to EN 15359:2011 (Solid recovered fuels — Specifications and classes). This characterization comprised de determination of: moisture content, volatile matter content, ash content, low heating value (LHV), bulk density, biomass content, elemental analysis (carbon, nitrogen and hydrogen) and metals (antimony, arsenic, cadmium, lead, cobalt, copper, chromium, manganese, mercury, nickel, thallium, vanadium, metallic aluminum).

2.2 Experimental apparatus

Gasification experiments were performed in a laboratory scale fixed bed pyrolizer with 36 litres of total volume. The reactor consisted of a vertical stainless-steel tube with an inner diameter of

15 cm. Temperature is provided by electric resistances positioned along the reactor which allowed to obtain a perfect thermic equilibrium in all the reactor components. Syngas condensation and purification unit consisted in a cooling box and a condensation chamber constituted by one acrylic column submerged in water that precipitate the resulting oil. The syngas formed before heading out into the atmosphere flows through three water columns where it is promoted the washing and purification of gases. The control unit is constituted by a volumetric flowmeter that measured the syngas formed during each experiment. The data acquisition (volume of syngas, temperature, time and pressure) was monitored in real time through a computer coupled to the system.

2.3 Refuse Derived Fuel (RDF) pyrolysis

All pyrolysis experiments were performed at approximately 120 minutes with initial feedstock of 450 g of RDF. The effect of temperature was evaluated in syngas composition, syngas heating value (kJ/m^3 N), specific syngas flow rate (m^3 N/kg RDF) and carbon conversion efficiency. The effect of reaction temperature was studied at 450, 600 and 750°C.

For each experiment, it was recorded the reaction time, mass loss, volume of produced gas, composition of final gas and solid char produced. The amounts of gaseous and liquid products collected were recorded along with the char remaining after the gasification process to check the mass balance. Syngas was collected in plastic bags and its composition were analyzed by gas chromatography. Lower Heating Value of syngas produced was calculated according to equation 1. Carbon conversion efficiency (CCE) (%) was calculated using Equation 2.

$$LHV = \frac{(CO * 126,36 + H_2) * 107,98 + CH_4 * 358,18)}{1000} \tag{1}$$

where, CO, H$_2$, CH$_4$, are the molar percentages of components of syngas.

$$CCE = \frac{(12Y * (CO + CO_2 + CH_4))}{22,4 \times C} \times 100 \tag{2}$$

where, Y is the dry gas yield (Nm^3/kg), C is the mass percentage of carbon in ultimate analysis of biomass feedstock, and the other elements are the molar percentage of syngas components.

3 RESULTS AND DISCUSSION

3.1 Refuse Derived Fuel (RDF) characterization

Refuse Derived Fuel (RDF) samples were characterized in order to evaluate their energetic potential and according to EN 15359:2011. Table 1 shows the chemical and physical characterization of RDF.

Results demonstrated that RDF presents high potential for energetic valorization because has low content of water (6.1%), high volatile content (77.2%) and high calorific value (24330 J/g). it is also possible to notice that carbon is the major component of this RDF (56.2%).

However, results also reveal that RDF has high concentrations of sulfur (2150 mg/kg) and chlorine (9120 mg/kg) which could represent operational problems on a large-scale exploration. In order to limit these problems, a cleaning step of the gas formed through the introduction of water columns was included in the execution of the experiments.

These results are quite similar to other authors, e.g., Genon and Brizio (2008) characterized RDF samples in a LCA study. These authors verified that RDF have also high calorific value (20000 J/g), high carbon content (53%) and also high sulfur content (5%) (Genon & Brizio, 2008). Dalai et al (2009) also observed that RDF from municipal wastes have high volatile content (84.7%), high carbon content (46.7%) and low moisture content (3.2%). In this case, authors only detected 0.5% of sulfur in RDF which could be related to the sample origin (Delai et al, 2009).

Table 1. Chemical-physical characterization of Refuse Derived Fuel (RDF).

Parameters	Unit	Results
Ash content	%	15.00
Moisture	%	5.70
Low Heating Value	[J/g]	24330.00
Bulk density	[g/cm^3]	0.10
Biomass content	%	39.70
Volatile matter	%	77.20
Elemental analysis		
Carbon	%	56.20
Nitrogen	%	7.14
Hydrogen	%	0.91
Metals		
Antimony	[mg/kg]	54.00
Arsenio	[mg/kg]	<0.80
Cadmium	[mg/kg]	0.30
Lead	[mg/kg]	42.00
Cobalt	[mg/kg]	2.00
Copper	[mg/kg]	2970.00
Chromium	[mg/kg]	520.00
Manganese	[mg/kg]	52.00
Mercury	[mg/kg]	<0.05
Nickel	[mg/kg]	15.00
Thallium	[mg/kg]	<0.20
Vanadium	[mg/kg]	5.00
Metallic aluminum	[mg/kg]	0.25

3.2 *Effect of temperature in RDF pyrolysis*

Temperature is one of the most relevant parameter in gasification process, since it modifies syngas composition and calorific value. In order to evaluate the effect of temperature in RDF conversion, gasification experiments were carry out at 450, 600 and 750°C. Figures 1–4 shows the effect of this parameter in syngas production, syngas composition, carbon conversion efficiency (CCE) and syngas heating value and specific syngas flow rate, respectively.

Results of Figure 1 shows a trend in syngas production with increasing temperature. In fact, syngas gas production increased from 24 liters at 450°C from approximately 120 liters of syngas at 750°C. Buah et al (2007) studied pyrolysis of RDF with a gross calorific value of 18.9 MJ/kg in a fix-bed reactor and found that the main weight loss within the lower temperature zone (240–380°C) was due to the degradation of the cellulosic fraction in the RDF, whereas the second peak of weight loss, occurring between 410 and 500°C was due mainly to degradation of plastic components. Chen et al (2008) adopted thermogravimetric analysis–Fourier transform infrared spectrometer (TG–FTIR) to investigate the pyrolysis of two different RDFs, and they found similar weight loss-temperature behaviors. The temperature zones were 220–430°C for biomass degradation, 430–520°C for plastics decomposition and higher than 650°C for carbonates.

These results represented also an increase of specific syngas flow rate (Figure 4) from 0.104 to 0.314 m^3 syngas/kg RDF at 450°C to 750°C, respectively. Regarding to syngas gas composition, Figure 2 shows the influence of temperature in this parameter. Results plotted in Figure 2 reveals also that the increasing of temperature promoted the formation of a methane rich gas. At 750°C methane yield increased to 24.3% from 4.3% at 450°C, mostly due to the cracking of larger molecules presented in oil products. In contrast, carbon dioxide yield decreased from 25.1% at 450°C to 11% at 750°C. The increasing of methane yield at 750°C resulted also in the increasing of calorific value of the syngas produced. At 750°C methane yield increased to 11.2 MJ/m^3 from 2.7 MJ/m^3 at 450°C (Figure 4). Hwang et al (2014) studied the pyrolysis of RDF from MSW at 500,

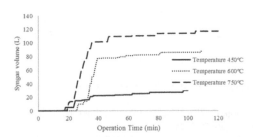

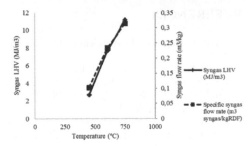

Figure 1. Effect of temperature on total syngas production.

Figure 2. Effect of temperature on syngas composition.

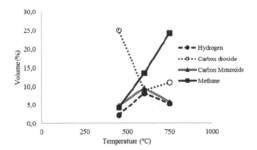

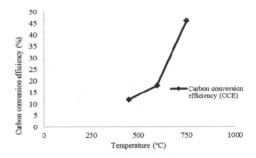

Figure 3. Effect of temperature on carbon conversion efficiency (CCE).

Figure 4. Effect of temperature on syngas heating value and specific syngas flow rate.

700 and 900°C. Authors also reported an increasing of specific syngas flow rate and gas LHV with temperature increasing. These authors obtained an increase of 0.13 to 0,43 m³ syngas/kg RDF at 500 to 900°C, respectively. The HHVs of the pyrolysis gas from this RDF were 10.4, 17.0 and 19.1 MJ/kg for temperatures of 500, 700 and 900°C, respectively.

Regarding to carbon conversion efficiency (CCE), results represented in Figure 3 shows that temperature also increased the carbon conversion from 12% at 450°C to approximately 50% at 750°C. Buah et al (2007) also found that that carbon conversion in pyrolysis increases with temperature increasing. These authors reported that at 400, 500, 600 and 700°C, the calorific values of the chars were 20.4, 16.7, 16.4 and 11.2 MJ/kg, respectively, with a yield decrease from 50% to approximately 31%.

4   CONCLUSIONS

The utilization of Refuse derived fuels or RDF as an alternative and renewable solid fuel can respond to a range of problems a robust endogenous resource can respond to a range of problems related to waste management. In fact, European policies in waste management give extreme importance to RDF pyrolysis. However, due to its high thermochemical complexity and implements costs, this technology is still far way of being implemented at global scale. For these reasons, and in order to overcome these technical issues, the study of its main reactions and products are necessary.

This study intended to evaluate the effect of temperature in syngas production, composition, heating value and carbon conversion efficiency. Results proved that pyrolysis is more efficient at 750°C. At this temperature, it was obtained a syngas of 11.2 MJ/m³ and a specific gas production of 0.43 m³ syngas/kg RDF. Results also proved that the increasing of temperature in pyrolysis reaction enhanced methane production and decreased carbon dioxide content.

REFERENCES

Agência Portuguesa do Ambiente (APA) 2014. *PERSU 2020 - Plano Estratégico para os Resíduos Urbanos*.

Buah, W.K., Cunliffe, A.M., Williams, P.T. 2007. Characterization of products from the pyrolysis of municipal solid waste. *Process Saf. Environ. Prot.* 85: 450–457.

Chen, J., Huang, L.W., Zhang, X.M. 2008. Pyrolysis analysis of RDF by TG–FTIR techniques. *Environ. Sci. Technol.* 31: 29–32.

Dalai, A., Batta, N., Eswaramoorthi, I., Schoenau, G. 2009. Gasification of refuse derived fuel in a fixed bed reactor for syngas production. *Waste Management* 29: 252–258.

Edo-Alcón, N., Gallardo, A., Colomer-Mendoza, F. 2006. Characterization of SRF from MBT plants: Influence of the input waste and of the processing technologies. *Fuel Processing Technology* 153: 19–27.

EN 15359:2011. *Solid recovered fuels. Specifications and classes*.

European Commission (EC) 2017. *Being wise with waste the EU approach to waste management implementation of the Circular Economy Action Plan*. COM (2017) 33.

European Environment Agency 2016. Annual European Union greenhouse gas inventory 1990–2012 and inventory report. *Publications Office of the European Union*, 2016.

Ferrão, P., Lorena, A., Ribeiro, P. 2016. Industrial Ecology and Portugal's National Waste Plans. *Taking Stock of Industrial Ecology*: 275–289.

Montejo, C., Costa, C., Ramos, P., Márquez, M.D.C. 2011. Analysis and comparison of municipal solid waste and reject fraction as fuels for incineration plants. *Appl. Therm. Eng.* 31: 2135–2140.

Pires, A., Martinho, G., Chang, N. 2011. Solid waste management in European countries: A review of systems analysis techniques. *Journal of Environmental Management* 92(4): 1033–1050.

Saveyn, H., Eder, P., Ramsay, M., Thonier, G., Warren, K., Hestin, M. 2016. Towards a better exploitation of the technical potential of waste-to-energy. *Science for Policy report by the Joint Research Centre (JRC)*, European Union.

Strezov, V., Patterson, M., Zymla, V., Fisher, K., Evans, T.J., Nelson, P.F. 2007. Fundamental aspects of biomass carbonization. *J. Anal. Appl. Pyrol.* 79: 91–100.

Velghe, I., Carleer, R., Yperman, J., Schreurs, S. 2011. Study of the pyrolysis of municipal solid waste for the production of valuable products. *J. Anal. Appl. Pyrol.* 92: 366–375.

Wang, N., Chen., D.Z., He, P.J., 2014. Reforming of MSW pyrolyis volatile on their char and the change of syngas. *In: Proceedings of the 5th International Symposium on Energy from Biomass and Waste*, 17–20 November 2014, Venice.

Werge, C. 2009. EU as a Recycling Society Present recycling levels of Municipal Waste and Construction & Demolition Waste in the EU. *European Topic Centre on Resource and Waste Management*, European Environment Agency.

WASTES – Solutions, Treatments and Opportunities II – Vilarinho, Castro & Lopes (Eds)
© *2018 Taylor & Francis Group, London, ISBN 978-1-138-19669-8*

Potential of exhausted olive pomace for gasification

C. Castro, A. Mota, A. Ribeiro, M. Soares, J. Araujo & J. Carvalho
CVR—Centre for Wastes Valorisation, University of Minho, Guimarães, Portugal

C. Vilarinho
*Mechanical Engineering Department, Mechanical Engineering and Resources Sustainability Center,
University of Minho, Guimarães, Portugal*

ABSTRACT: The exhausted olive pomace was fully characterized for the application as combustible in gasification. Nowadays, this waste is sold as an animal feed or fertilizer, losing its potential as a combustible for energy production. To analyse the potential as a fuel, a full physic-chemical and energetic characterization were performed. With the results was possible to conclude that: the waste was not dangerous, it is a potential combustible, releases 100.65 W g^{-1} at 496°C and have a full burning at 600°C. Considering all the results, was conducted a theoretical study, varying the amount of oxygen injected, in a range of 0.21–0.39 kmol kmol^{-1} of waste, allowing the selection of the best work point in terms of net energy. From the theoretical analysis, it was possible to obtain the syngas composition for all the points of the study. The results and directives for the gasification are presented in this paper.

1 INTRODUCTION

Olive oil is one of the most used products in cooking all over the world. Its characteristics make him a good choice in the preparation of new plates. Every day, the investigation has evolved in order to try to get a better product and to improve its production. The Europe is the largest responsible for the production and exportation of this product all over the world, particularly Spain, Italy, Greece and Portugal. In 2016, Portugal was responsible for the production of 757,373 hl of olive oil (INE, 2016). From the olive oil production results a great quantity of different wastes that needs to be treated and that depend on the production method. Some of these wastes resulting are dangerous for the environment, having impact on the soils, water and even in the air (Roig et al., 2006). One of these wastes produced is the olive pomace, which contains oil that is recovered in extraction factories that produce another waste: exhausted olive pomace. This waste, has a low amount of water (<10% w w^{-1}) and oil (<1%). Its field of application is vast, varying from animal feed to fertilizer and biofuel, etc (Paudel, 2009). However, in spite of being a combustible for small traditional furnaces, its combustion is generally inefficient. This inefficiency produces a lot of smoke, rich in carbon and carbon monoxide, resulting in the interference of "l'Agence Nationale de Protection de l'Environment" (ANPE) to prohibit the use of the exhaust cake as combustible (Masghouni & Hassairi, 2000). One alternative is the use of the exhausted olive pomace as a combustible in gasifiers. Gasification, in summary, can be defined as a thermochemical conversion of the product with the increase of the temperature. The type of gas injected in the gasification allows the component to be converted into a gas, known as synthesis gas (syngas), by different heterogeneous reactions. The syngas is composed by CO_2, CO, H_2, CH_4 and H_2O. At the same time it is possible to produce a small quantity of char and a tar (Belgiorno et al., 2003).

In this paper a full characterization of the olive oil pomace is presented, as well as a theoretical study created by the Equation (1), and some directives for the utilization of the waste as combustible in the gasification.

2 METHODOLOGY

2.1 *Waste characterization*

The olive oil exhausted pomace is a by-product of the olive pomace. According to that, a pre-treatment or a drying process of the waste is not necessary. To perform the theoretical study of the waste, a characterization in terms of elementary analysis was conducted, as well as physical and chemical analysis and the X-Ray Fluorescence (XRF) of the ashes (to make sure that the result compound of the burning process was not hazardous). The physical and chemical parameters and the elementary analysis were made in different methods and equipment's, such as:

- Gravimetry: dry matter, ashes, organic and volatile matter;
- TruSpec CHN Analyser: total carbon, hydrogen and nitrogen;
- CSN EN 15400: HHV (High Heating Value) and LHV (Low Heating Value);
- E.P.A. 200.7: metals and halogens;
- Intern Method: bulk density, pH, sulphur, TGA (Thermogravimetric Analysis) and DTA (Differential Analysis).

2.2 *Theoretical study*

In order to find what conditions were the best for the gasification process, it have been made a theoretical study (mathematical) to simulate the conditions needed. For the calculation of the syngas composition, there are one equation (1) that is the most important to consider (Mountouris et al., 2006):

$$Waste + wH_2O + mO_2 + 3.76mN_2 = n_1H_2 + n_2CO + n_3CO_2 + n_4H_2O + n_5CH_4 + n_6N_2 \quad (1)$$

where: waste = elementary analysis of the waste; w = quantity of moister of the waste; m = amount of oxygen intended to inject in the gasification process; n_{1-6} = constants calculated to stoichiometry.

Equation (1), represents the stoichiometry for the syngas composition depending on the waste elementary composition, water content (moisture) and, amount of oxygen that is inputted into the gasification process.

For this theoretical sheet the input parameters are the elementary analysis, the HHV, the amount of oxygen intended to inject and the temperature for the process. The outputs are the constitution of the syngas and the net energy (energy produced by the syngas—energy needed to the process).

3 RESULTS AND DISCUSSIONS

3.1 *Elementary, physical, chemical and X-Ray Fluorescence analysis*

By the analysis of the physical parameters presented in the Table 1, it is possible to conclude that the waste (exhausted olive pomace) has a low content of moisture (7.3%) and a low value of ashes at 815°C (6.1%). Comparing with the values of LHV of dry hood, which ranges between 15.48 MJ kg⁻¹ and 19.44 MJ kg⁻¹ (Repellin et al., 2010), the 20.3 MJ kg⁻¹ obtained

Table 1. Exhausted olive pomace physical and energetic parameters.

Parameter	Dry Matter 105°C**	Ashes 815°C	Volatile Matter 900°C	Organic Matter 550°C	Fix Carbon ***	HHV	LHV	Bulk Density
Unit*	%	%	%	%	%	MJ kg⁻¹	MJ kg⁻¹	kg m⁻³
Result	92.7	6.1	66.2	82.8	16.6	21.4	20.3	600

*Dry basis; **As it is; ***Calculated.

are promisor for the gasification process. In terms of organic matter, the 82.8% represents a good percentage of the material. In reference to the volatile matter, the 66.2% could be improved when compared with the 80% of the volatile matter of the wood (Demirbas, 2004), but that do not preclude the possibility to have a good syngas in the gasification process.

The elementary analysis presented in the Table 2 shows an amount of carbon about 56.9%, which represents a good quantity for a biomass and also a good factor in the prediction for the gasification. Another good characteristic presented, which can be founded in a biomass, is the lowest quantity of nitrogen and sulphur, 1.42% and 0.23% respectively. The hydrogen has an amount of 6.99% and, by calculous, the oxygen has a value of 28.36%.

In Table 3 it is possible to analyse the chemical parameters of the waste. As it can be seen, the main elements presented in the waste are the phosphor, potassium, zinc and copper. About these four elements, the two highlighted elements are the phosphor and the potassium, presented in an amounts of 3440 mg kg^{-1} and 24,800 mg kg^{-1}, respectively. The potassium is not a concern because is not dangerous for the environment, however, in the case of the phosphor, the situation is quite different. In gasifiers that do not work with flames to produce heat (the heat is from a power source, e.g. resistances) a high amount of potassium means that it is possible to have some combustion inside the gasifier. This lead to the production of CO_2, which is a nefarious gas in the composition of the syngas.

In Table 4 it is possible to conclude that the elements presented in significant relevance in the waste were almost totally burned with the increase of temperature. The most relevant are SiO_2 (silicon dioxide), P_2O_5 (phosphorus pentoxide) and K_2O (potassium oxide). The presence of the last two is explained by the high quantity of the elements in the waste composition. Is not possible to eliminate all, but it is possible to reduce their composition, even if the presence of these elements do not cause any inconvenient, since they are not a dangerous.

3.2 Thermogravimetric and differential analysis

In Figure 1 it is possible to analyse the thermogravimetric analysis (TGA) and the differential analysis (DTA) of the waste (in air conditions). A crucible loaded with around 20 mg of mate-

Table 2. Exhausted olive pomace elementary analysis.

Parameter	Total Carbon	Hydrogen	Nitrogen	Sulphur
Result* (%)	56.9	6.99	1.42	0.23

*Dry Basis.

Table 3. Exhausted olive pomace chemical parameters.

Parameter	pH (20.2°C)	P	K	Cd	Cu	Ni	Pb
Unit*	–	mg kg^{-1}	mg kg^{-1}	mg kg^{-1}	mg kg^{-1}	mg kg^{-1}	mg kg^{-1}
Result	4.76	3440	24,800	<0.4	27.3	4.9	<1

Parameter	Zn	Cr	As	Ba	Mo	Sn	Sb
Unit*	mg kg^{-1}	mg kg^{-1}	mg kg^{-1}	mg kg^{-1}	mg kg^{-1}	mg kg^{-1}	mg kg^{-1}
Result	31.5	3.1	<0.5	5.8	<0.4	<1	<0.5

*Dry basis.

Table 4. Ashes composition by XRF.

Parameter	Na_2O	MgO	Al_2O_3	SiO_2	P_2O_5	K_2O	CaO	TiO_2	Cr_2O_3
Result (%)	1.282	2.663	2.640	31.364	12.102	37.737	8.389	0.055	0.014
Parameter	MnO	Fe_2O_3	NiO	CuO	ZnO	Rb_2O	SrO	ZrO_2	BaO
Result (%)	0.058	3.340	0.046	0.185	0.071	<0.0005	0.034	<0.0005	0.016

rial is placed in the thermogravimetric analyser. The aim of the analyser is to measure and record the weight loss and the energy released of the sample, with the increase in temperature (Lu et al., 2013). In this study case, the temperature range set selected was 0–1000°C, and the heating rate was 10°C min⁻¹.

This study is important in a construction of a plan for pyrolysis or gasification because it is possible to obtain a full range of the behaviour of the material with the temperature rising. So, it is possible to stablish the working temperature in the gasification process. The best temperature to work it is when, in the TGA-DTA, the compound reaches the lowest value of weight and, when there is no more energy to release (≤ 0 W g⁻¹).

As it can be seen in the Figure 1, the TGA-DTA curves are divided in three stages. In the 1st stage the material loses some weight but do not release a lot of energy. Some of the energy released come from the moister content. The 2nd stage happens when the compound releases most of its energy and it is when de material reaches the pick of energy per gram of material (100.65 W g⁻¹ at 496°C). The 3rd stage (final stage) starts when the material do not release energy (negative energy). At the same time, the quantity of mass is almost null, which means that there are no more components to release.

3.3 *Gasification theoretical results*

As presented before, is was used a mathematical sheet to calculate all the variables needed for the gasification. By iteration, it was found the best value for the amount of oxygen needed to input in the gasification. The goal of this project is to have a gasification with the best net energy possible for the system. That point means the system is optimized, being more effective in energy production. Founded the best value for the net energy, a range between –30% and +30% of oxygen were selected to study the variation of the parameters. In Figure 2 it is possible to see the trend of the net energy and the LHV with the variation of the oxygen injected (in kmol) per kmol of waste.

By the analysis of the Figure 2, the LHV decreases with the rise of the oxygen. This is a typical situation in gasification. Normally, the gasification is planned to inject the lowest amount of oxygen to increase the LHV of the syngas (Ghassemi et al., 2014). Since the curve of the net energy reaches a pick, that point will be the reference to the amount of oxygen that will be inject in this project (0.3 kmol per kmol of waste).

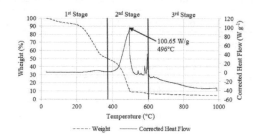

Figure 1. Exhausted olive pomace TGA-DTA, in air conditions.

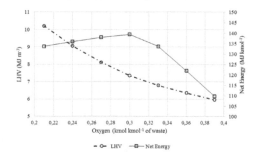

Figure 2. Syngas LHV and net energy obtained in different air flow conditions.

440

Table 5. Syngas theoretical comparison.

Parameter	H_2	CO	CO_2	CH_4	N_2	LHV
Unit	%	%	%	%	%	MJ m^{-3}
Exhausted olive pomace	23.90	34.10	0.10	1.12	40.52	7.35
Olive Pomace (Borello et al., 2015)	12.6	19.26	8.9	1.2	58.4	4.26*
Olive kernel (Skoulou et al., 2008)	7.78	4.81	19.47	2.99	64.95*	2.89
Olive Pomace (García et al., 2004)	9.3	8.4	19.0	1.9	60.55	3.3

*Calculated by the composition of the waste.

The LHV presented in the Figure 2 is calculated by the composition of the syngas obtained through the mathematical sheet. The values (in percentage) of each compound of the syngas for the best point (oxygen 0.3 kmol), presented in Table 5, are compared with the results mentioned in literature, namely the values of the olive pomace from Borello et al., 2015 and García et al., 2004, and the olive kernel of Skoulou et. al. 2008. Through the comparative analysis, it is observed that the value of LHV presented in this paper is better than the results of the others authors. Since the waste is different in the three cases means that the results will be different too. Moreover, the amount of H_2 and CO is better in this paper maybe due to the optimization of the calculus. Proceeding to a qualitative comparison, it is possible to conclude that all the rehearsals trend to the same path: a high amount of N_2 due to the injection of air and a higher amount of CO and H_2 in comparison with the amount of CH_4. The amount of CO_2 does not depend only on the combustion, but also on the type of gasifier used in the gasification (endothermic or exothermic—essentially). There are lot of variables that condition the syngas composition in a gasification that is the main reason why the results should not be compared in a quantitative way.

The syngas obtained presents a LHV of 7.35 MJ m^{-3} due to the quantity of H_2 (23.90%) and CO (34.10%), since the amount of CH_4 is irrelevant (pretended in the gasification) (Higman, 1950). In case of CO_2, the result indicates a lack of combustion (important in gasifiers that do not use the energy for the compound to work). The value of N_2 is consider irrelevant since it is an inert gas and it is filtered from the syngas before being used. Compared to other authors, the syngas obtained presents a better LHV, which represents a good trend for the gasification of this waste.

3.4 Gasification conditions

After all the study and all the analysis, it was possible to select what are the best conditions for the gasification. The TGA-DTA indicates the release of all the energy of the material at a temperature near to 600°C, to ensure the total decomposition of the material and that the working temperature for the gasification should be around 850°C (750°C is a good working temperature, the increment is just to ensure gasification temperatures in all parts of the gasifier). As presented before, the amount of oxygen needed for the gasification is 0.3 kmol kmol^{-1} of waste. Converting to air, the amount of air to inject is 1.26 m^3 per kg of waste. This amount of air should be controlled depending on the waste's stage time.

4 CONCLUSIONS

The exhausted olive pomace extracted from the olive oil process, is a possible waste to be used as a combustible for gasifiers, with the objective to produce energy and to increase the balance of energy of the producers. To validate the idea, there were made some analysis and calculus which are presented in this paper.

Through the results obtained in the full characterization of the waste it is possible to conclude that the waste is not dangerous for the environment (even after burning) and the weight loss occurs at 496°C (with a heat transfer of 100.65 W g^{-1}) stablishing the working temperature at 850°C to guarantee all the weight loss and compounds release.

The theoretical calculous obeyed an equation to syngas production (Equation (1)), obtaining its composition in percentage per compound and the net energy of the system. With all the conditions, it is possible to observe, theoretically, a decrease in the LHV value with the rise of the amount of oxygen injected and a pick in the net energy in a range of oxygen between 0.21–0.33 kmol kmol^{-1} of waste. The point selected as the optimum for the project was the one with the highest value of net energy (to have the most efficient process) obtaining a syngas with a LHV of 7.35 MJ m^{-3}, a net energy of 139.35 MJ kmol^{-1} and a syngas composition as shown in Table 5. That point occurs at injection of 0.3 kmol of oxygen per kmol of waste, which can be translated in 1.26 m^3 of air per kg of waste.

All of the results are a trend for the real gasification process and they will serve as a line of work. At the end of the gasification, an analysis to the syngas and the char remaining of the process should be performed to ensure the best efficiency of the process. In a practical rehearsal, it is possible to conclude the amount of gas in relation of the amount of air injected.

This could be a good way to use this waste, giving a better path, increasing the efficiency of the system and, more important, increasing the eco-friendly label.

ACKNOWLEDGEMENTS

The research work here presented is related to the activities included in the project proposal ELAC2014/BEE0364, with the acronym SUMO—Sustainable Use of bioMass from Oleaginous processing, insert in an ERANet-LAC program and financed by FCT—Fundação para a Ciência e a Tecnologia.

REFERENCES

Belgiorno, V., De Feo, G., Della Rocca, C., & Napoli, R.M.A. 2003. Energy from gasification of solid wastes. *Waste Management* 23(1): 1–15.

Borello, D., De Caprariis, B., De Filippis, P., Di Carlo, A., Marchegiani, A., Pantaleo, A.M., ... Venturini, P. 2015. Thermo-Economic Assessment of a Olive Pomace Gasifier for Cogeneration Applications. *Energy Procedia* 75(August): 252–258.

Demirbas, A. 2004. *Combustion characteristics of different biomass fuels* 30: 219–230.

Estatística, I.N. de. 2016. INE. Retrieved May 26, 2017, from https://www.ine.pt/xportal/xmain?xpid=INE&xpgid=ine_indicadores&indOcorrCod=0000709&xlang=pt&contexto=bd&selTab = tab2.

García-Ibañez, P., Cabanillas, A., & Sánchez, J.M. 2004. Gasification of leached orujillo (olive oil waste) in a pilot plant circulating fluidised bed reactor. Preliminary results. *Biomass and Bioenergy* 27(2).

Ghassemi, H., & Shahsavan-Markadeh, R. 2014. Effects of various operational parameters on biomass gasification process; A modified equilibrium model. *Energy Conversion and Management* 79: 18–24.

Higman, C. 1950. *Gasification*.

Lu, K.M., Lee, W.J., Chen, W.H., & Lin, T.C. 2013. Thermogravimetric analysis and kinetics of co-pyrolysis of raw/torrefied wood and coal blends. *Applied Energy*.

Masghouni, M., & Hassairi, M. 2000. Energy applications of olive-oil industry by-products: - I. The exhaust foot cake. *Biomass and Bioenergy* 18(3): 257–262.

Mountouris, A., Voutsas, E., & Tassios, D. 2006. Solid waste plasma gasification: Equilibrium model development and exergy analysis. *Energy Conversion and Management* 47(13–14): 1723–1737.

Paudel, S. 2009. Current Status of Wild Olive (Olea cuspidata Wall.ex G. Don) in Bajura District of Nepal, 41.

Repellin, V., Govin, A., & Rolland, M. 2010. Energy requirement for fine grinding of torrefied wood. *Biomass and Bioenergy* 34(7): 923–930.

Roig, A., Cayuela, M.L., & Sánchez-Monedero, M.A. 2006. An overview on olive mill wastes and their valorisation methods. *Waste Management* 26(9): 960–969.

Skoulou, V., Zabaniotou, A., Stavropoulos, G., & Sakelaropoulos, G. 2008. Syngas production from olive tree cuttings and olive kernels in a downdraft fixed-bed gasifier. *International Journal of Hydrogen Energy* 33(4): 1185–1194.

Analysis of foundry sand for incorporation in asphalt mixtures

L.P. Nascimento
CVR—Centre for Wastes Valorisation, CTAC—Center for Territory, Environment and Construction, University of Minho, Guimarães, Portugal

J.R.M. Oliveira
CTAC—Center for Territory, Environment and Construction, University of Minho, Guimarães, Portugal

C. Vilarinho
Department of Mechanical Engineering, Metrics—Mechanical Engineering and Resource Sustainability Center, University of Minho, Guimarães, Portugal

ABSTRACT: With the need to preserve natural resources, waste recovery and recycling are fundamental. However, it is essential to evaluate and characterize the wastes for a correct management namely by incorporation in industrial processes in order to avoid negative impacts. Foundry sand is a waste that can be reused as an aggregate and incorporated in asphalt mixtures. In this work, a mix design study of one asphalt mixture was carried out to assess the maximum amount of material that could be used (5%), together with an evaluation of its water sensitivity. Additionally the leaching behavior was evaluated. Due to the fluoride concentration, it was concluded that this waste should be sent to a non-hazardous waste landfill, however it does not constitute a limitation on its incorporation in asphalt mixtures which could be an effective process for foundry sand management from both the environmental and economic points of view.

1 INTRODUCTION

In environmental terms, the construction sector involves the extraction of natural resources, such as wood, water, minerals and natural aggregates, as well as high energy consumption (Algarvio, 2009). The construction industry accounts for 30% of carbon emissions, and the building park consumes 42% of the energy produced. At the world level, the construction industry consumes more raw material (approximately 3000 Mt/year) than any other economic activity (Torgal and Jalali, 2010).

Highway design and construction can cause numerous impacts on the environment, such as deforestation, loss of biological diversity, alteration of the natural drainage system and soil degradation. However, in an environmental perspective, the road construction process should seek to reconcile with environmental preservation using construction techniques and methods that avoid or minimize environmental degradation (Panazzolo et al., 2012).

Environment preservation is a basic factor that is related to humanity survival. Safeguard of natural resources and sustainable development play an important role in modern and fundamental building design practices (Oikonomou, 2005).

One way of minimizing the impacts caused by the paving industry is the use of recycled materials in the pavement structure, reducing the extraction of new resources and offering another destination for waste. In the construction of pavements, the most used materials are natural aggregates, which are granular materials extracted from quarries. However, these aggregates can be replaced (at least partially) by recycled aggregates obtained from waste produced in different industrial activities. Within these wastes, construction and demolition waste, rubbers, foundry sands and others wastes can be highlighted.

According to Yazoghli-Marzouk et al. (2014) foundry industry generates an extensive amount of wasted sand, known as foundry sand. This waste can be used in pavement layers, such as granular sub-base and base, or in bound layers like the surface course. Castro et al. (2004), estimated that some 50,000 tons of foundry sands are produced annually in Portugal, and that these may replace a small part of the smaller aggregates in asphalt pavements, depending on their gradation.

With regards to the foundry sand characteristics, it can be observed that the particle sizes are uniform and contain a large amount of silica or lake sand, being used in molds and cores for ferrous and non-ferrous components, e.g., iron, steel, copper or aluminum. It is generally reused several times by regeneration processes. However, when it can no longer be reused in the production of new materials, it is removed and becomes a waste that contains many heavy metals (Etxeberria et al., 2010, Siddique et al., 2009).

Yazoghli-Marzouk et al. (2014) analyzed the mechanical performance and leachate emissions from rigid pavements where 6% of foundry sand was incorporated, and the results were promising in the mechanical strength tests and also in the leaching tests.

In studies by Collins and Ciesielski (1994) and Benson and Bradshaw (2011) foundry sand was used as fine aggregates in hot mix asphalts, in order to substitute several portions of the conventional aggregate. The results showed that the proportion of 15% replacing the natural aggregate, has a satisfactory mechanical performance. When the proportion is higher than 15%, the asphalt mixture becomes more sensitive to the loss of aggregates, which can result in an early deterioration of the pavement.

Foundry sand recycling can save energy, reduce the need to extract virgin materials and reduce costs for producers and end users. Use of foundry sands as a fine aggregate in construction applications gives project managers the ability to improve sustainable green building by reducing their carbon footprint while qualifying for Leadership in Energy and Environmental Design (LEED) credits. The United States Environmental Protection Agency (USEPA) recently calculated that at the current level of recycling, 20,000 tons of CO_2 emissions are avoided, while 200 billion BTUs of energy are saved (Benson and Bradshaw, 2011).

Penkaitis and Sígolo (2012) studied the composition of foundry sands of a Brazilian industry that were deposited in landfills, where they carried out leachate tests, chemical analysis and a scanning electron microscopy analysis to characterize the waste. It explicitly showed that high concentration of metals (iron, manganese, boron and selenium) were also found in local groundwater. Also, elements considered as toxic (chromium, copper, cobalt, nickel, aluminums) were identified in the waste, although it is currently considered non-hazardous.

The present study was carried out to evaluate the water sensitivity of a high modulus asphalt mixture incorporating a certain content of foundry sand, and its ability to prevent the leaching of the foundry sand contaminants, when they are incorporated in the mixture.

2 MATERIALS AND METHODS

2.1 *Materials*

For the production of the high modulus asphalt mixture, granite aggregates (14/20, 6/14, 6/8, 4/8, 4/10 and 4/6 fractions) and limestone filler were used. A 10/20 penetration grade bitumen was used as the binder, following a mix design study previously carried out by Silva et al. (2009). A foundry sand obtained from a foundry company located in the north of Portugal, was used as an alternative to replace the fine aggregates.

2.2 *Methods*

2.2.1 *Chemical and morphological characterization of the foundry sand*
Prior to the mechanical performance analysis carried out to the asphalt mixtures, samples of foundry sand from the casting industry have been characterized for its chemical and morphological properties. The chemical composition was evaluated by X-ray fluorescence spectrometry

Table 1. Reference temperatures used for production of the asphalt mixtures.

	Reference temperature
	°C
Aggregates	175
Bitumen	17
Compaction	165

using a Philips X Unique II spectrometer. The morphological foundry sand particles were characterized in a SEM microscope (FEI Nova 200—FEG/SEM) and the chemical composition of the constituents determined by energy dispersive spectrometry (EDAX—Pegasus X4M).

2.2.2 *Production of the mixture with foundry sand*

The mix design of the asphalt mixture used in this study was based on a previous work (Silva et al., 2009). Thus, two mixtures with an equivalent mix design were produced for application in a pavement base layer (AC20 base): one with the incorporation of foundry sand and another with 100% new material (control mixture). For an adequate performance of the mixtures it is important to respect the production and compaction temperatures specified in EN12697-35 standard, as presented in Table 1.

After the production of the asphalt mixtures, using a laboratory mixer, the test specimens were compacted using the impact compaction method (also known as the Marshall compaction method) specified in EN 12697-30 standard.

2.2.3 *Water sensitivity*

The Indirect Tensile Strength Ratio (ITSR) test allows the determination of the water sensitivity of asphalt mixtures, based on the ratio between the average indirect tensile strength value of 3 water-conditioned test specimens and the average value of 3 dry specimens.

The indirect tensile strength (ITS) test method is performed using a uniaxial testing equipment to diametrically apply a load to a specimen until failure, according to EN 12697-12.

2.2.4 *Leaching behavior*

Samples of foundry sand have been collected in order to determine its leaching behavior for environmental evaluation. For this purpose a leaching procedure according EN 12457-4 standard was followed. This is achieved with a ratio Liquid to Solid (L/S) of 1/10, turned at 7.5 rpm for a period of 24 hours at room temperature. Once filtered, the leaching solution was analyzed for the most relevant parameters including distinct metals, chlorides, sulfates, fluorides, dissolved organic carbon (DOC) and total dissolved solids (TDS) and the results compared with Portuguese environmental legislation regulated by the Decree 183/2009 where the leachate parameters are defined for the different landfill classes.

3 RESULTS AND DISCUSSION

3.1 *Chemical and morphological characterization of the foundry sand*

The chemical composition of the foundry sand is presented in Table 2 and illustrates that the main components are silica (SiO_2) and alumina (Al_2O_3). Potassium, calcium, iron, magnesium and sodium oxides are also present in a lower content. Other compounds mainly sulfates are detected in this waste.

The morphological characterization by SEM shows a regular and uniform distribution of the sand particles and the corresponding peaks obtained by EDS analysis confirm that the waste is mainly constituted by SiO_2 and Al_2O_3 as shown in Figure 1. Based upon this results, foundry sand shows potential to be incorporated in asphalt mixture.

Table 2. Chemical composition of the foundry sand by XRF (wt%).

SiO$_2$	Al$_2$O$_3$	CaO	Fe$_2$O$_3$	K$_2$O	MgO	Na$_2$O	Other
71,4	17,0	2,37	2,36	1,00	2,20	1,99	Balance

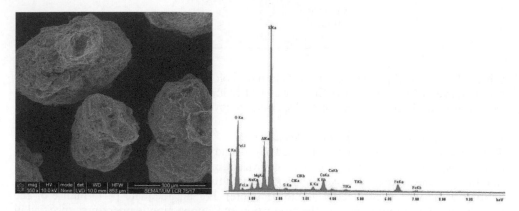

Figure 1. Morphological and chemical analysis of the foundry sand by SEM/EDS.

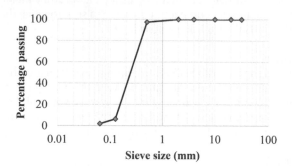

Figure 2. Particle size distribution of the foundry sand.

3.2 *Mix design and production of the mixture*

As previously mentioned, an AC20 base mixture studied by Silva et al. (2009) was used as a reference, and a new mix design was made to substitute a percentage of aggregates by foundry sand, fulfilling the grading envelope specified by EP (2014). Thus, in this study two mixtures were prepared, one conventional AC20 base and another AC20 base + foundry sand, which incorporated the largest amount of sand in the mixture that still complied with the envelope specifications. In order to do that, a particle size distribution of the sand was first performed (Figure 2) and its incorporation in the mixture was then adjusted as shown in Figure 3.

With the result of the sieve analysis presented in Figure 1, it is possible to see that the grains of the foundry sand are significantly small and uniform as observed in the graph, with the material concentrated between the sieves of 0.5 mm and 0.063 mm size.

In an attempt to obtain mixtures equivalent to the reference, in order to avoid the need to carry out an extensive mix design study, the grading curves of the mixtures with and without foundry sand were designed, according to the particle size distribution of each aggregate fraction. As can be observed in Figure 3, the incorporation of foundry sand makes it difficult to obtain a grading curve close to that of the reference mixture. Thus, the maximum percentage of foundry sand that was possible to incorporate was 5%. Taking this into account, the binder content used in the mixture was the same used by Silva et al. (2009).

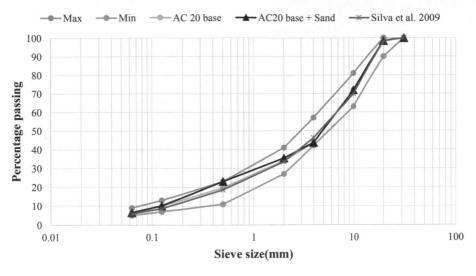

Figure 3. Grading curves of the studied mixtures.

Table 3. ITSR results of the studied mixtures.

Mixture	ITS dry kPa	ITS Wet kPa	ITSR (%)
AC20 base	2612	2382	91
AC20 base + Sand	2899	2405	83

3.3 Water sensitivity test results

The evaluation of the water sensitivity (ITSR) is essential to assess the performance and durability of an asphalt mixture, especially when recycled material is incorporated, since this material is not commonly used and needs an additional assessment of its behavior (Oliveira et al., 2013).

In the present study, 6 specimens of the reference mixture and 6 specimens of the mixture with incorporation of foundry sand were produced and the results are presented in Table 3.

It can be observed that the mixture with foundry sand presented a higher indirect tensile strength without water conditioning. However, its sensitivity to water was reduced by 8% in comparison with the reference mixture. Nevertheless, the mixture showed an ITSR value higher than the minimum value considered adequate for an asphalt mixture, which is 80%. Thus, based on the results obtained, it can be concluded that it is feasible to use up to 5% of foundry sand in asphalt mixtures for base courses (base AC20), without compromising its mechanical performance.

3.4 Leaching behavior

The foundry industry generates a significant amount of wastes including foundry sands.

The leaching test was carried out following the EN 12457-4 standard and the resulting leaching solution was analyzed regarding several parameters. The results have been compared with Portuguese environmental legislation regulated by the Decree 183/2009. In Table 4 the limits concerning landfill for inert wastes and the results of the foundry sand leaching behavior are presented.

Based on the results presented in Table 4, it is possible to observe that only the fluoride parameter doesn't respect the legal limits, enabling its classification as a non hazardous waste

Table 4. Results of leaching evaluation of foundry sand.

Parameters	Legal limits (mg/kg)	Result (mg/kg)
As	0.5	0.01
Ba	20	0.18
Cd	0.04	<0.04
Cr	0.5	<0.5
Cu	2	<0.25
Hg	0.01	<0.01
Mo	0.5	0.11
Ni	0.4	<0.3
Pb	0.5	<0.3
Sb	0.06	<0.01
Se	0.1	<0.01
Zn	4	0.17
Chlorides	800	182
Fluorides	10	50
Sulfates	1000	610
Phenol index	1	0.3
DOC	500	120
TDS	4000	1900

for disposal purposes. Sulfate and chloride concentration and metal dissolution level present low values, far below the legal limits for disposal. Based on this, in what concerns environmental aspects, no limitations for the process of foundry sand incorporation in asphalt mixture have been detected.

4 CONCLUSIONS

From the experimental results it can be concluded that the incorporation of foundry sand in asphalt mixtures is a viable alternative for the recycling and valorization of this waste, in order to reduce its deposition in landfill and preserve raw materials.

Its morphology, composition, particle size and uniformity enhances its use as a fine aggregate for asphalt mixtures. At the environmental level, no appreciable environmental effect is presenting. However, it is fundamental to analyze the leachate that this material can produce when incorporated in asphalt pavements, namely concerning the fluoride presence.

REFERENCES

Algarvio, D.A.N. 2009. *Reciclagem de resíduos de construção e demolição: Contribuição para controlo do processo.* Mestre, Universidade Nova de Lisboa.
Benson, C.H. & Bradshaw, S. 2011. *User guideline for foundry sand in green infrastructure construction.* In: Wisconsin, U.O. (ed.). Madison: Estados Unidos Recycled Materials Resource Center.
Castro, F., Vilarinho, C. & Soares, D. 2004. *Gestão de resíduos industriais por incorporação em materiais para construção civil.*
Collins, R.J. & Ciesielski, S.K. 1994. *Recycling and use of waste materials and by-products in highway construction.* Washington, DC: Estados Unidos Transportation Research Board.
EP 2014. *Caderno de Encargos Tipo Obra.* Volume V: 03—Pavimentação—Capítulo 14-03. Almada: Estradas de Portugal, S.A.
Etxeberria, M., Pacheco, C., Meneses, J.M. & Berridi, I. 2010. Properties of concrete using metallurgical industrial by-products as aggregates. *Construction and Building Materials* 24: 1594–1600.

Oikonomou, D. 2005. Recycled concrete aggregates. *Cement and concrete composites* 27: 315–318.

Oliveira, J.R.M., Silva, H.M.R.D., Abreu, L.P.F., Fernades, S.R.M. & Jesus, C.M.G. 2013. Pushing the Asphalt Recycling Technology to the Limit. *International Journal of Pavement Research and Technology* 6: 109–116.

Panazzolo, A.P., Frantz, L.C., Aurélio, S.O.S., Costa, F.L. & Muñoz, C. 2012. Gestão ambiental na construção de rodovias—O caso da BR-448 - Rodovia do Parque. *In:* PROMAB (ed.) *3° Congresso Internacional de Tecnologias para o Meio Ambiente.* Bento Gonçalves—RS, Brasil: Fundação Proamb.

Penkaitis, G. & Sígolo, J.B. 2012. Waste foundry sand. Environmental implication and characterization. *Geologia USP, Série Científica* 12: 57–70.

Siddique, R., De Schutter, G. & Noumowe, A. 2009. Effect of used-foundry sand on the mechanical properties of concrete. *Construction and Building Materials* 23: 976–980.

Silva, H.M.R.D.D., Sousa, R. & Oliveira, J. 2009. Fabrico de misturas betuminosas de alto módulo a menores temperaturas com betume 35/50 e parafinas. *XV Congresso Ibero-LatinoAmericano do Asfalto*, 2009. 375–384.

Torgal, F.P. & Jalali, S. 2010. *A sustentabilidade dos materiais de construção.* TecMinho.

Yazoghli-Marzouk, O., Vulcano-Greullet, N., Cantegrit, L., Friteyre, L. & Jullien, A. 2014. Recycling foundry sand in road construction-field assessment. *Construction and Building Materials* 61: 69–78.

Author index

Abreu, M. 389
Abreu, M.F. 125
Aderoju, O.M. 63
Akkouche, N. 93
Albuquerque, A. 87
Alegría, C. 203
Almeida, M.F. 209, 241, 277, 287, 293
Almeida, T.S. 265
Alves, A.C. 125, 409
Alves, C.I. 221
Alves, J.L. 23
Alves, R.C. 395
Alvim-Ferraz, C. 241, 287
Alvim-Ferraz, M.C. 293
Amant, V. 299
Andrés, A. 203
Antrekowitsch, J. 37
Araújo, C.A. 51
Araújo, D.J.C. 417
Araujo, J. 437
Araújo, J.M.M.G. 331
Araújo, N. 323
Araújo, S. 81
Arim, A.L. 215
Ayturan, Z.C. 147

Bacelo, H.A.M. 113
Baidarashvili, M.M. 345
Balistrou, M. 93
Baptista, A.J. 235, 305
Barbosa, B. 383
Barbosa, F.V. 331
Barroso, M.F. 31
Belo, I. 99
Bendjillali, K. 9
Benedetti, M. 305
Berardi, P.C. 209
Bernardo, M. 75, 359
Bessa, D. 283
Bessa, J. 365, 371
Boeykens, A. 31
Bonalumi, I.P. 265
Bongiovani, M.C. 57

Borges, J. 241
Botelho, C.M.S. 113
Boulekbache, B. 9
Brás, I. 81
Brito, R.S. 51
Budžaki, S. 241

Camacho, F.P. 259, 265
Campos, L. 283
Canavarro, V. 23
Carlota, S. 167
Carneiro, D. 277
Carneiro, J.R. 45
Carvalho, J. 431, 437
Carvalho, M.A. 409
Castanheiro, J.E. 167
Castro, C. 437
Castro, F. 403
Castro, F.A. 389
Castro, L.M. 221, 271
Catarino, M. 153
Cecílio, D.F.M. 215
Chaves, B. 283
Chemrouk, M. 9
Chen, W. 183
Chepushtanova, T.A. 191
Cifrian, E. 203
Ciríaco, L. 87
Coelho Pinheiro, M.N. 221
Coelho, J. 323
Colvero, D.A. 105
Cordeiro, A. 81
Corrêa-Silva, M. 323
Cosentino, S.L. 383
Costa, A.S.G. 395
Costa, E. 287
Costa, F. 139
Costa, J. 383
Cristelo, N. 161, 175, 323
Cruz, M. 287
Cunha, F. 365, 371
Cunha-Queda, A.C. 81

da Silva, A. 45
Dalmo, F.C. 1

Delerue-Matos, C. 31
Delvasto, P. 17
Denot, A. 299
Dias, D. 75, 359
Dias, G.A. 63
Dias, J.M. 209, 241, 245, 277, 287, 293
Dias, R.A. 271
Dias-Ferreira, C. 183, 235, 317
Díaz-Salaverría, J.V. 17
Dinc, G. 147
do Amaral, A.G. 57
Durão, F. 251
Durão, L. 425
Dursun, S. 147

Eisenlohr, L. 299
Estrela, M.A. 305
Evans, S. 305

Fangueiro, R. 365, 371
Faria, M. 81
Fernandes, A. 87
Fernandes, M.J. 245
Fernando, A.L. 383
Fonseca, I. 75, 359
Forero, B.J. 17

Gando-Ferreira, L.M. 133, 215
Gomes, A.P.D. 105
Gonçalves, M. 425
Gouveia, C.X. 17
Guimarães, C. 251
Guío-Pérez, D.C. 351
Guseynova, G.D. 191

Hachemi, M. 93
Hanke, G. 37
Himrane, N. 93
Holgado, M. 305
Hyppänen, T. 69

Itkulova, Sh.S. 191

Jensen, P.E. 183
Justo, A. 277

Kirkelund, G.M. 183
Klepacz-Smolka, A. 221
Klupsch, E. 183
Kolaitis, T. 425
Kůs, P. 377

Lacerda, P. 389
Lapa, N. 75, 359
Leão, C.P. 409
Llano, T. 203
Lobo, G. 81
Lopes, A. 87
Lopes, M. 99
Lopes, M.L. 45, 119,
   175, 209
Loubar, K. 93
Lourenço, E.J. 235, 305
Lourenço, L.F.H. 51
Luganov, V.A. 191, 339

Machado, A.V. 417
Machado, J. 389
Machado, T. 283
Mamyrbayeva, K.K. 339
Margarido, F. 251
Martins, C.I. 403
Martins, M.J. 277
Martins, R.C. 221, 271
Matos, I. 75, 359
Matos, M.A.A. 105
Mendes, B. 425
Merkibayev, Y.S. 339
Miguel, M. 359
Miranda, S.M. 99
Miranda, T. 323
Monteiro, D. 23
Morais, S. 31
Moreira, F. 125, 197, 409
Moreira, M.M. 31
Mota, A. 437
Mota, C. 365, 371
Motovilov, I.Yu. 191
Mourão, P.A.M. 311

Nabais, J.M.V. 311
Nascimento, D.B. 259
Nascimento, L.P. 443

Nebra, S. 1
Nikku, M. 69
Nobre, C. 425
Nogueira, C.A. 251
Nunes, M.A. 395
Nunes, O.C. 81
Nunes, S. 161

Oliveira, A. 403
Oliveira, A. 425
Oliveira, J.R.M. 443
Oliveira, M.B.P.P. 395
Oliveira, V. 317
Orazymbetova, S.D. 339
Ottosen, L.M. 183

Pacheco, M.J. 87
Pacheco, R. 389
Peças, P. 305
Peixoto Joele, M.R.S. 51
Perão, B.A. 259
Pereira, A.S. 99
Pereira, J.P. 235
Pereira, M.F.C. 251
Pereira, P.M. 119
Pesqueira, J.F. 277
Petriaev, A.V. 345
Pinheiro, C. 161
Pinheiro, C.T. 133
Pino-Hernádez, E.J.G. 51
Pinto, A.P. 167
Pinto, F. 75, 359
Pires, F.C. 245
Puga, H. 395

Quina, M.J. 133, 215, 221
Quinta-Ferreira, R.M. 271

Ramos, M. 153
Rangel, B. 23
Rani-Borges, B. 227
Ribeiro, A. 431, 437
Ribeiro, R. 23
Ribeiro, W. 75
Rico, C. 203
Rincón, S.L. 351
Rios, S. 161
Rizk, M.C. 259, 265
Rodrigues, D.P. 221
Rozumová, L. 377

Šafařík, I. 377
Sakharova, A.S. 345
Sant'Ana, P.H.M. 1
Santos, J. 203
Santos, J. 389
Santos, K.A. 105
Santos, S.C.R. 113
Schneider, R.M. 57
Sermyagina, E. 69
Sidella, S. 383
Silva, A. 389
Silva, E.J. 305
Silva, J.O.V. 293
Silva, M.E. 81
Simão, N.M. 1
Soares Dias, A.P. 153
Soares, E. 389
Soares, M. 437
Sousa, V. 317
Souza Matos, G.S. 51
Steinlechner, S. 37
Svatovskaya, L.B. 345

Tarelho, L.A.C. 105
Tavares, T. 139
Tazerout, M. 93
Tchikuala, E.F. 311
Teixeira, A. 23
Teixeira, J.C.F. 331
Teixeira, T. 389
Tibocha, D.A. 351
Topa Gomes, A. 323
Tsoutsos, T. 425

Vakkilainen, E. 69
Vaz, J.M. 317
Viana da Fonseca, A. 161
Vieceli, N. 251
Vieira, C.S. 119, 175
Vieira, J.M.P. 227
Vilarinho, C. 431, 437, 443
Vilarinho, M.C.L.G. 331, 417

Wegscheider, S. 37
Werberich, T. 57
Withouck, H. 31

Yesken, Zh. 339